MYSTIC, GEOMETER, AND INTUITIONIST

MYSTIC, GEOMETER, AND INTUITIONIST

The Life of L. E. J. Brouwer

Volume 1
The Dawning Revolution

DIRK VAN DALEN

Department of Philosophy
Utrecht University

CLARENDON PRESS · OXFORD
1999

OXFORD
UNIVERSITY PRESS

Great Clarendon Street, Oxford OX2 6DP

Oxford University Press is a department of the University of Oxford.
It furthers the University's objective of excellence in research, scholarship,
and education by publishing worldwide in

Oxford New York

Auckland Bangkok Buenos Aires Cape Town Chennai
Dar es Salaam Delhi Hong Kong Istanbul Karachi Kolkata
Kuala Lumpur Madrid Melbourne Mexico City Mumbai Nairobi
São Paulo Shanghai Taipei Tokyo Toronto

Oxford is a registered trade mark of Oxford University Press
in the UK and in certain other countries

Published in the United States
by Oxford University Press Inc., New York

© Dirk van Dalen, 1999

The moral rights of the author have been asserted
Database right Oxford University Press (maker)

First published 1999
Reprinted with corrections 2002

All rights reserved. No part of this publication may be reproduced,
stored in a retrieval system, or transmitted, in any form or by any means,
without the prior permission in writing of Oxford University Press,
or as expressly permitted by law, or under terms agreed with the appropriate
reprographics rights organization. Enquiries concerning reproduction
outside the scope of the above should be sent to the Rights Department,
Oxford University Press, at the address above

You must not circulate this book in any other binding or cover
and you must impose this same condition on any acquirer

A catologue record for this book is available from the British Library

Library of Congress Cataloging in Publication Data
(Data available)
ISBN 0 19 850297 4

Printed in Great Britain by
Biddles Ltd, Guildford and King's Lynn

*Dedicated
to the memory of
Hans Freudenthal*

PREFACE

> Writing a book is an adventure, to begin with it is a toy and an amusement, and then it becomes a mistress, and then it becomes a master, and then it becomes a tyrant, and the last phase is that, just as you are about to be reconciled to your servitude, then you kill the monster.
> Winston Churchill.

In recent years a number of biographies of mathematicians has appeared. This can be seen as part of an increased awareness of the relevance of history in scientific circles. In the contemporary curriculum, where more and more pressure is applied to improve the efficiency of university education, resulting in a teaching of the bare essentials, the demand for historical perspective is an encouraging experience.

The present book deals with a person and a period of crucial interest for the integrity and meaningfulness of mathematics. L.E.J. Brouwer and his ideas belong to an episode in the history of mathematics which has often been labelled with the terms 'crisis' and 'revolution'. The first half of the twentieth century was the scene of a foundational debate that rocked mathematics, but that in principle could have been solved in a couple of workshops and personal meetings. It is one of those exceptional historical accidents that at this particular point two strong-willed men were defending their opposing views, both with an inflated feeling of responsibility for saving mathematics. The constructivistic, human view of mathematics was represented by the Dutchman, Brouwer, a mathematical genius with strong philosophical views. His life story is full of friction and conflicts, but equally with moments of intellectual achievement and euphoria. His strong sense of justice, combined with an uncompromising nature, made him a champion of the underdog. As a rule, the ensuing fights heavily taxed his nerves.

If young temperamental and exceptionally gifted scientists recognize the mental and intellectual states described in the biography, the book will have had its use.

The author would have been shocked and incredulous had a gypsy fortune-teller told him at the beginning of his career that he was to write the biography of L.E.J. Brouwer. For a scientist is trained to advance his subject by looking forward, by adding new theorems, principles, experiments. The role of history was at that time rather modest in the sciences, only much later, when the negative consequences of an overly efficient teaching of a vigorously reduced curriculum became visible, the lessons of the past received the attention they deserved.

Be that as it may, I was only vaguely aware of Brouwer's personality when I started to teach. Brouwer was, for reasons that will become clear in Volume 2, considered both a great man and an embarrassment in Holland. The situation might be compared to that of a noble family hiding the eccentric old baron in an isolated wing: a man to be praised, but not to be seen.

Under the influence of friends and colleagues at MIT in the mid sixties, I started to view Brouwer with a keen curiosity. The stories that circulated in mathematical circles, cheerfully ignored and suppressed in Holland, suggested that here was something interesting to be investigated.

In 1976 I started in earnest to collect material for a Brouwer Archive. The Dutch National Science Foundation, ZWO, generously supported this activity by giving a one year grant to Walter P. van Stigt and Erik Heijerman.

The University of Utrecht has loyally supported me at a time when an eventual publication was still hypothetical, the Mathematics Department graciously put storage room at my disposal for storing the Brouwer *Nachlass*, and the Philosophy Department showed a laudable understanding for the time invested in the project and the regular visits to all kind of institutions and archives.

I am deeply grateful for the hospitality and support from the *Mathematisches Forschungsinstitut* at Oberwolfach, without the marvellous library, the helpful staff and the isolation from the outer world this biography would not have seen the light in this century.

When I set out to collect the necessary data and to write the biography I was blissfully unaware of the fact that a biography is a merciless mistress; it demands total devotion. In return it gives glimpses of fascinating people and events. I have benefitted from the project by acquiring insights into the life and work of a highly unusual person and into the scientific-historical developments of the first half of this century.

Looking back, my personal involvement in the Brouwer project seems an instance of predestination. My highschool teacher, B. Stroboer, was a student of Brouwer, and on the way to or from school I often accompanied him. During these walks he told me about the miraculous accomplishments of the mathematician Brouwer. As a result I enrolled at the University of Amsterdam, where I became the student of Brouwer's successor, A. Heyting. Thus I joined the tiny intuitionist company. The writing of the biography is the endpoint in the quest for intuitionists and intuitionism.

The life story of Brouwer is far more complicated than anything I had imagined. It consists of mysticism, politics, topology, philosophy of language, intuitionism, editorial boards, Göttingen, girlfriends, the pharmacy of his wife, and a hundred more things. The present volume and its intended successor are attempts to do Brouwer justice.

The bulk of the Brouwer material reached the archive via a detour. After Brouwer's death Mrs. C. Jongejan inherited the estate. She put Brouwer's scientific papers in the care of the Dutch Mathematics Society.

Freudenthal and Heyting made use of the material for the Collected Works of Brouwer. J. Dijkman, the chairman of the Dutch Mathematical Society at the time, donated the material from Brouwer's scientific estate to the Brouwer Archive. Subsequently a number of private persons and institutions put Brouwer letters, or copies, at the disposal of the archive. At present the archive contains a good part of Brouwer's *Nachlass*. I am, however, under no illusion of completeness. An unknown number of letters and manuscripts were lost as a consequence of a fire in Brouwer's cottage; furthermore personal letters were destroyed by the persons in charge of the estate. Finally, in the period between Brouwer's death and the opening of the archive there was a certain tendency of documents to disappear in thin air.

Some items of the Brouwer material have fortunately been located. A collection of notes concerning the Brouwer–Urysohn correspondence had been given (or lent) by C. Jongejan to P.S. Alexandrov, who was preparing Urysohn's Collected Works. Eventually they were discovered by Prof. A. Shiryaev in Alexandrov's former dacha. I am grateful for his assistance in making the material available; copies are in the Brouwer Archive. The son of Prof. Dijkman discovered after his father's death a box of Brouwer material that was taken home, and apparently forgotten at the time of the transfer of the material. This material significantly completes the collection.

Some important items are in private possession, for example Brouwer's own copies of '*Leven, Kunst en Mystiek*' and the dissertation. Fortunately we could obtain copies for the archive.

When I finally started to write the biography, I soon saw that it was impracticable, if not downright impossible, to stick to the chronological order. Brouwer was most of the time involved in any number of projects, conflicts and schemes. To treat them in parallel would result in stark confusion. To most observers during Brouwer's lifetime, this almost compulsive multiple activity was, however, almost invisible. As a rule, informants were greatly surprised to learn that at the time of the events reported by them, so many other things were going on simultaneously. Probably only Brouwer could keep track of this multitude of affairs. It is a platitude that nobody knows another person completely, but in Brouwer's case not only the innermost thoughts and emotions but also matters that were familiar in other circles escaped the observer.

But even though I am treating Brouwer's life according to topics, the reader will do well to keep in mind that the events of various chapters were taking place in parallel.

There are a number of major themes in Brouwer's life: mysticism, topology, intuitionism, philosophy of language, and a number of smaller ones: the Lebesgue quarrel, anti-nationalism—with his rejection of Denjoy's policy and the boycott of Germany, the Grundlagenstreit, the Menger fight, the Hungarian investment, the postwar tribunal, the opposition of the faculty and the loss of Compositio Mathematica. In almost all of these instances Brouwer was fighting a battle of sorts, and many of these battles ran for years and were carried out simultaneously. It still strikes me as curious that a person can get involved in so many disputes. There are two plausible explanations, and both apply: Brouwer was a high-strung nervous person, who could easily exaggerate matters when under stress. On top of that he had an extreme passion for justice; as Bieberbach put it: he was a justice fanatic (*Gerechtigkeitsfanatiker*). As a result he would counter injustice—no matter with respect to whom—with a state of total war. The biography will provide ample, and sad, examples. He combined his strong emotions with a display of ice-cold arguments, a combination that is not as unusual as one would think. His contemporaries, in particular his adversaries, could not always discern the emotions under the granite exterior.

Acknowledgements
The writing of this biography would have plainly been impossible without the assistance and information of a great many individuals and institutions. I am indebted most of all to the late Hans Freudenthal, who was part of the history which is the subject of this biography. His memory, archive and above all insight have been a constant inspiration. In spite of the unfortunate friction between him and Brouwer, he rose far above the bitterness of the past.

Brouwer's stepdaughter, Louise Peijpers, provided many details from Brouwer's past; she had a wonderfully clear memory. Most of her stories have received independent confirmation. She was bedridden at the time we talked to her, but from her bed she surveyed (and ruled) her world. Being a lady with a strong, albeit sometimes difficult, character she also had her conflicts with Brouwer. Nonetheless these did not influence the veracity of her information.

Walter van Stigt has been a good collaborator and friend; during the year sponsored by ZWO he was a most efficient and cheerful companion. His book *Brouwer's Intuitionism* is a lasting souvenir of that exciting initial period of the project. My student Erik Heijerman also contributed in a competent and imaginative way to the early historical and archival investigations.

L.E.J. Brouwer, the son of Brouwer's brother, Aldert, was the representative of the family who authorized the deposition of the Brouwer material in the Archive and who encouraged the publication of the historically and scientifically relevant material. I am most grateful for his cooperation and interest.

There is a long row of friends, students and colleagues of Brouwer who contributed to this biography with their information and impressions. I warmly thank them for their kind help. Many historians and archivists have helped me to find the right material and information. The Mayor of Blaricum, Mrs. Mr. A.J. Lecoultre-Foest, has provided assistance in matters of local history. The archives of the Universities of Amsterdam, Göttingen, Leiden, Delft, Paris, Groningen, Berlin, Harvard, Bonn, Oslo, Cambridge, and the ETH at Zürich have opened their doors to me and graciously permitted me to use the material. Furthermore, the archive of the Amsterdams Studenten Corps, the *Auswärtiges Amt* in Bonn, the Centrum voor Wiskunde en Informatica, the Frederik van Eeden Museum, the Heyting Archive, the Letterkundig Museum, the Boerhaave Museum of Science, the Smithsonian Institute at Washington, the Institute for Advanced Studies at Princeton, the Mittag-Leffler Institute (Stockholm), the Courant Institute (New York), the University Library of the Amsterdam University, the Gemeente Archief at Amsterdam, the University Museum at Amsterdam, the Koninklijke Academie van Wetenschappen, the Rijksarchief at Haarlem, the Rijksarchief at the Hague, the Rijksinstituut voor Oorlogsdocumentatie, the Ministry of Defence, the Ministry of Education, the Archive of Springer Verlag (Heidelberg), the Royal Society of London, the New Atlantis Foundation, the Einstein Archive at Boston, the Alexandrov Archive, and the Archive of the Hebrew University at Jerusalem have provided information and copies of relevant material. I am most grateful to the above mentioned institutions and their archivists, without their help the picture of Brouwer and his context would not have been as complete as it is now.

In the course of the project a number of persons have joined in the search for Brouweriana, above all my wife, Dokie; she took part in almost all fact finding missions, she was the photographer of the project, and she managed the worldly side of the operation with superior ingenuity. Without her the project would probably not even have started, let alone reached the reader. She also mastered the mysteries of TEX and saw the manuscript to its final form.

The technical advice of our systems experts, Karst Koijmans and Freek Wiedijk, has been beyond praise. In particular the latter was always prepared to solve the mysterious problems of TEX and its children.

The chapters of this volume have been read and criticized by the following colleagues and friends: Andreas Blass, Adrian Mathias, Jeff Paris, Wilfrid Hodges, Martin Hyland, Roy Dyckhoff, Craig Smorynski, J. Stillwell, I.M. James, Richard and Nancy Tieszen, Douglas Bridges, W.T. van Est, Ieke Moerdijk. My son, Harry, read all of the chapters and gave helpful hints. I would especially like to thank Fred van der Blij, who was willing to read an early draft of the present volume.

Their criticism and suggestions have significantly contributed to the final formulations, but I want to stress that any remaining shortcomings are wholly my responsibility.

The Oxford University Press has given me the much appreciated support in the process of transforming the manuscript into a real book. It is a pleasure to thank to the staff, in particular Julia Tompson, Elizabeth Johnston and Nandi Mahua.

A number of persons who have known Brouwer in one way or another have been interviewed, their information has substantially contributed to the life story of Brouwer, I list them in alphabetical order: Paul Bernays, Mrs. M.F. Brouwer-Rueb, C.L. ten Cate, Mr. Crèvecœur, Mr. and Mrs. R. Dudok van Heel, Casper Emmer, Max Euwe, Hans Freudenthal, Arend Heyting, Mrs. Y. Holdert-Brouwer, Johan de Iongh, F. Kuiper, Hannie and Gabi van Lakwijk-Najoan, H.J. Looman, Maarten and Frédérique Mauve, Mrs. M.A.G. Paulssen-Smeets, A.M. de Rijk, Mrs. B Sapir, Mrs. Stieltjes-Mauve, Dirk Jan Struik, Mrs. E.H. Versteegh-Vermey, L. Vietoris, Mrs. S. Volmer-Sapiro, Mrs. C. Vuysje-Mannoury, and B.L. van der Waerden.

Finally, I would like to express my gratitude to a long list of persons who in one way or another have contributed information or material, their support has been a significant factor in the historical reliability of the book. The following list may be, unintentionally, incomplete—I apologize for possible inaccuracies: Gerard Alberts, L. Balke, H. Behnke, S. van de Berg, L. Bieberbach, K.-R. Biermann,

R.H.T. Bleijerveld, D. Breedijk, E. Brieskorn, L. Van den Brom, N.G. de Bruijn, F.F. Burchsted, Catriona Byrne, Dirk van Delft, John Dawson, R. Delvigne, J.C. Doodkorte (St. Jobsleen), E.P. van Emmerik, Sjoerd van Faassen, Jens Erik Fenstad, R. Finch, Jan Fontijn, P. Forman, B. Fraenkel, Mrs. M. Fraenkel, H. Gispert, B. Glaus, A.D. de Groot, Vagn Lundsgard Hansen, Mrs. P.L. van der Laan-van Heemert, Dennis Hesseling, Mrs. F.J. Heyting-van Anrooy, Mrs. Y. Holdert-Brouwer, A.F. Jaegli, D. Kan, S.C. Kleene, Mrs. V. MacDermot, Saunders Mac Lane, Per Martin-Löf, Brian McGuiness, Herbert Mehrtens, H. Meschkowski, A.I. Miller, W.C. Mulder, W. Purkert, K. Ramskov, Mrs. C. Reid, R. Remmert, V. Remmert, David Rowe, Mrs. S. Schöffer-Burger, Walter Schmitz, Mrs. I. Schout-de Waal, Arjen and Lidy Sevenster, Mrs. M. Shillan, A. Shiryaev, Wilfried Sieg, R. Siegmund Schultze, A. Sklar, Ernst Specker, T.A. Springer, F.J. Staal, F. Stuurman, G. van Suchtelen, O. Tausky-Todd, A.S. Troelstra, Henk and Lies van der Tweel, Ralph Twentyman, G.F. Viets, Hao Wang, R. Weijdom, Janny van Wering, M. Weyl, B. Willink, L.C. Young.

Conventions
—A word of warning to the reader: this is a biography of a mathematician and philosopher, hence a certain amount of mathematics and philosophy will be found here and there. If you are not an expert in a particular topic, feel free to be content with a fleeting impression of the matter. A number of topics will be explained, but that does not make this book a text book. The treatment will, necessarily, always fall short of a precise exposition. There are enough didactic presentations around to supplement the sketches in the present book.
—The orthography of Dutch names is somewhat unusual; for example, Hugo de Vries is alphabetized under V, but if one just uses the family name one writes 'De Vries'. I will try to stick to the Dutch convention, but occasionally I may deviate from it.
—The list of documents and their sources will appear in Volume 2.
—Volume 1 and 2 of Brouwer's collected works are cited as CW I and CW II. Brouwer's papers are cited following the classification of chapter 11.
—In the book we use a minimal amount of logical and mathematical notations. The reader who is not familiar with them will be wise enough to skip the technical parts. For completeness sake I mention the logical connectives, which occur only sparingly: $\wedge, \vee, \rightarrow, \neg, \exists x, \forall x$ stand for *'and, or, if ... then, not, there exists an element x such that ... ', 'for all x the following holds ... '*.

Utrecht
September 1998
Dirk van Dalen

The second edition has given me an opportunity to carry out a number of improvements in the text. Many readers have provided me with suggestions and corrections. I am grateful in particular to Maurice Boffa, Dennis Hesseling, Hide Ishiguro, Joachim Lambek and Bonny Shulman for their kind help.
De Meern, July 2002

CONTENTS

1 Child and Student — 1
  1.1 School years — 4
  1.2 Student in Amsterdam — 12
  1.3 The religious credo — 17
  1.4 Friendship: Adama van Scheltema — 22

2 Mathematics and Mysticism — 41
  2.1 Teachers and study — 41
  2.2 First research, four-dimensional geometry — 46
  2.3 Marriage — 52
  2.4 Bolland's philosophy course — 56
  2.5 Among the artists and vegetarians — 59
  2.6 The Delft lectures — 64
  2.7 Family life in Blaricum — 77

3 The Dissertation — 80
  3.1 Preparations and hesitations — 80
  3.2 Under Korteweg's supervision — 87
  3.3 On the role of logic — 99
  3.4 Mathematics and the world — 102
  3.5 Observations on set theory and formalism — 105
  3.6 The public defense — 118

4 Cantor–Schoenflies Topology — 122
  4.1 The geometry of continuous change — 122
  4.2 Lie groups — 126
  4.3 Publishing in the *Mathematische Annalen* — 130
  4.4 Fixed points on spheres and the translation theorem — 133
  4.5 Vector fields on surfaces — 136
  4.6 Analysis situs and Schoenflies — 140

5 The new Topology — 151
  5.1 Invariance of dimension — 151
  5.2 The fixed point theorem and other surprises — 172
  5.3 The Karlsruhe meeting and the Continuity Method — 178

6 Making a Career — 197
  6.1 Financial worries — 197
  6.2 First international contacts — 203

	6.3 Climbing the ladder	205
	6.4 The shortcomings of Schoenflies' *Bericht*	208
	6.5 *Privaat Docent*	213
	6.6 Korteweg's campaign for Brouwer	218
	6.7 Schoenflies again	229
7	**The War years**	234
	7.1 Sets and sequences—law or choice?	235
	7.2 The International Academy for Philosophy	243
	7.3 Family life	250
	7.4 An offer from Leiden	252
	7.5 Significs and Jacob Israël de Haan	255
	7.6 Van Eeden and the International Academy	258
	7.7 Faculty politics	270
	7.8 The Flemish cause	275
	7.9 Air photography and National Defense	276
8	**Mathematics after the War**	284
	8.1 How to appoint professors	288
	8.2 The return to topology	290
	8.3 Brouwer, Schouten and the *Mathematische Annalen*	295
	8.4 The offers from Göttingen and Berlin	300
	8.5 The Academy—how Denjoy was elected	304
	8.6 Negotiations with Hermann Weyl	306
	8.7 Intuitionism and the *Begründungs* papers	312
	8.8 And Brouwer—that is the revolution	317
	8.9 Intuitionism, the Nauheim Conference	325
	8.10 The failure of the Institute for Philosophy	330
9	**Politics and Mathematics**	336
	9.1 The *Conseil* and the boycott of Germany	336
	9.2 The Nauheim conference and Intuitionism	341
	9.3 The Denjoy conflict	344
	9.4 Weitzenböck's appointment in Amsterdam	358
	9.5 Kohnstamm and the Philosophy of Science curriculum	361
	9.6 The New Chronicle	364
10	**The Breakthrough**	367
	10.1 The Signific Circle	367
	10.2 Intuitionism—principles for choice sequences	375
	10.3 Intuitionism in the *Mathematische Annalen*	387
	10.4 Beyond Brouwerian counterexamples	392
	10.5 Fraenkel's role in intuitionism	395
	10.6 Heyting's first contributions	401

11	Bibliography of Brouwer's writings	405
Contents of Volume 2		418
References		419
Index		426

1
CHILD AND STUDENT

It was into an optimistic country that Luitzen Egbertus Jan Brouwer was born—a country with a burgeoning science, literature and social awareness. After having been one of the backwaters of Europe, the Netherlands was finding its way back into the mainstream of culture and commerce. Since the golden age of Huygens and Stevin, science had been more a playground for cautious imitators than for bold researchers. At the turn of the century all that was changing. Physics was flourishing—Lorentz, Kamerlingh Onnes, Van der Waals and Zeeman were all putting the Netherlands[1] in the forefront of modern physics, Van 't Hoff was doing so in chemistry, Hugo de Vries in biology, and Kapteyn in astronomy. The fact that Holland counted a number of its leading scientists among the early recipients of the Nobel prize may be sufficient evidence of the quality of the research in the sciences in the Netherlands.

Mathematics was slower to pick up the new élan. The best known Dutch mathematician after Christian Huygens was Thomas Jan Stieltjes, who did not find recognition in Holland and who practised his mathematics in Toulouse. The mathematicians of the period were competent, but not on a par with their colleagues in Germany, France and Britain. Neither was their choice of subject very daring: the Dutch worked in the more settled parts of mathematics, from analytic-descriptive geometry to number theory and analysis. At the same time literature was freeing itself from the strait-jacket of nineteenth century conventions, and a wave of new authors had already transformed the literary landscape. The explicit aim of the leading spirits was to 'push Holland high up into the stream of the nations'.[2] Almost inevitably the new literary trend in Holland was closely bound up with developments in the social movements. Many of the leading poets and authors were involved in the promotion of a better social climate. A number of them played a significant role in the socialist movement, which eventually led to the birth of the Social Democratic Party and, subsequently, the Communist Party.

---

[1] We will indulge in some 'abuse of language' by using 'The Netherlands' and 'Holland' as synonyms. This is definitely incorrect from the standpoint of geography. Holland is the collective name of the two provinces North- and South-Holland. There used to be a county Holland under the Count of Holland, but under the Dutch republic it became a province like any other. The concentration of trade and government, however, lent Holland so much prominence, that one often used 'Holland' instead of 'The United Republic'. In spite of the efforts of the later kings of The Netherlands, the habit of using 'Holland' as a *pars pro toto* has persisted. The phenomenon is not uncommon, think of 'England' and 'United Kingdom' for example. The reader is warned.

[2] *Nederland hoog op te stuwen in de vaart der volkeren.*

The political scene at the turn of the century was mainly determined by the *Liberals*, who were the rightful heirs of the dominant movement of the nineteenth century, the *Protestants* (who were traditionally called the Christians) the *Catholics*, and the late-comers, the *Social Democrats*. Traditionally the Liberals made up the higher strata of society; the Protestants, Catholics and Socialists were instrumental in the emancipation of their respective sections of the population. History had equipped their respective parties with powerful charismatic leaders.

Generally speaking, Dutch society was experiencing a powerful upward thrust in its social structure. Progress had become the password of the day, and in this general movement there was a small but influential group that formed the backbone of the nation's new spirit: the schoolmasters. In the past they had been the favourite butt of pamphleteers and wits, for example, Multatuli, the reformist author, created the immortal schoolmaster *Pennewip*, as a caricature of the fossils that used to educate the nation's youth. At the turn of the century, however, a whole new generation of schoolmasters had come to populate the schoolrooms of the country. The new teachers were, for the greater part, idealistic promoters of a better future, equipped with an unshakable belief in the beneficial influence of knowledge. The nation was blessed at the time with a fine body of teachers and an effective egalitarian educational system. A side effect of the strong admiration of schoolmasters for learning was that they often sent their own children to institutions of higher education; a surprising percentage of the scientific Dutch community actually came from schoolmasters' families.

Luitzen Egbertus Jan (*Bertus*, as he was called) Brouwer, was one of these schoolmaster's sons that was to put his stamp on Dutch society and culture; he was born on the 27th of February 1881 at Overschie, son of Hendrika[3] Poutsma (15 September 1852 Follega–3 May 1927 Utrecht) and Egbertus Luitzens Brouwer (17 April 1854 Bakkeveen–3 May 1947 Blaricum) schoolmaster at a primary school at Overschie. Hendrika was of Friesian stock; her earliest registered forebear Tammerus Gerhardi (1579–1644) was a minister of the Dutch Reformed Church at Joure and IJlst in Friesland. The Poutsma family tree is adorned with a good number of parsons and schoolmasters, and towards the end of the nineteenth century there is a definite tendency to rise to the higher strata of the teaching profession. Two of Brouwer's uncles became teachers at the Barlaeus gymnasium at Amsterdam, the finest school of the city. One of them, Hendrik Poutsma taught English literature and language and wrote a classic textbook, he was awarded a honorary doctorate by the University of Amsterdam. The other uncle, Albertus Poutsma, taught Greek and Latin at the same school and eventually became its rector.

The forebears of Bertus were all Friesian. His father was born at Bakkeveen in Friesland, and he in turn was the son of Luitzen Luitzens Brouwer (born 20 April 1813 in Duurswoude), who likewise was a schoolmaster (*onderwijzer der jeugd*). The

---

[3] The spelling of names is not uniform in the registers of older archives, e.g. 'Hendrika' is also spelt as 'Henderika', 'Hendrica', 'Henderica'.

latter was the son of another Luitzen Luitzens Brouwer (born in Duurswoude 1756 or 1757), who was a farmer and shepherd.

The Friesians were (and still are) known for their reliability, and they were welcome additions to many a profession in the western part of the country, the part that traditionally is known as Holland. The position of the Friesians in the Netherlands is somewhat comparable to that of the Scottish in England. In view of the limited opportunities in Friesland, the more adventurous among them moved to Holland and settled there. For example, in the first half of this century the Amsterdam police force recruited a good number of its members from Friesland.

Bertus' parents were married on 8 April 1880. They immediately moved from Beetsterzwaag (Friesland) to Overschie, a small town which nowadays is part of Rotterdam, where their first child, Bertus, was born. The story goes that grandfather Brouwer came to see the baby.[4] He gravely looked at the child in the cradle, and spoke the memorable words: 'Let us hope that he can learn'.[5]

Bertus' parents were strict and honest people and none of the later extravagances of the sons seem to have visited them. Like all hardworking, sober Dutchmen they led a simple life in reasonable comfort, but without wasting money. Bertus' mother had indeed elevated saving to an art.

The most detailed information about the child and boy Bertus was provided by himself. At the age of sixteen he had to write a short 'auto-biography' as part of the initiation rites of *Amsterdamse Studenten Corps*.[6] The following lines are from this biographical sketch (September 1897):

> I was born on 27 February 1881 in Overschie. Here I lived for eight months, of which I naturally do not remember anything at all. I have never been back to that town, except for a few hours, when, four years ago, I went to Brussels, and stopped over in Rotterdam, skipping a couple of trains.[7] From there, one reaches Overschie in half an hour by horse tram along a road that curves through low, muddy, peaty meadows, crisscrossed with marshy drains and black bubbling ditches. The muddy cattle in those meadows feeds itself on the waste of the gin mills (which are centred in that area), the so-called 'swill' which is put into tub-like containers standing here and there in the field. The village of Overschie consists, like

---

[4]Told by Louise Peijpers , the daughter from the first marriage of Bertus' wife. Although Louise was no witness to the earlier events of Brouwer's life, she had a perfect memory for the family folklore. A large part of the information on Brouwer's early years is based on oral communications of Louise Peijpers.
[5]*Als hij maar kan leren.*
[6]The chief fraternity of the University of Amsterdam.
[7]The reader should note that this was rather exceptional; most Dutchman hardly ever crossed the borders of their country, let alone at the age of 12! One should, however, bear in mind that Bertus was a precocious child, who was at the age of twelve already in the third grade of the high school (HBS). Foreign travel later became a second nature to Bertus (and his brothers).

so many villages, of two rows of houses built on each side of a wide road. It is distinguished by a dirty street, an ugly town hall, ugly private houses, a couple of dunghills, and a drawbridge in the middle.

Within a year of the birth of Bertus, the family moved to Medemblik, one of the old towns in North-Holland that had flourished on maritime trading in the old days of the Zuiderzee. The Brouwers lived in Medemblik for eleven years, during which time two more sons were born, Izaak Alexander (Lex) (23 January 1883) and Hendrikus Albertus (Aldert) (20 September 1886). Bertus' father taught in Medemblik at an elementary school, the so-called *Burgerschool*.[8]

The arrival of the second child, Isaac Alexander, known as Lex, seemed to have upset Bertus. According to the oral tradition of the Brouwer family, he blamed Lex for driving him from the comfortable cradle. Whatever may be true in this story, a fact is that as a boy, Bertus harboured a thinly veiled dislike of Lex. The latter came in for a fair portion of refined or not-so-refined pestering.

The relationship with the youngest brother, Aldert, was much more friendly, although they were known to fight occasionally. Even in their old age they annoyed and amused the members of the Royal Academy with their quarrels.

1.1 School years

In Medemblik Bertus spent his early school years, and in 1890 he entered high school (the HBS) at the tender age of nine (which earned him posthumously an entry in the Dutch Guinness Book of Records). The student-biography also records his Medemblik years:

> My next abode did not rank much higher. I moved to Medemblik, where I lived for eleven years. In that town I learned to walk and speak, smashed a lot of things in the parental home, repeatedly fell down the stairs without —miracle, oh miracle—breaking my neck, and had the measles. All these things concern the first three years of my life. Furthermore I had at that time a great passion for umbrellas: when I saw one, I started to dance with excitement, clasped the thing in my arms while continuously bellowing 'pappetepé, pappetepé' [for *paraplu*]. I was most of all charmed by our house painter and our gardener, whose work I, as often as they were at our place, contemplated with the utmost attention, and I did not rest before I

---

[8]The Dutch educational system knew one elementary school and a number of secondary ones. The elementary school (preceded by the *kleuterschool* (kindergarten)) was called the *lagere school*, age 6–12; the secondary general education schools were MULO (or ULO), literally *(Meer) Uitgebreide Lagere Onderwijs* ((further) extended elementary education), HBS, *Hogere Burgerschool* (higher public school) and gymnasium. The MULO took 3 or 4 years, the HBS 5 years (there was also a shorter, three year variant) and the gymnasium 6 years. The gymnasium was in a sense the direct descendent of the old grammar or Latin school, and the HBS was the product of the new ideas of the mid nineteenth century. The gymnasium was the training ground for future academics, whereas the HBS was intended to provide young citizens with the necessary skills for trade and industry. The MULO was a simplified version of the HBS.

was allowed to use the paintbrush, or to employ the hoe, with the result that I once fell into the paint pot and at another time cut my own shins with the hoe. I have all my life suffered from such clumsiness.

Furthermore it was a precious thing for me to see a cow chewing the cud in the meadow, and I loved it if somebody took me for a walk and allowed me to sit in the grass at the side of a canal with a meadow with cattle at the other side. And also in winter, when the cows were in the cow shed, they offered me a colossal attraction; I loved the swinging of the tail and the resulting swarming up of mosquitoes.

As I got older my amusements naturally changed, and the old ones gave way to looking at pictures, and soon also to the reading of stories from Mother Goose. I rarely went to the elementary school. Most things I learned at home, and I have few recollections of the school. Only, I can still see how the schoolmaster pulled one of my classmates by his hair through the schoolroom so that the boy passed out, and how then all of us got a day off. But all the sharper are the impressions that I still have from the time that followed; how I commuted with a season ticket by local railway from Medemblik to Hoorn and attended the high-school in the latter place.[9] I remember still very clearly that often, waking up, I saw that the train was about to depart in 10 or 5 minutes, and how I hurriedly slipped on my clothes, got my books and sandwiches, ran down the stairs, dived into the ally at the side of our house, then covered the road to the station, then running like a madman and finally, out of breath, flinging myself through the station into the train,[10] where I usually still had to lace up my shoes, and to finish tidying myself up—in my haste I had omitted various necessary parts of this.

Although we lived fairly close to the station, it was quite a job— especially in the severe winter of 1890–1891—to be on time at the station, and the hour of departure of our train, which was at first twenty past seven, was moved forwards at every new timetable, until it finally reached half past six. Fortunately, I was not the only one to bear this cross, but I had three fellow sufferers. The journey took one hour, and this hour was, if necessary, devoted to the learning of our lessons, but otherwise, out of boredom often, to forbidden actions. We posted ourselves on the platform of the train, which was strictly forbidden, or we fiddled with the equipment, for example, the gas lamps; we were even so childish as to bother our fellow passengers, for example, by pricking them through holes in the backs of the benches with a pair of compasses. Thus we were obviously in disrepute with the passengers and the railway employees, and the employees took revenge in any way where they saw the slightest opportunity. For ex-

---

[9] Bertus was registered on the first of September 1890.
[10] The railway line Hoorn–Medemblik is nowadays operated by volunteers. In the Summer season one can make the trip in historic carriages drawn by an equally historic locomotive.

ample, if one of us forgot his season ticket, the guard gave no pardon and we had to pay the ordinary fare, and once the men pulled the following trick: in winter we usually stayed as long as possible around the stove in the waiting room, as it was ice-cold in the train, for we had ample time to board the train when we heard the whistle of the guard, since the train was always right in front of the waiting room. When our habit was noticed by the station master, the driver, and the guard, the train one morning rushed out as fast as it could without any announcement of its departure, and we were left behind. We did try to make a last minute jump on the running board, but the guards and the stoker prevented us from doing so. Of course, after this, the lingering in the waiting room was over. On the way to Hoorn we usually had little company to start with, but on approaching Hoorn the train slowly filled up with farmers, and at the end of the journey, especially on market days, we were packed like sardines. The proverb says 'the more the merrier', but we did not at all see our swelling company in that spirit. For all those farmers smoked like chimneys and at the same time they had a mortal fright of a draught, to the extent that they never allowed a window to be opened; and so we sat there amidst billowing smoke and in a terrible atmosphere. Thus we fostered a profound hatred for our fellow passengers, and so it is understandable that when we bothered them, there was a good measure of revenge involved.

During the first weeks we regularly spent our lunch break with the caretaker of the high-school, where we ate our sandwiches and got a glass of milk. After some time we all but stopped this routine, and instead went to the station, to which we had free access with our season tickets. There we could have lots of fun, for which we gladly gave up our glass of milk, and were willing to cover the fairly long road from the school to the station twice. We were the best of friends with the station master in Hoorn and all the staff of the station, and to that we owed the permission to push ourselves along the tracks on a trolley, or to make a ride on the freight train that left at a quarter past two to Purmerend while it was shunted; yes, even to act as a shuntman. Sometimes we played tag on shunting trains, jumped off rolling carriages, and up again, jumped from one carriage to another, climbed up on the roofs; in short, did all kinds of tricks which would have made our parents endure a thousand frights, could they have seen us there. Moreover, in winter we could skate without being bothered by other people on the smooth ice of a pond which belonged to the station and that could only be reached through the station. Usually, if it had snowed, we ourselves swept the rink before nine o'clock, on which we skated between school hours. Also at the station in Hoorn some mischief was practised, but here it never went so far as to spoil the good understanding between us and the staff of the station.

Bertus judiciously omitted to mention an incident that could have had a nasty ending: he once climbed onto the roof of the station in Hoorn, where he was seized with a spell of dizziness, so that he very nearly lost his balance.

Apart from Brouwer's own biographical sketch, little is known about his years in Medemblik. One would like to know about his childhood friendships, his development, his adventures in the quiet streets of the town and the expeditions into the countryside. Bertus' years in Medemblik are a closed book to us. There is just one minor but consequential detail that must be mentioned, as it plays a key role in Brouwer's later years.

In Medemblik there lived a family Pels with whom Bertus made friends. The daughter, Dina, was two years older than Bertus and she attended the same high school in Hoorn as Bertus. Dina entered the HBS in 1892, the same year that Bertus left for Haarlem, and she finished three classes. Subsequently she went to Amsterdam where she combined a job in a pharmacy with the training for the certificate of 'pharmacy assistant'. The training in those days was a matter of apprenticeship in one of the officially recognized pharmacies. Apparently Dina and Bertus met again when the latter enrolled in the university. As we will see later, the renewed relationship between the two had far reaching consequences.

In 1892 the Brouwer family moved again, this time to Haarlem, the capital of the province of North-Holland. Brouwer senior obtained a position as headmaster of the MULO at Haarlem. This move was the last step up in his career. Lacking an academic education or the supplementary qualifications (the so-called *middelbare acten*) he could go no higher in the teaching profession. Haarlem had more to offer to Bertus than the sleepy town in the north.

> As far as life in Medemblik itself was concerned, the older I got, the more I felt the unpleasantness of it. There were few boys for company, there was no surrounding countryside, there were no walks apart from the sea dike, sports were unknown, so that I had little recreation to go with my daily work. Thus, when we moved to Haarlem, I made a good switch as far as the town was concerned. In particular during the first year, there was a great deal for me to see; my lifestyle underwent a considerable change. The time reserved for learning was more and more cut back by other things. To begin with, walks took up a great deal of my time, because one can—if one is not overly prosaic—do quite a lot of walking in the countryside surrounding Haarlem before one starts taking it for granted. Already in *Den Hout*[11] it is possible to find ever new spots that catch the eye, even after having walked there for a hundred times. And then one can go to the dunes, where a hundred different hillscapes with ponds, villas and copses unwind before one's eyes.
>
> One can wander through the woods of Bloemendaal and Santpoort, and seek out the hollows in the dunes of Santpoort, and if one has had

---

[11] A wooded park in Haarlem.

enough of all those sceneries of nature, one can start botanizing, and find jewels of Dutch flora in the dunes. But apart from walks, also sports took up a great deal of my time, for soccer, swimming, cycling and cricket soon found a keen adept in me, and this gave me a lot of pleasure and a lot of excitement; for if one starts to practice a sport, it is easy to start racing, and many a soccer match or swimming contest deprived me of a night's sleep.

In the meantime I finished high-school (*HBS*); at the end of my high-school years, and the following year I learned Latin and Greek. Next I took an entrance examination for the sixth grade of the gymnasium, where I spent my last school year. This last year was not the least congenial year of my life; I was not overly pressed, I could devote a lot of time to sports, and I could get along perfectly with the boys of my class. There were eight of us, and now we have split up, four to Amsterdam, one to Leiden, two to Utrecht and one failed the final examination. And so a new period will now dawn for me.

I am a freshman (*groentje*), and I hope to become a student. Two thirds of the initiation has passed, and I have made many new acquaintances, talked a lot to them, and learned much from them. I have noticed how much I am lacking in general knowledge and moderation, and I have learned to respect men who, children of the same era but with more experience, could guide and advise me on the road which lies in front of me. Physically, I have, strictly speaking, not been bullied yet I have learned a great deal the hard way. This period is miserably exhausting for me, and I am glad that I have already had a long invigorating holiday, and that after one more week I can catch my breath again. One of the nuisances too, is all the work I get to do in this period: love letters, rhymes and proverbs in the various initiation journals, and not the least, the autobiography of four pages without a margin.

Fortunately no punishment has come on top of that, but I believe that if there had been, I would have dropped in my tracks. But let me put on a brave face; one more week and the barrier that separates me from student life has dropped.

In Haarlem the family lived in a new house on the Leidsevaart (the canal connecting Haarlem and Leiden); at that time the house was on the outskirts of the town. The HBS and the gymnasium were only a short distance from the house, so the travelling belonged to the past.

Mrs. Brouwer took boarders, two schoolboys: Fer and Lau van der Zee, whose parents were in the Dutch East-Indies (the present Indonesia). No records survive of the relationship between the Brouwer boys and the boarders. The only remarkable fact to relate here, is that Brouwer later made use of the pseudonym 'Lau van der Zee' in some contributions to the student weekly *Propria Cures* and the magazin of the Delft students, see p. 64.

Bertus always had a sweet tooth, he loved sweets and treats; as the eldest boy he was usually given an extra helping of the custard, nonetheless he was so fond of this dessert that he sometimes bribed his brothers and the boarders into giving him their portions.

The musical education of the Brouwer boys was taken in hand by their mother; she came from a talented musical family,[12] and she gave piano lessons to her sons, who did not always meet their mother's standards. When the brothers played abominably, or when they had neglected to study at all, she would occasionally play the rod of the gaslight on the backs of her darlings. On one occasion the rod broke; Bertus had it repaired and presented it to her as a birthday-present.

Father Brouwer had the reputation of being a gifted pedagogue, but just the same he had difficulties in handling his own offspring. Bertus, in particular, did not get along with his father. At the time that they were living in Haarlem, Bertus now and then fancied spending a night in the dunes, something his father would not allow. Once, when the urge had become too strong, Bertus managed to lock his father into the cellar. When he was released, Bertus was already safely in

The house in Haarlem.
(Photo Dokie van Dalen)

the dunes. The relationship with his father remained uneasy his whole life. When Bertus was a well-established citizen he used to pretend an indisposition as soon as his father visited him, taking himself to bed and moaning as if in agony. The intimi who were well aware of this play-acting spoke jokingly of Brouwer's *vluchtbed* (a bed to flee to).[13]

For all their intelligence, the Brouwer boys were no softies, in their exploits they came well up to the Tom Sawyer-mark. Aldert was an excitable, impulsive boy, prone to accidents. Bertus himself was no stranger to daring facts, he was an inveterate climber in trees, buildings, etc. Even at the age of seventy-two, at a picnic during a meeting in Canada, he upset his company by suddenly disappearing. To

---

[12] The Poutsma family produced a number of scholars and artists, among whom was Geesje Poutsma, an older sister of Bertus' mother, who gave singing lessons and was a concert singer herself. See also [Stuurman, F. 1995].

[13] Oral communication Mrs. F.J. Heyting–van Anrooy.

everybody's surprise he was discovered up a in tree.[14]

Family picture at the occasion of the copper wedding anniversary. The boys left to right: Lex, Aldert, Bertus.

Bertus' school career was highly unusual, his exceptional intelligence and, no doubt, a certain amount of private coaching by his parents had enabled him to cut short his elementary school years and to get into high school three years in advance of the normal entrance age. The phenomenon of pupils skipping one or more grades was not unusual in the old educational system; Bertus' progress was, however, remarkable. The more so, since his report cards, right from the beginning, show him at the top of the class—and he stayed there during all his school years. In his first year at high school he ranked first in arithmetic, geometry, algebra, history, Dutch, French, English, and German; in natural history and geography number 2; only in art (drawing) he ranked as number 26 (out of 27 pupils). In the second

[14] Oral communication J. Lambek.

grade of the HBS at Hoorn he ranked number 1 in all subjects except art. The mark that the Dutch educational system added for diligence (*vlijt*) is telling: whereas he scored in Hoorn the top mark for almost all subjects, he was graded a 3 out of 5 for diligence in mathematics, but a 5 for the other subjects. This seems a fair indication that the mathematics curriculum at the school in Hoorn did not have much to offer to Bertus.

In Haarlem he entered in September 1892 into the third class of the HBS; when he moved up to the fourth class, he again was number 1 in the class. Even for a clever boy this was something of a tour de force, as he had taken in the meantime the entrance examination for the local gymnasium. The official record says that on 13 and 15 January 1894 he passed his written and oral examinations for the admission to the first (sic!) class. The notice went on 'This candidate has to be a gymnasium student in order to be eligible for a grant of the St. Job's Foundation [see below]. He is admitted on the condition that he will promptly catch up in Latin and Greek.' And so Bertus learned his Latin and Greek while following classes at the HBS. In September 1894 he passed another entrance examination, this time for the third class of the gymnasium.

The school year 1894–95 was a busy year indeed, because Bertus compressed two highschool years into one year; at the end of the year he had mastered the total curriculum of the HBS and he took the final examinations (22, 23 and 24 July) with splendid marks (with exception of (again) arts, this time in the company of *cosmography*). Since he was not a regular student he had to take the examinations before a state committee (the so-called committee of experts (*deskundigen*)). The diploma was awarded on 9 August 1895. The following year he followed the lesson in class 4 of the gymnasium and at the end of the year he passed an entrance examination for class 6, so that he simply skipped a class. As a matter of fact he followed lessons in both parts, α and β—the literary and the science part—of the gymnasium (in the science part he was the only student!). Bertus did, as usual, very well, albeit that his father sternly admonished him with the words 'this must become a 4' in his report card, when Bertus had scored only $3\frac{1}{2}$ (out of 5) for German.

The gymnasium examination was conducted at the school itself by both the teachers and a committee of outside experts. The experts were as a rule university professors or lecturers, who spent part of their summer vacation travelling from gymnasium to gymnasium examining candidates. This examination spree was a traditional part of academic life; one was more or less expected to take part, and there was a modest fee. The system helped to maintain contact between the universities and gymnasiums, it was an implicit tool for control, and it enabled teachers to keep up their contacts with the professionals at the universities. Later in life Brouwer also regularly took part in this examination activity. There was a similar system with committees of experts for the HBS examinations. It is an interesting feature of the gymnasium examinations that they were primarily seen as entrance examinations for the universities and not as the crown on a high school career. This was literally and officially recognized in the *Nederlandsche Staats-Courant*, where we read in the issue of September 1897 that L.E.J. Brouwer had received a testimonial for admission to a study in the faculties of theology, law, literature and philosophy, and also of Medicine, Mathematics and Physics.

Since a gymnasium diploma was the normal requirement for admission to the university, the decision of Bertus' parents to send the boy to the gymnasium was dictated by their wish to give the boy a higher education. Another motive for the prolonged high school education may have been the age of Bertus at the time he finished the HBS: 14 years old, he would have been something of an oddity at the university. There is no doubt that Bertus' two extra years at the gymnasium were well spent. There are indications that he read and studied a lot in his spare time. Quite a number of more or less prominent Dutchmen had indeed been confronted by the fact that the HBS-diploma did not qualify for a university study; some outstanding Dutch scholars had nonetheless entered the University without a gymnasium diploma,[14] but this either required a special dispensation from the Minister of Education, or else one had to take an entrance examination. In 1917 the requirement of a gymnasium diploma for admittance to university was relaxed by law; the studies of medicine, mathematics and physics were opened to HBS graduates.

Judging from the marks that Brouwer earned at the HBS and the gymnasium, there was no specific field of study that was a priori excluded. Indeed, Brouwer had a deep love for languages and he cultivated his Latin and Greek during his whole career. His choice of the faculty of Mathematics and Physics could, however, hardly have been accidental, but the ultimate reasons for this felicitous decision remain somewhat vague, It can be said, however, that *a posteriori* his exploits fully justified the choice.

1.2 Student in Amsterdam

The sixteen–year–old Bertus enrolled at the University of Amsterdam, also called the Municipal University (*Gemeente Universiteit*). This university was an old institution in a new form. Traditionally the Netherlands knew only a couple of universities; the first university in the Low Countries was that of Leuven, in the present Belgium, founded in 1425. After the Reformation, the University of Leiden was founded in 1575 at the instigation of William of Orange, as a reward for the tenacious resistance of the citizens of Leiden during the siege by the Spaniards. The universities of Groningen and Utrecht followed in 1614 and 1636. Higher education in Amsterdam was provided by the Atheneum Illustre, founded in 1632; its elevation to '*hogeschool*' was effectively blocked by the University of Leiden. In 1877 the Atheneum was transformed into the University of Amsterdam ('UVA' for short). There was, in Brouwer's days, one more university: the Free University at Amsterdam, founded in 1880 by the Dutch Calvinists. In addition there was the

---

[14]Korteweg, Van 't Hoff, Lorentz, Kamerling Onnes, Dubois, Zeeman, Pannekoek, Zernike, ...

Polytechnic School at Delft, the former school of artillery of King William I[16] of the Netherlands. It was elevated to *Technische Hogeschool* in 1903, and today it is the Technical University at Delft. The universities of Leiden, Groningen and Utrecht were state universities, but the University of Amsterdam was a municipal university. As a consequence it was directly governed by the mayor, who was the chairman of the board of Curators, and by the council of the City of Amsterdam.

For a beginning student with aspirations in the sciences, the choice of the UVA was not an obvious one. The University of Leiden had gained a reputation in physics and astronomy with Kamerlingh Onnes and Lorentz as its star professors, who drew the international attention. The young UVA had done well for itself by engaging Van der Waals, Van 't Hoff and Hugo de Vries. For mathematics there were few reasons to prefer one university to another; as pointed out before, Dutch mathematics was awaking from a long slumber and outstanding authorities were not easily found. So the choice of Amsterdam is difficult to explain. One reason might be its proximity to Haarlem, another its reputation as an exciting cultural-political centre in The Netherlands. Moreover, some relatives (the Poutsma uncles) lived and worked in Amsterdam, and this may

Bertus.

or may not have been an extra argument in favour of the UVA. As a rule Dutch students were (and mostly still are) rather conservative with respect to their choice of university. Geographical arguments carried considerable weight, and mobility was markedly absent. Whereas students in neighbouring Germany usually changed universities before specializing, Dutch students tended to stick to their first choice.

Whatever the motivation may have been, on the 27th of September 1897 Brouwer registered as a student at the faculty of Mathematics and Sciences at the UVA. The young boy followed the example of most of his fellow students and joined the Amsterdam Fraternity, *het Amsterdamse Studenten Corps*. The 'Corps' was, and still is, split into a number of *disputen*, debating or social clubs, sometimes of a specialistic character. In Brouwer's days, and long after, the membership of the fraternity was almost obligatory. Non-members, called 'nihilists', were considered to lack the essential ingredients that were believed to make a student something more than just a person who studies. Bertus first signed up for the debating society NEWTON (23 October 1897). This was the society where the science students met and discussed, among other things, scientific topics. A month later he joined

---

[16] This was the first Dutch King of the House of Orange, to whom the crown was offered after the fall of Napoleon. To outsiders the history of The Netherlands may seem somewhat confusing. Until the French revolution the Princes of Orange were Governors of the Dutch Republic. Napoleon made his brother Louis Napoleon King of Holland; the Orange monarchy was introduced after Napoleon's fall. The famous William III was King of England, but Governor of The Netherlands.

another dispuut, PHILIDOR, the meeting place for chess players. A more important *dispuut*, however, may have been CLIO, a literary club, where Bertus met a number of fellow students who, in one way or another, were going to make their mark on Dutch society.

Although Bertus' intelligence was beyond dispute, one should not take it for granted that this by itself was enough to study at a university. The financial burden was far from negligible, and many a potential young scientist ended up as a shop assistant or bookkeeper. The income of a headmaster was scarcely sufficient to support a child at the university—let alone three! Bertus, however, was fortunate enough to obtain support from a private fund in Friesland, the *St. Jobsleen* at Leeuwarden.

This foundation is one of many private institutions that even today support worthy young students. Some of these funds offer grants to students of a particular geographic or religious background. The St. Jobsleen supported students of Friesian descent. The grant was awarded the first time to Bertus in 1894, when he enrolled in the gymnasium in Haarlem, and he received DFl. 450 a year; when he entered the university the foundation doubled the grant This was, by any standard, a generous amount, taking into account that a skilled labourer would consider himself well-paid with such wages. From this sum DFl. 150 was deducted to be paid out after the successful completion of the study.

The student fraternity brought Bertus into contact with a number of interesting fellow students, who helped him to extend his intellectual horizon. The most prominent among them was Carel Adama van Scheltema, the grandson of a clerical poet of the same name. Among the remaining members of Bertus' circle of friends, Jan van Lokhorst, Henri Wiessing, and Ru Mauve, stand out for one reason or another. The most elusive among them was Jan van Lokhorst, a mathematics and physics student, who exercised a considerable influence over Brouwer. This somewhat unusual person dressed as an eccentric; on one occasion he sported a yellow suit with matching shoes. Jan van Lokhorst had introduced Brouwer, when still a gymnasium student, into the company of artists, in particular of Thorn Prikker, a well-known visual artist, who spent his later life in Germany, and Boutens, an influential poet. According to Wiessing:[16]

> Only the above mentioned Jan van Lokhorst, one of his contemporaries who later switched to Leiden and died young (1904), had any noticeable influence on Brouwer's attitude towards life. In the last years of his life this Van Lokhorst was already venerated and consulted by considerably older and already well-established authors and visual artists.

The death of Van Lokhorst is the subject of one of the many legends surrounding Brouwer. His stepdaughter, Louise, related that when Van Lokhorst' death was approaching, Brouwer felt inexplicably drawn to his friend, whom he found dying, in a small hut.

---

[16] *Bewegend Portret*, [Wiessing 1960], p. 142.

A meeting of the fraternity CLIO. Brouwer is top left.

Henri Wiessing, an enterprising young man of Roman Catholic origin, became a close friend of Scheltema and Brouwer. After his studies he became a journalist with far left inclinations, and for some time he was the editor in chief of a progressive, left wing weekly *De Amsterdammer*, known affectionately as *de Groene* because of the green colour on its front page, see p. 288. Brouwer sent from time to time contributions, in the form of a small article or a letter to the editor, to Wiessing's weekly, and he stood by him in a number of literary-political affairs. At the time of Brouwer's undergraduate years, Wiessing was infatuated with Adama van Scheltema.

Ru Mauve[17] was the son of the painter Anton Mauve; he studied medicine for some time in Amsterdam, but decided to prefer a simple, idealistic life. In 1898 he exchanged the world of study for that of a craftsman; he took up a job with the famous architect Berlage but soon changed his mind again and joined the much discussed commune of Frederik van Eeden (see p. 60). Having experienced the pleasures and miseries of life in a commune, he departed for Florence. After studying architecture he eventually enrolled in Delft, where he left without a diploma. Mauve remained a lifelong friend of Brouwer.

---

[17] see page 62.

The poet, Carel Steven Adama van Scheltema, or 'Scheltema' as he was called for short,[19] was probably the most influential person in Brouwer's early life; the two met in CLIO and NEWTON.

Scheltema was four years Brouwer's senior, and already a man of the world. He was not as gifted, intellectually, as Brouwer; whereas the latter took the gymnasium in a gigantic stride, Scheltema had to struggle along. Among his fellow students Scheltema stood out by his striking personality. He had enrolled in the faculty of medicine in 1896, and although he soon discovered that he was not cut out to be a doctor, he duly took and passed the propaedeutic examination. Scheltema was a man of culture, blessed with a fair dose of charisma and authority. Already in 1897 he was elected to three important positions: member of the senate of the *Amsterdamse Studenten Corps*, editor of the prominent student weekly *Propria Cures*, and chairman of the Student Drama Society. Indeed, his fervent wish was to become a professional actor. After his performance of the title role in Richard II on the occasion of the lustrum festivities in 1899, he joined a theatre company. His retiring disposition made him, however, ill-suited for an actor's life, so he soon gave up the theatre.

After a short excursion into the world of art dealers, where he worked for the Van Gogh Art gallery, he once and for all gave up his quest for a regular occupation, and became a free-lance poet. He could afford to do so, mainly because his father, on his death in 1899, had left him enough to lead a modest but comfortable life.

Scheltema's father had died of a tumour in the brain; and the experience was so traumatic that Scheltema was haunted for the rest of his life by the fear of a similar fate.

Scheltema had taken a keen interest in the young student Brouwer and, in fact, became the self-appointed mentor of Bertus. The friendship that ensued is reasonably well documented by a collection of letters exchanged between 1898 and the death of the poet in 1924.[20]

During the first year of his study Bertus sampled some of the traditional activities of the university and the student societies. Wiessing described in his autobiography[21] his student friendships and provided some illuminating remarks on the young Brouwer, 'a young and very tall Friesian from Haarlem, about whom—although only a student and no more than sixteen years old—rumours circulated concerning his mathematical knowledge.'

In contrast to Scheltema, and Wiessing himself, 'this introverted '*éphèbe*' remained 'an obscure student'.[22] In the Corps he not only shunned prominence, he avoided members of the Corps, in particular the prominent ones. He had declined to join the social club, which had invited him after the initiation period, and with

---

[19]The family name is 'Adama van Scheltema', a typically Friesian name. The suffix 'a' usually indicates Friesian descent. We will stick to the abbreviated version 'Scheltema'.

[20]Cf. [Dalen, D. van (Ed.) 1984].

[21]*Bewegend Portret*, [Wiessing 1960].

[22]The traditional name for a student who seldom frequents the fraternity events.

the Corps he had, after a short spell in the rowing club *Nereus*, hardly any ties, except through the study club NEWTON.

1.3 The religious credo

As a young student, Bertus joined the *Remonstrantse Kerk*, a Protestant denomination that had its roots in a theological dispute in the seventeenth century. That particular dispute would have remained obscure, were it not for the circumstance that the religious rift among the Dutch Protestants (basically Calvinists) had its repercussions in social and political life. The Grand Pensionary Johan van Oldenbarnevelt (1547–1619), who favoured peace with Spain for the sake of economic expansion, had adopted the case of the 'Remonstrants', and, partly in reaction, Prince Maurits (son of William of Orange) had taken the side of the 'contra-Remonstrants'. The Remonstrants suffered defeat and remained ever since a small but refined religious denomination.

Brouwer's choice to join the *Remonstrantse Kerk* is somewhat puzzling, since his parents were members of the Dutch Reformed Church,[22] and presumably Bertus was baptized as such. There are no indications in his later life as to his religious affiliations, so there is no simple explanation of this step.

Before being solemnly accepted into the Church, Brouwer was, in accordance with the custom, requested to write a personal profession of faith. The text of this profession has survived; and it is highly significant in the light of Brouwer's later philosophical views. As a boy of 17 years old, he presented a coherent idealistic and even solipsistic view of his religious credo.

In the light of the following solipsistic reflections, it seems significant that already as a small child Bertus was occupied with the status of the ego. His mother used to tell that as a three years old boy, Bertus asked the question 'What is I?' According to the experts most children start to discover the self at an early age, but hardly ever to such a degree that they formulate it explicitly.

The confirmation took place in March 1898, and Dr. B. Tideman was the Minister. The candidate for confirmation had to write down his private views on a list of questions or topics provided by the minister; as a rule the confirmants reproduced as well as possible what they had learned in the Catechism class, not so Brouwer! The questions are unknown, but Brouwer's answers are interesting by themselves. A translation of the original text is reproduced below.[23]

> Point 14. What is the foundation of my faith in God? This is for me the main point of the profession of faith, and the only thing that may properly be called 'profession of faith of a *person*'. That I believe in God originates by no means in an intellectual consideration in the sense that I should conclude from various phenomena that I observe around me the

---

[22]'The one-time state religion.
[23]'The order of the answers is Brouwer's.

revelation of a 'higher power', but precisely in the utter powerlessness of my intellect.

For, to me the only truth is my own ego of this moment, surrounded by a wealth of representations in which the ego *believes*, and that makes it *live*.

A question whether these representations are 'true' makes no sense for my ego, only the representations exist and are real as such; a second reality corresponding to my representations, independent of my ego, is out of the question.

My *life* at the moment is my *conviction of my ego*, and *my belief in my representations*, and the belief in That, which is the origin of my ego and which gives me my representations, independent of me, is directly linked to that. Hence something that, like me, lives and that transcends me, and that is my *God*.

One should by no means read in the many words that I have used above, an intellectual *deduction* of the existence of God, for this belief in God is the bedrock, from which can be deduced, but that itself is not deduced. The belief in God is a *direct spontaneous emotion* in me.

Now I do think that this belief in God is of a somewhat different nature than the ordinary one, and this mainly because mine rests on a *Weltanschauung* that acknowledges only me and my God as living beings, of which I *know* myself, and sense my God, my master.

Furthermore, the representations which are given to me contain in themselves, for instance, also that there should be other egos, also with representations, but these are not real, they are parts of my representations, therefore they are *mine*. My representations are my life, thus at this moment I live in the representation that I think of my life, and write a profession of faith, but *in* that life I do not find my God, my God is under[24] and outside my life, only the fact that I live, makes me sense my God, it is not *in the way* I live that I find my God.

This view includes my *immortality*, or rather, cannot consider mortality. For the concept of *time*, like *space*, belongs to my representations, whereas my ego is completely separate from those concepts. My relationship to my God is a dependent trust in him, who *makes* me *live*.

But the life that my God gives me to live can be thinking about things that I observe and the state of things that I see around me and having opinions about various matters, also so-called *religious matters*, but then these are representations given to me by my God, who is outside and above it; they cannot encompass my God, for they originate from him.

---

[24] under, i.e. is the *Urheber* of my life, [Brouwer's footnote].

Only the *sensing of my God* belongs to my proper religion.

However, here the language is too awkward an instrument. The *sensing of God* and the *trust in God* is not a conscious thought, and hence a representation, for then it would again be situated outside God himself, but it is something that, as it transcends thought, cannot be thought, let alone written down; it is something that is tied to the unconscious ego, becomes conscious, that is receives representations, but is separated from those representations. Indeed, an image of it can come into being in a conscious thought, but only very vaguely.

This view of mine concerning the first point of the profession of faith renders the discussion of many other points superfluous.

[1–13] In the first place, a historic survey of religion can in my case contain nothing that guided me to my conviction and hence has no place here.

[15] Objections to the acknowledgement of a divine rule in this sense, that I would not see how to reconcile with various things that I observe around me, do not exist for me. For my perceptions are part of my representations and none of these can by their nature be an objection, all of them are in *their* existence a proof of *God's* existence.

[16,17] The characteristic feature of my conception of life in contrast to that of others has already been stated. I neither conceive life a burden I have to carry, nor a task to be fulfilled; no, my life is an accomplished fact, about which I cannot give an opinion. For to that end, I would have to view it objectively from the outside, but that I cannot do; for me life is the great unique It to which I cannot assign properties, because nothing can be compared with it.

This view does not at all imply that my life should be a dull, blind, will-less letting go. The life that my God gives me to live can be rich in hope, anxiety and aspiration, full of passionate pursuit of ideals, and my own free will can be strong; all this, indeed, belongs to the representations that can be given to me.

[18, 19] I already mentioned at point 14 my unlimited trust in God, and my conviction of my immortality.

[20] Among the representations which my God gives me are those that make me at some moments feel intensively his existence, this is then followed by a strong self-confidence and a joyous courage to live. Each time when that awareness forcefully thrusts itself upon me, stirring my inner life, I may speak of *love for my God*. For me such moments of contact do not have the character of prayer, because my wishes and sorrows do not play a role, but, on the contrary, have totally disappeared for me.

So far I have been able to connect the points with my religious conviction. The remaining part has a totally different meaning for me.

To wit, my God has also given me the ambition to make my life, i.e. my representations, as beautiful as possible.

[21] from this it ensues that I am struck by the loathsomeness in the world that surrounds me, and is part of me, and which I will try to eliminate; also as regards the world of men. I can hardly call this *love of my neighbour*, for I detest most people; hardly anywhere do I recognize my own thoughts and spiritual life: the shadows of men around me are the ugliest part of my conceptual world. So, in theory, I will never sacrifice myself for another person; God has, however, given me feelings such as compassion, which sometimes force me to act in that direction.

Through the unconscious pursuit of beautifying my representations I have of course opinions on the being or not being useful of institutions in the human world, therefore I can also write about these points, even though I stray more and more from my religious conviction.

I approve of a church because, even though we do not hear in it our own conviction, it can direct our thoughts to fields where by our own action and thought happiness can be found. Ecclesiastical rule and dogmatism are of course phenomena of degeneration; I approve of religious forms for the simple crowd, to be subdued, in a reverential non-understanding, by a church that wants to dominate.

Once again, I approve of the church as the one that points out our task to us; to find in a religious conviction a staff to go through life. And this is the credo of my religious feelings and convictions, of which I have now given an account for the first time, and which I have ordered and sifted even though the unity and force have suffered by an arrangement in points that was not mine. March 1898.

L.E.J. Brouwer.

This is not the place to give an extensive exegesis of the profession of faith, but let us just note a few interesting and important points.

In this document we can find Brouwer's views on his life, in his own formulation. The basic underlying idea is that life is just there; it is not within the competence of a person to put himself as a judge above it. Since the ego and life are almost synonymous, one cannot step outside life and view it from a higher position. He immediately goes on to reject a fatalistic view of life: 'the free will can be strong'. The views expounded by Brouwer are very similar to those of Indian philosophy and religion. At point [20] Brouwer describes the experience of feeling the existence of God. Here one recognizes what traditionally have been called mystic experiences. The last section but one treats the relationship with world and the fellow human beings. It describes Brouwer's feelings for his fellow men in surprisingly frank terms. Apparently the intervention of God is required to give him feelings of compassion.

The final views on the church and its role are rather cynical, to say the least, but given Brouwer's basic view on the ego and its relationship to God, not without a point. The minister must have been surprised at such a confirmant, but fortunately the Remonstrant Church had a reputation for open-mindedness and tolerance.

Before leaving the topic of the credo, it is worthwhile to pause for a moment, and reflect on its status. The question one would like an answer to, is 'how original are the basic ideas?'

Some of the material has the flavour of Schopenhauer. And one is also reminded of Cusanus, when Brouwer points out the impotence of our intellect in the face of the problem of God and his existence. The analogy with *De docta ignorantia* suggests itself.

There is no definite answer to the above question. On the one hand Brouwer was highly original and unorthodox in his thinking; he had an unusually penetrating mind, as his later works shows. So it was not beyond him to develop a solipsistic view all by himself. On the other hand, he was a avid reader, and a superior school like the Haarlem gymnasium may very well have exposed the young Brouwer to ideas and traditions that could easily escape the untutored student. One should also keep in mind that Schopenhauer was very much *en vogue* around the turn of the century.

It would certainly not be beyond a clever boy like Bertus to assimilate the ideas of Schopenhauer. In the absence of convincing evidence, I would be inclined to give Brouwer credit for the originality of the credo.

From the above profession of faith one gets a fairly accurate impression of the philosophical views of the young Brouwer. It appears that he had adopted a rigorous, Schopenhauer-like, view of the world, religion, and his fellow human beings.

The basic entity for Brouwer is his ego, and immediately after that there are the 'representations' (*voorstellingen*) of (or in) the ego. At this point Brouwer makes the radical choice for a strict idealism, there is absolutely no compromise with impressions from the outer world or representations of (or derived from) experience. The representations are inextricably bound up with the ego. Hence these impressions are autonomous in the sense that they cannot be checked against experimental or objective phenomena. The next step is not forced upon the ego, but rather a matter of free choice, namely the recognition of God as that which is the source of the representations and of the ego. As Brouwer stresses, God is not deduced from the ego and its representations, but the belief in God is a spontaneous act of the ego. One could almost say that it *happens* to the ego.

At the points [16] [17] we can already note some of the characteristic points that we will meet again in Brouwer's booklet *Life, Art and Mysticism*, namely that one should accept the world as it is; it not something one can complain about, it is (in the later terminology) part and parcel of one's Karma.

1.4 Friendship: Adama van Scheltema

The years at the university were far from smooth for Bertus; although the actual study did not present any problems, he suffered from nervous attacks that were to plague him his whole life. Nonetheless he fervently pursued a great number of activities. In the summer of 1898 we find him in the infantry barracks in Haarlem.

Now here is a riddle. What, one would ask, is a boy of 17 doing in the army? He is too young and he is a student, so he has better things to do than to play soldier. Or to put it more positively: his first duty is towards Athena and not towards Mars. There is no definite answer to the question in the absence of data. The most likely solution to the 'army problem' seems a coherent strategy on Brouwer's side to get his army obligations out of the way before the beginning of his real career. He joined the army as a volunteer in 1898 with the rank of *aspirant vaandrig* (reserve officer to be). Combining the information from the *Rijksarchief* and Brouwer's correspondence, we conclude that he was enlisted in the fourth regiment infantry quartered in Haarlem. A letter of 14 August 1898 to Scheltema shows that army life was not as pleasant as Brouwer had probably hoped. Brouwer was no softy, he enjoyed rough sports, had no objections to outdoor life, and he had survived very well at school, although he was invariably much younger than his classmates (recall that for young children age differences are far more important than later in life). So it is possible that he had underestimated the hardships in a world that was probably alien to him. In the army things were not done by Brouwer's rules. Even his extraordinary intelligence would work against rather than for him.

He entered the army on 6 July 1898, and he obtained leave (*groot verlof*) on 21 September. On 27 August he was promoted to reserve corporal, and that was as far as he would go in the military world.

The decision to get done with the army as soon as possible lent an illogical feature to his army career. At the time the lottery system was still in operation, that is to say, it was determined by lot whether one was conscripted. For those who could afford it, there was the possibility to 'buy a replacement'—a person who had not been drafted, and who was willing to take over the military obligations of the conscript, usually this was some ignorant, underpaid yokel. As a result the army was not exactly pleasant company, to say nothing of its efficiency. Most eligible men would wait for the lottery result, and then contemplate how to handle the situation. Not so Brouwer; he reported on 11 December 1900 at the 'lottery board' in Haarlem, having already completed seven-and-a-half months military training. He was not so lucky to draw a blank, but according to the record, he was exempted from military service on the grounds of voluntary service[25], that is to say the actual time served was to be deduced from the obligatory period. The record of the lottery board does not give much information. It listed his physical features: height 1.863 metres, oval

---

[25] *eigen dienst.*

face, blond hair and brows. The *Rijksarchief* records show that Brouwer was short-sighted at the right eye (0.5 dioptre) and that he was 1.848 meter tall at entering the military service. For some reason people usually thought Brouwer to be taller than he actually was, estimates of 2 meter are no exception. The explanation is probably the fact that Brouwer was extremely slender, thus creating an illusion of great height. Of course, it is very well possible that in the following years he added a few centimetres to his length; after all, he was not yet nineteen when his measures were taken.

The following letter makes it clear that even a short stretch of military service had been enough to fix Brouwer's view on army life for good.

> Your letter sounded me like the far away familiar tolling of bells, waking me from a torpor into which I gradually, dulled, broken and kicked, had sunk, and filling me with nostalgia for all the endlessly beautiful things that I miss [...]. The military service is capable of first poisoning and then killing a soul within a few months.[26]

There was one particular event that year that must have varied the daily routine a bit. As it happened, it was the year of the young Queen Wilhelmina's access to the throne. Her father had died when she was only ten years old, and her mother, Queen Regent Emma, had prepared her for her future duty as Queen of the Netherlands. On 6 September her inauguration took place in Amsterdam.[27] The route of the new Queen was lined with various parts of the army and volunteers from the student fraternities. One of the military men along the route was corporal Brouwer. Although he did not have fond memories of the army itself, the duty on the inauguration day filled him with some pride.[28]

To continue the story of Brouwer's military career, he was under arms off and on. Usually he spent part of the vacations in the army, to return to the university when the semester started again. In the *Rijksarchief* there is a record of the time spent in actual service: from 15 December 1898 to 14 January 1899, from 23 March 1899 to 5 April 1899, from 15 June 1899 to 30 September 1899, from 7 July 1900 to 26 July 1900, and finally from 20 December 1902 to 9 January 1903. On 1 July 1903 he was transferred (in his absence) to the tenth regiment infantry. We will see that his army periods were disastrous. They ruined his health and his nerves. It is not clear why and how Brouwer ended his military service. It is not unlikely that the authorities realized that the young man was not able to cope with life in the army.

The correspondence between Brouwer and Adama van Scheltema is an important, and almost the only, source, shedding light on the early years of Brouwer. It starts with a letter from Scheltema of 12 August 1898 and ends with a letter of

[26] Brouwer to Scheltema 14 August 1898.
[27] There is no coronation of the Kings and Queens of the Netherlands, but an inauguration. It takes place in the New Church on the Dam.
[28] Louise told that Brouwer used to recall the occasion with pleasure.

of Brouwer of 25 February 1924, almost two months before the tragic death of Scheltema.

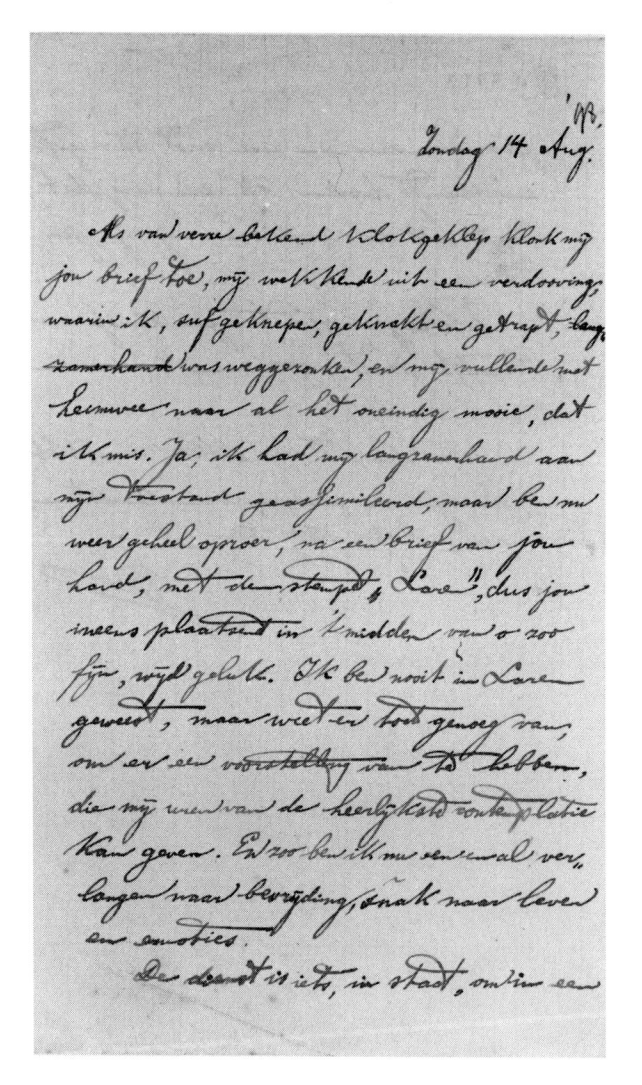

The first letter of the Brouwer–Scheltema correspondence (14 August 1898), Nederl. Letterkundig Museum.

The correspondence sets out on a slightly formal footing, the older student Scheltema addresses Brouwer as 'Dear chap' (*Beste Kerel*) and Brouwer writes *Waarde Scheltema*, a way of addressing that cautiously avoids formality on the one hand and familiarity on the other hand. The Dutch language had in the old days a refined spectrum of titles at its disposal, and '*waarde*' is the kind of opening that one used (and the more traditional still use) to address a colleague without offending him either by unsolicited familiarity or by haughty formality. Very soon, however, the tone of the letters changed into the informality of true friendship.

Bertus was fortunate to win the friendship and sympathy of the older and cultured student Carel, but Carel also gained a great deal from the company of his young friend. Carel, as the older and riper person, guided Bertus into the world of art and into the more worldly aspects of life; while at the same time he benefitted from the extraordinary philosophical and theoretical insights of Bertus. The friendship between Brouwer and Scheltema went largely unnoticed; of course their fellow fraternity members were aware of the close ties between the two, but their friendship, definitely, was not for public display. Although few who knew the young student Bertus would have foreseen it, he later acquired a large circle of acquaintances in all quarters of society. There was an extraordinary mixture of the cerebral scholar, jealous of his privacy, and the gregarious boyish man, with a hunger for company and talk. At no time in life was he deprived of friendships and relationships, but none of them equalled the friendship with Scheltema. This particular relationship had a tinge of the private and the sacred. As Brouwer himself expressed it at one occasion, his friendship with Carel Scheltema was a 'private friendship, one that was not introduced into any collective'.[29] As in all real friendships, there were conflicts and disagreements, but on the whole Brouwer's relationship to Scheltema was one of the few true ideal friendships in his life.

Scheltema, commenting on the friendships of his student life, wrote in his diary:

> I can be brief about Bertus—initially reaching him a *paternal* and appreciating hand, I started feeling a warm affection for him as an equal—and subsequently in almost all respects as a superior. Indeed I still surpassed him in understanding of life, wisdom and experience and strength of soul and determination in visibility— but I had to rank his abstract capacities in the realm of knowledge and beauty higher than mine. And finally I had met a man whom I had to place above me, which could not have been the case with anybody else, unless very temporarily and self-denyingly. In the mean time a mutual human sympathy developed through these spiritual appreciations. It did not decrease my pride in the fact that I, as the only one out of those one hundred members of the fraternity, had seen and been sensitive to this man as a most extraordinary person—and even more, that I gradually proved to be the only one Bertus could have continual contact with, yea, to whom he attached himself as the only *friend*—and that was not because he did not see others, or could not find spiritual contact. I have not had more pure, more fundamental and more penetrating discussions with anybody—with nobody else I am spiritually as tranquil, and so close to beauty in vision or analysis.
>
> This is by far the greatest human being that I have met so far, and I don't think to meet a greater one—he is *in all respects* the paradigm of a

[29][Wiessing 1960], p. 142.

man of exceptional genius—for being a *genius* he lacks the connection between his own mind and the world around him.

The essential part of our relationship was my truly tireless endeavouring to make him come closer to the material world.—

This may suffice to indicate Scheltema's appreciation of Brouwer. He recognized at an early stage that the young student had all the characteristics of the great man-to-be, and he drew an intense intellectual and emotional satisfaction from the contacts with Bertus. This deep sympathy and appreciation enabled him to deal with the less pleasant aspects of Brouwer's personality. Brouwer for his part was deeply fond of his older friend, and at the death of Scheltema he summed up his feelings in the following words:[30]

It was the reflection of your eyes, the inflowing grip of your hand, the warm engulfing of your voice, the peaty smell of your overcoat. It was the wild riches of your dream life, the confusing exuberance of your fantasy.

But around you roamed the compelling force of determinatedness, which you sensed and had to acknowledge, and you were determined and you wanted to understand, and to become a personality.

You have understood much, and you have become a personality. And a part of your flourishing rhythm has become common property.

One may well guess that Scheltema served Brouwer as a guide and mentor in the most diverse matters, and that Bertus gladly accepted Scheltema's advice. For instance, when Brouwer had fullfilled his first period in the army he did not hesitate to turn to Scheltema: 'Tomorrow my cage will be thrown open, will you help me find a room?'.[31] When Bertus was depressed or ill, Carel tried to comfort him and to give the sort of advice that a student of medicine can give. Indeed, the student Brouwer had a poor health, and the letters to Scheltema contain a litany of complaints, disorders, nervous breakdowns, etc. For the greater part the poor health of Brouwer was due to the nervous tensions of the highly-strung, brilliant boy. The complaints follow each other in persistent succession until the fall of 1903, but from then onwards they occurred less often. The correspondence provides a depressing list of, mostly vague, complaints:

- Now self-control and a diet for the convalescing patient, and only a wrinkle of chastening and immunity will remain (20 September 1898).
- Because of some infirmity I have to stay in bed, for that reason I did not come to you; please drop in later when I am again allowed to see people (4 May 1901).

---

[30] *Ter Herdenking van C.S. Adama van Scheltema* (In Memory of C.S. Adama van Scheltema). Note the reference to Homer, cf. [Wiessing 1960], p. 106, apparently the members of the fraternity practiced the 'handshake' that is referred to in the Iliad and the Odyssey by 'he grows him into the hand'.

[31] Brouwer to Scheltema 20 September 1898, written in the barracks, one day before going on leave.

- I remember little from before September, I stand completely reborn, weak and free; not tied by desire, not by memory. But the disease has to wear off first. I have already mastered it with my will, and I will make it the mother of my working power (5 December 1901).

- I am here sometimes very poorly—in bed for a week at a stretch, and then 3 days sleeping badly (11 June 1902).

- ..., maybe I will get rid of the pain in the back, that tormented me terribly in Haarlem (23 August 1902).

It may come as a surprise that this same Brouwer, who suffered the penalties of a delicate nervous system with accompanying physical phenomena, would occasionally carry out downright silly experiments. At one time he wanted to find out how long it took before the ice in the water formed itself around a person's legs. So on a freezing day he took the test himself by getting into the river Amstel and waiting patiently. The story does not tell whether he waited long enough. Curiosity always was a powerful stimulus for Brouwer.

From the Brouwer–Scheltema correspondence it appears that Brouwer did the sensible thing: often, when his tormented body refused to serve him, he withdrew from the city and the university. The letters hint at intervals of recuperation, either on the Veluwe, the thinly populated area in Gelderland renowned for its healthy forests, its heath and its sands, or at the seaside, Den Helder, the national naval centre, or in 't Gooi, the enclave of utopists, health freaks and artists, east of Amsterdam. One wonders how he found time to study at all; as a matter of fact it took him three and a half years to pass the first examination, the *candidaats examen*, certainly no accomplishment to boast of for a student of his brilliance.[33]

On 16 December 1900 Brouwer took this first hurdle of the academic course. He was examined by a committee of his professors, and the diploma was signed by J.D. van der Waals and chemistry professor, H.W. Bakhuis–Roozeboom. In view of his excellent grades and matching performance at the examination, the committee granted him the *candidaats* degree *cum laude*. In the summer of the following year Brouwer widened his horizon by a longer trip to Italy. Following in the footsteps

---

[33] The academic study in Holland basically consisted of two parts. The first was concluded by the '*candidaats*' examination and the second by the *doctoraal* examination. These examinations are roughly comparable to the BSc. and MSc. The titles connected to those examinations were *candidaat* and '*doctorandus*' (abbreviated as 'drs.' in titles). Some faculties had an extra examination at the end of the first year, the so-called *propaedeutisch* examination—endearingly called 'propjes'. The time schedule was flexible. Three years for each examination was the average, but clever students could compress the period somewhat. A doctorandus had the legal right to present a dissertation to the faculty, and after a successful public defence he was awarded the title of 'doctor'. A 'candidaat' could enter the teaching profession, but the diploma was basically considered an intermediate stage on the way to the final doctoral diploma. In the well-stratified society before the Second-World War each university diploma entitled its owner to an traditional civil title, a *candidaat* would (or rather could) be addressed as *Weledelgeboren Heer*, a doctorandus as *Weledelgeleerde Heer* and a doctor as *Weledelzeergeleerde Heer*. A professor was simply *Hooggeleerde Heer*. A female student/scholar had a similar title with '*Mevrouw*' or *Vrouwe* instead of *Heer*.

of countless predecessors he made this trip on foot.

After the *candidaats* examination, there was no marked improvement in health; but now Brouwer had decided to take action, to remain no longer passive under the cruel attacks on his body and mind. He had made up his mind to join the health fashion.

In 1902 he spent long periods in Blaricum in a pension that accommodated a loosely connected set of vegetarians, clean-living adepts, and the like. This was the locally celebrated Pension Luitjes, run by the couple Tjerk and Gerda Luitjes, which had already operated a vegetarian pension in Amsterdam before opening their establishment in Blaricum. Brouwer reported:

> Back from Blaricum, with the mixture of both sexes in vests with bare black feet with blue nails, the sunbathing of bare backs, and the gnawing of raw turnips and carrots; I am so miserable that I can not come tomorrow.[33]

Blaricum also appealed to his younger brother, Lex. In June 1902 Brouwer informed Scheltema that he was planning to share an apartment with Lex, but in September he reported that the plan was cancelled because his brother had decided to stay in Blaricum as a painter. Lex had given up his studies and had decided that the life of an artist was the right thing for him.[34]

Brouwer remained faithful to the practising of vegetarianism, albeit not in a fanatic way; he also became a lifelong adept of sun- or air bathing, fasting and the like. His poor health and nerves also drove him to visit spas. These institutions never reached the same popularity in Holland as they boasted in neighbouring countries. One short-lived spa, the *Bad Courant*, was in operation in Haarlem at the time the Brouwers moved from the town of Medemblik to Haarlem. So Bertus may have learned about the phenomenon in his hometown.

Aldert, Lex and Bertus in the garden in Haarlem.

---

[33] Brouwer to Scheltema 29 July 1902.
[34] Brouwer to Scheltema 4 September 1902.

The disastrous, if brief, spells of military service greatly harmed Brouwer's health, both physically and mentally. The army proved an ordeal to the young man, his fellow soldiers pestered him and his superiors did not take him seriously. A letter of 7 January 1903, two days before his definite farewell to the draft, showed an exhausted Brouwer.

> Dear Carel,
> I only want to send you just a single note. There is of course no time to write a letter while in military service, moreover I am still half sick. After the stomach problem and the chest complaint with blood tinged sputum, and now a rash in my face. After performing my daily duties I drowse for a moment in a chair, stoically contemplate the course of my life, and say 'bah' and turn in. In order to have ideals in life a physically sound core is required, and I have lost that. I only ask for a small corner to wither away. Nonetheless I perform all the duties towards my body in a dull constancy; maybe it will work out alright, so that I re-enter the world with a shining look; or it is finished, and with that too I am in accord. I do not know myself which of the two I desire (The Cool Lakes of Death).[35] Should your ideals pine away, seek that state too. Addio.
>
> Bertus

Scheltema reacted resolutely. He wrote from Paris:

> Something positive has to be done. Huet is ancient history—you should go to Winkler or Werthem Salomonson.[36] It is a totally wrong idea that nothing more could be done to a body like yours, especially since the physical condition is the main thing. For example, your cold comes from anaemia, which can be improved by a regular use of iron pills.[37]

From that point onwards the complaints became less frequent; Brouwer remained however an easy target for attacks of a nervous nature, with their accompanying physical disorders. As a result of the physical experiences of his student years he had turned into a health freak, with a strict and even eccentric way of living.

While Brouwer suffered from a long list of complaints, his friend was well on his way to becoming a successful poet. Scheltema had a strong feeling of social responsibility. In as far as his poetry was concerned, this inspired him to write poems for the working classes, simple and understandable. Politically, it made him side with the young socialist movement.

The Socialist Democratic Workers (Labour) Party in Holland, SDAP,[38] was founded in 1894. It attracted a great number of intellectuals and artists, among

---

[35] Novel by Frederik van Eeden, *Van de koele meeren des doods*.
[36] Winkler was professor in diseases of the mind and psychiatry, Werthem Salomonson held an extraordinary chair in diseases of the mind.
[37] Scheltema to Brouwer 17 January 1903.
[38] *Sociaal Democratische Arbeiders Partij*.

others Henriette Roland Holst-Van der Schalk and Herman Gorter. The first was the grand lady of socialist literature in Holland. She had married into a family of artists. Herman Gorter was one of the first naturalistic poets in Holland; in the course of time he had acquired considerable authority in Dutch literary circles.

The students, in particular in Amsterdam, traditionally the red city of the Netherlands, were quick to turn to the new movement. A number of them joined the Socialist Reading Society (*Socialistisch Leesgezelschap*), where the texts of the founding fathers of socialism were read and discussed, including the journals of the (in particular German) socialist movements.

A now almost forgotten, but at the time rather popular theoretical socialist, Joseph Dietzgen, was intensively studied by the students. Dietzgen was a socialist philosopher, born in Germany in 1828, who moved to the United States of America and died in 1888 in Chicago. By profession a tanner, he was a self-taught philosopher, author of *The essence of human brainwork. Excursions of a socialist into the domain of the theory of knowledge*.[39] Prominent Dutch Socialists–Communists studied and advertised his work, among them the astronomer A. Pannekoek and the above mentioned Henriette Roland Holst. Scheltema was particularly impressed by Dietzgen, and he was completely under the influence of his philosophy of socialism, somewhat to the chagrin of his friend Wiessing, who quickly developed an allergy for what he called 'the scientific Buddha of the despicable German social democratic bourgeoisie'. Wiessing characterized the particular brand of Dietzgen's socialist theory as 'the mysticism of disbelief'. Many of the students, including the members of CLIO, canvassed at the elections of 1901 for the SDAP, with its leader, Pieter Jelles Troelstra.

Scheltema's relationship to socialism was ambiguous, to say the least. He, an aesthete, far from the discomforts of the masses, felt absurdly uncomfortable in close contact with the less refined. Before his conversion to socialism, he wrote, for example, in 1897 from Visé in Belgium, to where he had retired to work, that the hotel was abominable, and that this was how it would be in a socialist state.[40] His opinion of his fellow men was at that time pronouncedly negative:

> [...] The friend above all; the sympathetic people, they constitute the world, and all those surrounding it are a shady bunch of scoundrels.

His eventual choice for socialism was that of the theoretician and the artist. The heyday of individualistic, impressionistic, naturalistic literature was past and Scheltema sensed a new perspective in the new ideals of social democracy; he formulated his conception for the first time in 1899 in *Propria Cures*, the Amsterdam student magazine, of which he was an editor. The literary output of Scheltema was twofold, he wrote poems and also books of a theoretical nature, including a travelogue of his Italian experiences. His poetry was written for a large audience, for workers, housewives, school children; it was extremely popular—at his death 65,000 volumes of his

---

[39] *Das Wesen der menschlichen Kopfarbeit. Streifzüge eines Sozialisten in das Gebiet der Erkenntnis Theorie.*
[40] [Bonger 1929].

poems had been sold! Contrary to what one might think, these sales did not make Scheltema a rich man. He refused on principle to make money out of the working classes. Typically, his poetry volumes would cost only one guilder. Among the more theoretical works there was a study *Foundations of a new poetry* (1908), with its telling sub-title 'Essay of a social theory of art contra naturalism, anarchism, the movement of eighty[41] and its decadents'. The study was sharply critical of the exponents of the movement of 'eighty' and also of some of the leading spirits of the socialist movement. This book, which so explicitly singled out its opponents, was attacked violently. Brouwer wrote a penetrating review and sent it to Kohnstamm[42] for publication in the *Tijdschrift voor Wijsbegeerte* (Journal for Philosophy). For one reason or another the editors decided not to publish the review, to the utter indignation of Brouwer, who wrote to Scheltema[43]

> The jerks Meijer and Pen[44] have made notes in the margin on my very paper, like pencil scratches on statues of Donatello. I did not want it to survive the insult, and I burned it.

In general Brouwer was an attentive reader, and at times a strong critic, of Scheltema's work:

> I have not been in the mood to write about your poems, although I have read them all—that golden fleet, it is beyond me, and I take the liberty to condemn it therefore, the elegy of Loke I find gross; and the stable of Jesus is vague in content, although it has some good lines; these are the only three that I would have wished to exclude. For the rest there is a soul in all of them, even so the reader gets a jolt towards the end: the shock of the sudden transition from life in contemplation in the first part to the end, that speaks in concepts; in the case of concepts the intellect is active, and there metre and sound, that accompany only affects of the soul, are ill-suited.
>
> If Kloos[45] argues in a sonnet, then that is permitted on the ground of the forceful effect of contempt or irony or whatever else supports the lot. If that disappears, and the usage of concepts in poems is such that the intellect works stronger, then I am insensitive to a poem; for that reason I sometimes find Goethe insufferable. [...] I will not go on with comments, for after all there is a richer soul in your verses than in those of our modern poets.[46]

[41] A literary movement in Holland that rebelled against the traditional school of the nineteenth century.

[42] Ph. Kohnstamm was a student of Van der Waals. He started his career in physics and switched eventually to pedagogy. He held chairs in Utrecht and Amsterdam.

[43] Brouwer to Scheltema 11 August 1908. There may actually have been more to the rejection of Brouwer's review than meets the eye, see p. 108.

[44] A disciple of the Leiden philosopher Bolland.

[45] Willem Kloos, poet of the *Movement of Eighty*.

[46] Comments on the volume 'Of Sun and Summer' (1902), Brouwer to Scheltema 11 June 1902.

In 1905 Brouwer wrote an exclusive literary criticism of the second of a series of three longer poems, *Londen*, *Amsterdam* and *Dusseldorp*. Like its successor above, it was rejected, but a copy survived in the archive of P.L. Tak, the editor of the *Kroniek*.[47] The review is interesting as Brouwer develops a sort of evolutionary schema in which socialist art *à la* Scheltema fits:

> The force of the new gospels is in their justified rejection of the one-sided positiveness of the old ones; but once they have made the old one perish, there remains only positiveness, which will cause in due time a reaction and they will be pushed back in turn.
>
> But the new will not force out the old if it does not first step into the forms of the old. Whoever presents something new in the forms of the old will make a career; but whoever makes people personally experience the new, will be reviled [...].
>
> A change of literary gospel is always preceded by an analogous change in the economy, where often the literary consequences of the old economy flourish on the illusion of the new one. The Dutch '80 of literature belongs to the '48 of the economy; before '80 the restoration in literary art was carried on by the liberals.[48] [...] And so the liberal literature is now carried on by the socialists Gorter and Roland Holst. Note that the old conceptless mood-poetry is now fed by the new economic gospel, and thus it provides this new gospel with old fashioned needs such as ideal-confirming art.

The implicit criticism of Gorter and Roland Holst, the figure-heads of socialistic art, was rather scathing—their poetry was seen by Brouwer (and by Scheltema!) as a belated exercise in bourgeois pre-socialist expressive art, and in this form supporting the new economic movement. Brouwer was inclined to see in the case of Gorter a strong visionary involvement with the new social gospel: 'strong and deep enough to break through the walls of bourgeois understanding; and so wild phantasms flower there on a ruin of intellect'; whereas in the case of Roland Holst 'the old logic, on the contrary, remained standing cleanly and squarely like a little house at the river Zaan'.[49]

Brouwer definitely had a point, but it was short of a chutzpah to express it at the time. In contrast to the older generation of socialist poets, Brouwer presented Scheltema as the real socialistic literator, who did understand the times and the modern world.

---

[47] A literary magazine. The review was posthumously published in [Delvigne 1985].

[48] The 'Movement of 80' was a Dutch literary movement which introduced in the Netherlands the concepts and styles of the leading European artistic circles. It imported successful innovations such as naturalism, impressionism and the like. The movement of '80 dominated the Dutch literature far into the twentieth century. Brouwer compared this movement with the political-economical revolutionary ideas of 1848. He saw Scheltema as one of the first literary artists escaping the influence of '80.

[49] The old heart of industrial North-Holland, reputed for its beautiful old Dutch houses, still to be seen in the local outdoor museum.

The first realization of socialism in literary art is Adama van Scheltema. Here no emotions are recognized, unless strongly felt as a necessary moment of interaction of his human nature with the living world around him and as a necessary parallel of his actions.

In the meantime he is, more than is good for him, ahead of his own time.

This last sentence can be understood when we realize that Scheltema, however popular as a poet of the common people, was not recognized by the literary world.

Brouwer's review addressed some admonishments to Scheltema as well: '... his prose will have to become less agitated, and ruled more sternly and strictly; it still has to loose more the character of the unbalancedness of the idle bourgeois, who is not part of the earnestness of creating.'

It should not come as a surprise for the reader that the review was rejected by the board of the *Kroniek*; the Gods at the Dutch Socialistic Olympus would not have been amused by irreverent remarks of a self-styled philosopher with mystical antecedents.

Objectively speaking, Scheltema will probably go down in history as one of the minor Dutch poets of our century, but that does in no way detract from his merits as a popular poet.

Carel Adama van Scheltema and Henri Wiessing. Nederlands Letterkundig Museum.

He has added significantly to the appreciation of poetry, and art in general, in the working and middle classes. His theoretical work, however, has never attracted the attention it deserved.

Brouwer's opinion of Scheltema was, we guess, coloured by his friendship, but even more by extra-poetical arguments, such as his interest in the theoretical foundations of art, society and communication.

In view of Brouwer's relationship with Scheltema, it is hardly surprising that he was influenced by the latter's political views. In the early years Scheltema, like so many students, was impressed by Dietzgen, and some of Scheltema's enthusiasm for Dietzgen rubbed off on Brouwer.

Rather aloof, Brouwer followed his friends on their rounds of the workers circuit and even went so far as to join the Socialist Reading Society.

The membership of that illustrious society, incidentally, played a trick on Brouwer. In 1901 on his walking tour to Italy, he was, in a mixture of eccentricity and practicality, dressed in a long cape, which he used when sleeping in the open air at night. One day in Italy he was arrested for vagrancy, and when the police found in his pockets the red membership card of the Socialist Reading Society, they were convinced of a good catch: a strange young man with a red card, printed in a cryptic nordic language, with the word 'socialistisch'. Brouwer was, however, released no sooner than he was arrested when a letter of the Italian mathematician Bianchi was found upon him, which contained the names of a number of famous scientists known to the local police magistrate. Brouwer used to narrate the story of his foot tour to Italy with justifiable pride.

One of the little adventures seems to come straight out of a classical adventure story: once, at the end of a day's hike, Brouwer had joined a small group of men who had found a suitable spot in the forest to spend the night. The company made a fire and prepared a meal. Brouwer was with proper hospitality invited to take part in the meal. He gladly accepted and joined in the general conversation at the fire. After a while he wrapped himself in his cloak and laid down to sleep. Listening to voices of his fellow travellers, he discovered that they were discussing how to rob the young foreigner. Waiting some time, he made his exit when everything was quiet. By the time he was missed, he was already far enough away to be safe.

The proper thing for a student at the beginning of the century was to join, if not the socialist party, at least the band of enthusiasts, who saw in socialism the road to a new and just society. Scheltema and Wiessing, each after their own inclination, became active supporters of the socialist cause, and for some time Brouwer joined them and frequented political meetings, albeit in a more contemplative spirit. He fostered for some time socialist sympathies. This is remarkable indeed, since the rigid materialistic views of, for example, Dietzgen contradicted everything he believed in.

The following quotation from his letter of 11 June 1902 to Scheltema gives some indication of his inclinations:

> *La société, c'est la blague*[51] and in this way it is actually rather funny—of course at the bottom there is the earnestness of life, and we are socialists, but that is only a dim basis for the grand totality that we try, according to our strength, to put somewhat in the right position.

But before two years had lapsed, he had cut his short liaison with socialism:

> My short-term socialist inclinations, dating back more than two years, have turned out not to be viable.[52]

---

[51] Society is a joke.
[52] Brouwer to Scheltema 18 January 1904.

Five years later his farewell to socialism was final. Korteweg, his Ph.D. adviser, was at that time apparently involved in the election campaign of the liberal party,[52] and Brouwer offered his assistance:

> Professor, I rarely think about politics, but my political sympathies are in the liberal, anti-democratic direction.[53] So if you can use the assistance of someone without experience in political work, then I will gladly help to support a liberal against a free- or social democrat.[54]

Later in life he became involved in local politics as an alderman for the Neutral Party in his hometown Blaricum.

The Brouwer–Scheltema correspondence shows us the melancholy process of the inevitable growing up of a young gifted, intelligent but impulsive young man; a process with all its agony, yearning, despair and elation. One of the first letters gives us the picture of a confused, unsettled young man:

> I am too drowsy to experience my freedom, my mind is not capable of any activity, not yet delivered from the oppression of that most dark revelation of Adam's curse.[55]

Slowly, through the years, his physical state improved, and with it his mental state. Although there were numerous relapses, he slowly gathered the strength to face the world. After the utter resignation in his illness during the military service in 1903, the complaints had become less frequent, and a certain defiance of the problems of the world could be discerned. Having done with his military service, he returned to his studies:

> —after an absence of two years it required some determination, especially where any trace of love for the subject [mathematics] was lacking. [...] My work is done without illusions, but with a feeling of joy for the activity in itself.[56]

In a later letter, looking back, he lamented the loss of innocence and susceptibility:

> What a lack of tenderness, of childlike innocence, of abandonment in the words that I write down; I know it, I would be ashamed if I met myself as I was five years ago; but just as one cannot stop the growth of one's beard, so one cannot stop the growth of the philistine-tissue through one's soul.

---

[52]'liberal' has to be understood in the classical sense of the nineteenth century, *laissez faire*, limited government interference. It was neither conservative nor Marxist.

[53]'There was a certain measure of anti-democratic feeling in the air at the time. The influential philosopher Bolland, for example, crusaded against democracy. It is possible that Brouwer also flirted with these ideas. Given the context of the letter, we may be sure that Brouwer referred to the Social Democrats.

[54]Brouwer to Korteweg 22 September 1909.

[55]Brouwer to Scheltema 20 September 1898.

[56]Brouwer to Scheltema 15 November 1903.

> Then let me be great as a philistine! And follow my course alone, unfeelingly among the dead stones to the fair End. And so leave my trace on the thus melancholy world. That is, Ambition is born within me, perhaps! But in any case, one that knows to control itself, and to collect quietly material, until its time has come. **I will have to remain obscure for a few more years, then my grasp will be felt.** [my boldface] Just because I feel the futility of all worldly things, no detraction or fear will disturb my course.[58]

That same year, on the sixteenth of June, he passed his final examination, the *doctoraal examen*, which earned him the title of *doctorandus*. Again, he passed with the highest honours, Cum Laude. He dryly informed Scheltema of this fact by a telegram containing just the word 'Cum'. In the years that lay between the first steps in science and the proud moment of being awarded the '*Cum Laude*', Brouwer had established himself as the equal, and partly the master of, his older friend. Their friendship had passed through all stages, reaching a stability that was not to be disturbed by the sometimes moody or even crude actions and remarks of Brouwer. The relationship between Brouwer and Scheltema had its touching moments, as well as its small unpleasantnesses. Their correspondence provides a revealing insight into the minds of the fast developing mathematical genius and the socialist poet. There are numerous small details that allow us glimpses of their habits, their reading, financial problems, ... For example, Scheltema strongly advised the 21-year-old Brouwer to read Flaubert's *Éducation sentimentale*[59] and a little later he asked him to return his copy of the *Contes* of Flaubert.[60] At another occasion Brouwer reported a visit to the theatre, and recalled the performance of Ibsen's *Wildente* (The Wild Duck). He wrote:

> Send me your new address. I hope that you will find solitary rest there; surround yourself with the books of your equals (*Ebenbürtige*) and congenials. I live with Pascal, Emerson, Madame Gimon and Montaigne. And do me a favour, if you do not yet know it, read the 'Journal of Marie Bachkirtoff'.[61] She has something of both of us, and she stands in between us.[62]

The fact that Brouwer had read the book of Marie Bashkirtseff is interesting. The author was a prodigy of Russian descent, who lived in Paris. When she died at the age of 23 she was greatly admired as a painter and author. Brouwer had recognized in her a kindred spirit—he was under no misconception where his own genius was concerned!

---

[58] Brouwer to Scheltema 18 January 1904.
[59] Scheltema to Brouwer 29 December 1902.
[60] Scheltema to Brouwer 2 May 1903.
[61] [Bashkirtseff 1888].
[62] Brouwer to Scheltema 18 January 1904. 'Bachkirtoff' is Brouwer's spelling. Scheltema misspelt the name.

One should not get the impression that Brouwer's reading was restricted to the exalted regions of literature. Much later, writing from his bed he confessed:

> I don't feel very ill, but if I leave my bed for longer than one hour, my heart starts behaving in a funny way. So I stayed in all week and read all the volumes of the adventures of Arsène Lupin.[63]

In those days the stories of the gentleman-burglar Arsène Lupin were popular, but hardly the reading matter that the more conventional would expect in the hands of a learned scholar!

In 1903 Brouwer had gathered so much confidence in himself that he considered himself on a completely equal footing with his former mentor Scheltema. It was time to cut the umbilical cord, and to face the world on his own. He proposed to seal a formal union, as between two kings, to be concluded on Ascension Day. Apparently there had been some friction between the friends, and it is not wholly impossible that Scheltema saw, to his dismay, the ugly duckling starting to turn into a swan, with a private and virtually inaccessible inner world. Brouwer clearly realized that the mentor–novice relation was over.[64]

Il faut savoir séparer (La Rochefoucauld) *C'est le privilège des grands esprits, de ne pouvoir se brouiller* (Voltaire).

> Carel, my rich poet, I have finished your book, but listen: In no realm are there two Kings, each has to live in his own country of subjects: that loneliness without their equals, that is what they are Kings for. But once a year they visit each other, and see their great contrasts, with nothing in common but the joint feeling of being royal, both being in the immediate grace of God; their associating can be nothing but showing each other the powers and glorification of their mutual domains.—And the only permanent grace, that, from the awareness of the kingship of each other, they can let flow to each other, consists of the rendering of obligatory courtesies and reporting outward appearances of their person and Kingdom.
>
> Carel, your realm is more summery than mine, and your people are more pacified—both our countries have been blessed by God with marvellous beauty.
>
> Well, after our discussion of Thursday night I believe that you are right, but the best idea for our having to live apart, is what I write down here.
>
> Let us meet every year on Ascension Day, and solemnly bathe in the cool sun of spring, and sup together, and exchange what the past year has brought us, and for the rest feel, invisible to each other, united in 'knowing the other to be a King'.

[63] Brouwer to Scheltema 25 February 1916.
[64] Brouwer to Scheltema 23 May 1903.

So, brother, do you agree with this?

Then, hail to You and Your Kingdom—Until 1904.
Bertus

Two days later Scheltema answered; he accepted the separation as a tragic fate. The royal metaphor had immediately sunk in: the idea of Kings toiling in solitude for the happiness of the people, or to put it otherwise: of the best and the noblest working for the redemption of their fellow men, was certainly not a rarity. One did not have to read Nietzsche or the like, to see that the best of the nation were destined to play an important role as prophets and moral guides. It tells us a good deal about Scheltema that, although a socialist by conviction, he whole-heartedly embraced the idea of the responsible, benevolent king. In a sense this is not so surprising; the socialist in Scheltema was partly, perhaps largely, the product of his intellectual efforts, a fact that he was himself very well aware of. In fact, his socialism was a tour de force of a theoretical nature, rather than a spontaneous manifestation of the heart. In spite of his heart-warming poems, he remained a theoretician.

> —Sometimes I am desperately mad about my own desperate soberness! Then I would like to kick my own soul and for once have a touch of anarchistic looseness.[64]

Far from being a game of pretending, the matter of the two kingdoms was taken completely seriously by both correspondents; it expressed in a metaphorical way their views on responsibility and society. As Brouwer put it, discussing the physical sorrows of the body:

> That is the misery of purification, [...], that holds the great soul back from soaring too high and that keeps it in the fear of God; of God, whom he has to serve in watching over God's children, asking nothing for itself—chosen are we—not for our pleasure in the world—we are the prophets, who, messengers between God and mankind, direct and inspire the development, the working, the growing, the flowering of it with the dewdrops that flow from our fingers—you walk earnestly and solemnly through your garden, and scatter them with firm and knowing gestures; I rush through my wilderness, and they roll, without my knowing it—indeed there are few who find them, but they are all the more valuable to them.[65]

A few weeks later he wrote again about the destiny of the exceptional person:

> —In the straight chain of the generations, where all the present is sacrificed to the future, nature sometimes allows itself to bear sidewards a barren beautiful flower, not connected with mankind that reproduces itself. Blessed is such a chosen one—mostly an eldest son, who in this way is

[64] Scheltema to Wiessing, 21 July 1903.
[65] Brouwer to Scheltema, 9 August 1903.

sacrificed to Minerva—as long as he is aware of his consecration, and not worried about not being directed in the presence of all the strong, blindly directed ones in the chain. The flower supports nobody in the greater context, it has no other duty than to be beautiful, sufficient in itself: for the open side, bent away, is turned towards heaven, which can refresh itself in the beautiful appearance; but only God sees the tears within.[66]

This theme is stressed more than once by Brouwer. It is not farfetched to conjecture that the above view on the exceptional, chosen son was a reflection on his own perspective of life. In his case, but equally well in that of Scheltema, the view had a prophetic ring. The line of the generations did not continue through them. In November he returned to the topic:

Ever working on, reading, and thinking—and in line with that thinking, harmonizing one's life more and more, carried by resignation and faith in God—it is the bliss here on earth. My house is homely, striking and modestly comforting. And it is sacred to me—I could do no evil in it and have no evil thoughts—in it I am even friendly disposed to everybody. If a boring person visits me—in my house I do not find him boring—if I walk out-of-doors the next day, I do not understand how I could have suffered him.

The ultimate harmonizing of our life seems most difficult, slowest, most laborious with people of our sort. It seems that in the progress of the generations, of all parents the oldest child may not simply be sent along in the mainstream of the 'striving and pairing', but must be offered to the gods of consciousness, of the infertile consciousness in the worldly motion, as an opening sacrifice—as a sideways shooting flower, which has no further purpose for the growth of the trunk—in compensation those gods will forgo their title to the other children. So, let those consecrated sacrificial animals be aware of their role, and let them not be jealous of the rough rye bread of the herd.[67]

The two-kings episode is interesting for two reasons. On the one hand it illustrates the spiritual intimacy of the friends and their mutual appreciation. Scheltema was well aware of the intellectual and emotional potential of the younger man, while at the same time Brouwer realized that Scheltema—be it with a tremendous effort—had found his place in the human society. When Scheltema pressed the young Bertus to accept the world:—'Try rather to see the reality than your own fantasies. The more you approach reality, the greater your chances of regeneration will be',[68] he well knew that utter isolation held destruction.

The idea of a select group of superior men, working for the salvation of mankind, was undeniably in the air: an intellectual elite as the new priests of the world. The

---

[66] Brouwer to Scheltema, 26 August 1903.
[67] Brouwer to Scheltema, 15 November 1903.
[68] Scheltema to Brouwer July 1901.

same idea occurred somewhat later independently to Frederik van Eeden,[70] who was immensely pleased to find out that Brouwer had conceived for himself the idea of a cultural, spiritual elite.

The relationship between Brouwer and Scheltema was one of deep respect and spiritual understanding, and its tragedy was the irrevocable passing of the unity and similarity of spirits. It was Brouwer who repeatedly urged his friend to accept the inevitable divergence of the courses that their lifes were destined to run. Both partners had to summon all their strength to accept the new relation—kings in their separate kingdoms, aware of their duty and destiny.

Scheltema, who had just read Nietzsche's 'Birth of Tragedy', characterized their differences in personality as : 'You are *Dionysus*, I am *Apollinius*, and the society we live in is *Alexandrinius*'.[71]

This particular friendship, that lasted for slightly more than a quarter of a century, was of tremendous importance to Brouwer, and presumably to Scheltema. It was guarded as a private treasure; few knew about it, and eventually a few isolated remarks were left. Were it not for the lucky preservation of a substantial part of the correspondence, the relationship between Brouwer and Scheltema would have remained nothing but a poorly substantiated rumour.

---

[70] See page 246.
[71] Scheltema to Brouwer, 6 August 1907.

2
MATHEMATICS AND MYSTICISM

2.1 Teachers and study

Let us go back to the starting point of Brouwer's university life and the years of study. When the young boy enrolled in the University of Amsterdam, that university treasured a few great men in the sciences, the most outstanding amongst them being the physicist Johannes Diederik Van der Waals and the biologist Hugo de Vries. In mathematics there were no stars of the same order, but on the whole the students were in competent hands. Lectures in mathematics were given by Diederik Johannes Korteweg, A.J. van Pesch and in physics by Van der Waals and Sissingh. Korteweg, in a manner of speaking, had saved nineteenth century Dutch mathematics from an inglorious historical record. When the zoologist Hubrecht was presenting a survey of fifty years of exact sciences in the Netherlands at the occasion of the Inauguration of Queen Wilhelmina in 1898, only two lines of his 11 page essay were devoted to mathematics, and it was Korteweg's work that was referred to.[1]

Brouwer was doubtlessly influenced most by his mathematics professor, Korteweg (31 March 1848 –10 May 1941), a man with a remarkable career, which was in a way characteristic of the first generation of scientists of the new era. The second mathematician in the faculty, Van Pesch, cut a rather poor figure compared to the impressive Korteweg. His lectures did not always measure up to the standards of his students. Wijdenes, one of Brouwer's contemporaries, told that when Van Pesch got into one of his muddles, Brouwer would get up, go to the blackboard, take over the chalk, and in his precise manner steer the lecture past the cliffs where Van Pesch had stranded.[2] He did, however, not have the audacity to take liberties with Korteweg.

The mathematics training at the universities in the nineteenth century was in the hands of well meaning professors, who followed at a safe distance the developments in the prominent centres. Apart from the well-known Stieltjes (who did not teach in Holland) there were no men of stature who could inspire the new generation of students. The rise of physics, and the sciences in general, in Holland, however, called for strong mathematics departments. But mathematics was still trying to catch up with the international developments. In the absence of outstanding pure mathematicians, who could have influenced the academic opinion, there was

---
[1] [Ritter 1898], p. 70 ff.
[2] Communication of L. van den Brom.

a general tendency to consider mathematics rather in her role of a handmaiden of the sciences, than that of the Queen of Science.

In this climate Korteweg started his studies at the Polytechnic School of Delft. As he was not particularly technically minded, he chose to break off his studies at the engineering school; not, however, without obtaining a certificate for teaching mathematics.

The situation in teacher training in Holland, before the wholesale reorganizations after World War II, requires some explanation. In Holland there were two roads to a teaching position at one of the nation's high schools (HBS) or gymnasiums: one could either obtain the normal degree of doctorandus at one of the universities, or one could study individually a particular subject, ranging from the languages to the sciences, and get a special teaching diploma. The examinations for these subjects were conducted by a state committee, the diploma was called the *middelbare acte (MO-acte)* (secondary certificate), and the subject and level was indicated by a code in letters and numbers. The first secondary certificate for mathematics was the *MO-KI* acte and the second and highest one the *MO-KV* acte.

With this *KV* diploma in his pocket, Korteweg found a teaching position, and from 1869 until 1881 he taught at high schools in Tilburg and Breda, towns in the southern part of the Netherlands. In the meantime he prepared himself for the academic entrance examination, in order to study mathematics at a university. In a fast succession he passed the entrance examination in 1876, the candidate's examination in 1877 at Utrecht and the doctoral examination in 1878 at Amsterdam, and without loss of time he defended in that same year his dissertation, *On the speed of propagation of waves in elastic tubes*; the doctorate was awarded 'cum laude'. At the age of thirty he was the first doctor of the young university of Amsterdam. His Ph.D. adviser was the physicist J.D. van der Waals. Three years later, in the same year that Brouwer

Diederik Johannes Korteweg.

his student-to-be was born, Korteweg was appointed a full professor in Amsterdam, holding the chair of mathematics, mechanics and astronomy from 1881 until 1913. From 1913 to 1918 he was an extraordinary professor.

Korteweg's mathematical production was impressive and wide-ranging, he published on such topics as theoretical mechanics, thermodynamics, the theory of voting, algebra, geometry, theory of oscillations, electricity, acoustics, kinetic theory of gases, hydrodynamics, astronomy, probability theory, actuarial science, philosophy, ... He was to a large degree the man who dealt with the mathematics behind the physical theories of his Ph.D. adviser, Van der Waals. Nowadays he is mainly known for the famous *Korteweg–de Vries equation* (1895), which he published together with his Ph.D. student, the mathematics teacher Gustav de Vries.[3] The equation describes the propagation of a solitary wave in a rectangular canal. The success of this equation should, however, not obscure his research on the folding of surfaces and on the Van der Waals surface.

As the chief editor of the collected works of Christiaan Huygens, (1911–1927) he combined his mathematical and historical interests; he solved the riddle of Huygens' *sympatisch uurwerk* (concerning coupled oscillators).[4] Korteweg was a noble and generous man, who played a central role in the national institutions of learning—the Academy, the Mathematics Society, the Senate of the University of Amsterdam, and, of course, the faculty of Mathematics and Physics. We will meet his name again, when we get to Brouwer's dissertation.

The second person to exert a profound influence on Brouwer's career was Gerrit Mannoury (17 May 1867 – 30 January 1956), a man who had also come to mathematics via the detour of a teacher's career. Mannoury was the son of a captain of the merchant navy. He finished high school in Amsterdam in 1885 and obtained his teacher's diploma three months after the final high school examination. For comparison: the regular study for a teacher's diploma took 4 years! In 1886 he got an appointment at an elementary school in Amsterdam, and in 1888 he moved to the *Instituut Schreuders*, a private educational institution at Noordwijk. Three years later he obtained a position at the *Openbare Handelsschool* (a high school that trained mostly for commerce) at Amsterdam, this position was combined from 1893 until 1902 with an appointment as a private tutor (*gouverneur*) of the son of Mrs. Henri Tindal (the widow of a newspaper tycoon). Between 1902 and 1905 the *Bloemendaalse Schoolvereniging* hired him; in 1905 he obtained a position as a teacher in Helmond. Finally, in 1910, he got a position at the high school in Vlissingen, where he taught bookkeeping, mathematics and economics. He also became the headmaster of the new evening school for commerce. On top of all that he worked as an accountant from 1894 onward.

While fully occupied as a teacher, he passed the numerous examinations that galled the life of many a school master (the so-called *acten*, discussed earlier). In particular he obtained the diplomas for teaching mathematics in secondary schools, the *KI* and *KV* diplomas.

---

[3] The equation also occurred in the dissertation of De Vries. A recent book of Willink cites evidence that the role of De Vries was modest, to say the least. cf. [Willink 1998].

[4] [Korteweg 1905].

Even before passing these examinations he published original papers in mathematics. His paper *Lois cyclomatiques* (1898) introduced the new discipline of *topology* in Holland; it was followed by two more papers in the same area.[5]

Mannoury and son.
Courtesy
J. Mannoury.

The paper treats a generalized form of the Euler-Poincaré formula. Mannoury proved in this paper a theorem which Van Dantzig has called 'Mannoury's duality theorem'. In Hopf's words, 'The theorem expressed by the [indicated] formulas, which you correctly call 'Mannoury's duality theorem', belongs completely to the area of modern duality theorems, and the fact that Mannoury knew it in 1897 shows how far he was ahead of his time. It is a pity indeed, that he did not continue this

---

[5][Mannoury 1898a], [Mannoury 1898b], [Mannoury 1900].

work. He was very close to the duality theorems of Alexander'.[6]

At roughly the same time he familiarized himself with the new symbolic methods of Guiseppe Peano.[7] The latter had introduced a symbolic language for mathematics. From 1888 onwards, Peano had studied and advertised a formalism that is fairly close to our present-day logical notation. Although Frege preceded him by almost a decade, Peano's notation was a great improvement in terms of readability. Peano's best-known publication of the symbolic language was his *Formulaire de mathématiques* (1898), and he went so far as to publish his result on the solution of differential equations in the symbolic notation. Brouwer later somewhat scathingly remarked that Peano's paper was not read until someone translated it back into common language! Mannoury quickly saw the theoretical value of Peano's language, but he was sensible enough not to write his own papers in Peano's formalism. Thus, curiously enough, this schoolmaster without a formal mathematical training not only introduced topology in the Netherlands, but also symbolic logic.

He enrolled at the University of Amsterdam to study mathematics; unfortunately, in view of his daily teaching duties, he could not attend the lectures, so the study was far from simple.

Korteweg, who recognized Mannoury's ability, gave him for some time private tutorials at home, on Sundays, and allowed him the use of his private library. Nonetheless, the limits of the combination of working and studying were reached before long; Mannoury gave up and never got a formal degree in mathematics. In view of his exceptional performance as a free-lance mathematician, Korteweg tried to further Mannoury's career. As a result he was appointed in 1903 *privaat docent* in the logical foundations of mathematics at the University of Amsterdam.[8]

The formal appointment to any of the positions of *privaatdocent, lector* (lecturer) or professor, required the appointee to give a public inaugural lecture. This was a formal occasion, attended by members of the senate in gown and cap. Mannoury presented his inaugural lecture with the title *The Significance of Mathematical Logic for Philosophy*, on 21 January 1903.

Brouwer has sketched the decisive role of Mannoury in his life, in the formal address, delivered at the occasion of the awarding of an honorary doctorate in 1946 to Mannoury:[9]

> As happens so often, I began my academic studies as it were with a leap in the dark. After two or three years, however full of admiration for my teachers, I still could see the figure of the mathematician only as a servant of natural science or as a collector of truths:—truths fascinating by their

---

[6][van Dantzig 1957], p. 7.

[7][Peano 1895].

[8]The position of *privaat docent*, similar to *Privatdozent* in Germany, brought the bearer of the title a nominal fee. Its main attraction was that it enabled one to keep a foothold in academic life, in the hope of a promotion.

[9][Brouwer 1946B2].

immovability, but horrifying by their lifelessness, like stones from barren mountains of disconsolate infinity. And as far as I could see there was room in the mathematical field for talent and devotion, but not for vocation and inspiration. Filled with impatient desire for insight into the essence of the branch of work of my choice, and wanting to decide whether to stay or go, I began to attend the meetings of the Amsterdam Mathematical Society. There I saw a man apparently not much older than myself, who after lectures of the most diverse character debated with unselfconscious mastery and well-nigh playful repartee, sometimes elucidating the subject concerned in such a special way of his own, that straight away I was captivated. I had the sensation that, for his mathematical thinking, this man had access to sources still concealed to me, or had a deeper consciousness of the significance of mathematical thought than the majority of mathematicians. At first I only met him casually, but I at least knew his tuneful name, which guided me to some papers he had recently published in the *Nieuw Archief voor Wiskunde*, entitled *Lois cyclomatiques*, *Sphères de seconde espèce* and *Surface-images*. They had the same easy and sparkling style which was characteristic of his speech, and, when I had succeeded, not without difficulty, in understanding them, an unknown mood of joyful satisfaction possessed me, gradually passing into the realization that mathematics had acquired a new character for me. For the undertone of Mannoury's argument had not whispered: 'Behold, some new acquisitions for our museum of immovable truths', but something like this: 'Look what I have built for you out of the structural elements of our thinking.—These are the harmonies I desired to realize. Surely they merit that desire?—This is the scheme of construction which guided me.—Behold the harmonies, neither desired nor surmised, which after the completion surprised and delighted me.—Behold the visions which the completed edifice suggests to us, whose realisation may perhaps be attained by you or me one day.'

2.2 First research, four-dimensional geometry

Brouwer's relation to mathematics remained ambiguous for a number of years, only when the success of his work in topology blocked his retreat, did he definitely resign himself to a mathematician's life. As the laudatio at Mannoury's honorary doctorate tells us, mathematics initially did not at all fulfil his expectations, mathematics as a clinical, sterile subject consisting of theorems and exercises did not in the least appeal to him.

At any rate, he decided to make the best of it; he had joined NEWTON, the club where students and their professors freely mixed. In 1899 he also became a member of the venerable *Wiskundig Genootschap*, the national mathematical society, founded in 1778, in the era of Progress and Enlightenment. Like all the venerable institutions of the Enlightenment the society had a motto: *Een onvermoeide arbeid komt alles te boven* (An untiring labour overcomes everything). The Mathematical Society, usually referred to as *WG*, traditionally met on the last Saturday of the

Brouwer (standing, left) at a diner of a student society, the bald man at the right is Korteweg.

month somewhere in Amsterdam. It was at these meetings that Brouwer fell under the spell of Mannoury, and it was there that he got to know the leading personalities in Dutch mathematics, if not personally then at least by sight. In spite of his earnest quest for the true living mathematics, he could not easily shake off his doubts. Nonetheless, he could not renounce his talent for mathematics. Even before his final examinations he had done original research in geometry. Wiessing reports that Brouwer gave a talk at a meeting of the science club NEWTON of the Student Corps:[10] 'only eighteen years old, he presented to the company at a meeting attended by Professor Korteweg, some theorems of four-dimensional geometry, found by him. Korteweg was completely confounded: 'I don't know what to think of it', the professor said reflectingly, 'it is a great and ingenious discovery, or it is a mystification!' '

The content of Brouwer's talk unfortunately is unknown, but it is fairly certain that it contained the germs of three papers that were submitted to the Dutch Royal

---

[10][Wiessing 1960], p. 142.

Academy by Korteweg:[11]

- *On a decomposition of a continuous motion about a fixed point O of S_4 into two continuous motions about O of S_3's,*
- *On symmetric transformations of S_4 in connection with S_r and S_l,*
- *Algebraic deduction of the decomposability of the continuous motion about a fixed point of S_4 into those of two S_3's.*

The above papers, and a considerable number of his later papers, were published in the Proceedings of the Royal Dutch Academy of Sciences. Most of these papers were in fact published twice, one version in Dutch and one in English or German. There were actually two series of publications, the *Verslagen van de Koninklijke Nederlandse Akademie van Wetenschappen* and the *Proceedings of the Koninklijke Nederlandse Akademie van Wetenschappen*. Names have changed since then; when Brouwer started submitting his papers, the Academy was called the *Koninklijke Akademie van Wetenschappen te Amsterdam* (Royal Academy of Sciences at Amsterdam), during World War II the name was changed to *Nederlandse Akademie van Wetenschappen* and after the war a synthesis was arrived at: the *Koninklijke Nederlandse Akademie van Wetenschappen*. The institution itself was founded by the first King of Holland, Louis Napoleon, the brother of the Emperor Napoleon. At that time there were already a number of 'learned societies', such as the *Hollandse Maatschappij van Wetenschappen*, the *Provinciaal Utrechts Genootschap*. These local academies were the fruits of the Enlightenment; they provided a platform for the scientific and commercial upper class of the Dutch Republic. When Louis Napoleon proposed to transform one of the existing societies into the national academy, he was met with protest and refusal, so he founded his own academy—which was considered an upstart by the older establishments. Eventually the Royal Academy superseded the local learned societies, the latter having remained to this day modest centres of the sciences and arts.

Before Brouwer was elected a member of the Academy, most of his papers were submitted by Korteweg, who, as a member, was entitled to present papers for publication in the Proceedings and the *Verslagen*.

The first of Brouwer's papers, mentioned above, was a treatise on rotations in four-dimensional Euclidean space. He showed by geometric means that a rotation in four-dimensional space can be obtained as the product of two rotations in three-dimensional space.

Freudenthal, in his discussion of the paper,[12] pointed out that the simplest way to treat the above transformations, is by means of quaternions. Brouwer first gave a geometrical proof and subsequently an algebraic one. Possibly Brouwer was not well versed in the geometrical applications of quaternions; whatever the reason may have been, he had opted for a laborious direct proof by geometric means.[13]

---

[11] Communicated respectively at the meetings of 27 February 1904, 23 April 1904, 23 April 1904.

[12] CW II, p. 22.

[13] Brouwer's theorem in modern formulation reads $SO_4 \cong SU_2 \times SU_2 / \pm (1, 1)$. Another geometric proof is in [Klein 1890], and a similar theorem can be found in [Cartan 1914].

These very first mathematical publications made Brouwer, unwittingly, an actor in a priority controversy. It was his first experience of that kind, but, alas, not the last one. The topic of Brouwer's paper—a study of the orthogonal transformation group of a four-dimensional euclidean space, had been the subject of investigations of a German mathematician, E. Jahnke,[14] a man of some weight, with a sizeable publication record.

Jahnke had spotted Brouwer's paper in the Proceedings almost immediately after its appearance; he wrote a rather condescending letter to Korteweg (15 March 1904) and magnanimously (and correctly!) assumed that Brouwer was unaware of Jahnke's publications on the subject, which, he said, already contained the results of Brouwer's paper. He acknowledged that Brouwer had obtained his results by new means, but he expressed his expectation that 'the author would use the opportunity to acknowledge in a short note in the same journal and if possible in the next issue my priority for the mentioned results'.

A letter from a man who had earned a reputation in applied mathematics, who was an editor of the *Archiv der Mathematik und Physik*, might have daunted a lesser spirit than Brouwer, but this young man was not to be silenced so easily. When Korteweg duly informed Brouwer of the claims of Jahnke, Brouwer carefully studied Jahnke's papers and concluded that he had in no way invaded Jahnke's priority rights. He wrote a cool, polite, but unmistakably provocative letter:[15]

> From your letter, kindly transmitted to me by Professor Korteweg, and the enclosed papers, I see that my treatise interests you, and that earlier investigations of yours are connected with it.

Brouwer went on to explain in some detail to Jahnke the contents of Jahnke's and his own papers:

> The reading of your papers suggested the following remarks, which will certainly be plausible to you.[16]

And after spelling out the geometrical meaning of Jahnke's method (or rather, the lack of it) and of his own method,[17] he closed with:

> Thus I hope to have shown that our papers under discussion have nothing in common, but that your final result is a by-product of my principle.

Considering the provocation, Jahnke's reaction was rather mild; he demanded from Korteweg the publication in the Proceedings of a rejoinder from his hand. The latter had more faith in his own pupil's insight than in that of his German colleague; he promised Jahnke a note in the Proceedings, while at the same time asking Brouwer to write an exposition of the matter for the Proceedings. Jahnke's note[18]

---

[14] At that moment *Oberlehrer* at the *Friedrich–Werderschen Oberreal Schule* and a *Privat Dozent* at the *Technische Hochschule Berlin*.

[15] Brouwer to Jahnke 20 March 1904.

[16] The contents of the letter are incorporated in [Brouwer 1904C2].

[17] See Freudenthal's commentary, CW II, p. 22 ff.

[18] [Jahnke 1904].

shows that he had still not grasped the geometrical meaning of the decomposition that Brouwer had obtained, but the affair ended quietly with Brouwer's algebraic derivation of the results.

Among Brouwer's papers there are some notes that comment on Jahnke's papers in a rather cutting way, he compared him to a man 'who has stumbled around without detecting anything but small traces, and who now sees that the thing itself has been found and that his traces have lost their value. Hence his hasty and anxious letter.' He went on:

- *A discovers that somewhere everything behaves just as in a magnetic field, and he even discovers that this field is remarkably simple. B finds the magnet and a very simple one at that; and says 'the matter is so and so'. Now A would not raise a priority claim against B, would he? At best one can say that Jahnke's researches suggested that there were two R_3's. I have indicated those R_3's (and not bothered with their properties, which Jahnke presents in full).
- I would say: if a man finally deduces by means of boring observations, which rest on, and are combinations of, equally boring observations of a predecessor (Caspary)[19] *remarkably* simple results from all those complicated things, and finally somebody comes along and says that no complicated things are going on, but that something very simple is the matter, then he is ashamed and he withdraws himself. At least he does not raise a priority claim.
- When Newton found the law of attraction, and deduced the laws of Kepler from it, Kepler would not have wanted to diddle the credit from him.

In spite of all his self-assurance, the fledgling scientist showed the hesitance that every beginner has experienced: Brouwer asked Korteweg how to send out his preprints;[20] he did not know how to find the addresses of mathematicians he only knew by name. 'And furthermore, can I send a copy to people like Klein and Veronese? Or would that be presumptuous without an introduction?'

Brouwer was enrolled for almost 7 years, not exceptionally long in those days, but nonetheless too long for such a promising student. So, what kept Brouwer so long at the university? The blame cannot be put on the fraternities, for after his first year Brouwer scarcely frequented them. The cause was rather the military service; according to his own statement[21] his studies were interrupted for a good two years. We have already seen how Brouwer was drafted (cf. p. 1.4), and his dislike of the experience. Indeed, his physical and mental health was seriously put at risk. The military service put so much stress on him that at times he felt utterly desperate.

---

[19][Caspary 1883].
[20]Brouwer to Korteweg 19 May 1904.
[21]Brouwer to Scheltema 15 November 1903.

The actual time spent in the army did not exceed eight months, but each bout of military training apparently upset Brouwer so much that the recovery asked time. The sensitive young man must have experienced military service as a kind of hell; it was not that he could not, or would not, cope with the physical hardships, as we have seen he had always enjoyed a good dose of rough soccer, and long marches did not tire him. It was rather the company that fed his distaste for the army. Even years later, during the First World War he recalled his national service with a shudder:

> My past service-time with the infantry is the darkest page of my life; from my equals I got little more than hatred, from my superiors little more than teasing and opposition; I repeatedly failed the examination for subaltern, and the consequence has been that after my military service for one and a half year I had a nervous disorder, and was not able to work; I recovered from it only very slowly.[22]

This letter, combined with the information of the Brouwer–Scheltema correspondence, confirms the picture of Brouwer as a man extremely susceptible to stress. The antique military establishment, not exactly known for its rationality or open-mindedness, clearly was not the environment to cherish an unorthodox—and probably contrary, character like Brouwer. The result was a prolonged physical and mental breakdown. Only in November 1903 did Brouwer get into the rhythm again. He wrote to his friend Scheltema:

> Of course you have excused me for remaining silent for so long. I have been busy; returning to my subject after two years of absence required some dedication, in particular where any love for that subject was missing. By now I have gradually succeeded, and I row with long strokes towards my doctoral examination. My work is done without illusions, but with a feeling of cheerfulness on account of the activity itself.

Once he had resumed his study, the subject matter offered no problems. One guesses that his publications on four-dimensional geometry could not have failed to impress his examiners. The doctorandus-diploma was awarded on 16 June 1904, (cf. p. 36), and—as a mark of excellence, with the predicate 'cum laude'! The diploma was signed by A.J. van Pesch and D.J. Korteweg.

It did not take Brouwer long to make up his mind on the matter to his future activity. He soon informed Scheltema of his plans to start to work on a dissertation.[23] He planned to absent himself for some time, perhaps for a longer period, in order to

> recover the clear relations, in which I have to position myself *vis à vis* the various persons and institutions within my narrow social horizon, in order not to be distracted from the cultivation of my power and the development of my clairvoyance in the service of God.

---

[22] draft of a letter to Lorentz 16 February 1918.
[23] Brouwer to Scheltema 4 July 1904.

Although the scales were by now definitely tipped in favour of mathematics, philosophy was still prominent in his mind:

> Next winter I will be either in Blaricum—where a cottage[24] is being built for me—working at a philosophical creed, that will be the prologue of my work—or in London, in the great British Library for my dissertation: 'The value of Mathematics' with the motto $O\upsilon\delta\epsilon\iota\varsigma\ \dot{\alpha}\gamma\epsilon o\mu\epsilon\tau\rho\grave{\iota}\kappa o\varsigma\ \epsilon\iota\sigma\iota\tau\omega$.[25]
>
> I thank you for your well-meaning admonition to me at the gate of the paradise of freedom. Did I wish a kingdom on earth, then it would perhaps be good to wall in myself in mathematics, and to have me crowned like a pope in the Vatican, a prisoner on his throne. But I desire a kingship in better regions, where not the goal, but the motive of the heart is the primary thing.

2.3 Marriage

This exalted message was followed, only six days later, by a letter that contained a short but weighty message:

> Carel, my friend, in my life a thaw has set in. I have exchanged marriage vows with Mrs. R.B.F.E. de Holl. Greetings and hail to you, comrade
> Your Bertus

Whereas from our modern point of view there is no reason why a student who has only just finished his final examinations should not get married if he or she wishes to do so, in the old days marriage was not an affair to be rushed into. Prudence and tradition required a couple to save a substantial sum, and for the bride to collect a complete trousseau, before one could even contemplate matrimony. Middle class morality, in particular, was quite strict and specific in matters of engagement, marriage, children, etc. Handbooks of etiquette spelt out all the rules to a nicety; the number of sheets, teaspoons, ... was precisely indicated.

The fact that less than a year ago Brouwer still viewed himself as the eldest son, destined to remain without offspring, either shows that nature is stronger than theory, or that in Brouwer's view marriage did not necessarily entail procreation. It may be remarked here that the marriage remained childless, so the prophecy was fulfilled in spite of the marriage.

There are only hints in the correspondence of Brouwer and Scheltema concerning the other sex. In view of the long-standing tradition of fraternities to introduce students to all aspects of life, including those of the flesh, it is not unlikely that either Scheltema himself, or one of the other members of his fraternity companions, took Bertus' practical education in hand. A visit to one of the traditional establishments may not have been an obligatory part of fraternity life, but it was not actually frowned upon.

---

[24] See p. 62. This cottage was referred to as 'the hut'.
[25] Let nobody enter without the knowledge of geometry (Plato).

References to females are scant in the Brouwer–Scheltema correspondence. At the time of his depression in 1902, Brouwer wrote to his friend, after discussing Carel's recent volume of poetry, *Of sun and summer*, that 'there is a richer soul in your poetry, than in that of our modern poets'. These words were followed by an urgent and heartfelt counsel:

> ... Now, yet, a woman for you, Carel, you will not reach your destination before that; she will open up so much with her magic wand, I felt that again only just now thinking back to last year. I, too, feel homesick for the arms of a woman and a kiss. I would, even a month ago, never have thought that this could haunt me so much. Yes, it is really easily said that a man should live with his reason; a burden of thousands of years sits in mind and body, which compellingly shows him what to strive for, how he has to be.[26]

Scheltema calmly replied that women had no place in his life as an artist:

> ... you forget that I love my own art above all and that my life with it, and my fighting against it, is the life with, and the fight against, the powerful muse—the most cruel of all women. [...] And the muse does not tolerate any love but the love of friendship—and in that I am rich enough!

In the same letter he turned the table on Bertus and called on him to find a woman:

> —be it either to the act of pairing that gloriously relieves me like a bath— or the love of her, of whom you had often spoken, and who will certainly want you back. Maybe she will now be able to help you better than anybody else!—But in this case don't act without Huet or your present doctor...[27]

Scheltema apparently referred here to some earlier relation of Brouwer, which might be hinted at in the letter of 5 December 1901: 'In order to quell a growing tragedy, I have to get away from here'.

Brouwer's wife-to-be was Reinharda Bernardina Frederica Elizabeth de Holl (Lize) born on 5 August 1870, eleven years Brouwer's senior. She was the daughter of Eelbartha Johanna Jacoba de Holl-Sasse, widow of Jan de Holl. The latter was a medical doctor, who had his practice on the Overtoom in Amsterdam. He died young and left his widow with 7 children. He also left her the pharmacy which he had run together with the medical practice. Mrs. De Holl had decided to keep the pharmacy; according to the regulations she had to hire someone with the proper qualifications in order to guarantee the required expertise.

Lize had married Hendrik Frederik Peijpers, a former army doctor and, incidentally, a full cousin of hers, when she was still a young girl. Peijpers was

---

[26] Brouwer to Scheltema 11 June 1902.
[27] Scheltema to Brouwer 12 June 1902.

sixteen years her senior; he had first been working in the pharmacy for his aunt, the widow De Holl, with whom he was lodged. The marriage was far from happy, Peijpers did not want any children, and did not hesitate to carry out an abortion if and when Lize became pregnant. When she was once more pregnant, Lize managed to circumvent her husband's intervention, and the child, Anna Louise Elisabeth, was born on 26 March 1893. Soon after that Lize got a divorce and in the meantime she and Louise had moved in with her mother. It should be kept in mind that in those days a divorced woman with a child was in an extremely uncomfortable position. Her social status was far from enviable.

There are no elaborate accounts of Brouwer's courtship,[28] but the following story, as told by Louise, is undoubtedly authentic.

Bertus had renewed his friendship with a girl he knew from Medemblik, the earlier mentioned Dina Pels.[29] Dina, who was somewhat older than Bertus, had found herself a place in the pharmacy of the widow De Holl, where she combined a full-time job with a training as pharmacy assistant.

Lize, before her marriage to Bertus.

The two made long walks and exchanged their experiences (a salient detail, reported by Louise, is that Bertus insisted on carrying Dina's purse, to return it safely at the end of the walk). Dina told about the proprietor of the pharmacy, the widow De Holl, and the routine at the shop. She also mentioned that a young divorced daughter with her child lived with Mrs. De Holl. Bertus' interest in the daughter and the pharmacy was soon aroused, and he devised a plan of action in the best romantic tradition: he climbed onto a roof in the neighbourhood of the pharmacy at the Overtoom,[30] in order to watch the object of his curiosity. The inspection must have led to a favourable conclusion, for a meeting was arranged, and on 10 July 1902 the two met. What Lize thought of this curious student is not known. It is reported that she had her doubts about the wisdom of marrying a man eleven years younger than herself. Brouwer, anyway, did not hesitate long. He marshalled all his charm and power of persuasion to win the heart and hand of the young divorcee, whom he often and fondly praised for her Memlinck[31] face. The campaign was successful. In spite of the negative advice of some of her friends, Lize accepted Brouwer's proposal. The two formed a striking couple, Bertus, over 1.87 meter tall,

---

[28] Most of the information on the courtship and the marriage is from oral communications of Louise Peijpers. Confirmation of the information on Dina Pels was provided by Mrs. J. Schout-de Waal.

[29] See p. 7.

[30] In other accounts of the same events the roof is replaced by a tree.

[31] Fifteenth century Flemish painter.

towered over Lize, who measured no more than one-meter-fifty. Two years later the marriage took place, two months after Brouwer's doctoral examination, on 31 August 1904—the birthday of Queen Wilhelmina (*koninginnedag*).

And so Dina Pels played a brief but decisive role in Brouwer's life. She later married a medical doctor in Alkmaar by the name of Formijne. She died at the age of fifty. Brouwer kept up the relationship with the Pels family—among the gratulations on the fortieth anniversary of his doctorate in 1947 there was a letter from a member of the family. In 1953, on his tour of the United States Brouwer visited one of the Pels relatives, who had emigrated to the States. It may be stressed that Brouwer had a very strong loyalty to his friends and relatives; there are numerous testimonials to this fact. He always enjoyed a quick visit to those whom he had admitted in his personal circle, irrespective of status or gender. On the other hand he had no patience with potential bores; his motto was 'if you won't have anything to do with someone, pick a fight right away—it saves a lot of time'.

The marriage was a simple affair. Bride and groom took a streetcar to the city hall of Amsterdam, where the civil wedding took place. Both brothers Lex and Aldert had, on this special occasion, accompanied Bertus to the house of the bride; the three of them played leap-frog on the way.

At the dinner, which followed after the official part of the wedding, the uncle of the bride, Reverent De Holl, held a diatribe on the topic of the marriage of two students—one of them, moreover, the mother of a child!—who did not have a penny, and who had nonetheless commissioned the building of a house for themselves, unscrupulously borrowing money![32] The twelve year old Louise quietly snatched some dainty morsels from the table and stole away. In her memory the atmosphere at the dinner was stifling.

Scheltema, who in some way took part in the wedding spoke bitterly of the occasion:

> I have not understood anything of your wedding and in particular of the embarrassing ceremony, and the, for me insulting, invitation of Poutsma[33] and the whole collection of people, one and all, that I found disgusting! Really Bertus, that day was a great sacrifice for me! I have made it without demurring because you insisted, and seemed to have your reasons ...

After the marriage the young couple moved into the rooms over the pharmacy at the Overtoom, waiting for the completion of their Blaricum cottage.

The daily routines of the couple were not much changed by the marriage. Brouwer had started to work on his dissertation, and Lize was fully occupied as a pharmacy student. Mrs. de Holl had, as we said before, no licence to run the pharmacy. Therefore she had to hire a licensed pharmacist to run the professional part of the shop, although the management remained in her hands. Since the salary of such a *provisor* (as he or she was called) presented a serious drain on the finances

---

[32] Cf. the letter from Brouwer to Scheltema 4 July 1904.
[33] Presumably this was one of the uncles, a teacher at the Barlaeus Gymnasium.

of the pharmacy and the family—not to mention the space problems when a provisor happened to live in—Mrs. de Holl had considered the possibility preparing Lize for the supervision of the pharmacy, with the intention of eventually letting her take it over. As a result Lize had enrolled as a student in the University of Amsterdam, and diligently studied pharmacy. Roughly a year later the pharmacy did indeed change hands and became the property of the young couple, cf. p. 198.

2.4 Bolland's philosophy course

Brouwer, clearly, was of two minds about his future scientific career; as we have seen above, he had not yet made a definite choice between mathematics and philosophy. If the text of his profession of faith had not been preserved, this would have come as a complete surprise, for there was at that time no other visible sign of interest in, or familiarity with, philosophy. The letter of the fourth of July to Scheltema underlines that, notwithstanding his early success in mathematics, Brouwer was totally serious about the role of philosophy in his work. His interest in philosophy was probably the explanation for his involvement in the following short episode, which was of such significance, that it was the immediate cause of his mystical-philosophical monograph *Life, Art and Mysticism*, cf. p. 66.

Philosophy in Holland, around the turn of the century was largely dominated by G.J.P. Bolland, a man with an incredible career. He was basically a self-taught philosopher, a fast learner and an even faster user of new knowledge, a powerful protagonist in every sense of the word. After a colourful but tragic youth, he became a teacher in the Dutch Indies. Absorbing in a high tempo the philosophy of predominantly German thinkers, he soon developed into a formidable character in Dutch philosophy. His life looked like one long series of conflicts, most of which were of his own making. He acquired a certain notoriety by his extreme anti-Catholicism. In spite of his unusual reputation, he was appointed to the chair of philosophy in Leiden, where he preached the philosophy of Hegel. Bolland was one of those legendary professors who tyrannized his audiences, his students, and probably his colleagues as well. He could, at his lectures, request specific members of the audience ('that person with the unborn face') to leave the hall because 'otherwise I cannot do my work'.[34] Nevertheless, he was a popular and inspiring lecturer, who did not hesitate to give his opinion on any subject.

His reputation as a forceful and interesting speaker made him much in demand for courses and talks. In view of Bolland's success at other universities, the Amsterdam students decided to invite the great man for a series of lectures. Brouwer joined the committee, and he soon was the major force behind the invitation.

On 15 January 1904 Brouwer wrote his first letter to Bolland. Even Brouwer must have felt some awe, for the letter is unusually timid. Bolland's part of the correspondence has not been preserved, so we can only try to interpolate his reactions. Apparently he consented to give the lectures, but not without his conditions. From

---

[34] In this respect Bolland was not an exception. Cf. [Wiessing 1960], p. 241.

Brouwer's third letter[35] we gather that only those participants were welcome who had bought Bolland's book 'Pure Reason'.[36] The letter shows that Brouwer was seriously worried about the size and the quality of the prospective audience:

> A great concentration of 'serious' listeners has not been secured. Many joined at first, who 'wanted to hear Bolland' for pleasure, but already the condition of the purchase of the books—which was seen, in addition to the assumption of a considerable advance knowledge, a demand for activity, instead of pleasure—caused the number of prospective participants to dwindle to some twenty-five.

In spite of these discouraging messages, Bolland agreed to give his course. Brouwer went so far as to invite Scheltema to Bolland's lectures, although Bolland was a notorious anti-socialist. Brouwer told Scheltema that 'I so often heard the voice of Dietzgen in his words last night that I have to let you know'.[37] Scheltema resolutely declined the invitation: 'it must be most unpleasant to hear the Hegelian philosophy from this abominable man'.

The lectures were far from successful, as one can read in the student magazine *Propria Cures*. Whereas everywhere else Bolland was greeted by his audiences with hardly suppressed awe, the Amsterdam students shared no inclination to flatter Bolland's ego. When he made his entree he was met with smothered laughter and giggling. Bolland reacted predictably. He mercilessly attacked his audience—'you are nothing special, and I don't expect anything from you ... '

Propria Cures reported the lectures in some detail; the issue of 15 October contained an introduction to the thoughts of the master in tones that echoed some of Brouwer's ideas (actually, he may have been the writer, but there is no certainty about that) including some mocking remarks about science, socialism and females, followed by a panegyric and a satire.

The lectures continued for some time, but the atmosphere in the lecture hall did not meet the speaker's expectations. For one thing, Bolland was used to be treated with reverence; he did not suffer from modesty, and he considered himself the most important philosopher in Holland, and so he expected his audience to treat him accordingly. His handling of disagreeable listeners (and it did not take much to be considered thus) was crude but effective. If the manners or behaviour of members of the audience were not to his liking, he could suddenly interrupt his lecture—'If those foetuses will not dispatch themselves, I cannot allow myself to continue', or words of a similar import.

He characterized his Amsterdam audience in various unflattering ways, and finally gave up the course, writing that he could only present his 'Collegium Philosophicum' to an audience of 'students' and not to one of 'spectators'. The *Propria Cures* copy of 16 December 1904 published Bolland's letter of cancellation, which (quite correctly) stated that the 'Collegium philosophicum' was offered to a circle

---
[35] Brouwer to Bolland 5 March 1904 (Boll. B 1904, 29. Leiden University Library).
[36] *Zuivere Rede*, [Bolland 1904].
[37] Brouwer to Scheltema 8 October 1904.

of 'students', but not as a public performance for a fee, let alone for free. He definitely refused to lecture for an Amsterdam public that wanted to be amused and that did not even bother to procure the obligatory book, as they showed by their empty hands.

The issue of the 'Bolland Lectures' continued for some time to occupy the columns of *Propria Cures*; a number of comments and reader's letters were published that in turn defended or attacked the great man; there was even some excitement in Roman Catholic circles on the assumption that Bolland had abused the catholic hearers or vice versa.[38] Brouwer also took part in the discussion: he published a number of notes under the pseudonym *Lau van der Zee*.[39] In his contribution with the title 'Grounds of consolation' he concluded, after a flowery and convoluted commentary on Bolland, that 'Bolland's lectures have stopped at the point, where his path and the paths of his audience who were about to liberate themselves, parted ... And thereafter he stayed away, just in time.'

Brouwer's relationship with Bolland is far from clear. The philosophical tradition in Holland eventually turned away from, and even against, Bolland. True, there was a hardy band of faithful followers, who devoted their undivided loyalty to the works and thoughts of Bolland, but gradually they were outnumbered and eventually forgotten.[40] Slowly the name 'Bolland' became a synonym for 'weird and unscientific', hence the present generation can no longer imagine the spell that Bolland cast over Holland. But anybody who takes the time to peruse the books of Bolland will find quite sensible thoughts (next to obscure passages). There is, for instance, a small monograph, *Intuition and Intellect*,[41] which contains quite sensible ideas, next to unfounded speculations. In particular, it showed that the mathematical layman, Bolland was not as uninformed about mathematics, as later commentators suggested.

Comparing Bolland's and Brouwer's writings, one can see that they share certain ideas, but whereas Bolland's text may be compared to the confused sounds of an orchestra that is tuning, Brouwer's philosophy is the crystal clear music of a transparent symphony. At some point, Brouwer studied Bolland's writings, together with P.C.E. Meerum–Terwogt, a contemporary of Brouwer, who had become a Bolland adept. He also visited the master personally.[42] There is no doubt that Bolland acted as a catalyst for the young Brouwer, but once the latter had become a philosopher in his own right, with his own program, Bolland was no longer of any influence.

Nowadays there is little appreciation for Bolland's philosophical views, but during his life he exercised a considerable influence on his followers and adversaries.

---

[38] Bolland was a virulent anti-papist. No love was lost between him and the Dutch Catholics.

[39] One of the lodgers of the Brouwer family went by the name of Lau van der Zee, cf. p. 8. In the Brouwer archive there is a manuscript in Brouwer's handwriting, signed by Lau van der Zee and subsequently published in *Propria Cures*. So the identity of this 'Lau van der Zee' is well authenticated.

[40] The reader may find more about Bolland in a recent biography [Otterspeer 1995] (Dutch).

[41] [Bolland 1897].

[42] Communicated by Mrs. N. Kapteyn–Meerum–Terwogt, the daughter of the above mentioned Meerum–Terwogt. No details of the conversation between Bolland and Brouwer are known.

Almost all Dutch philosophical publications in the beginning of the century in one way or another paid a tribute to the recognized master and tyrant of philosophy.

This short episode is of some importance, as it shows that Brouwer was actively involved in matters of philosophy. He did not just want to dabble in philosophy, but wished to pay serious attention to the developments in that field. Already in his short observations in the student magazines one can discern his own private views on the basic issues of (in particular) moral philosophy. In the *Propria Cures* issue of 19 November 1904 Brouwer published under the name Lau van der Zee, a short note *On Morality (Excerpt)*. This note is almost a short preview of Brouwer's later lecture series *Life, Art and Mysticism*.

The main theme is the loss of the original innocence—'In passivity the world is a garden of marvel and joy and silence. There is no separation, no reality and one wants nothing.' The loss of this primordial equilibrium, on Brouwer's view, is caused by man's concentration on certain phenomena, the active directing of attention. 'The source of *All* is lost, one has been born.' Brouwer then goes on to indicate the modes of liberation, that is means to regain paradise. These modes are, strangely enough, distinct for man and woman. 'Moral' is to be found in the quest for the lost primordial state. It is remarkable to note that the female road to liberation is rather negative compared to the male one. This short note can, like the profession of faith, be seen as an ouverture to the mystical-moralistic book of 1905.

2.5 Among the artists and vegetarians

The year 1904 was an eventful one. Not only was it the year of the doctoral examination and the marriage, but also the year of Brouwer's settling in Blaricum, a small town (village) not too far from Amsterdam. The Brouwer–Scheltema correspondence contains a number of glowing references to Blaricum and its general surroundings. Brouwer evidently was infatuated with Blaricum, and in order to appreciate this phenomenon, we have to take a closer look at the town and the surrounding area.

Blaricum had been a desperately poor village populated by farmers and shepherds. Its soil was sandy and could not be expected to yield more than a scant crop.

At the end of the nineteenth century things started to change, the more affluent citizens of Amsterdam had discovered the charms of country life in an area where prices were still reasonable, and where the air was clear and healthy: the commuter had been born. At the same time *het Gooi*, a geographic unity, comprising Laren, Blaricum, Bussum, Huizen etc. attracted a number of artists. Somewhere in the eighteen-seventies the painter Jozef Israëls discovered the picturesque charms of Laren; in his wake more artists followed, Johannes Albert Neuhuys settled in Laren in 1883, and in 1886 he was joined by the painter Anton Mauve. The latter became famous for his paintings of the landscapes of Het Gooi, the flocks of sheep with their shepherd, the heath, and the small farmhouses and huts of the local population. He was so much identified with 't Gooi that one spoke of 'the land of Mauve'.

Gradually Laren and the neighbouring Blaricum became a well-known centre for painters; the list of resident painters of whom some had a more than local fame contains too many names to include them all. We must be content to mention a few: Jacob Kever, Frans Langeveld, Wally Moes, Jan Veth (who was an author as well), F. Hart Nibbrig, Arina Hugenholtz, Evert Pieters, F. Oldenvelt, Willem Dooijewaard, William Singer, Herman Heijenbrock. In the world of painting, Laren became known for its *Larense School*. A special role in the history of Het Gooi was played by William Singer, the son of the American steel giant William Singer from Pittsburgh. He had chosen to become an artist and to forego his rights as a successor to his father's steel industry. After some wandering through Europe, he alternatingly lived in Laren and in Norway. In Laren a magnificent house was built for him, which today is the town hall. After the Second World War, the widow of William Singer donated the funds for the founding of a memorial foundation and for the Singer Museum, which now attracts art lovers to Laren.

Art was vigorously promoted in Laren by an enterprising hotel keeper, Jan Hamdorff. No account of Laren would be complete without the mention of this enterprising individual, who governed the local art world as a benevolent autocrat, with a keen eye for the interest of his artists and of himself.

Much later Mondriaan and Van der Leck worked for some time in Laren. Laren and Blaricum not only attracted the adherents of the visual arts, but also considerable numbers of the Dutch literary society spent a part of their life in the idyllic villages and in the neighbouring towns. The poet Herman Gorter, the authors P.L. Tak, Frans Coenen, Victor van Vriesland, Carry van Bruggen, the couple Henriette and Richard Roland Holst lived in Het Gooi, and last but not least the famous, but somewhat controversial author, psychiatrist, philosopher, philanthropist Frederik van Eeden. We will meet the latter again in connection with the so-called 'significs'.

At roughly the same time there was an invasion of a totally different kind: the advance of the communes and of the health fanatics. A number of these communities, usually called *colonies*,[43] founded on idealistic, mostly socialistic and/or religious bases, have made history; they have influenced life in het Gooi to no small degree, although nowadays they are considered just a curiosity in the local history of the region.

The best-known colony, was *Walden*, founded by the above mentioned Frederik van Eeden (1860–1932). Van Eeden was well-known, and not only in the Netherlands; he had studied medicine and through his own efforts he had become the first psychiatrist in Holland. He was a sensitive man, the author of a number of books, dramas, and poems, with a keen social conscience, rejecting, however, Marxism as an acceptable basis.[44] His colony was named after Walden Pond in Concord, Massachusetts, where Thoreau carried out his famous experiment. Thoreau's book had fired Van Eeden's socio-romantic imagination. In 1898 he bought some land

---

[43]*kolonies*. Cf. [Boersen 1987], [Heyting 1994].

[44]There is a two-volume biography of Van Eeden (in Dutch) by Fontijn, cf. [Fontijn 1990, Fontijn 1996].

in Bussum from an ex-patient, and started to master the practical and theoretical problems of running a colony.

Following Henry Thoreau, he had a cottage built for himself, followed by a number of cottages for members of the colony. In 1899 Walden opened its doors. It attracted a mixed group of people, consisting of a number of disciples and patients of Van Eeden and some farmers. Walden suffered from the usual defects; idealism and love of mankind are no substitutes for organisation and leadership. Van Eeden, who was often absent, was not cut out to be the practical leader of a group of colonists. The enterprise was a financial disaster, in particular for Van Eeden. In 1904 the colony, as an official institution, ceased to exist.

The project had, however, caught the imagination of the Dutch people. Walden became something like a catch word; socialists and communists condemned the idea on principle, and ethical idealists treasured its memory—much as the true romantic adores ruins.

The second movement was centred around another charismatic personality—Professor Jacobus van Rees. Van Rees was the son of a social-liberal historian, Professor Otto van Rees, and the father of the painter Otto van Rees. Inspired by Tolstoi, he became a religious-anarchist, active in the fight against military service, the killing of animals, alcohol and tobacco. In 1899 he helped to found the colony of the International Brotherhood in Blaricum. The colony was an agriculture enterprise, which functioned for a brief period. A lack of expertise, discipline, and the barren soil eventually finished off the colony.

The original inhabitants of Blaricum and Laren were not altogether pleased with the presence of, what they called, 'reds' and 'grass eaters' (*plantenvreters*). The colonists dressed and behaved in objectionable ways, and they were, in the eyes of the hard working indigenous population, lazy and ignorant. Nonetheless, there was a good deal of tolerance, after all, the communes brought the shopkeepers business. Violence only erupted during the big railway strike in 1903, when it was rumoured that the colony people would stop the *Gooische Stoomtram*, the connection of het Gooi with Amsterdam.

The adherence to the fundamental Christian-anarchist principles of the 'International Brotherhood' gradually eroded, and even more so after a number of Friesian socialist farmers had joined, they clearly wanted to combine socialism with successful farming—'they needed livestock for dung and they had bought along guns to shoot the rabbits that ate the cauliflower'. Eventually the Brotherhood was dissolved in 1911.

Already before that time, Brouwer had bought a strip of land from Professor Van Rees, situated

Ru Mauve. Courtesy Mrs. M. Stieltjes-Mauve.

along the Torenlaan in Blaricum, and it was at this spot that Brouwer, while still a Ph.D. student, had a hut built. It was designed by his friend Rudolf Mauve, the son of the painter Anton (see page 59).

At the end of October 1904 the wooden cottage was ready to receive its occupants; it was a charming construction of a modest size, basically one room plus a kitchen, and a bedroom upstairs. It had a thatched roof and was situated in a wooded lot. The location exactly answered Brouwer's dreams; a secluded spot in the middle of a romantic landscape. The privacy was later increased by an enclosure of rush-mats.

In the seclusion of his private domain Brouwer enjoyed the pleasures and rites of a healthy life. He practised a number of traditional health activities, such as vigorous exercises to improve the circulation of blood and oxygen, open-air bathes, mostly in the nude, sleeping—weather permitting—in the open air, swathed in wet sheets.

The eating habits and the food were subjected to a strict regime. In this respect the couple was well-matched: Bertus and Lize both practised vegetarianism. Lize was, as all sources confirm, well-informed about vegetarian diets, traditional herbal cures, and the like. The marriage of Brouwer certainly could not have been more felicitous with respect to his lifestyle. Lize was known in the village for her knowledge of herbs and as a pharmacist she used to prepare bottles of herb cures; even in old age, she could be seen stirring a huge cauldron with a brew of all kinds of herbs. Under her guidance the pharmacy also dispensed homeopathic medicines. Less kind characters spoke jokingly of her as 'the poison mixtress'[45]

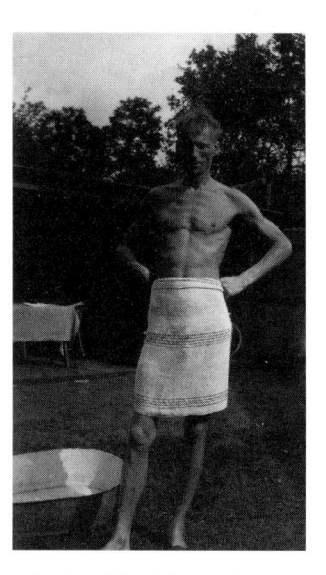

The healthy life in Blaricum. Brouwer in his garden.

Not that she was capable of harming any living being, but the image of an earnest, tawny old lady going about her business of preparing potions of herb drinks irresistibly evoked the thought of old fairy tales.

Bertus had acquired his knowledge of the vegetarian kitchen in a German health clinic of Doctor Just, in the town Jungborn in the Harz. He had been a long-time sufferer of complaints of the nose; his father saw this as the cause of his son's long drawn out studies, and he ordered the boy to have his nose treated, but after it was flushed, things became even worse. The doctor proceeded to prescribe him seven goose eggs, daily, and one and a half pound of steak, in order to shore up his general condition. The result was that Bertus felt more sick than before. Finally he went to Just's health clinic, where he was introduced to the secrets of diets, vegetarianism,

---

[45] *gifmengster*.

open air baths, exercises, etc.[46] He adhered to the vegetarian diet the rest of his life, albeit for pragmatic reasons. He was not dogmatic enough to resist the temptation of an occasional bite of meat or chicken.

Bertus and Lize in their Blaricum garden.

Among Brouwer's papers there are some notes that illustrate the eating habits of the Brouwers. An undated list, probably from the early years in Blaricum, gives detailed instruction for the daily diet; there is an enumeration of wild herbs for each month, for example, 'April: scurry grass; lady's smock; wild sorrel; stinging-nettle; dandelion; plantain; lamb's lettuce; onions and the like; carrots, lemons, … September: acorns; beech-nuts, cabbage-lettuce, cucumbers; endive; onions c.s.; French beans; lemons ….' The procurement of food products was a matter of serious consideration. The choice of rice, for instance, was not left to chance: 'brown rice (Van Sillevolt, rice-huskers, Rotterdam)'. The note gives general rules, based on a list of very detailed instructions for the choice of fruits and nuts, arranged per day for the various seasons. In combination with Brouwer's adherence to vegetarianism, the items give some insight into his daily routine. The self-chosen Spartan lifestyle is illustrated by the following rules:

- In case of momentary fainting one takes according to one's need a juicy fruit, or milk, or milk and bread.
- No more departures [of the rules] allowed as a guest
- No more cleansing baths.
- Never eat by artificial light.
- Always: Once per week swimming in the open.
- Once a week play football or practice another intense physical exercise (preferably with danger and fights).
- To bed only after fasting for 3 hours.

[46] Oral communication Louise Peijpers.

- In case of fever immediately the fruit diet.
- Sleep at least from dusk to midnight.
- Rise as early as possible.

The reader may perhaps wonder if Brouwer tackled the food matter in dead earnest. It should be borne in mind that his student years were one prolonged misery of medical problems. He had, clearly, decided to fight the physical weakness of his body by a systematic regime. And it may be said that the method proved successful. Although he had his breakdowns and illnesses from time to time, he boasted a wiry, lithe body without a trace of fat.

The whole atmosphere of Blaricum and Laren and the congenial housing in the hut must be viewed as the ever changing and yet permanent background of Brouwer's life. No offers from famous universities were able to uproot him. Blaricum was his irreplaceable home. After his fame had spread, he did not even have to leave the village; the established and the newcomers pilgrimaged to his hut.

2.6 The Delft lectures

During the years 1904 and 1905 Brouwer suddenly displayed an interest in the cultural side of student life in Delft. He published a number of short notes in the Delft student weekly, and ended by giving a series of lectures.

His first note was a short comment on Frederik van Eeden's book 'The Joyous World'.[47] In this note Brouwer opposes the view that an improvement of the economic circumstances will result in a morally and ethically better world, 'A bad father beats his child; to improve the father by anaesthetizing the child, and thus taking away the pain, proves hopeless. Therefore it would be necessary to raise the ethical level, then, slowly, the Joyous World will grow. And for that reason the book of Van Eeden is the deathblow for any Marxist.'

The note on Van Eeden's 'Joyous World' was soon followed by an ecstatic exhortation to attend Bolland's lectures, which were to be held in Delft.[48]

As we have seen, Brouwer was involved in the organization of Bolland's lectures in Amsterdam. Although Bolland's performance was not exactly successful in the nation's capital, Brouwer saw no reason not to promote him in Delft, where an active student association had a tradition for organizing cultural events.

Brouwer send an exalted letter, under the pseudonym of Lau van der Zee, to the student weekly:

> ... Then a shining star will appear, the joyful sign of hope, then you will rise to higher regions, to God's glory, although this will not be attained in this life.
> Bolland sees that star clearer than you will learn to see it; the veils disappear before his eyes. And thus he can lead you on the difficult shining way to God's throne.

[47] *Studenten-weekblad* 6 October 1904.
[48] *Studenten-weekblad* 17 November 1904.

The note was (correctly so, I would say) judged incomprehensible, so that in a next issue an article appeared in which the author calmly outlined arguments for the role of philosophy in Delft, adding that 'We cannot forego the occasion to show the author of this exhortation, Lau Van der Zee, our appreciation for his laudable efforts, although we consider his way of operating most dubious.'

It is certainly surprising that Brouwer showed such an interest in the promoting of philosophy in Delft. There are two partial explanations: in October 1904 he was still making propaganda for Bolland in Amsterdam, and he may have decided to give his fellow organisers in Delft a helping hand, and furthermore he had connections in Delft. In fact, his brother Aldert had registered as a student in Delft in 1903.

There is a third possibility: Brouwer was preparing a philosophical exposition ('a philosophical creed that will be the prologue of my work', cf. p. 52) as a part of his dissertation. It is not unlikely that it did not take him long to realize that the material was not exactly suited for a faculty of mathematics and physics. As we will see (p. 96) even a modest philosophical motivation did not survive the axe of the adviser. So he may well have considered Delft as a suitable platform for his philosophical message. Any of these reasons or a combination, may have sufficed for his involvement. It is not known how Brouwer got himself invited to give a series of lectures in Delft. It seems plausible that his brother introduced him in the local student association '*Vrije Studie*' (Free Study), and that Brouwer sufficiently impressed the governing body to get himself booked for a series of lectures, the first of which was held on 29 March 1905.

The lectures were well attended and apparently quite successful. The organizers were somewhat sceptical about the text that was going to appear in print—'The oral lecture was necessary indeed for a proper understanding, I think that in print much will seem incomprehensible'.[49]

Brouwer, although no doubt seriously trying to get his message across, could hardly suppress a quiet amusement at the behaviour of the audience (which, by the way, listened patiently for 3 hours(!) on end). He reported to his friend Scheltema that:

> The lecture will be printed at the request of the Delft public. When I send you the booklet, you will read it, won't you, and you will not lay it aside unopened in fear or disgust? Why then didn't you mix with the public as a solitary, darkly watching enemy among all those others, who were either stupidly frightened or admiring, or did not understand, or got angry. If you had seen how a couple of girls, in the second break, cried that they could not bear it any longer and demanded to be taken home, your nostrils would have flared, you would have snorted of hatred.[50]

The promised book itself was written and produced at an incredible speed, the publisher already advertising it in the weekly of 8th of June:

---

[49] *Studenten-weekblad* 6 April 1905.
[50] Brouwer to Scheltema 7 April 1905.

BROUWER. Leven, Kunst en Mystiek.

with the chapters

    I   The sad World
    II   Introspection
    III The fall caused by the intellect
    IV The reconciliation
    V   The language
    VI Immanent Truth
    VII Transcendental Truth
    VIII The Liberated life
    IX Economics

The language of the treatise is partly borrowed from the majestic language of the bible, and partly it uses the expressive emotional language of the literature of the turn of the century, albeit with the personal Brouwerian flavour. If the reader feels that Brouwer's language is unusual and convoluted, he certainly is right. Brouwer had no mercy on his readers, even his Dutch is hard to read, and the extraordinary length of his sentences, with many subordinate clauses, was notorious.

There is a number of main themes, with which the book is concerned, all of those, however, can be retraced to the central theme.

The main point of *Life, Art and Mysticism* is the truly mystical doctrine that man's ultimate goal and challenge is total introspection—a turning into oneself (*zelfinkeer*). All the remaining points and chapters are elaborations of that particular task. Before Brouwer turned to this central issue, he first presented in a 'pedagogical' first chapter the disastrous influence of man on the world and nature in general. It is called 'THE SAD WORLD', as a wordplay on Van Eeden's 'The Joyous World'. The violations which Brouwer describes, will nowadays generally be recognized as such, but at the time of writing little understanding could be expected. On the contrary, the things that Brouwer condemned, would be greeted by most as miracles of progress. Practically speaking, in Brouwer's young days the crimes against man and nature had only just begun. The wholesale poisoning of complete areas and seas, the destruction of the natural balance in the system of our rivers, just to mention a few topics, has reached nowadays a dimension that one could not have imagined in 1904. Hence, one may well assume that messages, as contained in 'THE SAD WORLD'—if they would find readers at all!—would be brushed aside as totally irrealistic, and as scare mongering. One cannot do better than read Brouwer's text, of which parts are reproduced here. A translation, unfortunately, does not do justice to Brouwer's prose, which is extremely solemn.

> The Netherlands came into existence and was preserved by the deposit of silt of the rivers; a balance between the dunes, the delta, the tides and

the discharge of the water was established—a balance in which temporary floodings of parts of the delta were incorporated. And in that land a strong human race could live and endure. Meanwhile, people were not content, they built dikes along the rivers to regulate or prevent the flooding, changed the river courses at will in order to improve the drainage or the shipping routes, and in the meantime cut down the forests. Small wonder then, that thus the subtle balance of the Netherlands was undermined, that the Zuiderzee was eaten away, that the dunes were slowly but inexorably washed away. And that nowadays ever harder labour is required to protect the country from total destruction. And does it not seem curious to observe, how this self-inflicted labour is not only accepted in resignation, but that it is even lent a lofty cachet of a task imposed in the name of God or Inexorability?

The people originally lived separated, and each tried to preserve for himself his balance in the supporting environment of nature, amidst sinful seductions; thàt filled their lives, no interest in each other, no worry about the morrow. Hence, also, no work and no grief; no hatred, no fear; also no pleasure. Meanwhile, one was not content; one sought power over each other, and certainty about the future. Thus the equilibrium was destroyed, ever more sore labour for the suppressed, ever more infernal conspiracies for the rulers, and all are the suppressed and the rulers at the same time; and the old instinct of separation lingers on as pale envy and jealousy. [...]

It is part of the balance of eternal and omnipresent life, that everyone will be called from this life on earth, when his time has come; and until that time [he is] physically and spiritually ill, as befits his evil mood of thrift, thirst for power, vanity and fear; once more, one is not content with this, one tinkers with the body by means of medicines and prescribed ways of life, and with the souls by hypnosis and suggestion, thus disturbing the purgatory of the lusts, and destroying the balance between psychological responsibility and physical constitution; the body is degenerated from the morale to such an extent, that one can indeed no longer be held responsible for one's crimes, for one's actions in this world. Although medical science boasts in recent times of the prolonging of the (incidentally, far too short) span of human life, what is the value of it? It is as sad to leave this life after one's time, as before one's time—and death? 'Nature never destroys without returning something better for it'. [...]

The life of the individual is illusion, the pursuit of goals by means of heavy labour and—disillusion; towards his death, which he awaits, unprepared in full unawareness, the insight, to have lost his life, startles him, were it not that his intellect soothingly enshrouds him with the thought that *without* illusions life would really have been nothing at all, or that in any case he will take a good measure of experience as a surplus with him into the grave.

> Yes, those arrogant elderly, they deceive themselves that experience, that injury and disgrace, that a long life of sin, stamped on their frozen features, which are stripped from all naiveté, and shining from their lifeless eyes, that thàt only leads to wisdom, and they cry out against the young, when it comes to saying what is nearest for man.
>
> The life of mankind as a whole, is an arrogant eating away of its nests all over the perfect earth, a meddling with her mothering vegetation, gnawing, spoiling, sterilizing her rich creative powers, until it has gnawed away all life, and the human cancer withers away over the barren earth.
>
> They call the folly in their heads, which accompanies thàt, and which turns them insane: 'Understanding the world'.

'THE SAD WORLD' sketches in dramatic tones the degeneracy of man, who has exchanged his natural stability for the sinful state of never ending subjugation of nature and his fellow creatures. The chapter serves as a contrast to the next one, which sketches the possibilities and virtues of introspection. By turning one's attention away from the world, to the inner world of the self:

> ...the passions become silent; you feel yourself pass away from the old exterior world, from time and space and all other manifold things. And the eyes of a joyous silence, which are no longer tied, open up.

This chapter contains descriptions of the mystic experience, that seem to put it beyond doubt that Brouwer was no stranger to the experience. It paints in the words of a visionary the victory of the introspecting Self over the sad World. This chapter and later ones contain a number of quotations of Meister Eckehart and Jakob Böhme; this shows that Brouwer was well acquainted with the old European mystics.

The inner world, that the Self can obtain access to, is a boundless, chaotic mixture of fantasy worlds.

> And in that merging sea of colours, without separation, without permanence and yet without movement, that chaos without disorder, you know a Direction, which you follow spontaneously, and which you could just as well not follow. You recognize your 'Free Will', in so far as it was free to withdraw itself from the world, in which there was causality, and then remains free, and yet only then has a really determined Direction, which it reversibly follows in freedom. [...]
>
> The phenomena follow each other in time, bound by causality, because you yourself want, shrouded in clouds, the phenomena in that regularity.

The passages in 'INTROSPECTION' do not yet have the preciseness and conciseness of the later explanations of 'move in time' and other notions, but in a poetic way the nucleus of Brouwer's foundational credo is expressed here.

In Brouwer's opinion the sorry state of the world, including man himself, is caused by the interference of the intellect. Chapter III, THE FALL THROUGH THE INTELLECT, deals with the phenomenon of man's apparent effectiveness in matters of domination of the world.

> Intellect renders men in the Life of Desire, the diabolic service of the connection goal–means between fantasies. While in the hold of the desire of one thing, the intellect hands them the pursuit of another thing as a means to that end; thus for the shifting of the riverbed: the making of a dam; giving vent to one's jealousy on another: setting his house on fire;
> …

Here Brouwer formulates for the first time his *end–to–means principle*, which was going to play an important role in his overall philosophical considerations.

Whereas, the intended domination that is implicit in this 'leap from end to means', is already in itself objectionable to the introspective person, Brouwer points out a serious inherent shortcoming of the end–to–means transition. The transition, he says, is always slightly 'off key', so that repeated use of it, eventually leads to effects that were not desired.

> The act, which seeks the means, now, always somewhat overshoots the target; the means has a direction, which makes an angle, albeit a small one, with that of the target; it thus works, except in the direction of the target also in other dimensions; an effect, that, if the attention were not isolated from it, could perhaps be experienced as very harmful; but more: the attention gradually looses sight completely of the end and henceforth only sees the means. And in the sad world, where together with the Intellect, Drilling and Imitation are born from Fear and Desire, and nobody any longer surveys the whole human bustle. Many come to know that, what originally was a means, only as an end in itself; they pursue, let us say, an end of second order; with which perhaps again a means will be discovered, and that again makes a slight angle with its corresponding end. If the alluring leap from ends–to–means thus is repeated several times, then it can easily happen that eventually a direction is pursued, that apart from its deviations in other dimensions, makes moreover an obtuse angle with the very first direction, and so counteracts it.

The chapter provides a whole catalogue of disastrous consequences of this practice, some of them are now generally recognized, whilst some—even today—would be considered irrealistic exaggerations. For example, Brouwer's views on nature are forerunners of the present ecological tenets:

> Does not industry originally deliver her products with the end to create in nature an environment of maximally favourable conditions for human life? In that connection, it was neglected that those products were themselves manufactured drawing on nature, in which, for this end, interventions were made in a disturbing manner. The balance of the conditions of human life was violated to a greater detriment, than the industrial products could ever benefit us. All the required wooden material, for instance, has led to the disappearance or degeneration of so much forest, that in the

temperate zones hardly any crop for human consumption grows spontaneously. And more: we started to view the generating of industrial products as an independent goal And in the pursuit of that goal, created as a means, a new industry of instruments that facilitate the old industry: a further blow to the old balance. In addition we recklessly started to collect the raw materials in remote countries, giving rise to trade and shipping, with all their physical and moral horrors, and the mutual suppression of nations.

In principle the same lines could have been written by modern environmentalists and reformers, and they were certainly in the minds of nineteenth century utopists and reformers. So far, Brouwer's indictment of the intellectual human imperialism differed from the moral programs of his predecessors and contemporaries mainly by the fact that his philosophical-ethical principles were far deeper and more radical. His rejection of the human pursuit of domination of the world, nature and fellow humans, was total and well-argued. He did, however, not stop at castigating society for the crude exploitation of economic, social and political powers, but went on to draw the ultimate conclusion. Namely, that even in the domain of the intellect and the mind, man was perpetrating the iniquities of the destruction of the natural balance. He extended the theme outlined above by adding science to the list of culprits perpetrating the abomination of iterated jumps from ends to means, to means, to means, and so on. Science is introduced as a means to further industry, and in turn becomes an independent subject; it then is followed by the 'foundations' of the science under consideration, which in turn is followed by 'epistemology'—'but the embarrassment ever increases, until all heads are reeling'. As for the scientists who take part in this regression, Brouwer comes to the merciless conclusion that

> Some of them give up quietly in the end; having thought, for example, for a long time about the intangible link between the intuiting of consciousness, which evolves with life out of that Anschauungs-world, and the Anschauungs-world, which itself only exists by and in the forms of the intuiting consciousness—an embarrassment, stemming from one's own sin of establishing an Anschauungs-world—then they put the 'I' which was self-created just like, and simultaneous with, the Anschauungs world in the opening, and say: 'Yes, there should, of course, remain something incomprehensible, because it is 'I' who must understand.

In a sharp indictment Brouwer accused an immense catalogue of established practices—science, the industry of stimulants and of pleasure, the misuse of art and religion, the medical industry and profession ('The medical industry was in the right hands with barbers and quacks, ... ').

Clearly, the conflict between the life of the mystic in self-contemplation and the ambitious world of improvement and domination, presented Brouwer with a real and significant problem. A sincere person like him could not just expose the undermining of worldly life; he had at the very least to consider solutions. The chapter, 'RECONCILIATION', offers such a solution in a remarkably mature way; it avoids the pitfall of action for transforming the sad world into a better one—'each attempt to eliminate the non-balance only causes a shift of the non-balance'.

The solution which Brouwer offered, consisted of a reconciliation with the straying world, a resigned life in which pain, labour, desire and fear belong to one's fate. One should not frivolously add to the burden of one's karma, but one should neither wish to be better than one is—'that would be a voluntary following of evil desire'. Nor should one wish to improve the world beyond what it is—'that would be evil lust for power'.

These considerations may offer an answer to the vexing question, how one can live in a world like ours, without an abject betrayal of all that is good and sacred. The ideal of detachment, as preached by mystics and Bhuddists alike, points a way out of the horrible dilemma 'collaborate or resist'. Brouwer, doubtlessly, must have considered the problem of reconciling the contemplative life of a mystic and that of the academic scientist. He was far too sincere, just to ignore this fundamental problem. We should at the same time keep in mind that at many occasions nature was stronger than principles. We will see Brouwer rush off to rescue the innocents and to fight noble battles for justice. Only the unimaginative live by the book!

It should not come as a surprise that, given Brouwer's views on inner life, communication and language were secondary notions. There is a special chapter 'LANGUAGE' to which many of his later philosophical insights can be traced. The basic claim is that there is no communication between souls.

> No two persons will experience exactly the same feeling, and even in the most restricted sciences, logic and mathematics, which can properly speaking not be separated, no two [persons] will think the same thing in the case of the basic notions from which logic and mathematics are built. Yet here the will is parallel in the two, for both there is the same forcing of the attention by a small insignificant area in the head. [...]
>
> But the use of language becomes ridiculous, where one deals with the finer gradations of will, without living in that will; just as when so-called philosophers or meta-physicians among themselves discuss morality, God, consciousness and free will; people who [...] share no finer movements of the soul, ...

It seems tempting to conclude that Brouwer must have advocated an abstinence from communication; there is not much supporting evidence, however. He practised great care in his scientific communications, even to such an extent that they became difficult to read. Moreover, in daily life he was an inveterate conversationalist. In view of the earlier remarks on 'Reconciliation', there is nothing paradoxical in this. On the contrary—a person who is aware of the weaknesses of communication will probably take extra care in his use of language. Later developments will shed more light on this topic.

The Chapters VI and VII deal with IMMANENT TRUTH and TRANSCENDENT TRUTH; the first truth 'points in the world at the consummated Karma of the world', the second points, in the world, at the personal life: 'Immanent truth clarifies, transcendental truth makes devout'.[51]

The chapter on Immanent Truth has acquired a measure of notoriety because of its view on women. It is a theme that in the underground folklore belongs to the *chronique scandaleuse* of an otherwise respectable science. The chronicles of intuitionism have always passed over *Life, Art and Mysticism* in an embarrassed silence; the mystic views of the founder of mathematical intuitionism were thought to be a liability that might very well detract from the objective virtues of intuitionism. The resulting picture of intuitionism showed a somewhat flat pragmatic practice, which—whatever one may think of the mathematical subject—did not do justice to its historical roots. At one point there was, however, justified cause for reticence—the topic of the female. Brouwer's conception of the role of women in the world is rather dated; this may surprise those who think of Brouwer as the revolutionary innovator—but, whereas this characterization is certainly apt in relation to his topological work, he may be considered a conservative in a number of philosophical matters.

The fact that Brouwer's intuitionism was considered new and revolutionary, can be simply explained by the observation that people had neglected their inheritance of idealistic philosophy, so that after a spell of the fashionable formalism (in mathematics) and neo-positivism (everywhere else), Brouwer's doctrines seemed to the less informed the newest thing, instead of a return to nineteenth century idealism. In fact, Brouwer was basically a conservative; *Life, Art and Mysticism* was definitely a protest action against the prevailing optimism of Progress.

His views on women can be classified as equally conservative, although one must understand that his views are part of the total mystic view that is being propounded.

The chapter *Immanent Truth* deals with the aspects of the resigned life in the world with all its conflicting desires and interests (as opposed to Transcendent Truth, which deals with life, disengaged from the influences of the world) and treats among other things the influences of various art forms and the burdening of the karma by 'avarice, ambition, and ... the illusion of woman.'

There is nothing new in the repudiation of the female by mystics and hermits. The history of the church provides numerous examples of saints who had a keen eye for the dangers of female company. In fact Christianity is not the only religion to take (or at least, to have taken) a rather defensive view towards women. So, in itself, the claim that the female burdens the karma of the man is quite in line with the tradition of ages. Brouwer, however, went beyond this observation (which, of course, carries an unmistakable danger sign: beware!) by classifying woman as a creature of a separate level. Her true function is to ward off disturbing influences from the man's karma, although she is, paradoxically, the greatest temptation to

---

[51] "If Truth points in the world to the personal life, free from the ties of fear and desire, where the bliss and wisdom and the quiet rejoicing of the timing in upon oneself flourish on modesty, poverty and quiet fulfilment of duty in this life on earth, which is one's own accomplished karma, then it is Transcendent Truth."

him. The role of the woman is a serving one:

> Humble she will be, humbly she will wish to take all ignoble work from his hands, all work other than the pure indulgence of the faculties of his body, in which he walks the earth; without the wink of an eye she will give her life, to save his balance.
>
> Serene will be her eye, tenaciously and patiently she lives on, and does what serves the beloved. Her body will be unwrinkled, motionless, without passion to seduce, not conscious that it seduces and yet so unbearably seductive in its taunting composure, that no man can endure it.
>
> The Venus of Milo shows in a clear, pure way that karma of the woman, of the quiet, desireless, unconscious, and yet so infernally seductive woman.

According to Brouwer, a woman can also burden her Karma, for example, with 'male activities'. Noble institutions will degenerate when women intrude, so the work of man, when taken over by women will necessarily be degraded. By the usurpation of parts of the prerogatives of man, the woman acts against her karma. In a long catalogue he lists all the shortcomings of the woman. If she is independent, she looses her femininity and burdens her karma, and if she is truly female, she is a shadow of her beloved and is guilty of naiveté:

> In worldly matters and worldly convictions she will naively follow the beloved, and defend views, unthinkingly copied, as objectively indisputable axioms against all objections from third parties; in disputes with such a woman the ridiculousness of language as a means of reaching an agreement clearly appears in the form of the notorious 'feminine logic'.

Whereas the philosophical and mathematical conclusions that Brouwer would eventually be drawing from his mystic insight and convictions were original and even revolutionary (perhaps in a counter-revolutionary way) his views on women, it seems, were rather modelled after the prevailing views of the nineteenth century. Many of the cliché's that turn up again and again in the treatises on the weaker sex,[52] appear in Brouwer's essay; women are temptresses, endangering male purity—pale, without expressive lines. Even Ophelia, the preoccupation of the Victorian period, is called as a witness.

The passages on women are certainly not the strongest in Brouwer's book, but at least he was not guilty of building a scientific argument, which was not uncommon in this particular area.

His message was mainly one of a moral nature, a warning from an ascetic mystic to the world. To the young man with the exalted ideals of the introverted seeker, women personified the dangers to man's karma. No doubt his personal emotional life and history had influenced his philosophic outlook. As a result, woman was not only the spectre of the fall from the ardently sought inner peace, but she was furnished with all the paraphernalia of the temptress and of the weak. Woe to him that succumbs to the distracting charm of the female:

---

[52]Cf. [Dijkstra 1986].

> But truth in art shows the distinct lines: man should avoid, ignore woman; but the woman should live in the man, holding herself insignificant, powerless and worthless, and sacrificing everything to the beloved. A real woman is pale, supple, without expressive lines, with dull, dreamy eyes: she has no muscular strength, and cringes from nothing. And a man who turns to a woman, has lost his life.

The choice of images and of words suggests that Brouwer was acquainted with the nineteenth century literature on the role of the woman. Many of themes and descriptions have a familiar ring to the connoisseur of the Victorian era. It is tempting to conjecture that his stern views on the role of the weaker sex may have been the result of female attacks on the bastion of Bertus' himself. Given the lack of facts, this has to remain what it is: a conjecture.

Many have wondered about Brouwer's theoretical aversion to women and his rather progressive daily outlook; in the twenties he was one of the first mathematicians (maybe the first) in Holland to engage a female assistant, and he admired his female colleagues in the academic world. He was on good terms with the renowned Emmy Noether and with Olga Taussky, and he certainly did not avoid female company, neither in he context of science, nor in his private life. Presumably he conceived the sermons on the distracting female in the framework of the avowed goal of the mystic: the unconditional introspection, whereas—as we have seen above, the actual life in the world required the sincere mystic to suffer the ways of the world, under pain of loss of karma through pride.

There is also a brief mention of science and truth:
> Furthermore immanent truth breaks through in science as well. It has separated the observed things from the ego, and placed it in a Anschauungsworld, which is thought to be independent of the ego, and which has lost the connection with the Self, which alone feeds and directs. Thus it builds outside life a mathematical-logical substrate, a chimaera, and within life, a tower of Babel with its confusion of tongues.

Here the adoption of an independent outer world appears as an attribute of immanent truth.

Life, Art and Mysticism contains a number of further remarks on society and its organization. Brouwer's conception of work as something noble and lofty clashed forcefully with more progressive ideas. We have already mentioned the inroads of women into the labour market and its negative consequences:
> The gradual usurpation of certain forms of work by women will go inexorably hand in hand with a degeneration of that work into an ignoble state.

The graduate student, who only a few years ago attended socialist and communist meetings in the company of Scheltema and Wiessing, apparently had realized that certain features of socialism were incompatible with the world of the mystic.

It should be fairly evident that socialism, with its preoccupation with the world and its socio-economic features, has no call for a mystic credo, and conversely the mystic would consider the materialistic socialist as one of those unfortunate, unavoidable parts of his Karma. Brouwer viewed the consequences of socialism with some mild horror:

> Until quite recently the state, and public life, were viewed as something honourable, even metaphysical; and a position in society was considered a noble task; [...] But the socialistic movements have in the last century washed away that aspect of 'honourable', [...].
> When, as the endpoint of the socialistic degeneration, the state will have turned into a well-oiled automaton, well—then the administration will perhaps be completely left to women.

The final chapter, 'ECONOMICS', must be seen as a logical conclusion of the preceding chapter. The proper attitude in life towards the theoretical contemplation of life and society, of the human who has taken to heart the call to introspection and detachment, is one of renunciation.

> There is one more thing, which the free life should be careful not to get tainted with, as long as its ties with society last: economics.

For, inherent to economics is, according to Brouwer, the idea that 'foolishness and injustice' are essential—otherwise economics would be superfluous. Thus the intellectual study of the ways and laws of misfortune and injustice will not attract the free man. Desire of property evidently turns the attention outward;

> ..., for he who views something as desirable or deplorable, views it as something outside himself, as part of a world which exists independently and persistently, as part of a fixed inalienable possession, that one can cultivate, take care of, clean, raise, as one can with one's flowers or chickens. Exerting influence outside oneself, be it for improving the world or for one's own power, is: blinding, vanity, thirst for power.
> The Free rather view their fellow beings as delusions which solicit compassion, disturb the path of life, which are to be borne like guilt, for their freedom does not suffer them around. And the Free cautiously slip past them.

Society is, in this perspective, an artificial web of power and domination, complete with its moral justification in abstract terms of 'suppression', 'justice', 'rights', ... Here, Brouwer appears in the cloak of the radical anarchist, denouncing the social theories that legitimize the powers that be. The theorists are vigorously criticized:

> They are talking about 'Human Rights' as if man brought rights into his life, and more than miserable duties, as a punishment for being born.
>
> They are talking about 'labour', its necessity, and the happiness it brings. As if the labour of mankind were something else, but a blind convulsion of fear for what is no evil, and of desire for what brings misery.

> They are talking about the 'poor oppressed' and about 'oppressed classes'; as if anybody would here be born in a state of oppression which he does not deserve!
>
> They also talk of talents and of joy of living; that would not have a fair chance in the poor oppressed classes; as a matter of fact, joy of living does not exist, it is only desired; for life is joyless.
>
> Or they talk about 'Justice', and cry in the joyful vista of childish vanity, that eventually it will reign on earth. But is justice anything (else) but the petrification of that union of men, which expresses their separation without independence?
>
> The economists and leaders of the people also love to talk about a 'future state of the deliberately co-operating people'; this would be possible for people without fear and desire, but those would not work, and a world of such persons would not exist.

Thus this final chapter once more confirms the basic tenets of the earlier ones: meddling with the organization of the life of human beings is doomed from the start. The final words sum up the lesson:

> he, who knows not to possess anything, not to be able to possess anything, not to attain stability, and who resigns in resignation, who sacrifices everything, who gives everything, who no longer knows anything, wants anything, wants to know anything, who lets everything take its course and who neglects everything, to him will be given everything and to him is opened the world of freedom, of painless contemplation, of—nothing.

The Delft lectures and the subsequent book went largely unnoticed; there was a review in one of the daily newspapers, a rather devastating one, and there it stopped. Brouwer sent complimentary copies to his friends and colleagues. Two reactions have been preserved; his friend Scheltema disagreed on principle:[53]

> I received your booklet shortly after my letter and I have started to read it. So far I did not read with the aversion I expected, and I wish many a Philistine this literature as a refresher—but you know how 'heartily' I disagree with you, how, for example, the conscious social democrat immediately rejects the premise you start from (that is that the original animal-like human life is the happiest imaginable one, and at least should be worth pursuing most) on the contrary, stresses that the happiness of the human society will only *begin* after the *conclusion* of the era of barbary in which we still live ...

A surprisingly mild reaction considering Brouwer's volleys at socialism! Korteweg, Brouwer's Ph.D. adviser, was more critical:[54]

---

[53] Scheltema to Brouwer 16 May 1905.
[54] Korteweg to Brouwer 13 May 1905.

That I am greatly interested in you and hence appreciate the sending of your slim volume, therein you are certainly not mistaken.

Whether I shall read it? I thumbed through it, but it is not the reading that I wish or that is good for me. It is true that there are abysses very close to us, but I do not like to walk on the brink of them. It makes me dizzy and less able for what I need to do. Whether it is good for you, I don't know. So much is certain, that I would rather see you walk along other paths, even though I find it sometimes also hard to follow you there, where you cut so deeply through fundamental matters.

Life, Art and Mysticism earned its author a reputation of eccentricity, although rather by an obscure oral tradition than by direct acquaintance with the text.

The question that immediately comes to mind is, was Brouwer serious about it all? Considering the available evidence, the answer is probably 'yes'. He may have exaggerated here and there to provoke the audience, but by and large the mysticism was genuine. In addition to his pronounced views on life, the complacency of progress and the progressives probably was a valuable source of stimulation.

2.7 Family life in Blaricum

To Louise, who was only 12 years old at the time of *Life, Art and Mysticism*, the whole matter seemed mysterious. She was attending in Amsterdam the school of a certain *Master Gerhard*, which was conducted on socialist principles. The fee for the school was forty cents a week. With this education, based on the principles of clean, healthy, idealistic socialism at the turn of the century, she must have been puzzled, to put it mildly, by her stepfather's rather unusual, gloomy views.

Like any child, she would go exploring in the house when she was left by herself. She recalled that she was drawn by a magnetic power to the cupboard in which the manuscript of *Life, Art and Mysticism* was stored. She cautiously climbed on a chair and read bits and pieces of it, closing the cupboard and jumping down as soon as she heard a key in the lock. Mysterious as the contents were to her, she understood enough to guess that this book should not get into the hands of Master Gerhard, and so she told Brouwer that she knew about the manuscript, asking him to promise that he would not show it to her teacher. Eventually Master Gerhard got wind of the book, as a result poor Louise was expelled from the school on the grounds that her parents did not conform to the socialistic ideals.

Since she was no longer at school after this incident, Louise was made responsible for the housekeeping in Blaricum—something she hated. The hut had little comfort, just one bedroom upstairs. Louise had to sleep on a wicker chair downstairs. Brouwer worked downstairs or in the garden, often reclining in his characteristic pose in the same long wicker chair.

Lize stayed in Amsterdam during the week, managing the pharmacy. Brouwer stayed home to work and went over to Amsterdam for his teaching, visits to the library, occasional meetings with fellow mathematicians, and for his regular concerts in the Concertgebouw. When it was more convenient to stay in Amsterdam, he joined Lize in the apartment over the pharmacy.

The hut in Blaricum. Photo by Dokie van Dalen.

Once a week the mathematician Hendrik de Vries came to visit. He usually brought his violin, which he played well, and Brouwer accompanied him at the piano. Louise's role was to play the conductor. The lessons of mother Brouwer had born fruit. Bertus became an ardent amateur pianist. He loved Beethoven, and according to Lize, his favourites at the time were the piano scores of the Beethoven Symphonies.

When Brouwer got it into his head to order a day of fasting, Louise was not allowed any food at all, not even yesterday's leftovers. She soon discovered, however, that Brouwer dodged the fast by nibbling nuts from the drawer of his desk. From then on she took her preparations, and bought enough biscuits[55] to survive the day.

As part of her education, Brouwer decided to teach Louise how to cook. The first lesson was how to prepare rice; it ran as follows: 'rice is a product from the Indies; it is carried by a ship and there negroes sleep in it. Therefore I want you to wash the rice ten times before you boil it' (Brouwer knew very well how lazy Louise was). Needless to say Louise did not follow the instructions. There are many little stories and incidents that show the uneasy relation between Brouwer and Louise. To mention one example, Brouwer told Louise to eat by herself in the kitchen when Lize came over for dinner.

Louise also had to walk from time to time to her school in Amsterdam. And sometimes Brouwer got it into his head to teach her French; sitting at a terrace in

---

[55] frou-frou, a special kind of wafer.

Amsterdam, he would quite sternly take her through the drill, not hesitating to slap her if her attention relaxed.

The relationship between Brouwer and Louise was problematic from the beginning. Brouwer was easily irritated by Louise and considered her lazy, stupid and stubborn. She certainly was far from industrious, but Brouwer's treatment of her did little or nothing to improve her attitude. As long as Brouwer lived, they had fierce clashes about everything. In the beginning Brouwer had the advantage of his age, but gradually Louise began to free herself from the pressure of her stepfather. And eventually the two battled on equal terms. Whereas Brouwer had made it a principle to avoid unpleasant and obnoxious people, he could not avoid Louise. For Lize the situation was extremely painful, because Louise was her only child, and she did not want to sacrifice the happiness of Louise to that of her husband, but neither did she want to hurt Bertus. The surviving correspondence between Lize and Louise shows how difficult it was for Lize to steer a safe course between daughter and husband. Louise told how, when she was still living at home, Brouwer sometimes introduced her to his visitors with the words: 'and this is my silly daughter'. The sad thing was that it was not intended as a joke. In return Louise thought that her stepfather was himself as mad as a hatter; in her opinion all the fuss at the house in Blaricum only served to keep him quiet. If no prophet is recognized in his home country, then certainly he is not recognized at all in his own home![56] Only very late in life, long after the death of her stepfather, she changed her views on him.

Even in her old age Louise's memory was wonderfully clear, she provided many facts that at first sight seemed curious. A complicating factor was that she was very confused about religious matters. The details she provided on daily life and the personal history of Brouwer and the crowd around him have, however, mostly been born out later by independent confirmation. By extrapolation I have come to view her information as generally reliable.

---

[56]When in 1981 a conference was dedicated to the centenary of Brouwer's birth, a short biographical article appeared in the Dutch weekly *Vrij Nederland*, [van Dalen 1981]. Reading this biography, Louise suddenly realized that Brouwer was not a fool after all. As she regularly communicated with the spirits of the departed, she noted that Brouwer's spirit had found rest after this public recognition.

3

THE DISSERTATION

3.1 Preparations and hesitations

While the excursion into philosophy and mysticism was going on, Brouwer quietly continued his mathematical research. He read much and tried to get acquainted with modern developments. Although his principal teacher, Diederik Johannes Korteweg, was a man of quality and of good taste, his own work was almost exclusively in the domain of applied (or at least, applicable) mathematics. There is no doubt that Korteweg offered a solid mathematical education, but on the whole, the choice of topics was rather conservative. The marvellous and exciting innovations from Paris and Göttingen were not taught or discussed. Students were definitely not brought in touch with the newest developments in pure mathematics.

We have seen that Brouwer had unusually strong philosophical views, which were, if anything, far from progressive or fashionable. Although he had come to terms with the world in the sense that a crusade for the spiritual liberation of mankind would only result in a burdening of his karma, and that thence a 'live and let live'-policy was the proper choice, he could not just shrug off the preaching habits he had acquired during the *Life, Art and Mysticism* period (which, by the way, partly coincides with the research for the dissertation).

There are a number of documents relating to the preparation of the dissertation, in the first place a series of notebooks, in which Brouwer jotted down his ideas and comments, and in the second place a synopsis he made for a first version of the dissertation. In the synopsis a selection of the material in the notebooks is sorted into chapters.

The notebooks contain a mixture of all sorts of topics. It is interesting to see that when he started out, the mystical–philosophical considerations were still uppermost in his mind. Gradually the non-mathematical remarks give way to purely mathematical topics, to re-appear suddenly somewhere in full force. In particular at the end the philosophical content increases again. The notebooks are not dated, so it is difficult to make specific guesses about timing and progress.

Here and there in the notebooks one can find schedules, plans, outlines for reading, research and writing. They are instructive in the sense that they shed some light on Brouwer's reading. They also illustrate his self-discipline. The following schedule, written on the inside of the cover of the first notebook, may serve as an example:

  1 week Russell
  1 week non-Euclidean geometry (Klein, Lie)
  1 week Dedekind and Cantor

1 week Poincaré and *Revue de Métaphysique*

—-—

Elaborate notes on Sundays

Like so many passages in the notebooks, this schedule is struck out, probably because the work had been done. Did Brouwer stick to the schedule? It is not impossible. Brouwer is known, in particular in his early years, as a hard worker, with explosive bursts of activity; nonetheless, four weeks for this amount of reading and the corresponding compiling of notes seems a tall order.

On page 1, in the margin, another list of things–to–do for the dissertation is given:

1. Foundations, Russell; Couturat, algèbre de la logique; D. Hilbert in Heidelberg Congress; Hilbert, found. of geometry, Teubner.[1]
2. non-Eucl., Klein, Lie
3. Numbers, Cantor, Dedekind.
4. Metaph., Kant, Poincaré, Couturat
5. Projective views on vector– and potential field
6. Work out notes
7. Hankel complex numbers
8. Hertz. Cf. Poincaré.

Only Hankel and Hertz have not been treated in the notes, the rest is discussed fairly extensively.

It is clear from the notes that Brouwer still was occupied with his mystic-philosophical views. The form these views take in the context of mathematics are not all that different from what we have seen in the previous chapter.

As we have seen, 'the highest attainable', state in life is the return to the deepest home of the self, the ultimate introspection (p. 66). The exodus from the original state of the soul[2] is what Brouwer calls 'sin'. The manifestations of sin, in this sense, are manifold. To mention a few: the seeking power over nature, the domination of fellow beings, the exploitation of one's talents. Since mathematics was, in Brouwer's eyes, party to the domination of nature, it was one of the tools of sin— unless practised for more lofty purposes, that is the free unfolding of the self in playful development.

The notes are pervaded with a reticence to take part in a sinful practice, or rather, to collaborate (no matter how pure his personal motives were) in a worldly design that he could not approve of.

We will reproduce a number of Brouwer's moral comments here:

---

[1] Brouwer cites in his dissertation the *Festschrift* version of the *Grundlagen der Geometrie*, but he privately used the French translation in *L'Enseignement Mathématique*.

[2] One would be inclined to say 'mind', but in Brouwer's later work 'mind' is given specific meaning, cf. [Brouwer 1933A], [van Stigt 1990]; hence it is better to avoid the term here.

> Whoever practises mathematics with a conscience (even though that is often maintained by fear, or undeserved respect for others and by suggested theories) finds few things, although these few may perhaps make a brilliant impression. He can produce much more if the place of the conscience is taken over by the mentality of the businessman who just 'markets' things.
>
> The activity of modern mathematics is forbidden, much as all excesses [...], and publishing, just as much as the marketing of meat extract.
>
> Let the motivation behind mathematics be the craving for the good, not passion or brains.
>
> The joint work with the ideals of mathematicians (thus not as merchandise) is based on the understanding of, and the resignation in, each others' wickedness.
>
> The role of foundational research must be: *given* the temptations of the devil, who is the world and its categories, to appreciate the true value of the world, and to relate it constantly to God.
>
> One should refuse to do mathematics, but since this point has been reached, one should refuse to do the next step, that is mathematical logic.
>
> My own, spontaneously observed, life has no fixed laws, but it is a miraculous play of chance. Just as one of the elements of this world of chance, there floats the dark cloud of the mathematics practising rabble which acts in this world in such a way that it can only react in laws (that is something from the underworld).

The above mentioned notion of 'sin' occurs frequently in Brouwer's early writings, in particular in his notes and letters. Later in life, when he became, whether he liked it or not, part of the academic establishment, the moral overtones disappear, to reappear suddenly in his great post-Second World War address, *Consciousness, Philosophy and Mathematics*. The keywords in his early description of sin are 'centralization, and 'externalizing,'[3] both of which indicate the passage from free, undirected contemplation in the self to the concentration on certain aspects, and to the positioning of the contemplated (or experienced) concepts in an independent outer world. In other words, sin is a form of treason to the self (or ego) and not something that concerns the church or society.

The contents of the notebooks and the subsequent resumé make it clear that Brouwer had not shed his convictions of the year before, when he presented his views on the world and on mankind in *Life, Art and Mysticism*. Indeed, the moral aspects are far from neglected in the notes. Sprinkled through the notebooks are remarks on the topics that he valued above all. He must have felt the paradoxical position he was forced to adopt: he had to write a dissertation on a topic in mathematics, and thus to become an accessory to the sinful power game of science, while

---

[3] *centraliseren* and *veruiterlijken*.

at the same time preserving his loyalty to the detached life of the mystic.

The dissertation episode is in a way the watershed in Brouwer's relationship with the world. He did not give up his mystic ideals or his inner convictions, but he realized that he had to establish a pact between himself (his ego) and the world. This is not uncommon in students who reach the end of the blessed period of uncompromised innocence, but in Brouwer's case it was a really painful transition. The Brouwer–Scheltema correspondence betrays the growing pains and the sadness of the loss of independence. In a letter Brouwer refers to 'the coarse mansion of society', where they would have to 'light its chandeliers, and grace its door-posts'[4]

Not so long ago, Brouwer had still been making his often biting comments on the topics that occupied his thoughts. There are undated notes, which possibly belong to the same period as *Life, Art and Mysticism*, which illustrate the difficult process of transformation from defiant student to respectable scholar. They are the aphorisms of the young mystic, untainted by collaboration with the corrupting powers of the world. A few examples may give the reader an impression of the audacious views of the young man before his surrender to convention:

> One could see as the goal of one's life: Abolition and delivery from all mathematics.
>
> Improve the world, or make it beautiful? Aren't we humans the earth's bacteria of decay, which help and speed up the process of decomposition for which the time had come?
>
> Most people become socialists in order to join a suitable milieu and make relations in it.
>
> The vulgar ones do not know that language–truth and understanding–truth do not exist; thus their passion concentrates itself on 'wanting to twist the truth'.
>
> Understanding between 2 people is gradual; but mathematical understanding is something like 'yes' or 'no' just like sleeping is something like 'yes' or 'no'.
>
> That mathematics and its application are sinful follows from the intuition of time, which is immediately felt as sinful.
>
> The rich and the poor are the complementing elements of a world fallen in sin. He who knows to back out of it will be neither rich nor poor.
>
> The feeling of dignity and being honoured is analogous to the erroneous feeling of ease, of power, of being armed, and other feelings, which tempt one to fall asleep.
>
> The degeneration of the practical mathematical activity can be its

---

[4] Brouwer to Scheltema 8 July 1907.

desire to see mathematical systems (theoreticians) and the desire to realize mathematical systems (tinkerers, agitators and intriguers). The normal organizing talent has not yet to be considered a degeneration.

Sorrow is the inability to find in one's head the mathematical system that gives rest. Only he who wants rest therefore knows sorrow.

By learning too much science, that is common thinking, you forget how to think for yourself in the struggle with others.

He who views people as an animal species lower than himself, no longer fights duels, just as little as one agrees to use, in the fight against animals, a weapon on the basis of a choice made in mutual understanding.

Mathematics justifies itself, needs no deeper grounds than moral mysticism.

The space of animals and trees is not Euclidean, does not even have a group of motions.

Isolated remarks of this sort also occur in the notes for the dissertation. In a way the genesis of the dissertation is the story of temptation, as experienced by all hermits and saints. No matter how much Brouwer fought the evil influence of the world, the fascinations of mathematics proved stronger than his Spartan views. Brouwer's views were permeated with the awareness of the sin of externalization with its intended domination of the world and nature. It is no wonder that trade and business figure prominently on the list of sins; even the Jews, the traditional exponents of small and big business, are mentioned in passing: 'How the Jews dominate the farmers with the help of mathematics, and how the farmers do the same to the animals. Mathematics as a part of the technique of culture, put on the market'.[5] It is as if we witness Brouwer's losing battle against the temptress, Mathematics. His notes contain fierce attacks at the sins of externalization, but they make the rather sad impression of rear guard actions. The final, printed version of the dissertation is completely neutral. The anger of the young Brouwer had found an outlet in the fierce indictments of his contemporaries Russell, Hilbert, Couturat and others. The parting shots at the sinful world are a curious mixture of sadness and indignation.

The notebooks show that Brouwer had an extensive reading program; that he went through a great many publications, with the following central topics:

- Axiomatic Geometry and Foundations of Geometry
- Non-Euclidean Geometry
- Foundations of Mathematics, Logic and Set Theory
- Potential Theory.

[5] From the summary of Brouwer's notes for the dissertation.

Mathematische Gesellschaft 1902. The picture shows a number of persons that were to become prominent in one way or another in Brouwer's career. Front row, left to right: Abraham, Schilling, Hilbert, Klein, Schwarzschild, Grace Young, Diestel, Zermelo. Second row: Fanla, Hansen, C. Müller, Dawney, E. Schmidt, Yoshiye, Epstein, Fleisher, F. Bernstein. Third row: Blumenthal, Hamel, H. Müller. Niedersächsische Staats- und Universitätsbibliothek Göttingen.

It is interesting to read how Brouwer found by himself a way through the undergrowth of new developments and contradicting positions of the mathematics of the turn of the century. From the dissertation and the notes, one can more or less piece together his sources, and see by whom he was influenced.

The influence of Hilbert is clearest in the treatment of the Lie group section of the dissertation and in the discussion of formalism and axiomatics. At the time Brouwer wrote his dissertation, Hilbert's *Festschrift* (*The Foundations of Geometry*) had not yet dominated geometry to the extent that the memory of Hilbert's predecessors had become obliterated. This can be seen in Brouwer's notes and the dissertation; indeed Brouwer quotes carefully the axiomatic researches of Klein, Pasch, Pieri, Schur, Vahlen, and Veronese. Of course, Lie's work is taking a central place in the dissertation, but that is no surprise, as Lie groups are explicitly treated. The influence of Hilbert's *Grundlagen der Geometrie* has been so massive, that later generations were under the impression that the axiomatic treatment of geometry was Hilbert's personal achievement. In the beginning of the century this optic distortion was not yet common. Hilbert's *Grundlagen der Geometrie* clearly derived a large part of its popularity more from its forceful presentation, which fitted the mood of the times, rather than from its novelties.[6]

The alternative approach to geometry was picked up by Brouwer directly from the works of Lie and Klein.

In the foundational part, Brouwer mostly dealt with Russell's *An Essay on the Foundations of Geometry*, Hilbert's Heidelberg address *On the Foundations of Logic and Mathematics* and a variety of papers of Poincaré. He also read the papers of Bernstein, Cantor and Zermelo.

It is easy to see that in the case of the foundations of mathematics, Brouwer stuck to his own philosophical–foundational views, and that he measured, so to speak, the authors in the field by his own yardstick.

For that reason, it is somewhat disappointing that Brouwer did not join the discussions of the day. For example, with respect to Zermelo's axiom of choice, he only remarks that the axiom cannot be right, because if we ask a person to make choices from a collection of (non-empty) sets, it is unlikely, that if the person is presented the same set twice, he will make exactly the same choice. Hence Brouwer objects to the functional character of 'choice'.

The potential theory did not find a place in the dissertation; it was published in a series of papers in the proceedings of the Academy.[7]

The most direct information concerning the dissertation is provided by the correspondence between Brouwer and his Ph.D. adviser, Korteweg. The choice of supervisor was not difficult; the only alternative to Korteweg was the mathematics professor Van Pesch, a scholar who practised mathematics at a safe distance from the frontline of contemporary research.

Korteweg, on his part, had no hesitation to accept the young man as his Ph.D. student; already before Brouwer had finished his regular studies at the university, three papers of his on transformations in four-dimensional space (Cf. p. 48) had been communicated by Korteweg to the Academy, so that the latter was eminently aware of the quality of the young man.

The preparations for the dissertations were carried out in parallel with the research on vector distributions and potential theory. The last two topics were safely out of the way when in September 1906 Brouwer started the actual writing of his

---

[6] Cf. [Freudenthal 1957].
[7] [Brouwer 1906A2, 1906B2, 1906C2].

dissertation in earnest. The last paper of the series was communicated to the Academy on 29 September 1906.

The series consisted of two papers on potential theory, *The force field of the non-Euclidean spaces with negative curvature* and *The force field of the non-Euclidean spaces with positive curvature*, and one paper on higher dimensional vectorfields, *Polydimensional vectordistributions*.[8] In the latter Brouwer proved, among other things a generalization of Stokes' theorem to arbitrary finite dimensional Euclidean spaces. In 1919 Brouwer returned to the subject, pointing out that the proof of his 1906 paper also established the non-metric form of the higher-dimensional Stokes' theorem. In the original paper no references were given, but in [Brouwer 1919Q2], Poincaré is mentioned as having enunciated the theorem already in 1899, 'without a proof however'. In a footnote Brouwer referred to earlier publications of Poincaré in which the simultaneous vanishing of both sides of the Stokes equation is stated.[9]

The paper *The force field of the non-Euclidean spaces with negative curvature*, contained a novelty that has escaped the general attention: on page 5 Brouwer introduced the parallel displacement (without giving it a name) years before the notion officially entered into the literature. Brouwer must have attached some value to this invention, since he indicated the definition by a line in the margin in all his reprints.[10] Struik told that he stumbled on Brouwer's definition more or less by accident, when checking some literature.[11] He informed Schouten, who was surprised that the notion had been introduced (in a special case) before he and Levi-Civita had formulated it (independently). It was Struik's impression that Brouwer was not aware that his formulation of the notion of parallel displacement preceded that of Levi-Civita and Schouten, before somebody called his attention to the footnote in Schouten's *Ricci-Kalkül* mentioning Brouwer's priority (at least for the case of constant curvature).

The potential theory papers were communicated by Korteweg to the Academy between 26 May and 29 September 1906. This productivity shows that Brouwer, in spite of the pressure of time, managed to think of other things than the coming 'promotion'.

3.2 Under Korteweg's supervision

Korteweg's role as a Ph.D. adviser was, I guess, restricted to competent and critical comments. The topics were Brouwer's own choice, and there was little overlap between the contents of the thesis and the work of Korteweg. Korteweg must have been somewhat cautious with this remarkable student. He agreed with the choice of the topics, provided 'enough mathematics was left in'. In a letter[12] Brouwer

---

[8] [Brouwer 1906B2], [Brouwer 1906C2], [Brouwer 1906A2].
[9] Cf. [Poincaré 1899] (*Méchanique céleste III*), [Poincaré 1887], [Poincaré 1895].
[10] Cf. *CW II*, p. 58, 69, 71, 78, 83, 86.
[11] D.J. Struik to Van Dalen 16 May 1992.
[12] Brouwer to Korteweg 5 November 1906.

reminded Korteweg of his permission, defending himself against the, real or imagined, scepticism:

> You know well, that when I selected my topic two years ago, it was not a matter of inability to tackle a more 'common' one, but only because I felt an urge towards this subject: it originated spontaneously in me'.

Fortunately for history, a fair portion of the correspondence between Brouwer and Korteweg has been preserved, so that we can reasonably well reconstruct the last stages of the writing of the dissertation.

The first letter that mentions the 'Promotion',[13] that is the official defence of the dissertation and the awarding of the doctorate, is the one of 11 January 1906. It deals with the grant of the St. Jobs Foundation. Since Brouwer was already twenty-five years of age, he was reaching the limits stipulated in the regulations. He argued, however, that one of the rules of the Foundation allowed an escape, so that a promotion at the age of 26 years would not disqualify him for another year of support. The Foundation was fortunately wise and liberal enough to grant him an extra period until 12 November 1906; Korteweg's strong recommendation had swayed the opinion of the regents of the board.

Korteweg was putting a lot of trust in the capacities of Brouwer, although his student was exceptionally brilliant and intelligent, it was a daring gamble. With less than a year to go, and not a syllable on paper, Korteweg was promising more than the average Ph.D. adviser would be willing to guarantee. To make things worse, Brouwer was tied up with his research papers until the last minute. As a matter of fact, Brouwer was still in the weeks before the summer vacation of 1906 making the final improvements of his above mentioned paper, *Polydimensional Vectordistributions*, in which he wanted to correct some mistakes in the preceding Dutch version. He had already received the proofs, and the desk editor claimed that changes were against the rules. Eventually Korteweg's help had to be enlisted in an effort to get the corrections done; he really had to put his foot down to protect his student. In his words, 'One does not force a man to go round with his tie in disorder, even if he gets company—because he tied it the wrong way in the morning!' Moreover, he said, Brouwer was already greatly upset, and it could easily set him fretting, 'It is difficult for a person of a conscientious disposition if he has to leave mistakes in his work, and we must be somewhat careful with Brouwer. The last few years he has been alright, but not so that it doesn't worry me any more.'

The desk editor grudgingly gave in, 'I don't want to cause an illness of Brouwer—and so it has to be done'.

Finally, on 7 September, Brouwer[14] reported that he had stopped reading new material, and that he had started to order his notes into chapters.

---

[13] The formal ceremony of the defence of the dissertation and the awarding of the doctor's degree is called the *promotie* in Holland. The tradition of the public defence of the dissertation before the faculty has, with minor changes, been preserved until to-day.

[14] Letter of 7 September 1906 to Korteweg.

The draft listed 8 chapters:[15]

- Axiomatic Foundations
- Examples of Axiomatic Uniqueness Proofs
- Genetic in the Mind, and Empirical Foundations
- Mathematics and Liberation of the Mind, and Philosophical Valuation of the various Exact Sciences
- Mathematics and Society
- The Construction itself (the Building proper[16])
- So-called Philosophical Foundations
- Criticism of Set Theory

By that time it was just plainly impossible to finish the dissertation before 12 November; presumably Korteweg again exerted his influence to protect Brouwer's income, for in the letter of 18 October Brouwer thanks Korteweg for approaching the St. Jobs Foundation.

Korteweg, indeed, had persuaded the regents that in this particular case an exception should be made. The person was clearly more than ordinarily gifted, and financial problems would probably put him under severe stress. So, he proposed to end Brouwer's grant on 11 November, and to pay the deducted DFl 1.500,–[17] after the completion of the Promotion. This proposal was almost immediately accepted by the board.

On 16 October Brouwer submitted a tentative list of chapters to Korteweg— it rather differed from the above draft:

1) The construction of mathematics.

2) Its origin in connection with experience.

3) Its philosophical meaning

4) Its founding on axioms

5) Its value for society

6) Its value for the individual.[18]

Within a month this list was further reduced to three chapters which constituted the final version of the dissertation:

---

[15] *Axiomatische Grondslagen,* • *Voorbeelden van Axiomatische Uniciteits-bewijzen,* • *Genetisch in de Geest en Empirische Grondslagen,* • *Wiskunde en Geestesvrijmaking en Philosophische Waardering der verschillende Exacte Wetenschappen. (Verband tussen Levens-Logica en Wiskunde),* • *Wiskunde en Samenleving,* • *De Opbouw zelf,* • *Zogenaamde Philosophische Grondslagen,* • *Aanmerkingen op de Mengenlehre.*

[16] Brouwer used the terms *'opbouwen'*, *'bouwen'*, *'wiskundig gebouw'*. in his Dutch publications; the literal translation is 'building', 'mathematical building'; we will stick to the more usual (but also less colourful) 'construction'.

[17] The result of the yearly deductions of DFl 150,– since 1897, cf. p. 14.

[18] *1) De opbouw van de wiskunde. 2) Haar wording in verband met de ervaring. 3) Haar philosophische betekenis. 4) Haar grondvesting op axioma's. 5) Haar waarde voor de samenleving. 6) Haar waarde voor het individu.*

I   The Construction of Mathematics
 II Mathematics and Experience
 III Mathematics and Logic

The whole episode of the preparation of the dissertation is typical of a dignified and noble academic past, a time when letters could be exchanged the same day, when professors had time to read drafts of dissertations at short notice, when publishers produced hand-composed proofs within days. Considering that in September 1906 Brouwer had not written any part of the dissertation worth mentioning, and that in October he was still arranging the material, when he had already told the publisher to start printing, it is a small miracle that he eventually finished in February 1907. In the middle of October he had finished chapter I, and he more or less counted on Korteweg's kind co-operation, so that he could send it to the printer in a week's time.

The second chapter was, however, an altogether different matter; it dealt with the place of mathematics in the outer world, and the first version must have been something of a shock to Korteweg. He had read, or at least glanced through, the philosophical opus *Life, Art and Mysticism*, but probably had hoped and expected that Brouwer would keep his philosophical activity and his scientific activity strictly separated. In Brouwer's case the philosophy was, however, the basic ingredient that made the mathematics work. Here Brouwer and Korteweg had interpreted their agreement in different ways. Brouwer had understood Korteweg's condition 'as long as there is enough mathematics left' as a lower bound on the mathematical content, without stringent restriction on the philosophical part. Korteweg on the other hand wanted mathematics, and possibly philosophy, but only if it was of the traditional sort—say Kant, Poincaré, Frege, Russell. Brouwer expressed his view in a letter:

> [You] suspected that [the topic] would strongly drive me into philosophy, which it indeed did. To the extent that I even, at times, completely lost sight of mathematics. But what I brought you now, exclusively treats *how mathematics roots in life*, and how, therefore, the points of departure of the theory ought to be; and all special subjects of the dissertation derive their meaning in relation to this fundamental thesis.[19]

Brouwer had no intention of leading a double life, mathematics during working hours, and philosophy as a gentlemanly occupation for the leisure hours. On the contrary. In chapter I he had already put the *ur-intuition* on the stage and in chapter II he intended to present an eloquent description of the role of science, that is mathematics, in the domination of the environment by man. Indeed, the introduction to chapter II, as originally conceived by Brouwer, was a clear and coherent elaboration of the moral issues in *Life, Art and Mysticism*. 'The struggle for the domination of nature and fellow humans', according to him, 'is strikingly different from the brutal assimilation and destruction practised by other creatures. The secret of the success of man, lies in his potential for objectivization of the world.'

---

[19] Brouwer to Korteweg 5 November 1906.

The primeval phenomenon is simply the intuition of time, in which the iteration of 'thing–in–time, and one more thing' is possible, but in which (and this is a phenomenon outside mathematics) a sensation can resolve into constituent qualities, such that a single moment of life is lived as a sequence of qualitatively distinct things. One can, however, restrict oneself to the simple observation of those sequences as such, independent of the emotional content, that is from the various gradations of frightfulness and desirability of that which is observed in the outer world. (Restriction of the attention to intellectual contemplation). Then the strategy of the goal-oriented activity of humans, is to substitute the means for the goal (the preceding one for the succeeding one in the intellectually conceived sequences) when the instinct deems the chances of the means better in the struggle. [......] However, in general the strategy, consisting of the observation of causal sequences,[20] and in connection with this, the shift from goal to means, is successful, and gives mankind its power. Indeed, if the capacity were not effective, it would not be there, just as a lion would not have paws if they were not effective.

One succeeds in discovering regularity in a restricted domain of phenomena, independent of other moments of life, and latent under intellectual observation. In this way one succeeds to find a weak spot in nature, to render in this way an enemy helpless in some essential part of life. In order to maintain the certainty of an observed regularity as long as possible, one often tries to *isolate* systems, that is to keep away that what is observed as interfering with the regularity; thus man creates much more regularity in nature, than originally occurred spontaneously in it. He desires that regularity because it makes him stronger in the struggle for existence, because it enables him to predict, and to take his measures. (Rejected parts)[21]

In the above lines Brouwer describes his key concept, the 'leap from end to means'.[22] It involves by itself a measure of externalization: observed sequences are detached from the observing ego, they are lumped together in 'similar' or 'identical' sequences, which are known as *causal sequences*. These causal sequences are essential in the shift from goal to means. Because causal sequences are no longer incidental, one-time, observation (or sensation) sequences, but rather more or less stable sequences which we recognize on the basis of earlier observations, we may,

---

[20] In his later writings, for example [Brouwer 1929A,1933A], Brouwer introduces the term 'causal sequence', which denotes the 'equivalence class' of sequences under the identification by the subject. One may safely conclude from the use of 'sequence' here, that Brouwer had this sort of sequence in mind. The individual, isolated sequence has to be supplemented with a notion of identification in order to receive a certain stability. We will anticipate the terminological practice of Brouwer's mature intuitionism in so far that we will freely use 'causal sequence' from now on.

[21] [Stigt 1979], [Stigt 1990], [Dalen, D. van 2001].

[22] Brouwer indeed uses the Dutch *sprong* ('jump' or 'leap'), probably to indicate that the phenomenon is not a gradual continuous shift, but a discontinuous transition. In later publications Brouwer calls this the *mathematische Handlung* [Brouwer 1929A], and *cunning act* [Brouwer 1949C].

when we wish to attain an event B, and we have learned that in a particular causal sequence another event A invariably precedes B, try to realize the means A, and trust that the causal sequence will take us to B.

The first, and partly rejected, version of chapter II was, like *Life, Art and Mysticism*, based on strong moral views. Mathematics, in the wider sense intended by Brouwer, is a prime tool for the subjugation of nature—and fellow humans—and as such it is despicable: 'He who dominates, is already cursed, and they are cursed qualities that help to dominate', and 'Since the assimilation of the environment removes it [mankind] ever further from the natural situation that originally supported mankind, each conquered and adapted environment becomes ultimately intolerable for mankind'. (Rejected parts)

The topic 'language' was—understandably—also treated; this time Brouwer's views were somewhat milder than in *Life, Art and Mysticism*. Language is presented as an imperfect tool of communication, which could never guarantee the evocation of the same sensations in different individuals, but which could at least direct the mathematical actions of the listeners in the direction desired by the speaker. The mystical writer, however,

> will carefully seek to avoid everything that smacks of mathematics or logic; otherwise weak minds might easily be led to mathematical believing or mathematical acting, outside the domain where either society, or their personal struggle for life requires it, and thus come to all sorts of foolishness. (Rejected parts)

Chapter II had to render an account of the pleasant, but often thought unreasonable, success of mathematics in handling the world, say physics and the other natural sciences. Brouwer did not duck this responsibility. He showed how his philosophy provided a natural explanation of the sciences, and the place of mathematics in them.

In his man-based system, physics was no exception in the general framework; the physical phenomena, were to be handled as causal sequences of a specific kind.

> Now, should it not be surprising, that one really succeeds, not only to observe causal sequences that reappear again and again, but that so many groups of phenomena, that affect the naive senses in a totally different way, can be brought under some general viewpoints, which correspond with simply constructible mathematical systems? That would be a miracle indeed, but let us keep in mind, that the physicist only deals with the projections of the phenomena on his measuring instruments, which are all manufactured following a similar technique from fairly similar *rigid bodies*, and that it is not surprising that the phenomena are forced to share a similarity with either similar 'laws', or no laws. The laws of astronomy, for example, are no more than the laws of our measuring instruments, when they are used to follow the course of the heavenly bodies. (Rejected parts)

The Ph.D. adviser (*promotor* in Dutch) Korteweg, had the challenging task of guiding his student in the proper academic directions. Teacher and student had regular discussions, and (fortunately for us) a frequent exchange of letters. After Brouwer's letter of 16 October, they must have had a fairly heated discussion, for on 18 October Korteweg writes that he spent his time rereading the first five pages of the manuscript. He admitted that after the oral explanation '...everything appeared in a different light, and even seems very well and clearly formulated', and he regretted his hasty judgement.

—'I am sorry that, when we talked I still was of another opinion, so that I felt it necessary to tell you a few things which must discourage you, although that was not my intention. So now temper my regret of that action by working calmly but steadily at your second part.'

Korteweg and Brouwer did not quite see eye to eye on the foundational topics of the thesis. They exchanged a good many letters and probably had even more discussion sessions.

On Sunday(!) 4 November there was another discussion between student and teacher at the home of Korteweg. Apparently Korteweg had again expressed his doubts as to the admissibility of a number of topics, for the next day Brouwer wrote a long letter[23] defending his approach.

In the emotions of the moment, Brouwer even forgot the civil opening, and plunged at once into the discussion. The first sentence contains the barely veiled suggestion that his adviser might be out of touch with the important developments in mathematics: he sent Korteweg the issue of the *Göttinger Nachrichten* that contained the famous problems which Hilbert had presented at the International Conference of Mathematicians in Paris in 1900.[24] These problems, which became known as *Hilbert's problems*, were intended as a challenge to, and homework for, the mathematicians of the twentieth century. And indeed, they have occupied a host of mathematicians; the problems often opened up new and important avenues in mathematics.

Brouwer, in his letter, claimed the (partial) solution of three of the problems; we will see to what extent he did solve any of them, but it is quite clear that he did not wish to become a doctor on the strength of some routine research. He immediately went for the highest honours!

The problems dealt with are the first, second and fifth. Problem 1, the *Continuum Problem* went back to Cantor (who at one time thought he had a solution); it is usually presented as the *continuum hypothesis*: Cantor had conjectured that *infinite subsets of the continuum are either countable or have the same power as the continuum*. Problem 1 asked to prove or disprove the Continuum Hypothesis.

Problem 2 posed the question of the *consistency of arithmetic*, that is it asked for a proof that no contradictions can be derived from the usual axioms of arithmetic (as

---

[23] Brouwer to Korteweg 5 November 1906.
[24] [Hilbert 1900],[Browder 1976], [Gray 2000].

formulated, for example, by Peano). And, finally, problem 5 asked to rid the theory of Lie groups of the traditional differentiability conditions.

Somewhat defiantly, Brouwer wrote, 'I send you this book, because I thought to perceive that you had some doubt whether the topics in my dissertation were, after all, worth the effort.'

Apparently, Korteweg had demonstrated a certain lack of confidence in the choice of topics. The letter of Brouwer eloquently defended the contents of the thesis. It is most likely that Korteweg was worried about the philosophical passages; he might, with good reason, feel some misgivings about the reception of this work in the faculty. The reference to Hilbert's problems was clearly designed to put him at ease. Brouwer was, however, not the person to give up his philosophical views without a struggle, and so the greater part of the letter is devoted to a defence of the foundational parts. In reply to Korteweg's questioning of the place of Kant in a mathematical dissertation, he pointed out that Russell's *Essay on the Foundations of Geometry* and Couturat's *Les Principes des Mathématiques* were largely devoted to discussions of the Kantian principles, 'and Poincaré points out that the present foundational controversies are a continuation of the old mathematical–philosophical struggle between Kant and Leibniz. 'Why avoid the name of Kant,' he said, 'when his ideas are discussed in mathematical texts, for example, of Russell and Couturat?'

It was, however, not just the pure philosophical content that bothered Korteweg; even the views of his student on the topic of physics—usually deemed on safe middle ground—were disturbing.

Korteweg must have commented on that particular point too, for Brouwer explicitly defended his views by referring to Poincaré, who had expressed similar opinions. Poincaré had in *Science et Hypothèse* declared that the only meaning of 'the earth rotates' is that 'in order to classify some phenomena in a simple way, it is most convenient to assume that the earth rotates'. Brouwer wholeheartedly supported Poincaré's view:

> And I think that something like that is far from absurd; on the contrary, that it convinces anyone at first reading. The system of the heavenly bodies is nothing but a mathematical system, freely built by us, of which people are proud, only because it is in this way effective in controlling the phenomena.[25]

The philosophical credentials of Brouwer had been questioned the day before by Korteweg, who had maintained that he was not certain at all that his student had studied Kant thoroughly enough to be entitled to an opinion. Brouwer felt that he could not let this pass: 'I can, of course, not give you that certainty, but I can anyway *tell* you that I have read all of the *Critique of Pure Reason* and studied many parts (among which those that are relevant to my dissertation) repeatedly and seriously'

---

[25] Brouwer to Korteweg 5 November 1906.

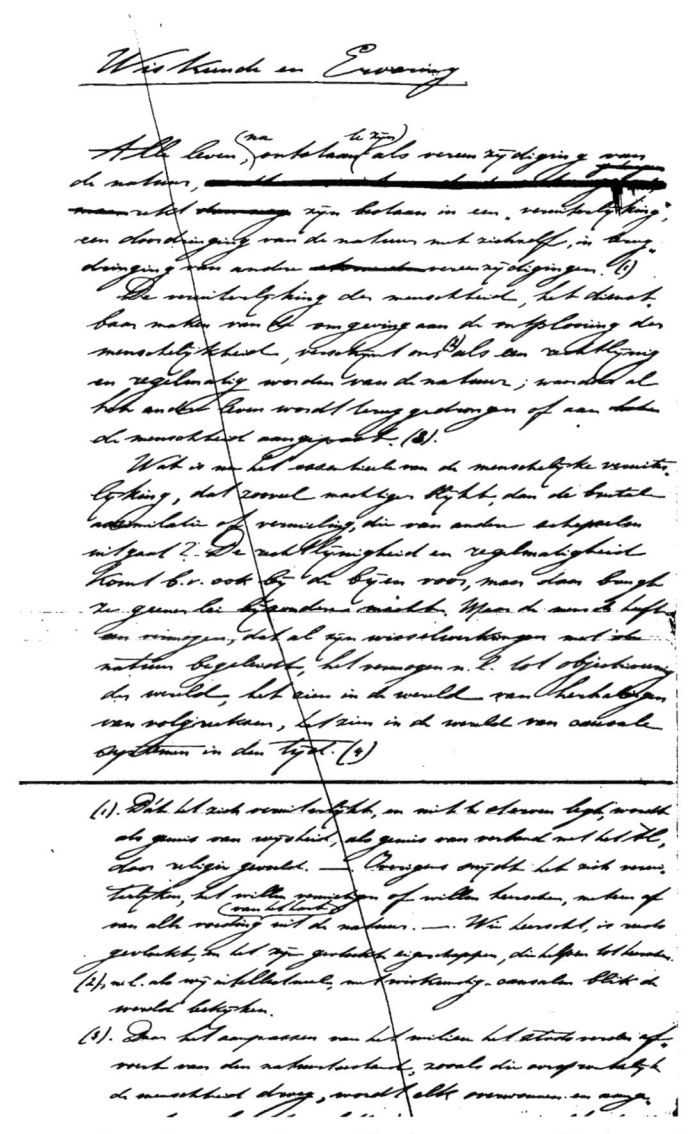

A page from the rejected part of the dissertation—'Mathematics and Experience'

Korteweg, wisely, took his time; he replied Brouwer that same day[26] that he would answer after he had read the whole manuscript, adding that 'I did not for a moment doubt the thoroughness of your mathematical preliminary studies.'

---

[26] Korteweg to Brouwer 5 November 1906. Note that in those days one could dispatch a letter and receive the reply the same day!

Brouwer could not, however, bring himself to wait for Korteweg's comments; he dispatched another extensive defence on 7 November. 'The second chapter', he said, 'essentially dealt with two topics: (a) how the mathematical experience accompanies the essentially-human actions, (b) to what extent the mathematics of experience can be a priori.'

The material dealing with physics and astronomy had only been inserted as an example of (a), and if it had to go, Brouwer would rather drop the rest of chapter II as well. He also motivated his extensive treatment of Russell's *Essay on the Foundation of Geometry*:

> he is the only one who writes about the philosophical foundation of mathematics, and (at least most of the time) uses an *exact* language, against which one can be up in arms. Hegel, Schopenhauer, Lotze and Fechner don't do that (I just drop a few names) if I discussed them I would get on purely philosophical terrain, which you don't want, nor do I wish it. Russell is, as far as I know, the only one who tackles the problem of the *a priori*, equipped with mathematics. And, criticizing him, I could remain on *my own ground*, which, in my eyes, is all the time mathematical, and in this criticizing I found opportunity to emphasize my standpoint in various directions. And that was for me the main point, not the book of Russell, which I find important for its character only, but which I find for the rest an absolute failure.

On the whole Brouwer was, in his dissertation, critical of Russell in a polite, academic way, as the reader can check for himself. But in the résumé, which Brouwer made before the actual writing of the text, Brouwer found it difficult to control himself. He repeatedly speaks of 'Russell's nonsense' and at one place he cannot stand it any longer: 'To put something higher than mathematics, you must feel a non-mathematical intuition; Russell has none of that, and yet he starts to bullshit.'

On 11 November Korteweg gave his opinion on the manuscript so far—in particular chapter II.

> After receiving your letter I have again considered whether I could accept it as it is now. But really Brouwer, this won't do.[27] A kind of pessimistic and mystic philosophy of life has been woven into it, that is no longer mathematics, and has also nothing to do with the foundations of mathematics. It may here and there have coalesced in your mind with mathematics, but that is wholly subjective. One can in that respect totally differ with you, and yet completely share your views on the foundations of mathematics. I am convinced that every supervisor, young or old, sharing or not sharing your philosophy of life, would object to its incorporation in a mathematical dissertation. In my opinion your dissertation can only gain by removing it. It now gives it a character of bizarreness which can only harm it.

---

[27] *Maar waarlijk, Brouwer het gaat niet*, 11 November 1906.

Korteweg had clearly been rather piqued by Brouwer's hints as to his expertise. His retort was gentle but firm:

> You inform me of all sorts of matters which could not possibly be unknown to me, as a regular reviewer of the *Revue de Métaphysique et de Morale*, as if they were things that I would not know. You thought to have understood that you were not allowed to use the name of Kant, even where it concerned opinions of Kant on mathematics, and you thought that I found the view 'that astronomy is nothing but a convenient summary of causal sequences in the reading of our measuring instruments' absurd. No, not *that* view; I admit that one can present the matter in that way, although in my opinion the general law of gravity has little to do, indeed, with our measuring instruments which led to its discovery, than that these make measurement *überhaupt* possible; but that the similarity of the laws which are valid in very different parts of physics would find its origin in the similarity of the used instruments; it was that claim that appeared absurd to me.

Although Korteweg insisted that the passages dealing with Brouwer's philosophy of life should be struck out, he did appreciate the comments on Kant, and he saw no objections to the more traditional treatment of the philosophical issues in mathematics. In spite of his firm stand concerning the more mystical parts, the letter was rather conciliatory. In particular, he wrote, his questions did not in any way imply superficiality on Brouwer's part; they were exactly what they should be: requests for elucidation.

Brouwer's reply, two days later, is interesting because it contained an elaboration of his views on the role of mathematics in physics. Little of this discussion has reached a wider audience, thus contributing to the popular misconception that Brouwer's theory could not, and did not, provide a foundational basis for applied mathematics.

The following passages from the letter of 13 November 1906 could be read independently of their philosophical (that is Brouwerian) background; one could even imagine a positivist writing it. It is good, however, to keep in mind that it is an integral part of Brouwer's idealist philosophy.

> You think that [...] the general law of attraction has preciously little to do with the instruments, which have led to its discovery; but are laws anything but summaries of phenomena by induction, means for the control of phenomena, and existing only in the human mind? In itself the law of attraction exists only with respect to the Euclidean space, and that only exists through an efficient, but arbitrary extension of the domain of motion here on earth. Without solid bodies on earth, the law of attraction would not exist, and the relation between both is laid down by the astronomical measuring tools. The law of attraction exists in relation to the astronomical phenomena, as molecules in relation to the state equation; both appear to summarize efficiently a group of phenomena, and to be

effective as a tool for prediction; only the law of attraction wins in simplicity from the molecular theory. But once again: the law of attraction is a hypothesis; the distance from the earth to the sun is just as much a hypothesis.

With respect to the issue of 'similar instruments yield similar laws', Brouwer pointed out that:

> There is no difference between the electromagnetic field of a Daniell-element, projected on our measuring instruments, and that of a Leclancher-element; but if we consider it without bias, we must expect that there must be as much difference between the fields, as between copper sulphate and ammonium chloride; but on our counting- and measuring instinct, operating by means of certain instruments, they both act in the same way; it turns out that a similar mathematical system can be applied to both, but it is only the lacking of suitable instruments, which so far has prevented us to find other mathematical systems which can be applied to the one field and not to the other. In each phase of development of physics, the measuring instruments—which 'have been found suitable', form a restricted totality, relative to the totality of measuring instruments which 'could be found suitable to govern all kinds of other, as yet unknown, phenomena'. Parallel to this, we have that 'the 'mathematical systems which have already been applied to nature' form a restricted totality relative to the totality of mathematics, that 'would be applicable to nature, if physics had been sufficiently extended'. —And where now every restricted group of mathematical systems has its invariants, it is to be expected that every restricted group of physical phenomena has its invariants, exactly on the grounds of those restrictions, in the form of laws or principles which are valid for all phenomena of that group.

The letter continues to elaborate in a convincing way the role of mathematics, in which, after all, the physical laws are formulated. Brouwer argues that physical measurements yield outcomes in suitable rigid mathematical groups (for example the group of rotations in the plane, when measuring torsion), and hence that

> Each physically measurable quantity is eventually reduced to a measurement in a rigid group, and it is *the laws of those measurements which are looked for in all sorts of different circumstances.*

Some people have questioned the role of Korteweg in the process of Brouwer's development; Brouwer's best-known student, Heyting, was for example completely unaware of Korteweg's influence on Brouwer; he was, on the contrary convinced that Korteweg had played no role whatsoever in Brouwer's scientific development. In the sense of modifying Brouwer's philosophical views or even determining the topics of the thesis, this is certainly correct, but in the sense of guiding the young man, Korteweg doubtlessly played an important role—as the cited correspondence bears out.[28] By asking the right questions, he forced Brouwer to refine his arguments and formulations. Moreover, a dictatorial Ph.D. adviser would have put off Brouwer. It is unthinkable that one could have forced him to forsake his personal ideals in mathematics.

In spite of his personal convictions, Brouwer could see Korteweg's point that the highly subjective philosophical sections would not enhance his credibility among his fellow scientists, and so he dropped them—but he kept the manuscript among his papers.

He realized of course, that without the basic motivation chapter II had lost some of its force:

> After their sudden appearance on the foreground as a replacement of their late leader, it was not immediately possible, to dress them all, so that they together could by themselves save the performance.[29]

It is true that the reading of the dissertation without the benefit of the 'rejected parts' leaves one with the uncomfortable impression that Brouwer's arguments were a bit *ad hoc*.

3.3 On the role of logic

Although the agreed date of the defense of the dissertation came unpleasantly close, adviser and student still continued their exchange of arguments and questions. The final topic to be discussed in their letters was that of logic. In his letters of 18 and 23 January, Brouwer answered questions and objections of Korteweg. Korteweg must have concluded from Brouwer's philosophical discourse that logic was seen as a part of, or possibly an application of, psychology—a not uncommon misconception at the end of the nineteenth century. Frege and Husserl had already demolished the bastion of psychologism, but it seems fairly certain that Brouwer was unaware of those particular writings.

In his answer to Korteweg[30] he dispelled any doubts on this point: 'From your characterization of theoretical logic as a part of psychology, I understood that I had expressed myself rather vaguely, for it was my express intention to show theoretical logic, although it is a science, has under no circumstance significance for psychology'.

In the next letter[31] he elaborated his views on logic. In complete opposition to the then prevailing beliefs, he argued forcefully that mathematical reasoning is *not* logical reasoning. Mathematical reasoning, in his view, consisted of (mental) constructions, whereas logical reasoning took place in the realm of language. Thus,

---

[28] The correspondence was after Korteweg's death deposited in the Amsterdam University Library by Mannoury, but this, apparently, was not known to Heyting. That the Korteweg files were unearthed was the result of a hint of Bastiaan Willink, a nephew of Korteweg, who was aware of the existence of the Korteweg archive.

[29] Brouwer to Korteweg 11 January 1907.

[30] Brouwer to Korteweg 18 January 1907.

[31] Brouwer to Korteweg 23 January 1907.

he stated in the dissertation that 'Mathematics is independent of Logic' (Diss. p. 125) and 'Logic depends on Mathematics' (Diss. p. 127). On the whole his view of the role of logic was rather dim. He considered it as a secondary activity dealing with linguistic figures that could on principle not be expected to take the place of mathematical constructions, and in those cases where one could easily supply the missing constructions, as in the syllogism 'all men are mortal, Socrates is a man, hence Socrates is mortal', the result is not very informative.

Brouwer was downright pessimistic about the use of logic in mathematics: 'I show in the beginning of the chapter that mathematical arguments are *no* logical arguments, that it only makes use of the connectives of logical argumentation out of poverty, and that it will keep the accompaniment of the language of logical argumentation alive, long after the human intellect will have outgrown the logical arguments.' In spite of this harsh indictment, he apparently did reconsider the role of logic after he finished the dissertation, for a year later, in the 'Unreliability'-paper, he presented a milder view.

Above all, his objections were directed against the shift of the attention from mathematical arguments (that is mental constructions) to the study of regularities in the accompanying *language* of mathematics. Even when the system of mathematical–logical language is shown to be free of contradictions, its subject matter is 'language', and there is no immediate connection with mathematics proper, for example arithmetic (Diss. p. 132).

A more argued case against logic was presented in the letter of 23 January to Korteweg:

> One knows nowadays very well, that if one derives by logical reasoning something concerning the outer world that was not so immediately *a priori* clear, it is for this very reason totally unreliable; for one does not believe any longer the underlying postulate that the world must be a finite, albeit very large, number of atoms, and that each word should represent a (hence also finite) group, or group of groups, of atoms. In other words, one knows very well that the world is not a logical system, and that it does not lend itself to logical reasoning; one knows very well that each debate is really humbug, and that it is only to be settled for mathematical problems, but then *not through logical reasoning* (although this may seem to be so in an inadequate language; how misleading this appearance is, may be seen in the case of axiomatic foundations and transfinite numbers) *but through mathematical arguments.*
>
> Theoretical logic does not teach anything in the present world, and one knows this, at least sensible people do. It now only serves lawyers and demagogues; not to instruct others, but to fool them. And that this can be done is because the vulgar unconsciously argue as follows: that language with logical figures exists, and so it will presumably be useful— and they meekly will let themselves be deceived by it; just as I heard some people defend their gin drinking with the words: 'why else is there gin?'. Whoever has illusions of improving the world can just as well crusade

against the language of logical reasoning, as against alcohol and no more is it a 'strange company' that does not drink alcohol, than it is a 'strange company' that does not argue logically. Although I believe that perhaps no abuse is entrenched more, than that which is blended with the most popular parts of language. (Rejected parts)

It is instructive to compare the above paragraph to Brouwer's mystic writings. The tone is more tempered (in spite of the reference to gin drinking), and there is an argument, not just indignant rejection.

Brouwer, in his letter, calmly and clearly defends the priority of mathematical, that is constructive, argument over logical proof. He uses an example of Euclidean geometry: an isosceles triangle has an acute angle. The logical interpretation is that the set of isosceles triangles is a subset of the set of acute angled triangles (Brouwer goes so far as to reduce the problem to subsets of $\mathbb{R}_6$) but the mathematical interpretation is that one concludes from the construction of an isosceles triangle that it has an acute angle.[32] The passage also contains a remark to the effect that the result can be obtained by reductio ad absursum as well—the doubts concerning the principle of the excluded third had not yet materialized!

The part about the demagogue, reveals a glimpse of the emotions underlying Brouwer's scientific activity. Realizing that the contents of the letter were rather unconventional, Brouwer apologetically added : 'Perhaps I have after all expressed my intentions more clearly in this somewhat wildly written letter, than in the subdued text. But, maybe, the text will appear in another light after this letter. That would please me very much.'

The explanation above of the shortcomings of logic, in particular the stressing of the non-finiteness of the universe, repeatedly occurs in later writings, including the dissertation itself; the finiteness assumption underlying the belief in logic is mentioned on page 130 of the dissertation.

The dissertation was finished in the nick of time; in spite of some printing problems, Brouwer managed to meet the faculty deadline. It reflected Brouwer's ambitions by dealing with a multitude of topics. The first chapter contained the 'real' mathematical material. Basically it contained the partial solution to Hilbert's problem no. 5, the treatment of Lie groups without the differentiability conditions, and problem no. 1—the continuum problem. Brouwer showed that 'independent of the differentiability, [...] there is only one construction for the one, resp. two-parameter continuous, uniform groups' (Diss. p. 35), while he left the three parameter case as an open problem.

After an investigation of the *Helmholtz–Lie Raumproblem*, Brouwer turned to the continuum problem (see page 93). The first chapter closed with a brief treatment of non-Archimedean uniform groups on the continuum, and the non-Archimedean geometries.

The topological part of the dissertation was, so to speak, Brouwer's ticket for the mathematical community. An improved version was submitted to the *Mathematische*

---

[32] Here a crude version of the proof interpretation of the implication may be conjectured.

Annalen, and it was the topic of one of Brouwer's talks at the international congress in Rome in 1908. His work on Lie groups immediately drew the attention of the experts (see p. 130).

There is one particular feature of chapter I which we have not mentioned so far, and which even in 1907 must have puzzled the readers. At the end of the nineteenth century it had become an accepted insight that the real numbers should be introduced as a derivative of simpler number systems—the rationals, the integers and ultimately the natural numbers. This was the result of the efforts of Weierstrass, Cantor, Dedekind and others, the so-called arithmetization of analysis.

Brouwer did not follow this new trend, he firmly declared that the continuum and the natural numbers were both given as parts of the *ur-intuition of mathematics:*

> the substratum of all perception of change, which is divested of all quality, a unity of continuous and discrete, a possibility of the thinking together of several units, connected by a 'between', which never exhausts itself by the interpolation of new units. (Diss. p. 8)

Thus the continuum is immediately given by intuition, and not reduced to the discrete. Over this intuitive continuum a scale can be constructed of a denumerable set of discrete points (for example the dual (or binary) scale). Points of the continuum may be approximated by approximation sequences—'which can, however, never be regarded as finished, thus have to be considered as partly unknown.' Brouwer recognized the possibility that the scale need not automatically be everywhere dense (that is to say, the intuitive continuum need not be Archimedean) but he contracted, by 'brute force', segments in which there was no point of the scale. Or to put it positively, two points are viewed as distinct,[33] when their dual approximations differ after a finite number of digits. The continuum will turn up again, when we discuss the continuum problem.

3.4 Mathematics and the world

Chapter II, which had severely suffered from Korteweg's axe, expounds Brouwer's views on the relation of mathematics to experience. In spite of Brouwer's rather solipsistic views in the Profession of Faith, the formulations in this chapter are fairly neutral, that is an outer world is introduced without comments. One may assume that at this point Brouwer did not want to be drawn into ill-timed philosophical discussions. It should be observed that a neutral terminology by no means excluded an idealist—nor a realist—reading.

The causal sequences take an important place in this chapter, and Brouwer argues that here mathematics plays its role in the struggle for life.

The basic scheme of Brouwer's grand design was the handling of causal sequences, in such a way that regularities occur, and that and predictions can be made.

Here the individual makes clever use of the end-to-means jump, (which is no longer diabolical in the thesis). After remarking that, there is no guarantee that in a

---

[33] In later writings Brouwer would use 'apart' for this positive notion.

causal sequence the end will *always* be obtained after realizing the means (the rule may no longer apply) Brouwer holds forth that

> in general the tactics, consisting of the consideration of the causal sequences and on the regression from the end to the means, where it is easier to intervene in the means, appears to be efficient, and provides mankind with its power. One succeeds to discover regularity in a restricted domain of phenomena, independent of other phenomena, which can thus remain completely latent in intellectual consideration.
>
> In order to maintain, as long as possible, the certainty of the regularity, one tries to *isolate* systems in the process, that is to keep away that what has been seen to disturb the regularity. Thus man makes much more regularity in nature than originally occurred spontaneously in it. He *wishes* that regularity, because it strengthens him in the struggle for existence, because it enables him to predict, and to take his measures. [...]
>
> The intellectual consideration of the world widens its scope, because one builds from the ur-intuition of the intellect abstract mathematics, independent of direct applicability. And thus one has a stock of virtual causal sequences at hand, which are only waiting for an occasion to be projected on reality. (Diss. p. 82)

So, the external world (one may think of the physical world) is viewed as a huge configuration of causal sequences, and the individual will try to optimize his environment (the 'struggle for life'). It is therefore of importance to make the end–to–means jump more and more perfect. For this purpose, the individual tries to eliminate events and phenomena that might jeopardize the transition from means to the end. For instance, one drops a ball in vacuum, or one observes the sky from an isolated mountain top. The mathematical treatment will have the best chance of succeeding in those cases were isolation can be realized. Of course, there are innumerable cases where no regularities can be called forth.

Brouwer's next observation is that things may get so complicated, that the steps to be taken to obtain isolation, cannot immediately be seen, or it might not be clear what the suitable means is for a given end. It is here that pure mathematics intervenes. The individual starts to generate virtual causal sequences, that are not directly observed (or not the direct result of sensations). These sequences may be easier to handle, and the mathematical treatment will then yield results that can be applied in the case of concrete, actual causal sequences. In this way the individual develops a stock of mathematical theories, that can be called on at any moment. Euclidean geometry is an example of such a successful mathematical theory (that is the stock of man-made imaginary causal sequences) it suffices for a large number of applications.

The most useful instances of this kind of mathematical manipulation is to be found in those cases where 'a large number of causal sequences are subsumed under one viewpoint by means of a system built using mathematical induction, that is called *law*. The difference between two sequences falling under it then only depends on the difference in values of the parameters occurring in the law.' (Diss. p. 84)

We see that causal sequences and the end-to-means jump thus become the cornerstone of Brouwer's universe and its properties. In the course of time he refined the arguments, but he remained faithful to the original idea.[34]

The first part of chapter II deals with this notion of causal sequences, laws and predictions (for example the mechanistic explanation of nature). The last part is concerned with philosophical matters, mainly in the context of Russell's *Essay on the Foundations of Geometry*. The discussion of Russell's *Foundations* is preceded by Brouwer's own views on *objectivity* and the notion of *a priori*.

On Brouwer's view 'objectivity' of quantities or laws simply means the 'invariance under the simplest or the most common interpretation'. For example

> one usually calls mass objective, and thinks of its indestructibility; we have seen, however, that masses are nothing but coefficients which simplify through their introduction, the mathematical image of nature, and which remain invariant under mathematical transformations, which represent the natural phenomena. Were one to find, however, natural phenomena which can most easily be represented by assuming the masses to be variable, then one could only keep calling them objective, on the grounds of their invariance under a *very important group of phenomena* in the image of nature. (Diss. p. 95)

So, objectivity gets a specific and well-determined meaning on Brouwer's view. Both physical time and space are, under this interpretation, highly objective. The matter is different for the notion of *a priori*; if one understands by *a priori* 'existence independent of experience', then on Brouwer's view, all of mathematics is *a priori*, Euclidean as well as non-Euclidean. If one understands by it 'necessary condition for the possibility of science', then since

> scientific experience finds its origin in the applications of intuitive mathematics to the real world, and since there is, except for experimental science, no other science than just those properties of that intuitive mathematics; we may call nothing *a priori* but that one thing, which is common to all mathematics, and which, on the other hand, is sufficient, to build all mathematics on—the intuition of many-one-ness, the ur-intuition of mathematics. And since this coincides with the awareness of time as change per se, we may say: *The only a priori element in science is time.* (Diss. p. 98)

Thus Brouwer sets himself to 'rectify and update' Kant's views. On pages 114 and 115 he spells out the inadequacies in Kant's arguments for the *a priority* of 3-dimensional Euclidean space.

He not only rejects Kant's assumptions (we cannot get external experiences except in the empirical space; empirical space is the 3-dimensional Euclidean space) but also the argument leading to Kant's conclusion.

---

[34][Brouwer 1929A, 1930A, 1933A1, 1949C].

We note that Brouwer added the observation that 'Properly speaking the building of intuitive mathematics *per se* is an action (*een daad*) and not a science (*een theorie*) ....' The reader will recognize here Hermann Weyl's later dictum: 'mathematics is more an action than a theory' (*mehr ein Tun als eine Lehre*), cf. p. 323.

The discussion of Russell's *Essay* consists of a series of criticisms on methodological and mathematical issues (for example Russell's claim that a variable constant of space curvature is unthinkable).

3.5 Observations on set theory and formalism

Chapter III, MATHEMATICS AND LOGIC, is a criticism of the developments of the last decennium. As we have seen, on Brouwer's view mathematics was the construction of 'mathematical buildings', and as such it was a language-independent activity. Logic, which was not (yet) exclusively identified with formal mathematical logic, dealt with descriptions of mathematical states of affairs, and according to the majority of the mathematicians, mathematics could not do without logic in order to establish its theorems. Brouwer turned this round, and boldly asserted that *Mathematics is independent of Logic, and Logic is dependent on Mathematics*. (Diss. p. 127)

Brouwer's criticism was directed against four recent developments in the foundations of mathematics:

1. The founding of mathematics on axioms.
2. The theory of transfinite numbers of Cantor.
3. The logistic of Peano–Russell.
4. The logical foundations of mathematics according to Hilbert.

The axiomatic method was criticized on fundamental grounds, that is to say, the properties and the study of the axiomatic systems, with their notion of consequence, cannot replace the intuitive mathematics of (mental) constructions. Furthermore, the basic desideratum of the axiomatic method, which says that one can always find a *mathematical* interpretation for a consistent logical system, is not substantiated. Brouwer was, however, cautious enough not to go too far. He did not state that there were cases where such a mathematical interpretation ('building' in his terminology) could not be provided; he merely stated that the existence

> was not established by the axiomaticians, not even for the case that the given conditions include that the thing that is looked for is a mathematically constructible system; for example, it has nowhere been proved, that if a finite number has to satisfy a system of conditions, of which it can be shown that they are not contradictory, that such a number then also exists. (Diss. p. 142)

Brouwer added in a footnote, that therefore it was not certain at all that 'every particular mathematical problem must necessarily admit an exact solution, be it that the answer to the given question can be given, or that the impossibility of the solution, and thus the necessity of the failure of all attempts, is established'. This conviction, *Hilbert's dogma*, was put forward by Hilbert in his lecture *Mathematische*

Probleme, with the remark that every mathematician is at heart convinced of its correctness. Brouwer summarily declared it 'unfounded' in his Thesis no. 21.[35] In the light of his doubt of Hilbert's dogma, it is rather surprising that a few pages earlier Brouwer had acknowledged the *principle of the excluded middle* (PEM) as vacuous or trivial. This principle, also going by the names *principium tertii exclusi*, principle of the excluded third, and *principium tertium non datur*, was one of the basic principles of Aristotle's logic, together with the *principium contradictionis*. The latter asserts that one cannot at the same time have A and *not* A (i.e. $\neg(A \wedge \neg A)$),[36] and the former states that one has either A or $\neg A$, and there is no third possibility (i.e. $A \vee \neg A$). Brouwer had accepted the principium contradictionis as evident.

On his view, the words of a mathematical argument of a logician are nothing but an accompaniment of the wordless building process. When the logician concludes a contradiction, the building mathematician observes that the building process is blocked. The latter does not need the principium contradictionis for this observation (Diss. p. 142). The principle of the excluded middle, he claimed, was correct but uninformative:

> 'A function is either differentiable, or not differentiable' tells us nothing, it expresses the same thing as the following: If a function is not differentiable, it is not differentiable. (Diss. p. 131)

This statement will surprise the modern reader; how could Brouwer accept PEM and reject Hilbert's dogma?

Although a conclusive answer will be hard to give, a possible solution to this riddle may be conjectured. C. Bellaar–Spruyt was teaching philosophy in Amsterdam when Brouwer took his courses. Bellaar–Spruyt did not publish much, but he influenced Dutch philosophy by his firm defence of Kant, and his equally firm attacks on the then predominant materialistic philosophy, embodied by the Utrecht philosopher Opzoomer.

There is no conclusive evidence that Brouwer took the courses of Bellaar–Spruyt, but it is plausible that he did. In the Brouwer Archive there is a picture of Bellaar–Spruyt, which Brouwer kept for all those years, and—more relevant, Brouwer's own philosophy has evident Kantian features. A philosophically interested, curious student would certainly attend the courses of his local philosophy professor, in particular if he had the reputation of an independent thinker, like Bellaar–Spruyt.

The logic course in Amsterdam was also in the hands of Bellaar–Spruyt, and his lectures have been published. Here one finds a hint towards the solution of the above mentioned problem. PEM is explained by Bellaar–Spruyt by means of an example: 'If you deny that Alexander was a great man, well, then you have to

---

[35] A dissertation was in Holland always supplemented with a list of theses, which were presented without proof, and which had to demonstrate the proficiency of the candidate in a wide area. The theses also functioned as a concession to those members of the faculty who had to take part in the examination, but could not spare the time to read the dissertation.

[36] From now on we will follow the conventions of logic, and write $\neg A$ for *not A*.

acknowledge that he was not a great man. Both opposite judgements, A. was a great man and A. was not a great man, cannot both be false.'[37] Replace 'Alexander is a great man' by 'the function is differentiable', and you get Brouwer's example. It is plausible, and even likely, that Brouwer learned this particular form of PEM from Bellaar–Spruyt. Needless to say that this reading does not pass the criteria of the constructivist.

There may have been more influence of Bellaar–Spruyt on Brouwer than just this case of PEM. Some Brouwerian themes already occur in Bellaar–Spruyt's lectures. Here are some examples: (i) logic is different from psychology. Brouwer would wholeheartedly agree, but where Bellaar–Spruyt sees logic in its normative role, he only views logic as means of recording the original mathematical constructive activity. (ii) In the section on 'The pernicious influence of language on the clarity of our thoughts', Bellaar–Spruyt discusses a phenomenon that very much looks like a case of 'the leap from end to means': 'It is indeed a familiar psychological law that if a state of mind A is often followed by B, the arising of A leads to the arising of B'.[38]

Although it is still a big step from a simple observation to the systematic use in the context of causal sequences, one may guess that the young Brouwer amply received impressions in the lectures of Bellaar-Spruyt.

Summing up, one is tempted to see Brouwer's view of PEM as a thoughtless adoption of the formulation that was given in the course. It could, of course, also be that he hesitated to destroy one of the pillars of Aristotelian logic—after all, a more than two-thousand-year old heirloom!

Among his private notes there is some reference to PEM, that seems to indicate that he already mistrusted the principle. To jump ahead of our story, soon after he finished his dissertation, he hit on the right reading of PEM:

> this requires that each assumption is either correct or incorrect, mathematically formulated: that for each assumed incorporation of systems in some way into each other, either the termination [success], or the encounter of an impossibility can be constructed. The problem of the validity of the principium tertii exclusi is thus equivalent to the problem of the *possibility of unsolvable mathematical problems*.[39]

In general, he argued, 'the principium tertii exclusi is, as yet, not reliable in infinite systems'. The paper also contained the first instance of his celebrated 'unsolved problems', that were at the basis of his later counterexamples, now known as *Brouwerian counterexamples*:

- Is there in the decimal expansion of π a decimal that in the long run occurs more often than others?

---

[37] [Bellaar-Spruyt 1903], p. 18.
[38] [Bellaar-Spruyt 1903], p. 62.
[39] [Brouwer 1908C], p. 5.

- Are there in the decimal expansion of π infinitely many pairs of consecutive decimals which are equal?

In both cases we (still) have no clue as to the truth of the statements, hence PEM cannot be considered to hold for them.

The above reflections were embodied in a paper, *The unreliability of the logical principles*, published in a semi-obscure Dutch philosophical magazine[40] (the same journal that had rejected Brouwer's review of *The foundations of a new poetry* of Scheltema.[41]) As a consequence it was read by few colleagues, even in Holland. The paper was received with a certain mistrust; after long deliberation the editorial board consented to publish it. Its hesitation may have had something to do with the fact that Brouwer was a selfmade philosopher, and with the friction between Brouwer and the philosophical community, resulting from an attempt of Brouwer and Mannoury, two years before, to found a philosophical journal.

Maas & van Suchtelen, the publisher of Brouwer's dissertation, had approached Mannoury with a proposal to establish the first philosophical journal in Holland. Mannoury, after some consideration, answered that he was prepared to set up such a journal provided Brouwer was to join him. Subsequently, the two explored the Dutch philosophical landscape in order to find out if the project was viable. They soon met a determined opposition from the side of the professional philosophers, who probably did not want to see the Dutch journal in the hands of philosophical amateurs, so the various philosophical societies suddenly felt the urge to start a journal themselves. When Maas & van Suchtelen was confronted with the situation, it decided to make the best of it by joining sides with the professionals. Kohnstamm, who was the spokesman for them, made the condition that Brouwer and Mannoury would not be asked to join the editorial board.

The editor in charge of the paper, the physicist Kohnstamm, made it quite clear that his firm stand in the meeting of the board had saved the day. He also used the opportunity to tell that, in his opinion, Brouwer had completely missed the point of the PEM, 'I cannot see that your observations on unsolvable mathematical problems violates the PEM. It seems to me that the matter is of the same sort as the question whether a square circle should be considered round or angular.[42] This would not be the last manifestation of incomprehension that Brouwer was confronted with in his long struggle for a renewed mathematics.

Whereas Brouwer's objections to logic and its uses in mathematics were of a general philosophical nature, his criticism of Hilbert's formalism had quite specific technical points. As the role of Hilbert and his formalism will become prominent in the nineteen-twenties, it is worthwhile to go into Brouwer's arguments in the dissertation. The criticism in the dissertation was directed at Hilbert's Heidelberg address,[43] *On the Foundations of Logic and Arithmetic*, which had appeared in 1905.

---

[40] *Tijdschrift voor Wijsbegeerte* (Journal for Philosophy) [Brouwer 1908C].
[41] Cf. p. 31.
[42] Kohnstamm to Brouwer 3 January 1908.
[43] [Hilbert 1905].

Hilbert had set himself the task to complete what he had started with his *Foundations of Geometry*' (1899), that is to show the consistency of number theory—on which the consistency of geometry rested. The Heidelberg talk could be considered as a try-out of a new method for consistency proofs. The systems treated were only small fragments of logic and arithmetic, but it was the method that counted. Hilbert had indicated means to show that in the formal systems under consideration one could not derive the formula expressing a contradiction. The reader who is not familiar with this type of argument may consider the following problem: is it possible that by using the usual rules of elementary algebra, one ends up with a formula with more left than right brackets? The answer is 'no', and one easily proves this by means of complete induction. So here one proves something about objects of the language of mathematics (that is formulas, strings of symbols), instead of the usual objects of mathematics, such as numbers, functions, triangles, ... Hilbert's aim was to use this technique to show that in a fully formalised part of mathematics (say arithmetic) one could not derive a contradiction (say '$0 = 1$'). The subject of 'deriving' was in this context to be taken in its formal sense, as, for instance, expressed in Hilbert's Paris address of 1900:

> When we are engaged in investigating the foundations of a science, we must set up a system of axioms which contains an exact and complete description of the relations existing between the elementary ideas of that science. The axioms so set up are, at the same time, the definitions of those elementary ideas; and no statement within the realm of the science, whose foundations we are testing, is held correct unless, it can be derived from those axioms by means of a finite number of logical steps.[44]

Brouwer viewed Hilbert's scheme to free the foundations of mathematics from intuition as misguided. He argued that if one proves by means of our intuition that in the formal-logical description of this intuition one cannot derive a contradiction, then one has not made any progress:

> Who will prove a mathematical theorem, by deriving it once more on the strength of the theorem itself, and will then say 'now the assumption is justified as well'? (Diss. p. 172)

Popularly speaking, we can paraphrase Brouwer's objections as: you need intuitive mathematics to justify the (consistency of) its formalization (in particular, you need induction to justify induction) and consistency of the formalization does not justify the underlying intuitive arguments (constructions). This insight was lacking in Hilbert's paper. Some thirty years later, Gödel drove home the message by means of technical logical arguments.

In addition to the criticism of Hilbert's meta-mathematics, Brouwer gave a meticulous analysis of the various levels involved in the proof theoretical treatment of mathematics, (Diss. p. 173–175). He distinguished 8 levels in Hilbert's exposition

---

[44][Hilbert 1900].

of 1904, a classification that in precision out did both Hilbert–Bernays' and Tarski's later language–meta-language distinction.

He gave an 'enumeration of the phases which were mixed up in the logical treatment of mathematics':

1. The pure construction of intuitive mathematical systems, which, if applied, are externalized by viewing the world mathematically.
2. The language parallel of mathematics: mathematical speaking or writing.
3. The mathematical consideration of the language: logical language constructions, built according to principles from ordinary logic or its extension to the logic of relations, are observed, but the elements of those language buildings are linguistic accompaniments of mathematical constructions or relations.
4. No longer thinking of a meaning of the above mentioned logical figures; and the copying of the construction of those figures by a new mathematical system of the *second* order, for the time being without a language which accompanies the constructions; it is the system of the logisticians, which easily becomes susceptible to the figure of contradiction at the least free generalizing extension, unless Hilbert's precautions are taken against it. And these precautions make up the real content of Hilbert's treatise.[45]
5. The language of logistics, that is the words which accompany and motivate the logistic building activity; Peano, indeed, takes, as much as possible, care to tie the accompanying thoughts to symbolic signs. Nonetheless the system remains decomposable in the proper construction, and the principles according to which the construction develops itself. Even though those principles are symbolically formulated, such formulations must be considered as heterogeneous with respect to the further formulas, to which the first ones are applied—not as formulations, but as intuitive acts of which the adjoined formulations are only accompaniments.

    Hilbert needs those intuitive language acts, thus also the accompanying language, more than Peano, because he wants to prove the non-contradiction of his logistic system *in itself*, something Peano does not care about.

    The verbal content of Hilbert's treatise up to page 184, V. belongs to the fifth phase.
6. The mathematical consideration of that language. The explicit performance of this step is something essential with Hilbert, in contrast to Peano and Russell; he notes, looking back at his own words, logical figures, which develop according to logical and arithmetical principles, including among other things, the theorem of complete

[45][Hilbert 1905].

induction. The elements of these logical figures, such as the words *mehrere, zwei, Fortsetzung, an Stelle von, beliebig*, etc., are a linguistic accompaniment of construction acts in the above mentioned mathematical system of the second order.

7. No longer thinking of a meaning of the elements of above mentioned logical figures; and the copying of the construction of those figures by a new mathematical system of the *third* order, for the time being without accompanying language.
Hilbert carries the transition from **6** to **7** out in his mind loc. cit. page 184 and 185 under V, first paragraph.

8. The language accompaniment of the mathematical system of the third order which motivates and shows the non-contradiction of it.
This phase is, in the *words* of the above mentioned paragraph loc. cit. page 184, 185, the last one found with Hilbert.

One could go on even further, but the mathematical systems of even higher order would all be roughly copies of each other; it thus makes no sense to pursue the matter further.

For that matter, the previous phases, from the third one on, are not of mathematical interest either. Mathematics only belongs in the first one; it cannot remain free of the second one in practice, but that phase remains a non-mathematical unconscious act, be it guided and supported by *applied mathematics* or not, but never gaining priority with respect to the intuitive mathematics. (Diss. p. 173 ff.)

The promised solution of Hilbert's problem no. 2, the consistency of arithmetic, is not given the prominence that one would expect. This is perhaps not surprising, for, from Brouwer's point of view, the effective construction of the mathematical system automatically protected it from contradictions. He pointed out that the early axiomaticians (including Hilbert) were quite satisfied with the procedure: '...the advancing of mathematical systems as existence proofs for the logical systems, implies that one still saw that the mathematical system required no further existence proof than its intuitive construction' (Diss. p. 137).

Brouwer's views become visible, so to speak, in his discussions of the axiomatic method and of Hilbert's formalism; he probably considered the proper consistency methodology (that is model construction) so natural and obvious, that he did not want to spell out the underlying details.

It is virtually impossible to do justice to all parts of Brouwer's dissertation, so for the moment we have to be content with a short final glance at Brouwer's reaction to set theory. Set theory was the hot topic of the turn of the century, it tickled the curiosity of the mathematical community, and although it had a long incubation period before it was accepted as a useful and self-evident part of daily practice (and mathematical training) it occupied some of the finest minds of the period. The founding father was Georg Cantor, professor in Halle. Cantor had, after some experimenting with sets that played a role in 'real mathematics', that is sets that occur in number theory, algebra or analysis (including the new discipline, topology), opened up an inexhaustible universe of sets, sets of sets, etc. His fundamental theorem on power sets held that there are essentially more subsets of a set, than it has elements. For finite sets this was not particularly shocking, a direct calculation shows that $\{0,1\}$ has 4 subsets, namely the empty set, $\{0\}, \{1\}$ and $\{0,1\}$, and, in general $\{0, 1, 2, \ldots, n-1\}$ has 2^n subsets. For infinite sets there is a problem: it is no use to count them. So instead, Cantor introduced the concept 'as large as', or 'equivalent' (*gleichmächtig*) Two sets are equivalent if they can be brought into 1 *to* 1-correspondence.[46] This is a reasonable generalization of 'as large as' for finite sets: $\{0, 2, 7\}$ is as large as $\{\pi, 12, \sqrt{2}\}$ and there are obvious 1–1 correspondences. Cantor's famous theorem tells us that the power set $\mathcal{P}(A)$ (that is the set of all subsets of A) is always larger than A. To be precise: A is equivalent to a subset of $\mathcal{P}(A)$, but not to $\mathcal{P}(A)$ itself. As a result there is no such thing as a 'largest set' in Cantor's universe. The set of natural numbers, $\mathbb{N}$, is infinite, and it is smaller than $\mathcal{P}(\mathbb{N})$, which in turn is smaller than $\mathcal{P}(\mathcal{P}(\mathbb{N}))$, etc. Indeed, $\mathbb{N}$ is 'the smallest infinite set', we call sets equivalent to $\mathbb{N}$ *denumerable*. Elementary set theory tells us that the continuum is equivalent to $\mathcal{P}(\mathbb{N})$.

Furthermore, Cantor generalized the notion of counting beyond the existing practice, he introduced *transfinite numbers*. The first transfinite number is ω (that is, ω comes after all the natural numbers) and the first few transfinite numbers (ordinal numbers) look like $\omega, \omega+1, \omega+2, \cdots, \omega+\omega(=\omega.2), \cdots, \omega.3, \cdots, \omega.n, \cdots$ These ordinal numbers are all denumerable. Cantor took the bold step to consider the set of all denumerable ordinal numbers (traditionally called *the second number class*); this set itself is another ordinal number: the first non-denumerable ordinal ω_1. Cantor showed that there is an unbounded number of transfinite numbers exists in his universe.

At the time Brouwer's dissertation was written, this material was all fairly new; lots of mathematicians considered Cantor's creations as if they were inmates of some asylum rather than useful, let alone necessary, citizens of the mathematical kingdom. An influential mathematician like Poincaré was publicly sceptical of set theory; all the same he was excellently informed about the new developments, and he gave a beautiful talk on set theory in Göttingen.[47] Set theory was rocked around the turn of the century by the so-called paradoxes. Cantor, Burali-Forti, Zermelo and Russell had shown that uncritical handling of certain set-forming operations led to contradictions. The 'set of all sets', the 'set of all ordinal numbers', and the like, spelled disaster. When the dust had settled somewhat, it was Zermelo who got the debate on set theory under way again by publishing his miraculous proof of the well-ordering theorem,[48] which states that each set can be well-ordered, that is put into an *ordering like an ordinal*. Technically speaking, a set is well-ordered by an

---

[46] That is, if there is a bijection (1–1 mapping) from one onto the other.
[47] [Poincaré 1910].
[48] [Zermelo 1904].

ordering relation $<$, if each non-empty subset has a first element. Well-ordered sets are exactly in the form that corresponds to the transfinite 'counting' of sets. Cantor had posed the question if all sets could be well-ordered; the more conservative mathematicians were inclined to doubt this, they usually pointed out that there was not even the faintest clue how to well-order the continuum. Brouwer, in his dissertation, refutes the well-ordering theorem by pointing out that in the case of the continuum most of the elements are unknown, and hence cannot be ordered individually—'So this matter also turns out to be illusory.' (Diss. p. 153)

Zermelo's proof was based on a new principle, the *axiom of choice*, which postulated that given a set of non-empty sets, there was always a (choice-) function that picked an element from each of these sets. The matter caused quite a stir in mathematics; objections of various kinds were raised by eminent scholars, Peano, Borel, Poincaré, Schoenflies, ...

Zermelo's second great achievement was the formulation of an axiom system for set theory. It turned out to be so adequate that, up to minor improvements, it is still the main basis for modern mathematics.

With respect to set theory, Brouwer maintained that the larger part of the creations of Cantor was beyond the realm of the mentally (and thus mathematically) constructible. Examples of (according to Brouwer) meaningless word play are the *second number class* and the higher *power sets*. In Brouwer's opinion, mathematical objects had to be constructed; one can thus easily imagine that considerable parts of Cantor's universe had to be jettisoned. Brouwer had no problem accepting the countable ordinals, but he balked at the second number class as a whole (indeed his supplementary thesis no. XIII boldly asserts 'The second number class of Cantor does not exist'). Cantor introduced the second number class as the totality (*Inbegriff*) of all countable ordinals, which according to Brouwer 'cannot be thought, that is to say, cannot be built mathematically'. Brouwer argued that constructions based on the 'and so on' are legitimate only if the 'and so on' refers to an iteration of the same thing at most ω' times. In the case of Cantor's second number class there evidently is no ω-iteration, nor, says Brouwer, is there a repetition of similar things, (Diss. p. 146). 'Thus Cantor is here no longer on firm mathematical ground.'

Similarly, Brouwer rejects the unrestricted exponentiation of sets.[49] Brouwer's argument is simple: we cannot imagine (think), for example, 2^{2^N}. It is worth noting that he does not mention the power sets $\mathcal{P}(A)$.[50] It would be a matter of wishful reading to conclude that he had already realized that power sets and exponents are, constructively speaking, different things. It probably was the influence of Cantor, but it is equally well possible that Brouwer just did not care for 'the collection of all subsets'. It would not fall under his generation principles of sets anyway.

A conceptual analysis of the mathematical construction process led Brouwer to the insight that:

---

[49] A^B is the set of all mappings from B into A.

[50] Cantor himself systematically discussed subsets in terms of their characteristic functions (*Belegungen*).

we can create in mathematics nothing but finite sequences, and further, on the ground of the clearly conceived 'and so on', the order type ω, but only consisting of *equal elements*,[51] so that we can never imagine the *arbitrary* infinite binary fractions as finished, hence as individualized, because the denumerably infinite number of numerals behind the (decimal) point cannot be viewed as a denumerable number of *equal* things, and finally the intuitive continuum (by means of which we subsequently have constructed the ordinary continuum, the *measurable continuum*). (Diss. p. 143)

The explanation of 'and so on' may seem somewhat puzzling. On a literal reading a denumerable sequence like '1, 2, 3, and so on' could only be viewed as '1, 2, 3, 3, 3, 3, . . .' But it is the *operation of successor* that is repeated, so one does get '1, 2, 3, 4, 5, . . .' As a consequence the 'and so on' here requires an operation or a law. Hence one cannot expect to get arbitrary sequences (think of the choice sequences which Brouwer introduced later).

In spite of his critical view of Cantor's set theory, Brouwer was no bigoted foe of the topic as such. As his topological papers were to show, he was fully aware of the usefulness of sets and their machinery. In the dissertation he did not preach abolition of set theory, but rather a thorough overhaul of the subject.

We have seen that he claimed in a letter to Korteweg to have solved the continuum problem. His proposed solution was based on an analysis of the possible set creating operations on the domain of point sets.

As the continuum problem was supposed to be a central issue in understanding the nature of the continuum, it is clear that definition or construction is of the utmost importance. In classical mathematics (say, Cantorian set theory) there are various precise, and equivalent, definitions, so that a treatment of the problem can at least be entertained on the basis of an exact specification. In Brouwer's mathematics, this is problematic indeed. Since the continuum was given outright by an act of the individual intuition, there seemed to be little hope for the treatment of the problem.

The intuitive continuum was, so to speak, an amorphous mass out of which individual points could be picked, but totalities of individualized elements could only be created in denumerable quantities (Diss. p. 62). In his words: 'The continuum as a whole was intuitively given to us; a construction of it, an act which would create as individualized, 'all' points of it by means of mathematical intuition is unthinkable, and impossible.'

Brouwer's view of the continuum shows a suggestive similarity to the mystic experience of the initial chaotic state of the individual. In this state, also, there are no sharp bounds, everything is flowing and amorphous. From that point of view it is indeed implausible, if not impossible, that all mathematical structures are made up of individual, sharply distinguished, elements. So the continuum, being traditionally the flowing, continuous medium, must from Brouwer's point of view, have

---

[51]Diss. p. 143. [Brouwer's note]: Where one says: 'and so on', one means the arbitrary repetition of the same thing or operation, even though that thing or operation may be defined in a complex way.

been the structure *par excellence* created by a human faculty analoguous to the original mystic state.

On the basis of the ur-intuition (see p. 102), Brouwer distinguished three set generating principles for subsets of the continuum:

1. Combinations of finite sets, sets of the order ω and sets of the order η (that is finite sets or sets that are similar to the natural numbers or to the rationals).
2. The 'completion to a continuum' of a set of order type η, which is dense in itself.
3. The complement of a dense in itself subset of type η, with respect to the continuum.

These three construction principles, according to Brouwer, yielded only countable sets and sets equivalent to the continuum, and hence the continuum problem was answered in the affirmative (Diss. p. 67). Actually, Brouwer was more cautious then the above suggests; he wrote 'seems to be solved'. In chapter III, Brouwer returned to the continuum problem (Diss. p. 149), see below. this time to show that the problem was not well-posed. For he argued that neither 'the set of points of the continuum', nor 'the set of all numbers of the second number class' exist as mathematical objects, so their equivalence is unthinkable. On the other hand one can turn the whole problem into a logical problem: introduce the relevant sets as hypothetical objects and show that the continuum hypothesis is non-contradictory.

The reflections on the possible point sets, and their generation principles had led Brouwer to consider another kind of 'pseudo' sets, or cardinalities. Evidently there are totalities that cannot be exhausted by successive selection of points, the continuum is an obvious example. The new notion was based on the idea that, although one could not exhaust such totalities, one could, with luck, always find, after an exhaust-attempt, another point in a systematic way. In fact Cantor's diagonal method yields such a systematic method for the continuum.

Brouwer coined a new term for this phenomenon:

> We mean by a *denumerably unfinished set* one, of which only a denumerable group can be indicated in a well-defined way, but for which then, according to some previously defined mathematical process, new elements may be derived from such a denumerable group, which belong to the set in question. (Diss. p. 148)[52]

The notion of 'denumerably unfinished' makes the impression of an interesting digression, almost an afterthought in the context of the continuum problem. It plays, however, no role in Brouwer's actual mathematics. The notion occurs a couple of times in later papers, to disappear, once the choice sequences have made their entry. After that it is mentioned in passing, but without significant or helpful comments. The idea behind this innovation was to apply the predicate 'denumerably unfinished' to 'collections' that could not be handled as sets according to the

---

[52] Observe the similarity with Dekker's productive sets, cf. [Rogers 1967].

above mentioned construction principles. Since the denumerably unfinished sets were not really sets, Brouwer decided, as a compromise, to treat them as a 'manner of speaking': 'We can, however, introduce these words as arbitrary expressions for a known intention'.[53] As examples of denumerably unfinished sets he listed: the totality of all ordinals, the totality of all definable points of the continuum, the totality of all possible mathematical systems. The comparison of denumerably unfinished sets, according to Brouwer, was simple: every two of them were equivalent. On the other hand, he admitted in a footnote, (Diss. p. 149), that one could just as well consider denumerable and denumerably unfinished sets equivalent, as each denumerably unfinished set may be mapped onto the ordinal ω^2. This seemingly paradoxical statement was clarified by pointing out the difference between 1–1 mappings between two enumerable sets and those between a denumerable and an denumerably unfinished set, the latter are themselves 'unfinished'.

On the basis of the above considerations, Brouwer recognized the following cardinalities:

1. the various finite ones
2. the denumerable infinite
3. the denumerably unfinished infinite
4. the continuous.

In this more refined context, Brouwer reconsidered the Continuum Problem. In the tradition of set theory there were basically two (equivalent) formulations of the continuum problem. One did not mention ordinals and cardinals: *each infinite subset of the continuum is either denumerable, or equivalent to the continuum*. The other one made use of cardinals and ordinals: $2^{\aleph_0} = \aleph_1$, in words *the continuum is equivalent to the second number class*. The second formulation was problematic for Brouwer. As neither the second number class, nor the 'set of all points of the continuum' exist in the proper sense as sets of individual points, the problem had no meaning in this form. The problem could, of course, be turned into an exercise in logic by abstracting from the meaning and trying to show the consistency of the continuum hypothesis (Diss. p. 150). That is to say, since the set of *all* points of the continuum does not make sense, one can consider the set of all definable numbers of the continuum. This is a denumerably unfinished set and hence equivalent to the equally denumerably unfinished second number class. Hence, under this interpretation, the equivalence turns out to be trivially true.

In 1908 Brouwer attended the International Mathematical Congress in Rome, where he gave two talks, one of them with the title 'The possible Cardinalities'.[54] In this talk Brouwer expressed himself more cautiously. The phrase 'continuum problem' had disappeared. But the above classification is by and large upheld. He concluded that there is only one infinite cardinality—that of the denumerable sets, but he mentioned the *denumerably unfinished* cardinality, which is a *method* rather

---

[53]This is similar to the notion of 'class' in Zermelo–Fraenkel set theory; a class is not a set, but it is a convenient fiction.

[54]*Die Moegliche Mächtigkeiten* [Brouwer 1908A].

than a set, and the *continuous* one, which is something finished, but only as a *matrix* and not as a set.[55]

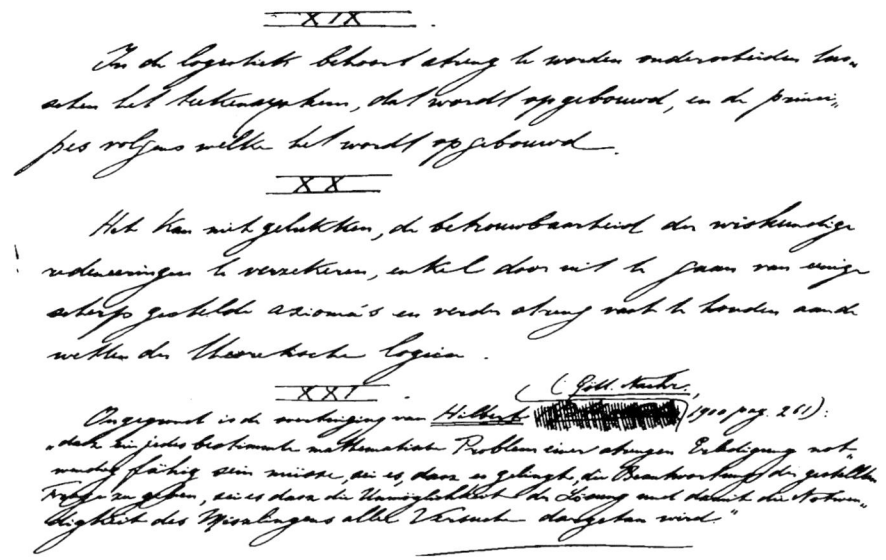

[handwritten manuscript text]

The last three of the obligatory theses of the dissertation.

Finally, Brouwer upbraided in his dissertation Poincaré (with whom he agreed on most issues, and whom he greatly admired) for 'not adopting the intuitive construction as the sole foundation of his critique'. Brouwer refers here to the famous (or notorious) quote 'in mathematics the word exist can have only one meaning; it means freedom of contradiction'. (Diss. p. 177).

The dissertation ended with a summary, which opened with a bold declaration, vaguely reminiscent of Cantor's slogan *The essence of mathematics is in its freedom*:

> Mathematics is a free creation, independent of experience; it is developed from a single *a priori* ur-intuition, which can both be called *constancy in change* and *unity in multitude*.

The summary only deals with the foundational issues of the dissertation. It recapitulates three fundamental points:

> (i) The application of mathematics, in the form of projecting mathematical systems, is also a free act of the human being.
> (ii) Mathematical definitions and properties are not to be studied mathematically. They are to serve as support of our memory, and in communication.
> (iii) A logical construction of mathematics, independent of mathematical intuition, is impossible.

---

[55] [Brouwer 1908A], p. 571.

The main body of the dissertation was followed by the obligatory theses; they were partly concise formulations of fundamental ideas of the dissertation, and partly claims about topics in mathematics, not connected with the themes of the dissertation. In all there were 21 of them. The first draft of the Theses contained 39 items. The number was, most likely under the guidance of Korteweg, reduced to the above mentioned 21.

There is a draft of a letter from Brouwer to Professor J. de Vries (Utrecht University), in which Brouwer gave a short résumé of the contents of the dissertation, together with some elucidations. In this letter Brouwer comments on Cantor's set theory. The following quote is interesting, but puzzling:

> With respect to the question whether actually infinite sets exist, and if 'yes', which ones?
>
> Here I neither subscribe to the opinion of Poincaré, who rejects the actual infinite out of hand, nor do I accept all the transfinite sets of Cantor, but I acknowledge denumerably infinite sets, and with a restriction, the continuous cardinality, and finally, with another restriction, a new cardinality, which I call denumerably infinite unfinished. I expose however, all the higher cardinalities of Cantor as a logical chimera. At the same time I try to strip transfinite set theory of its parasite parts, such as transfinite exponentiation, the theorem of Bernstein with its applications, and more; all of which result from the false logical foundations of set theory. In this connection I can formulate:
>
> 1. Actual infinite sets can be created mathematically, even though in the practical applications of mathematics in the world only finite sets occur.

This view on the actual infinite is a plausible consequence of Brouwer's concept of 'ur-intuition', for the continuum as a whole is immediately given to the individual, and it undeniably is infinite. The denumerable sets can, however not be considered as actual (that is 'finished') infinite.

This particular statement on 'actual infinite' did in fact occur explicitly in the original list of theses as no. 38. Although this particular thesis was dropped, there is an explicit acknowledgement of the existence of actual infinite sets in the dissertation (p. 176), unfortunately Brouwer did not elaborate the matter.

3.6 The public defense

The grand apotheosis took place on 19 February 1907. On that day Brouwer defended his dissertation in public before a delegation of professors from the faculty. His dissertation had, as tradition and the rules decreed, been printed and distributed. The publishers, Maas & Van Suchtelen (Amsterdam–Leipzig), and their printers, the steam printing shop Robijns & Co (Nijmegen), had done a marvellous job. The font and the choice of paper would have been the delight of any bibliophile.

In the old days the formalities of the 'promotion' took place in the auditorium (*Aula*) of the University at the Oudemanhuispoort, a solemn hall, reminiscent of a church hall, where the doctor-to-be occupied a small lectern in front of the imposing rector's lectern in the shadow of a statue of Pallas Athene. The actual examination and defence took three quarters of an hour; the candidate had to defend his dissertation including the theses against the objections of the faculty members (professors only).

The candidate was accompanied by two helpers, *paranimfs*, dressed in tails, like the candidate. The paranimphs were placed on both sides of the lectern, where they stood like statues during the ceremony. The professors filed in, preceded by the beadle with his staff decorated with medals, the soft clinking spreading an almost devotional atmosphere. Once the professors were seated, the examination was opened, according to the local Amsterdam tradition, with the so-called *opposition from the floor*: persons from the public were allowed to attack the dissertation and the theses. This sounds more liberal than it actually was (and still is). These opponents from the floor were picked in advance by the candidate and the adviser, and the approval of the rector had to be obtained before any such opposition could take place. As a consequence, this part of the 'opposition' was often a set up. The candidate could easily arrange the questions beforehand with the opposition, and most of the time this indeed happened. In Brouwer's case, the opposition from the floor was carried out by Mannoury and Barrau, (a fellow student, who was a bit older than Brouwer). Both had informed Brouwer of their questions[56] and Brouwer had, as appears from some sheets in the Brouwer archive, meticulously prepared his defence. Mannoury's objections were directed against the use of irreducible concepts in mathematics (Diss. p. 180) and of infinite (albeit potentially infinite) sets, in particular the use of 'and so on'. Barrau questioned the notion of the continuum as a matrix, and wanted to see the discrete as the basic notion in mathematics. It may be assumed that Brouwer got the better of these opponents from the floor.

The promotor also used the occasion to give vent to his doubts. In his opinion Brouwer's wholesale desertion of Kant's a priority of space was ill-motivated. The intuition of time was, according to him, not rich enough to allow geometry of all dimensions. Thus he proposed to allow for an intuition of displacement (transformation) in addition to the ur-intuition of time. He observed that although the candidate had no use for an extra intuition in addition to the intuition of time, it by no means followed that therefore it did not *exist*. This extra intuition of space had, according to Korteweg, to be locally Euclidean. Brouwer's reply has not been recorded, but presumably he would not have had much difficulty in warding off this blow too.

And so at the age of 25 he acquired the degree of doctor in mathematics and physics, with the predicate Cum Laude. He was assisted by his paranimphs Carel Adama van Scheltema and Ru Mauve, two friends of the first hour.

---

[56] Brouwer to Korteweg 16 February 1907, Mannoury to Brouwer 13 February 1907.

When the examination was over, Korteweg could show his delight at the success of his pupil, addressing the 'young doctor' with words of warm praise. Given Brouwer's preference for the deep foundations, he was not surprised that his student came to know

> ... those most recent investigations of men like Cantor, Hilbert, Poincaré and Peano, where they occupy themselves with investigations in those subterranean parts of this building, where for many of those who are used to work on the higher stories in clearer light, there seems to be a twilight that seems almost impenetrable. And, certainly, those researches would not have been valued so highly, if they had not been conducted by men, the guidance of whom one had learned to trust by their mathematical work of a different nature.

Korteweg was perfectly aware of the gigantic task that Brouwer had taken upon his shoulders. He knew that his student had written his dissertation as a mathematical credo. He expressed in his speech the hope that Brouwer might be in the position to elaborate his ideas. In spite of his recognition of the foundational program of his Ph.D. student, Korteweg could not help wondering why Brouwer had not simply continued the research that just preceded the dissertation, but he accepted and admired Brouwer's choice. Nonetheless he urged Brouwer to devote some attention to the work at the higher floors, for which the talent was certainly not lacking. Prophetic words indeed! As we will see later, Brouwer took the implicit message to heart.

There is no mention of the traditional celebration after the promotion; usually the 'young doctor' invites his Ph.D. adviser, the paranimphs, and a selection of his friends and relations for a dinner and possibly a party after the 'promotion'. It is likely that Brouwer observed the tradition, but one cannot be certain. Scheltema, one of his paranymphs, fled from the city immediately after the occasion to the solitude of the Veluwe. Like Bertus, he was not made for boisterous parties. A week later the two friends automatically resumed their traditional exchange of birthday wishes (Scheltema's birthday was actually the day before Brouwer's). Scheltema had observed his friend well enough to see the changes in his life. He wrote, 'may the next year bring you closer to reality than you were—you are already somewhat closer than before'. He added, 'and, please don't give up on your plan to visit those professors abroad.'

The family was probably impressed by the learned member of the Brouwer–Poutsma clan. One of Brouwer's aunts presented him at the occasion of his doctorate with a slim volume of Eastern mysticism, *Le Livre de la Voie et de la ligne-droite de Lao-Tse*, by Alexandre Ular.[57]

The reception of Brouwer's dissertation in Holland is hard to judge; it is a reasonable conjecture that it was rather mentioned and quoted than understood. Brouwer was way ahead of his fellow countrymen; even Mannoury, who was with-

---

[57] [Ular 1902], The book of the way and the straight line of Lao-Tse.

out doubt the most competent reader, misunderstood Brouwer on certain crucial issues. In a couple of reviews, one for the mathematics journal and two for a wider readership, he discussed the basic issues of the dissertation. The latter, more philosophical in nature, ended with a downright repudiation of Brouwer's claims of the reliability of (his!) mathematics:

> No, Brouwer, logicians do not ensure the reliability of the 'mathematical properties', but neither will you through your continuity intuition, for the simple reason *that it does not exist*. Mathematics is a human artefact, a human conception, in which there is no truth but that which relates to human language, intention and society. Your book is an action of thought-courage and a consequence of an acquired higher insight, but that thought-courage and that insight, ... they are 'unfinished'. Dissociate yourself (indeed *completely*) from all conventions and agreements, from the language and all word-constructs, and I am certain that you will arrive at the acknowledgement (which is the only true foundation of mathematics): there is no immutable truth and no immutable measure for truth—there is no absolute unity, no absolute space and no absolute time, there *is* no mathematics.[58]

These were hard words to swallow for Brouwer, even (or maybe, in particular) coming from a friend. Nonetheless he replied in a mellow mood:[59]

> Is there no mathematics? And what about your objective criticism which is therefore so mathematical? Designed in accordance with norms and hence understandable? And in particular: why was there the age-long uncomfortable feeling with respect to the axiom of parallels, and not with respect to the fundamental property of arithmetic?[60] Because the first is mathematized reality, and latter mathematized mathematics, free from reality, which here means 'not experienced by me'.

The above skirmish did in no way detract from the friendship between Brouwer and Mannoury. They remained the best of friends for the rest of their lives.

---

[58] [Mannoury 1907].
[59] Brouwer to Mannoury 1 August 1907.
[60] [that is all countings of a finite set yield the same number].

4

CANTOR–SCHOENFLIES TOPOLOGY

4.1 The geometry of continuous change

The evolution of geometry may, if one wishes to do so, be viewed as an everlasting struggle to free the subject from the shackles of convention, to allow the study of more and more general figures and properties. Where originally straight lines and circles were studied, soon conics were introduced and gradually more general curves entered the domain of geometry, the spiral, the conchoid, the cycloid, the sine-curve, curves with no tangents at all, etc.

This expansion of the domain of geometry went hand in hand with a more liberal attitude with respect towards the notion of 'the same' (or 'similar'): are only congruent triangles the same, or are all triangles the same, are circles with the same radius the same, are all circles the same, are circles and ellipses the same, are all conics the same? The yardstick of this notion 'the same' is the notion of 'comparison'. Euclid compared triangles with the yardstick of 'congruence', that is two triangles are the same if they are congruent, but in the theory of perspective figures are considered 'the same' if they can be projected onto each other from a point (the eye of the beholder). One might go even further and consider, for example, figures on a rubber balloon. These figures are 'the same' if they can be obtained from each other by squeezing the balloon, or by twisting it or poking at it. The strange shapes lines on a balloon can assume, when suitably twisted, are nothing compared to Cantor's result that, from a suitable viewpoint, a line segment and a square are 'the same'. In other words, one can fill up a whole square using all the points of a line segment—a shocking fact indeed! What one calls 'similar' depends, of course, on what kind of transformations one is willing to allow. Should one demand that straight lines remain straight, or that angles do not change?

In the nineteenth century mathematicians started to realize that it is the comparison that counts, or to put it a bit more formally, the transformations one allows determine the properties belonging to the geometry under consideration. This viewpoint was laid down in Felix Klein's *Erlanger Programm* (1872).

The subject of the present chapter, topology, was determined by the yardstick of transformations which are one–one[1] and continuous both ways; the technical term for these mappings is *homeomorphisms* or *topological mappings*.

The geometry on a rubber balloon is a fine example of topology; one may stretch, distort, inflate the surface as long as one does not cut or tear it. It is clear

---

[1] Such a transformation f can be reversed. To be precise: there is another transformation g, so that f followed by g, and g followed by f take points to themselves.

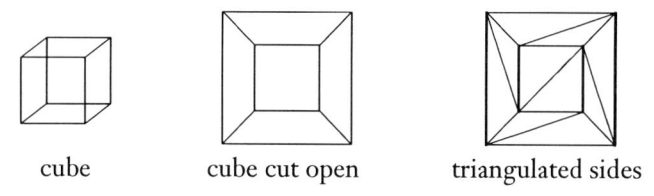

FIG. 4.1. Euler's formula for the cube

that notions like 'straight line' or 'angle' lose their meaning in this example. There are, however, quite a number of geometric properties that remain invariant; for example, the number of intersections does not change, and an O cannot be turned into an F.

Let us illustrate how one works in topology, by looking at Euler's theorem (which, by the way, can be traced to Descartes). We will look at a simple form.

Consider a convex polyhedron (that is a surface that is built up from flat pieces, with intersection lines and points, without dents or holes; think for example of a cube). There happens to be a fixed relation between the number of faces (F), edges (E) and vertices (V): $V - E + F = 2$.

We immediately see that if the polyhedron is stretched, bent, dented, etc., this relation between F, E and V remains the same. This insight is used to give a simple proof of Euler's theorem.

Let us demonstrate the method on a simple polyhedron, the cube, Fig. 4.1.

We cut away the front face and stretch the remaining part of the cube until it lies in a plane. With a bit of extra stretching we can get the edges straight again. The figure now looks like the middle figure of Fig. 4.1.

Instead of proving the above relation, we now have to show $V - E + F = 1$, because one face has disappeared.

The next trick is to split the faces in triangles, see the right hand part of Fig. 4.1. We note that given a polygon, the introduction of a single triangle introduces one extra edge and one extra face. Hence in the formula the $+1$ and the -1 cancel each other out. Repeating the process, one obtains a plane graph with only triangles. If for this graph the relation holds, it also holds for the original plane figure.

For one triangle we get $V - E + F = 3 - 3 + 1 = 1$, and by adding another triangle we add two edges, one vertex and one face, hence $V - E + F$ does not change. Now repeat the argument a finite number of times and we are done.

The reader will have noted that the procedure essentially uses a number of tricks that Euclid would have frowned upon, that is angles and lengths are completely ignored, and straight lines were bent and stretched ad libitum, etc. This is one of the tricks of topology; forget about Euclidean notions, but deform your figures continuously (and one–one).

After Euler the subject remained dormant for almost a century; in 1833 Gauss observed 'Of the Geometria situs, conjectured by Leibniz and into which only a few geometers, Euler and Vandermonde, were allowed a glimpse, we know after one and a half century little more than nothing'.

The name '*topology*' was introduced by Listing in his *Vorstudien zur Topologie* (1847–48); the alternative name *analysis situs*[2] survived for more than half a century, and was used by prominent topologists like Poincaré and Schoenflies, but gradually it has become a historic curiosity.

After Listing the interest in topology gradually increased; it was noted that many geometrical phenomena were not really dependent on the standard notions of traditional geometry. Geometry itself was relaxing its attitudes towards its proper contents. The insight that there was more than Euclidean geometry, and the systematic use of transformations in projective geometry, affine geometry, and so on, culminated in Felix Klein's above mentioned *Erlangen Program*, which characterized the various geometries as the objects and notions invariant under specific transformation groups.

This program opened the way for a more systematic approach to topology, in all its forms. The subject had at that time also received vigorous impulses from the combination of analysis and geometry, in particular from Riemann's work on the foundations of analysis, differential geometry and Riemann surfaces—originally designed for the study of the global behaviour of analytic functions. Riemann's geometric investigations were also the point of departure for the particular line of research in the foundations of geometry, that nowadays is listed under the heading of *Riemann–Helmholtz Raumproblem*, a project of determining the Euclidean and non-Euclidean geometries by means of their transformation groups. This particular direction in topology and geometry was followed by Sophus Lie, Killing, Elie Cartan and others; the theory of Lie groups is the mathematical offspring of the Riemann–Helmholtz project.

In France the subject of topology was taken up by Camille Jordan, his name is most of all connected with a particular kind of curve in the plane; this so-called 'Jordan curve' is obtained by applying a topological transformation to the unit circle. In popular terms, one considers the unit circle as a rubber band, which is placed in the plane without cutting the band and without introducing self-crossings.

Jordan stated the following basic theorem: a Jordan curve divides the plane in two components, that is there are two parts such that one cannot traverse from one part to the other via a (continuous) path without crossing the Jordan curve. The theorem will appear completely trivial to the outsider; however, when prodded, he will probably not even know how to begin to prove it. Jordan's own proof (1887) did not quite pass the test of correctness, and eventually Oswald Veblen gave the first rigorous proof in 1906.

A Jordan curve.

The key-word in topology is 'continuous'. The notion of continuity had been

---

[2]Actually a misnomer, cf. [Freudenthal 1954].

scrutinized in the nineteenth century by a number of mathematicians. After a quiet existence as a rather obvious concept, something like 'a curve is continuous if one can draw it without lifting one's pencil from the paper', it became a central notion in analysis and subsequently in geometry. Precise definitions were furnished (along the lines of the ϵ-δ-tradition) and most of the basic properties of continuous functions on the real line, or the plane, were established. A well-known example is the intermediate value theorem of calculus: 'if a continuous function f is negative in 0 and positive in 1, then there is a point a between 0 and 1 such that $f(a) = 0$'. The theorem may not seem terribly exciting at first blush, but to value its significance one must have a clear insight into (i) the nature of the real line, (ii) the nature of continuity.

At the time that Brouwer turned his attention to topology the grand master in the field was Henri Poincaré, a man with a wide ranging interest. He had come to topology through his investigations of differential equations, the qualitative behaviour of functions defined by such equations required radically new tools—which Poincaré found in topology. The spaces that were investigated by Poincaré, Lie, Élie Cartan and others were still closely related to Euclidean spaces of arbitrary dimensions. They were obtained by judiciously pasting together pieces of Euclidean spaces after suitable treatment (such as bending or twisting); the resulting spaces were called manifolds (or varieties, *Mannigfaltigkeiten* in German).

A man like Poincaré, who commanded almost the whole of mathematics, brought a considerable part of the mathematical arsenal into play when studying subjects from topology (or analysis situs, as he called it), for example, he usually supposed surfaces, etc. to be differentiable (or more). A new generation, with Brouwer in the forefront, would set topology free from those strong extraneous conditions; But not only would the study of Euclidean spaces be liberated from too stringent conditions, the spaces themselves would also be subjected to drastic generalization.

In the eighteen-seventies topology had begotten a curious child: set theory. Georg Cantor had introduced certain transfinite arguments in his study of Fourier series, and after studying sets of real numbers and sets of points of Euclidean spaces, he started to consider arbitrary sets and functions. In 1908 Zermelo provided the final touch: an axiom system for set theory. Set theory, in its turn, enriched topology with an abstract setting in which neighbourhoods could be postulated, so that the traditional definition of 'continuous' could be mimicked (Fréchet). This so-called 'set-theoretic topology' allowed a considerable generality that was very convenient in diverse fields of application.

Brouwer entered the stage when most notions were still rather rudimentary and not universally adopted; indeed, his topological investigations played an important role in setting better standards for the discipline. He was self-taught, and thus it is not surprising that his familiarity with the subject of topology showed some gaps; he was, however, a fast learner. Although he had marked preferences for certain parts of topology—for example the topology of planes, spheres, and manifolds in general—he knew how to handle topology in the broadest sense. We will consider here his contributions in the early period, before the First World War.

4.2 Lie groups

Brouwer's first encounter with topology was in an area that was already well understood at the end of the nineteenth century: the theory of the Lie groups, named after Sophus Lie, the Norwegian mathematician who uncovered their properties. Lie groups consist of transformations of finite-dimensional manifolds (the reader may think of Euclidean-like spaces), which depend continuously on a finite number of parameters.

'*Group*' is the technical term for some set with a binary operation (usually called 'product' after the multiplication of numbers) such that

(i) there is an identity element, that is an element e such that $a \cdot e = e \cdot a = a$ for all elements a,

(ii) the product of two elements again belongs to the group,

(iii) the product is associative, that is $a \cdot (b \cdot c) = (a \cdot b) \cdot c$,

(iv) every element a has an inverse b, that is $a \cdot b = b \cdot a = e$ (b is usually denoted by a^{-1}).

There is no lack of examples of groups: the integers, for instance, with addition and subtraction form a group. Addition is associative: $m + (n + p) = (m + n) + p$, the identity element is 0, that is $0 + n = n + 0 = n$ and finally $-n$ is the inverse of n, that is $n + (-n) = 0$.

Similarly, the positive real numbers, with multiplication and inverse, form a group: $a \cdot (b \cdot c) = (a \cdot b) \cdot c$, $a \cdot 1 = 1 \cdot a = a$ and $a \cdot a^{-1} = a^{-1} \cdot a = 1$. Along the same lines, vectors form a group under vector addition.

Groups play an important role in almost all areas of mathematics, in particular in geometry, where they naturally occur as transformations. For example, the congruences of elementary geometry (those transformations of the plane that do not change length) form a group. The main idea of composition and inverse in the case of transformations is that 'what one can do in two steps, one can also do in one step' and 'each step can be reversed'; it explains, for example, why congruence is an equivalence relation: if triangle Δ is congruent with triangle Δ', and Δ' with Δ'', then Δ is congruent with Δ''; if Δ is congruent with Δ', then Δ' is congruent with Δ, and Δ is congruent with itself.

In the original approach to the theory of Lie groups, not only the algebraic features of the groups played a role (that is those concerning the operations of the group by themselves) but also properties that belong to the area of analysis (or calculus), such as continuity and differentiability. Lie groups, in the style of the nineteenth century, were transformation groups of locally Euclidean manifolds, depending on a finite number of parameters; Lie had required that the parameters of the product of two transformations were twice continuous differentiable functions of the parameters of the factors. He proved that under those conditions, these functions were even analytical. An easy way to express this is: the operations are analytical (similarly, 'continuous', 'differentiable' etc.)

This differentiability condition gradually came to be viewed as rather artificial, and when David Hilbert presented in 1900 his list of mathematical problems to the International Conference of Mathematicians, the elimination of differentiability conditions from the theory of Lie groups was on the fifth place of the list.

In modern terminology, a Lie group is an analytic manifold with a group structure so that the operation xy^{-1} is analytical. Hilbert's problem number 5 asked to replace the latter condition by 'xy^{-1} is continuous'.[3]

One guesses that Brouwer, when looking for suitable topics for his dissertation, looked through Hilbert's list and decided that this fifth problem suited his taste and capacities. There is among Brouwer's early papers none that leads up to this particular problem, so the choice must have been more or less deliberate. As we have seen, Brouwer felt that his Ph.D. adviser, Korteweg, did not fully appreciate the contents of the dissertation, and, trying to convince him, he simply referred to Hilbert's problems, claiming that he had solved three of them, including the fifth one. Indeed, he treated the one-parameter case in the dissertation and derived the differentiability, rather than postulating it. As a result the basic results of Lie applied also to this instance of a more general setting.

A year after obtaining his doctor's degree, Brouwer attended the International Conference for Mathematicians in Rome, where he gave two talks, one on Hilbert's fifth problem and one on the possible cardinal numbers in constructive mathematics.[4] In the first talk, *The theory of finite continuous groups, independent of the axioms of Lie*,[5] he carried out the elimination of the differentiability (with the help of an auxiliary argument, spelled out in *About difference quotients and differential quotients*[6]) and gave a geometric-topological classification of the groups. The uniform differentiability notion, introduced by Brouwer, surpassed from a methodological point of view the traditional 'continuously differentiability'; it did not find its way into practice.[7] It is nowadays adopted in constructive analysis.

In view of the importance of the subject and the prior involvement of Hilbert, it is not so surprising that in the same year a substantial paper of Brouwer, *The theory of finite continuous groups, independent of the axioms of Lie, I*,[8] (which extended the dissertation and the Rome talk) was published in the *Mathematische Annalen*. The second part appeared a year later, also in the *Mathematische Annalen*, and dealt with the two-parameter case. At the end of the paper Brouwer remarked that he now had all the tools for enumerating the groups of two-dimensional manifolds, and promised to do so in a next communication. This third part, however, never appeared, although he mentioned in a letter of 9 April 1924 to Urysohn that

---

[3] The problem was eventually solved in 1952 in the affirmative by A.M. Gleason, D. Montgomery and L. Zippin. As a consequence one may now define a Lie group as a topological group over a locally Euclidean manifold.

[4] [Brouwer 1909B], [Brouwer 1908A].

[5] *Die Theorie der endlichen continuierlichen Gruppen unabhängig von den Axiomen von Lie.*

[6] [Brouwer 1908D2].

[7] Freudenthal, CW II, p. 101.

[8] *Die Theorie der endlichen continuierlichen Gruppen unabhängig von den Axiomen von Lie. I* [Brouwer 1909C].

> Due to numerous distractions, the manuscript of my third communication on *The theory of finite continuous groups* (*Die Theorie der endlichen kontinuierlichen Gruppen*) unfortunately still waits in my drawer for the finishing touches, which indeed I hope to be able to make in the near future.

Just looking at the dates of publication, one might get the erroneous idea that the sequel to Part I was a matter of routine. As a matter of fact, Brouwer suffered a serious setback—which eventually proved a boon. In a letter of 14 May 1909 to Hilbert he reported that a couple of months ago ('last winter'), when he was about to send Part II to the *Mathematische Annalen*, he had discovered that Schoenflies' investigations on Analysis Situs,

> on which I had based my work in the most extensive way, cannot be vindicated in every part, and that also my group theoretic results became questionable.

And thus Brouwer decided to occupy himself thoroughly with Schoenflies' topology. The result was the celebrated paper *Zur Analysis Situs*,[9] which 'rejected or modified some parts of Schoenflies' theory, and completely rebuilt other parts'. We will return to this paper on page 143.

After this substantial excursion into the realm of topology of the plane, he wrote to Hilbert on 26 July 1909 that the manuscript of Part II would follow in a month's time.

It is quite certain that the two papers did not exhaust Brouwer's results on Lie groups. According to Freudenthal,[10] he was aware of the difficulties in the three-dimensional case, such as presented by Antoine's set[11] and the horned sphere of Alexander.[12]

In the summer of 1909 Brouwer finally met Hilbert in the flesh. He had missed him at the Rome Conference, where (according to the list of participants) Hilbert was absent, but now Hilbert spent part of the summer vacation in the sea-side resort Scheveningen. The meeting, a precious gift to a young mathematician, made a lasting impression on Brouwer. He wrote in elated terms to his friend Scheltema:[13]

> This summer the first mathematician of the world was in Scheveningen; I was already in contact with him through my work, but now I have repeatedly made walks with him, and talked as a young apostle with a prophet. He was only 46 years old[14], but with a young soul and body; he swam vigorously and climbed walls and barbed wired gates with pleasure. It was a beautiful new ray of light through my life.

---

[9] On Analysis Situs [Brouwer 1910C].
[10] CW II, 117.
[11] [Antoine 1921].
[12] [Alexander 1924].
[13] Brouwer to Scheltema 9 November 1909.
[14] 45 years, actually.

The famous mathematician and his young admirer must have found enough to talk about. Brouwer's ideas about Lie groups and topology were of course close to Hilbert's interest. Later in life Brouwer often referred to this first meeting. He used to point out that foundational matters also came up in their conversation; in particular he explained his various mathematics, language and logic levels to Hilbert. It is rather plausible that he told Hilbert his objections to the Heidelberg talk.[15] After all, why else mention the levels? Olga Taussky reported that Hilbert visited Brouwer in his Blaricum cottage.[16] As far as we know, the 1909 trip to Scheveningen was Hilbert's only visit to Holland, so the conversation of the two men must have been continued in Blaricum.

As Brouwer had mentioned in his dissertation, Hilbert had already in 1903[17] determined the Lie group of the Euclidian plane independent of differentiability conditions; Hilbert's arguments were briefly summed up in Brouwer's dissertation. When he reworked his results for the *Mathematische Annalen*, he probably returned to Hilbert's paper, and checked it in detail. In a cordial letter of 28 October 1909[18] he explained to Hilbert, that the 1903 paper required a number of corrections and addenda, and went on to spell out the details. After his excursion into Cantor–Schoenflies topology, the topological details required for the shoring up of Hilbert's paper presented no problem. It is not known if and how Hilbert reacted to this flood of good advice, no answer to this letter has been found. Hilbert revised the paper, which was republished as Appendix IV to the *Foundations of Geometry* (from the third edition (1909) onwards). The revised version avoids the difficulties pointed out by Brouwer, but whether in consequence of Brouwer's letter or independently, cannot be ascertained. Hilbert, apparently did not further discuss the matter with Brouwer, for as late as 16 June 1913 the latter wrote a letter to Hilbert:

> I recently read that a fourth edition of your Foundations of Geometry is to appear. Have the remarks concerning Appendix IV [that is the paper from the *Annalen* 56], which I sent to you in the fall of 1909, been taken into account? I would in any case be glad to assist you in putting the relevant sections right if you should wish so, and if the imprimatur has not yet been given.

He mentioned the subject again in his note on the history of dimension theory in 1928, at the time when the relation with Hilbert was already beyond repair: in a comment on the proper dating of mathematical contributions (which was at issue in that paper) he pointed out that it was beyond dispute that one should put the date of the set theoretic foundation of geometry concurrent with "Hilbert's original paper of 1906, and not with the reprint of this note, in which an evident oversight in the underlying axioms was corrected on the basis of indications in a letter of mine of October 1909". The fact that Brouwer kept the draft among his papers, seems to

---

[15] See [Brouwer1928A3].
[16] Taussky to Van Dalen 1991.
[17] [Hilbert 1902], [Hilbert 1909] and later editions.
[18] Brouwer to Hilbert 28 October 1909, reproduced in CW II, p. 102 ff.

indicate that he attached some importance to this piece of topology and geometry; the circumstance that it concerned one of the greatest mathematicians of the day will have added to its value.

There was a long and friendly correspondence between Hilbert and Brouwer. The latter admired the great statesman of mathematics as no other. The above letter of 28 October 1909 illustrates Brouwer's reverence for the elder and wiser man, whom he described in 1909 to his friend Scheltema in glowing words.[19] Only the letters of Brouwer in Hilbert's archive, and some drafts, have survived. They all testify to an unreserved and cordial relationship between Brouwer and Hilbert. In view of the later conflicts between Hilbert and Brouwer, there have been conjectures as to earlier irritations and frictions. No traces can be found in the pre-*Grundlagenstreit* correspondence.[20]

Brouwer sent reprints of his papers to Hilbert with warm dedications. Some of these can still be found in the archive of the Mathematics department of the Nagoya University. Here are a few examples: the first Annalen paper on Lie groups bears the inscription 'in warmest admiration from the author' (*in wärmster Verehrung vom Verfasser*); invariance of dimension paper—'To professor Geheimrat Hilbert in warmest admiration from the author' (*Herrn G. R. Hilbert in herzlichster Verehrung vom Verfasser*); the second paper on Lie groups—'To his dear and admired professor Geheimrat Hilbert, L.E.J. Brouwer' (*Seinem lieben und verehrten Herrn G. H. Hilbert, L.E.J. Brouwer*). The dedications at the papers are certainly no empty phrases. Brouwer indeed deeply admired Hilbert; he was totally sincere in his love (which is the right expression, in spite of its melodramatic reputation). The later rejection of Brouwer by Hilbert must have been all the more painful.

4.3 Publishing in the *Mathematische Annalen*

Brouwer's first paper on finite continuous groups, *Die Theorie der endlichen kontinuierlichen Gruppen, unabhängig von den Axiomen von Lie, I*[21] published in the *Mathematische Annalen*, got a rather patronizing review in the review journal *Jahrbuch über die Fortschritte der Mathematik* ,[22] and this led to a curious correspondence between the author and the reviewer, Friedrich Engel, the close associate of Sophus Lie, and co-author of the three volumes *Theorie der Transformationsgruppen*. The surviving correspondence has been reproduced by Freudenthal in the *Collected Works II*. The letters show a striking conceptual rift between the new and the old generation. Engel, as one of the founding fathers of the subject, was so strongly entrenched in the traditional analytic approach, that Brouwer's general set theoretic

---

[19] Brouwer to Scheltema 9 November 1909.

[20] The only hint that could throw doubt on the good relationship is a letter of Bernays to Freudenthal, relating to the above mentioned letter of Brouwer to Hilbert. In this letter Bernays more or less expressed his surprise that Hilbert and Brouwer were still good friends at that time. There are, in my opinion, no historical facts that would support this. But it is, of course, possible that Hilbert expressed in private conversation criticism of Brouwer.

[21] [Brouwer 1909C].

[22] The volume for 1909 appeared in 1912.

approach was in effect unreadable for him.²³ The review opened with Engel criticizing Brouwer's definitions as being somewhat vague and too restrictive, moreover, Engel thought that there were too many loose ends. 'But', he remarked, 'it would be unjust to demand that such a difficult problem would really be completely solved at a first go'.

A review of this sort was just the thing to provoke Brouwer. From the formulation it can be guessed that Engel did not fully grasp Brouwer's definition, and that he was in some confusion as to the local behaviour of the transformation in the neighbourhood of the identity and its global group theoretical behaviour. Brouwer reacted immediately with a cold and slightly patronizing letter, asking Engel for chapter and verse:²⁴

> Concerning the first point,²⁵ I would be grateful if you would be so good as to tell me which extended version of the problem you have in mind; for, I did not succeed in getting a clear idea from your hints.

The reader should keep in mind that in the time between the publication of the paper and the review, Brouwer had turned from an unknown beginner into one of the foremost experts in topology. The review was somewhat condescending in tone, and Brouwer must have felt that he could not leave it at that. At the time of this correspondence Brouwer was in a somewhat ambiguous position: his mathematical talent was well-recognized in the world, but in Amsterdam he was still at the mercy of the board of the university and the city council, as just another extraordinary professor. Nonetheless, I guess, he would have reacted like he did under any circumstances. Even as a beginner he would have taken up the gauntlet and accepted the consequences.

In reply to Brouwer's letter, Engel frankly admitted that he could not keep up with Brouwer's set theoretical formulations, but he insisted that the clarity of formulation left much to be desired:

> Maybe they are clear to the set theoreticians incarnate (*eingefleischte Mengentheoretiker*), but I must say 'the expressions of the system sound obscure to uncircumcised ears'.

Engel's letter²⁶ illustrates the fact that Brouwer and Engel were worlds apart; their terminology and way of thinking, the first one topologically and the second one analytically, is so different that a common view was virtually excluded.

The first two letters in the correspondence make a somewhat tense impression,

---

²³In Freudenthal's words: 'Engel, [...] who could not grasp a group except in its analytic setting, and Brouwer, who had shaken off the algorithmic yoke and from his conceptional viewpoint could not comprehend his correspondent's difficulties. Manifolds and one–to–one mappings as substrate and action of Lie groups instead of cartesian space and many-valued mappings was indeed a great step forward, though for older contemporaries of Brouwer it was too much' (CW II, p. 142).

²⁴Brouwer to Engel 21 January 1912.

²⁵that is the excessive restrictions.

²⁶Engel to Brouwer 28 January 1912, CW II, p. 144.

Brouwer barely hid his annoyance at being lectured by an elder statesman, for a—in his eyes—perfect paper; the style of the letter is polite, bordering on the cynical. Whereas the letter may indeed display deep reverence, it could equally well be construed as mischievous. The closing sentence, 'I would highly appreciate to reach an agreement concerning the above, with a group theoretician of your authority', might have been written 'tongue in cheek', but it probably was sincere. Engel, in turn, seemed to sway between appreciation and mild sarcasm, writing that he had not intended any disapproval, but rather had, in the words of Lessing, evaluated the product of the Master, 'doubting in admiration, and admiring in doubt'. At the time of the exchange of letters Brouwer's fame had already reached the corners of the mathematical world, so that Engel must evidently have been aware of the merits of his correspondent, therefore it is quite likely that Engel was prepared to accept Brouwer's mathematical authority, but not without questioning. He closed his letter with the remark that he could not quite see the gain of investigations like Brouwer's in relation to the ingenuity and toil that were invested, expressing the hope that 'you would now also occupy yourself in group theory, for there is much to be done'. As mentioned above, the main point of confusion in Engel's understanding was the relation between local and global. The following letters returned to this matter. The tone of the letters, however, had greatly improved, but Brouwer still insisted that, even though Engel's objections were mainly the product of a difference in mathematical culture, readers of the review would get the impression that the paper contains actual gaps.[27] He felt that he was entitled to a rehabilitation and therefore he kindly asked Engel to correct possible misconceptions in the review of the second part of the paper.[28] The letter also contained a birds-eye view of the basic topological notions, including an elegant definition of connectedness.

Engel, indeed, set out on his journey to Canossa—without actually reaching it; the review of the second paper contained a somewhat reluctant retraction of the earlier statements, followed by a complaint that 'In general, I cannot dissemble that I am beset with a mild horror in the face of such colossal generality of the investigation, and the large number and variety of required inferences'. The second review only confirms the impression of the first one, namely that the master of the traditional school of Lie groups was separated by a wide gulf from the newcomer with his abstract topological methods.

It seems not unlikely that the wish, both of them expressed at the end of this correspondence—to talk things over in person, was fulfilled at one time or another. Whether this indeed happened and how profitable it was, we shall probably never know.

There is one more paper in this series dealing with Lie groups, it was the text of an address to the Dutch Physical and Medical Society,[29] it provided an elegant

---

[27] Brouwer to Engel 6 March 1912.

[28] *Die Theorie der endlichen kontinuierlichen Gruppen, unabhängig von den Axiomen von Lie, II.*, [Brouwer 1910H].

[29] *Het Nederlandsch Natuur- en Geneeskundig Congres.*

and short characterization of the various geometries of Riemann's program.[30] The paper shows the hand of the master in mixing geometrical, topological and group theoretical arguments.

The technique of eliminating differentiability conditions was applied once more in the paper *On a theory of measure*,[31] in which Brouwer sharpened the results of G. Combebiac on a certain functional relation.

4.4 Fixed points on spheres and the translation theorem

The investigations on Lie groups took place at the same time as work on continuous mappings on spheres, work on vector distributions and the above mentioned research on the topology of the plane.

On 27 February 1909, Korteweg presented to the Academy of Sciences the first paper by Brouwer of a series 'Continuous one–one transformations of surfaces in themselves'.[32] The question that Brouwer asked himself was 'whether this [that is the one–one continuous mapping of the sphere into itself] is possible without at least one point remaining in its place'?

The method used in this paper is wholly based on the available techniques of plane topology, it involves a detailed study of the behaviour of a family of concentric circles and their images. The best-known results of the paper are:[33]

> A continuous one–one transformation in itself with invariant indicatrix [that is orientation preserving] of a singly [that is simply] connected, two sided, closed surface possesses at least one invariant point.

and its companion

> A continuous one–one transformation in itself of a singly [simply] connected, closed surface leaves at least one point invariant.

As one can easily see, the two-sided simply connected closed surface need not have a fixed point under a one–one continuous orientation inverting mapping[34] (think of a mapping that takes each point of the sphere to its antipode). Brouwer's papers on the topology of surfaces are connected to his research on Lie groups, the fixed point theorem for the sphere is, for example, used in [Brouwer 1910H] (p. 193).

---

[30] Characterization of the Euclidean and non-Euclidean motion groups in R_n, CW II, [Brouwer 1909E] (originally in Dutch).
[31] [Brouwer 1911K].
[32] [Brouwer 1909F2].
[33] [Brouwer 1909F2] p. 10, 11.
[34] Popularly speaking one fixes the orientation on a surface by indicating a direction on a little circle, the orientation at other places is determined by shifting this circle all over the surface. If this is possible, that is if one never can get via different routes two circles circulating in opposite directions around one point, the surface is called orientable, and it has an orientation. The sphere is an example of such a surface. The Möbius strip is an example of a one-sided non-orientable surface (take a rectangular strip of paper, twist it and glue the ends together).

The paper on surface mappings was the first in a series of eight papers in the Proceedings of the Academy, the last one appearing in 1920. The pre-war papers are of the same nature, they use elementary techniques; the post-war papers use the new tools that Brouwer had developed himself before the war.[35]

The second paper in the series deals with arbitrary two-sided surfaces; it ends with Brouwer's first, and (as he soon realized) incorrect, version of the *translation theorem*: a 1–1 continuous mapping of the plane without fixed points is a continuous image of a translation.

The translation theorem was, so to speak, the natural supplement to the stream of fixed point theorems, it spelled out what one could expect in the absence of a fixed point for transformations of the plane (where, of course, no fixed points are guaranteed, think of an ordinary shift). At the time, the translation theorem was something of a dashing exploit, fit to be told in small groups of professional mathematicians, huddled in the bar after a strenuous day at the conference. The theorem has attracted the attention of topologists ever since; there are a number of new and clever proofs around nowadays, see for example [Franks 1992].

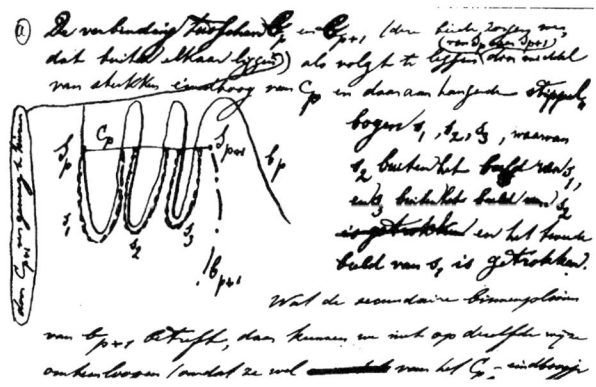

An example of Brouwer's private notes on topology.

The papers following this communication, were all concerned with continuous bijections of surfaces. Unfortunately, the early ones suffered from one and the same defect: Brouwer had relied on Schoenflies' monograph *The development of the theory of point sets II*,[36] often affectionately referred to by the insiders as the *Bericht*. Schoenflies' book was, together with the Encyclopaedia paper of Dehn and Heegaard, the latest word on topology; in particular Schoenflies dealt with all the items that were relevant to the foundations of analysis, such as the basic notions of convergence, open, close, dense, perfect, ..., including the notion of 'curve'. As it happened, Brouwer's work made generous use of properties of curves. Much to Brouwer's disappointment, a good deal of the treatment on curves by Schoenflies showed serious defects. As late as 1912 Brouwer somewhat crossly remarked[37] that the survey, *On one–one, continuous transformation of surfaces into themselves*,[38] which

---

[35] When the publisher, at the end of the series, mixed up the numbering, the numbers 7 and 8 were erroneously published as 6 and 7. Brouwer patiently corrected the numbering by hand in the off-prints!

[36] *Die Entwicklung der Lehre von den Punktmannigfaltigkeiten II*, 1908.

[37] [Brouwer 1912K2], p. 360.

[38] *Über eineindeutige, stetige Transformationen von Flächen in sich*, [Brouwer 1910F].

he had presented at Hilbert's request to the *Mathematische Annalen*, was based on unsubstantiated claims of Schoenflies:

> In carrying out the ideas sketched in the second communication[39] on the subject, I found out that in some points in the course of the demonstrations indicated there, the Schoenflies theory of domain boundaries, criticized by me,[40] still plays a role, so that the theorems 1 and 2 formulated on p. 295 and likewise the 'general translation theorem' based on them, and enunciated without proof in *Mathematische Annalen* 69, p. 178 and p. 179, cannot be considered as proved,[41] and a question of the highest importance is still to be decided here. The 'plane translation theorem' stated at the end of the second communication and similarly in *Mathematische Annalen* 69, p. 179 and p. 180, has meanwhile been proved rigorously by another method.[42]

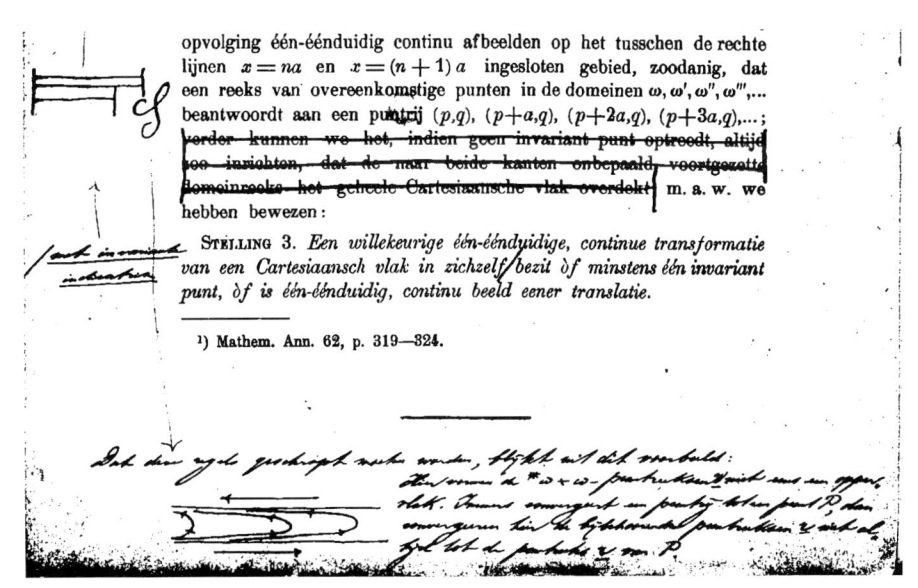

The handwritten corrections at the end of Brouwer's paper *Continuous one-one transformations of surfaces in themselves*

The correct proof of the translation theorem in the *Mathematische Annalen*, is a

---

[39] [Brouwer 1909H2].

[40] Cf. [Brouwer 1910C].

[41] (Brouwer's note) Already the property of p. 288 that the transformation domain constructed in the way indicated there, determines at most two residual domains, vanishes for some domains, incompatible with the Schoenflies theory.

[42] [Brouwer 1912B].

complicated piece of plane topology in the traditional style. It is remarkable that Brouwer formulated the theorem in rather sweeping terms, so that a reader could (without reading the proof) easily be misled. In 1919 he published a short note with the proper formulation, and a counter example to the superficial reading.[43] This, by the way, was already inserted in handwriting by Brouwer in his reprints of the second communication (Dutch version). In all likelihood those corrections were inserted soon after the publication.[44]

The proper formulation of the translation theorem requires more topology than our exposition allows. Let it suffice to say that a continuous 1–1 mapping of the plane without fixed points splits the plane in a number of strip-like parts disjoint from their images, so-called 'translation fields'. Freudenthal, in his editorial comments in [Brouwer 1976] (p. 218), wrote that 'In the twenties and thirties Brouwer stated orally his belief that the original proof could be salvaged. It would be worthwhile to try it because its idea, in particular the construction of a transformation field at one stroke, is more attractive than that of Brouwer's final proof'.

A reader of Brouwer's papers will be struck by the conscientious manner in which corrections and emendations were provided. Brouwer, apparently, was striving for perfection in his papers, even after they had appeared. A number of private copies of reprints carry handwritten corrections or additions, that often appeared as supplements to later papers, or as individual corrections, even when only trivial slips or misprints were concerned.

4.5 Vector fields on surfaces

One month after the communication of the first paper on the continuous 1–1 mappings of the sphere, Korteweg presented another paper of his student: *On continuous vector distributions on surfaces*.[45] The paper was the first of another series of papers, all published in the proceedings of the Academy. It dealt with vector distributions on spheres in a generalized setting, that is to say, on homeomorphic images of spheres. Brouwer's approach was based on Peano's existence theorem for differential equations;[46] he investigated singularities of vector fields by means of differential equations. There is a close connection between this series of papers and the fundamental papers of Poincaré on the qualitative theory of differential equations (*Mémoire sur les courbes définies par une équation differentielles I, II, III*, 1881, 1882, 1885).[47] Poincaré had introduced a whole new aspect to the theory of differential equations. Whereas traditionally one studied solutions locally, that is in the neighbourhood of a point, he initiated the study of global behaviour:

> Finding out the properties of differential equations is a task of major importance. A first step has been made by studying the proposed function in

---

[43] [Brouwer 1919M2].
[44] Cf. CW II 220, and Freudenthal's comments on 219.
[45] [Brouwer 1909G2].
[46] [Peano 1890]. Brouwer also cited the simplification of [Arzela 1896].
[47] [Poincaré 1881, Poincaré 1882, Poincaré 1885].

the neighbourhood of one of the points of the plane. The time has come to go further and study the functions in the whole plane [...]

The study of a function thus includes two parts: (1) The qualitative one, as it may be called, where the curve is studied geometrically, (2) the quantitative one, where values of the function are calculated numerically.[48]

It appears that Poincaré's papers were somewhat slow to be absorbed into the daily routine of mathematics; Brouwer, anyway, had not read them when he embarked on the investigations of vectorfields. But he had heard the presentation of Poincaré's talk at the Rome conference. During the conference Poincaré fell ill, and someone else had to read his address. Poincaré had planned to give at this conference his views on 'The future of Mathematics', with 'Differential equations' as one of its topics. Poincaré had warmly advocated the 'qualitative discussion of curves defined by a differential equation'; in view of Brouwer's enthusiastic report of Poincaré's talk,[49] it seems plausible to look here for the motivation of the choice of a topic of Brouwer's subsequent research.

Jacques Hadamard.
Courtesy M. Loi.

Although Brouwer was an admirer of Poincaré, and had read quite a number of his publications (in particular the more philosophical ones), he had somehow missed the topological œuvre of Poincaré, to the detriment of both Brouwer and topology, one might say. There are two letters in the correspondence between Hadamard and Brouwer which show that Brouwer was not aware of Poincaré's work in the area. Hadamard, thanking Brouwer for two reprints—the above paper on vector fields and possibly one of the papers on mappings of the sphere—suggested an easy solution to the fixed point theorem on the sphere by applying the main result of the paper on vector fields. In the same letter he drew Brouwer's attention to Poincaré's *Mémoire* of 1881. Brouwer's reply of December 24 1909 contained a refutation of Hadamard's suggestion and a grateful acknowledgement of Hadamard's reference to Poincaré's work. In the same letter he announced another approach to a proof of the fixed point theorem on spheres, 'Reading the memoirs of M. Poincaré cited by you, I got another idea'.[50]

Brouwer's investigations were, however, no duplications of Poincaré's earlier work, his viewpoint was a truly topological one, whereas Poincaré assumed some

---

[48] [Poincaré 1881]. Translation by Freudenthal. CW II, p. 282.

[49] Cf. p. 204.

[50] Cf. [Johnson 1981], p. 154. The letter from Hadamard was undated, but it seems likely that it was answered promptly by Brouwer.

algebraic and analytic conditions. The use of Peano's existence theorem is telling, since it requires in the differential equation $\frac{dy}{dx} = f(x, y)$ only the continuity of f.

The results of Brouwer were of a more general nature than Poincaré's, in particular he did not assume any uniqueness conditions (which are implicit in Poincaré's paper), since Peano's theorem did not require any. The price to pay was, however, a degree of complexity that robbed him of potential readers. In the long run Poincaré's results became the standard reference for the qualitative theory of differential equations. It is not wholly impossible that the place of publication, that is the Proceedings of the Amsterdam Academy, was less than fortunate. It did not have a circulation among mathematicians comparable to that of ordinary journals; as a rule periodicals of the sort of academy proceedings, which (at least in those days) published papers from a variety of disciplines in the sciences, were stored in the general sections of central libraries, and not in the mathematics sections. Also, for whatever reason, no comprehensive presentation of the material of the vector distribution series was submitted to the *Annalen*, or any other journal.

Freudenthal, in his comments in CW II (p. 282), points out that where Poincaré used uniqueness conditions and assumed algebraicity and simple singularities, 'Brouwer on the contrary chose the utmost generality and even in specializing he admitted more general types than Poincaré had done. Thanks to a simpler design, Poincaré's work has been influential in the long run, but it is strange that Brouwer's contributions have been entirely overlooked'.

The main results of Brouwer's first paper on vector fields were

> THEOREM 1. *A vector direction varying continuously on a simply connected, two-sided, closed surface must be indeterminate in at least one point.*

And from this follows directly:

> THEOREM 2. *A vector distribution anywhere univalent and continuous on a singly* [i.e. simply] *connected, two-sided, closed surface must be zero. or infinite in at least one point*

If we project the complex plane stereographically on the sphere, a complex function becomes a vector distribution on the sphere. So we can also interpret the above result as follows:[51].

> THEOREM 3: *A univalent, continuous function of a complex variable being nowhere zero or infinite and without singular points cannot exist.*

Freudenthal observed that 'theorem 2 is usually ascribed to H. Poincaré, 1881, though in fact Poincaré asserted and proved it under quite restrictive conditions.[52] Brouwer generalized it to higher dimensions and proved it by more adequate methods.'[53] Brouwer's first theorem became popular as the 'hairy ball'—one cannot

---
[51][Brouwer 1909G2] p. 856
[52]CW II, 282.
[53][Brouwer 1911D], p. 112.

comb a hairy tennis ball without getting a crown.

If Poincaré's basic papers were slow to attract attention, Brouwer's work on vector fields remained buried in the pages of a poorly read journal—to the disadvantage of the mathematical community.

The first paper was followed by two more with the same title;[54] they contain Brouwer's closer analysis of the nature of finite sets of singular points. The second communication contains the structure theorems for the singular points and for the behaviour of the field in the neighbourhood of the singular points. Apart from the classification of singular points, the paper contains a few novelties such as the introduction of the winding number by purely topological means, and its use in the study of the structure of zero points in relation to the behaviour of the field in its neighbourhood,[55] as illustrated by

> THEOREM 5. *The total angle which, by a circuit of a simple closed curve enveloping one point zero, the vector describes in the sense of that circuit, is equal to $\pi(2 + n_1 - n_2)$, where n_1 represents the number of elliptic sectors, n_2 the number of hyperbolic ones, which appear when a vicinity of the point zero is covered with tangent curves not crossing each other.*

Furthermore the paper contains the first appearance of homotopic changes of vector fields.[56] The third communication continues the study of finitely many singular points under special conditions, for example, in the absence of simple closed tangent curves (Brouwer introduced in this paper the picturesque and suggestive term 'irrigation fields').

Following a suggestion of Hadamard,[57] Brouwer applied the winding number to the (finite) set of singular points, obtaining generalizations (in an exclusively topological setting) of Poincaré's results from the *Mémoires* quoted above.

Brouwer apparently, and not unreasonably, felt that there should be a connection between the fixed point theorem on the sphere[58] and the existence of singular points of vector distributions. The obvious trick that comes to mind first, that is the construction a vector field out of a continuous one–one mapping via the great circle connecting them, was suggested in Hadamard's letter[59] and rejected by Brouwer; he spelled out his arguments in the third paper on vector distributions,[60] and gave a correct procedure to get the fixed point theorem on the sphere in the orientation preserving case from the theorems on singular points.[61]

---

[54] [Brouwer 1910A2, 1910D2].

[55] [Brouwer 1910A2].

[56] Cf. Freudenthal, CW II, p. 302.

[57] Cf. [Johnson 1981], p. 153.

[58] [Brouwer 1909F2].

[59] Hadamard to Brouwer 24 December 1909.

[60] [Brouwer 1910D2], CW II p. 314.

[61] §3 of [Brouwer 1910D2]. See also Hadamard's acknowledgement in his appendix to the second edition of Tannery's *Introduction à la théorie des fonctions II* (dealing with topological applications of the Kronecker index), [Hadamard 1910].

Looking at those papers through modern eyes, one cannot fail to be impressed by the approach (and the stamina) of Brouwer—he undertook the investigation of the properties of singular points with his bare hands.

All the methods involved are available to any student with a basic knowledge of differential equations and a sniff of general topology. The advance with respect to older research, for example Poincaré's, consists of a consistent exploitation of the topological viewpoint. One will find in Brouwer's papers certain arguments that have gone out of fashion, for example the use of the second number class and infinitesimal circles and the like, but on the whole the papers are very much readable.

Five years later Brouwer published in the journal of the Dutch Mathematical Society an application of the above theory to the projective group: 'On the orthogonal trajectories of the orbits of a one parameter plane projective group'.[62] At that moment his first topological period was over. His attention was already fixed on foundational matters.

4.6 Analysis situs and Schoenflies

The name of Schoenflies has already come up earlier in these pages in connection with the development of topology around the turn of the century. It is time to have a closer look at him and his work.

Arthur Moritz Schoenflies was born in Landsberg an der Warthe (now Gorzow, Poland) on 17 April 1853. He studied with Kummer in Berlin and after his Ph.D. and *Habilitation* became an extra-ordinary professor in Göttingen (1892) for applied mathematics. In 1899 he was appointed full professor in Königsberg (Kaliningrad), and in 1911 he moved to Frankfurth am Main where he had a chair until 1922. He died there on 27 May 1928.

Schoenflies was a geometer at heart, he started his career in mathematics with a number of investigations of a fairly traditional kind; but in the wake of Jordan he entered into a more modern era with his study of the groups of motions (*Bewegungsgruppen*). In this area, he carried out an extensive classification. This line of research was close to another of his activities: the study of crystal-structure by group theoretic means. The latter was laid down in a comprehensive book *Crystal Systems and Crystal Structure*.[63]

The bulk of Schoenflies' mathematical production is however concerned with the set theory of Cantor. He published extensively on the subject and was at the turn of the century recognized as the outstanding expert in the field. The DMV, the German Mathematics Society, asked him to prepare a survey (*Bericht*) of set theory (including point set theory, that is topology). Part I, the development of the theory of point sets, appeared in 1900 and Part II in 1908. For some time both

---

[62][Brouwer 1915B].

[63]*Krystall Systeme und Krystallstruktur* (1891); a new edition was published in 1984!

Schoenflies' monographs were the only comprehensive texts on set theory in the wider sense—Hausdorff's book appeared in 1914.[64] It is true that many textbooks on analysis or function theory contained sections explaining the basic notions of set theory, but these usually stuck to the immediately applicable parts of the subject.

Both in set theory and topology, Schoenflies had produced a considerable œuvre. He was, however, in the dubious position of a man on the borderline of two cultures; he belonged to the first generation of set theorists, and in ideas and method he was very close to Cantor. He lacked, however, the penetrating insights required to push the subject beyond its initial phase. More imaginative mathematicians, above all Zermelo, created the framework and concepts that made set theory the subject we know today. Schoenflies, the active champion of the new disciplines, was allowed to see the promised land, but he was not the one to lead mathematics into it. Where Zermelo showed set theory the way, Brouwer gave topology its ticket for the twentieth century. Already in 1905, in the animated discussions following Zermelo's proof of the well-ordering theorem, Schoenflies had been taken to task by Zermelo: 'his paper, however, contains further errors and misunderstandings that cannot be ignored here'[65]—and, again, in 1910 he was the victim of a vigorous dressing down, this time by Brouwer.

The new methods and concepts introduced by the latter, changed the subject almost beyond recognition. One could well say that after the introduction of the new Brouwerian methods, Schoenflies' approach became something of a curiosity.

Nonetheless, Schoenflies' role in topology should not be underestimated; his survey monographs are examples of a painstaking sorting and collecting of facts and conjectures in a new area, where the wheat had not yet been separated from the chaff. Looking back, we can say that Schoenflies took, in preparing the surveys of set theory and topology, the role of midwife upon himself. More so, one would be inclined to say, in the case of topology, than in the case of set theory.

From 1899 onwards Schoenflies had written a series of papers with the intention of providing a systematic treatment of the basic concepts that were beyond Cantor's and Jordan's work. In particular his *Contributions to the theory of point sets I, II, III*[66] are clear, easily readable expositions. The great survey of 1908, *The development of the theory of point sets, II* collected all hitherto published material in one volume.

Schoenflies formulated his goal in the first of the above Contributions:

> As one of the most general problems of the theory of point sets, we can point to the task of formulating and establishing set-theoretically the fundamental theorems of analysis situs, and setting forth the relationships which exist between the set–theoretic–geometric and the analytic modes of expressing these concepts and theorems. The paradoxical results, as they occur, for example in the one–one mapping of continua and in the Peano curve, have completely destroyed the naive ideas of analysis situs.

---

[64] There were other texts, for example, [Young–Young 1906] but they had only a modest influence.
[65] [Zermelo 1908], [Schoenflies 1905].
[66] [Schoenflies 1903, Schoenflies 1904, Schoenflies 1906].

All the more, we must demand that set theory provide a substitute and define the basic geometrical concepts in a way that returns to them their natural content, characteristic of analysis situs. Even if the much maligned intuition, is no source of proof, it still seems to me that it is a goal of research to reconcile the content of geometrical definitions with the content of intuition—at least in the domain of analysis situs.

As to the proper subject of research, Schoenflies was quite clear. Following Klein's Erlangen Program, he stated, in justifying his new definition of the notion 'connected' that Analysis situs can be considered as the science which studies notions invariant under univalent and continuous mappings (obviously he had topological mappings in mind).

The *Beiträge*-papers deal with the topology of the plane, and Schoenflies treats the major topics of the day, for example properties of plane curves, polygons, the structure of perfect sets, Jordan curves.

It is not exactly known when Brouwer started to study Schoenflies' topology, but both the topics of topological mappings of surfaces and of vector fields were presupposing a lot of Schoenflies' material. The earliest mention of Schoenflies is in Brouwer's 1908 Rome talk,[67] where he indeed prided himself on being the first person to do something useful with topology:

> Only the groups of the two-dimensional manifold can be determined with the help of the results so far of Schoenflies, and I believe that thereby for the first time an application of these results is given.

Clearly Brouwer was not slow to absorb Schoenflies' papers; in his dissertation he referred to the first volume of the *Bericht* and he had read and digested the 1908 survey of Schoenflies in the year of its appearance; and so it is fair to say that he knew the latest in topology (but, as we have seen, by no means all the classical papers).

At the meeting of the Dutch Mathematical Society of 30 October 1908, Brouwer gave a talk with the title *On plane curves and plane domains*.[68] Although the manuscript with the same title in the Brouwer archive is undated, there is little doubt that it is the text of the talk, because Brouwer refers to the Schoenflies proof of the invariance of the Jordan curve 'earlier this year'. Brouwer presented an exposition of the various notions involved in plane topology, such as Peano curve, closed curve, Jordan curve, domain, approximating polygon, and accessibility. He pointed out the mistakes in Schoenflies' recent proof and gave a proof of the Jordan–Schoenflies theorem.

The text shows that Brouwer had learned his plane topology, and also that in ease and insight he was second to none.

His investigations of topological maps of spheres and vector fields had brought home to him the questionable state of Schoenflies' foundation of topology, so he

---

[67] [Brouwer 1909B].
[68] *Over vlakke krommen en vlakke gebieden*.

set himself the task of getting the topic back on its feet before proceeding to the topics that were really attracting him. In a letter to Hilbert, he gave the motivation for his investigation of the basic notions of topology:[69]

> To regain clarity, it was first of all necessary, to check thoroughly the theory of Schoenflies involved here, and to ascertain precisely, on which of its results one can build in full faith. This was the beginning of the enclosed paper, which has serious consequences for various parts of the theory of Schoenflies, and more or less recreates some parts.
>
> It seemed fitting to me that, if possible, it should be published at the same place, where Mr. Schoenflies had originally published his paper, therefore I send it to the editorial board of the *Mathematische Annalen* for publication, and at the same time I will send a copy to Mr. Schoenflies.

The paper that Brouwer sent to Hilbert, was his pioneering *Zur Analysis Situs*,[70] which changed the landscape of elementary topology. It consists of two parts, the first one contains counterexamples to a number of Schoenflies' statements and the second one redevelops some parts of the topology involved. Apart from its scientific value, it has a certain curiosity value, for it is the only paper in the *Mathematische Annalen* with coloured illustrations! These illustrations were, by the way, a source of worries to the printers, the editor and to the author; they were discussed repeatedly.

Although Brouwer clearly was annoyed at having relied on doubtful material, he went out of his way not to affront Schoenflies—after all he still was a newcomer without a position in the academic world. After stating in the paper that 'several of his [that is Schoenflies'] results were false, and others, 'indeed correct, but insufficiently supported', he cautiously added in a footnote: 'I stress specifically that this paper does not try to diminish in any way the high value of the researches of Schoenflies'.

Some of the paper's counterexamples to claims or theorems of Schoenflies are still relevant, but the evolution of the basic concepts in topology has made some others less interesting. Nonetheless, one should keep in mind that Brouwer's counterexamples themselves played an important role in the cleaning up of the foundations of topology.

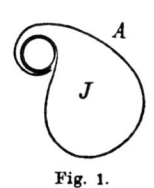

Fig. 1.

For instance, in the *Bericht* Schoenflies stated that 'One easily recognizes the correctness of the following theorem: VI. The boundary of a domain (*Gebietsgrenze*) is a set which is closed [...]'.[71] Brouwer's counter example is simple enough to please, even to-day a beginning student: The domain, J, 'is obtained if one continues a simple curve arc to both sides, and wraps both extensions, without letting them meet, like a spiral around one and the same circle.'

---

[69] Brouwer to Hilbert 14 May 1909.
[70] [Brouwer 1910C].
[71] A domain is a connected open set.[Schoenflies 1908], p. 112.

The best-known construction of the paper is that of a curve which splits a square into three domains. That particular example gave rise to an intensive study of so-called indecomposable continua[72] (i.e. continua that cannot be cut into two proper sub-continua), a tradition in topology that reached well into the forties.

To appreciate the difficulty of the problem, it will be helpful to introduce a few notions. The simplest kind of curve we know is a line segment, and since we are doing topology, it seems plausible to call the result of bending and stretching still a curve. If the result is not self-intersecting, we speak of a simple open curve. In precise terms: a simple open curve is the homeomorphic (or topological) image of a line segment. A closed curve is the homeomorphic image of a circle. Now Jordan's theorem tells us that a closed curve (or a Jordan curve) splits the plane in two parts, the interior (which is bounded in size) and the exterior. Both the interior and the exterior are connected, and one cannot go from a point of the exterior to a point of the interior without crossing the curve, which is the common boundary of both parts. The converse of this theorem, established by Schoenflies tells us that if a closed bounded subset C of the plane determines two open domains, such that each point of C is accessible from both domains, it is a Jordan curve. So it looks as if closed curves and a splitting in two parts are hand in glove. Schoenflies used the property of the above set, C, as the definition of a closed curve. One can imagine the general surprise when Brouwer showed that there are closed common boundaries to *three* domains. It just shows that our normal geometric intuition has a lot to learn!

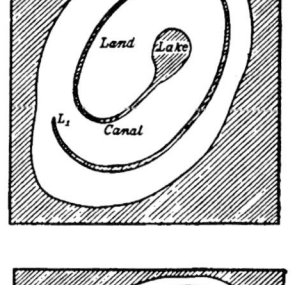

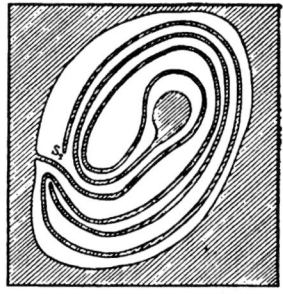

Brouwer's construction was given an amusing form by Wada,[73] who visualized the procedure as follows: consider in a saltwater sea an island with a

Wada's island and canals.

freshwater lake. One starts digging canals in a systematic way such that alternatingly one extends the fresh water system and the salt water system:

On the first day a canal is dug starting at the lake, and not meeting the sea, such that the distance from the points of the shore to the fresh water supply is $< \frac{1}{2}k$ (where k is some unit of measure).

On the second day, a canal is dug from the sea, not mixing with fresh water, so that the distance from all points at the border of the lake or the first canal to the salt water is $< \frac{1}{4}k$. On the third day the digging is resumed at the end of the fresh water canal and an extension is made such that the distance from the shore and the

---

[72] A continuum is a compact connected set with at least two points.
[73] [Yoneyama 1917].

borders of the salt water canal to the extended first canal is $< \frac{1}{8}k$, etc.

This process is continued countably many times; the result is (eventually) a curve which separates the fresh and the salt water. One can also start with two lakes, say with blue and red water). The result then is one curve that separates three 'domains' of water.

The final part of Brouwer's paper contained a critical examination of the main topological notions and their properties as treated in the last chapter; notions such as *domain, accessibility, curve*, etc., *connectedness, Jordan curve*, etc., and indications how to repair the lacunae.

The paper closed with a brief catalogue of false statements followed by two claimed, but not satisfactorily proved facts, including the elusive invariance of the closed curve.

The last mentioned problem was settled by Brouwer himself in a later paper.[74] He had, as a matter of fact, already mentioned this open problem in his inaugural address of 1909. The 'Analysis Situs'-paper did not only mark the end of a period in point-set topology (aptly called *Cantor–Schoenflies topology* by Freudenthal), but it was also a turning point in Brouwer's career. It drew attention to a promising, ambitious mathematician, who could use some support in his attempts to find a position in the Dutch academic world.

Brouwer was, deservedly so, proud of the above mentioned exploits in set theoretical topology (in the plane). An extra bonus was the attention paid by the incumbent grand master of topology, Schoenflies. It definitely had the pleasant effect of strengthening Brouwer's position on the job market (see p. 209). Schoenflies accepted the collapse of his Magnum Opus with grace, albeit reluctantly. He insisted to publish his comments in the same volume of the *Annalen*, following Brouwer's paper.[75]

The correspondence about the gaps in Schoenflies' *Bericht*, of which only Schoenflies' letters have been preserved, shows a confused and hurt Schoenflies.

It is painfully clear that he had not seen the pitfalls of his own speciality, and even after Brouwer's observations, he remained convinced that only marginal corrections would do.

In his he letters he tries to save as much as possible, and to minimize the impact of Brouwer's criticism. The first letter, 27 May 1909, from Schoenflies, is still rather optimistic about the extent of the damage. After thanking Brouwer for the copy of the manuscript submitted to the *Annalen*, he gave vent to his disappointment, 'My delight that you have carried the study of my *Bericht* to such depths is, alas, not without a bitter flavour.'

He admitted that a number of geometrical shapes of curves, under his definition as 'boundary of domain', had escaped him. He then went on to fill the remainder of the letter with corrections and 'refutations' of Brouwer's criticism.

But this was just the beginning; letter after letter was exchanged—without much

---

[74] [Brouwer 1912L].
[75] *Bemerkung zu dem vorstehenden Aufsatz des Herrn L.E.J. Brouwer.*

effect. Schoenflies had great difficulty in following Brouwer's arguments. Apparently Brouwer soon lost his patience, for we find in the letters of Schoenflies signs of exasperation: 'Would it then really be impossible for me to convince you in writing?' (14 August 1909); 'It pains me, so to speak, that you could think, that I mean by the outer boundary of a domain, what you mean by it. I have given you no cause for a thing like that. Moreover I had believed, on the basis of all my papers in this area, to be protected from that!' (12 September 1909)

After this letter Brouwer must have lost control of his patience altogether, for Schoenflies opened his next letter with the words 'To my strong disappointment, I must start to say that I must decidedly insist to be spared letters of the kind of your last letter. You have neither reason nor occasion for that' (13 December 1909).

By now the first couple of revisions of the proofs had already been handled by the *Annalen*; it became rather embarrassing to go on much longer. Nonetheless, Brouwer must have protested once more, as can be inferred from Schoenflies' reply (19 December 1909): 'If you think that you "in a sense have the right to demand the deletion of the words indicated by you", and furthermore, that I had acted against our agreement, I must defend myself most emphatically. I can in no way acknowledge this'.

The exchange stopped essentially here, and so Schoenflies' reaction to Brouwer's paper was in final form roughly seven months after Brouwer submitted his paper.

Although Brouwer had cleared the troubled waters of Schoenflies' topological work, the two reluctant companions were not through yet. Destiny would bring them together once more.[76]

After this 'animated' correspondence, the harmony between Schoenflies and Brouwer was restored through Hilbert's mediation. In the end the two authors agreed to send out combined reprints. (see also p. 210).

Brouwer continued to work in this area, although it had to give way to more challenging problems when Brouwer discovered the new topology, described in the next chapter. Among other things, there is a lovely, elegant proof of the Jordan Curve theorem,[77] Brouwer's proof was adopted by Hausdorff in his *Grundzüge der Mengenlehre*. In 1923 Erhard Schmidt also presented an elementary proof, which could compete in elegance with Brouwer's proof. Schmidt's proof is completely elementary and uses only the tools of point set topology. This paper also contains the theorem that subsequently became known as the 'Phragmen–Brouwer theorem', which states that the frontier of a connected component of the complement of compact connected set is connected. The terminology is poorly chosen; Phragmen's only paper in topology contains just the theorem that a closed point set with no connected subset, does not split the plane.[78]

---

[76] Cf. p. 230.
[77] [Brouwer 1910E].
[78] [Phragmén 1885], cf. Freudenthal, CW II, p. 383.

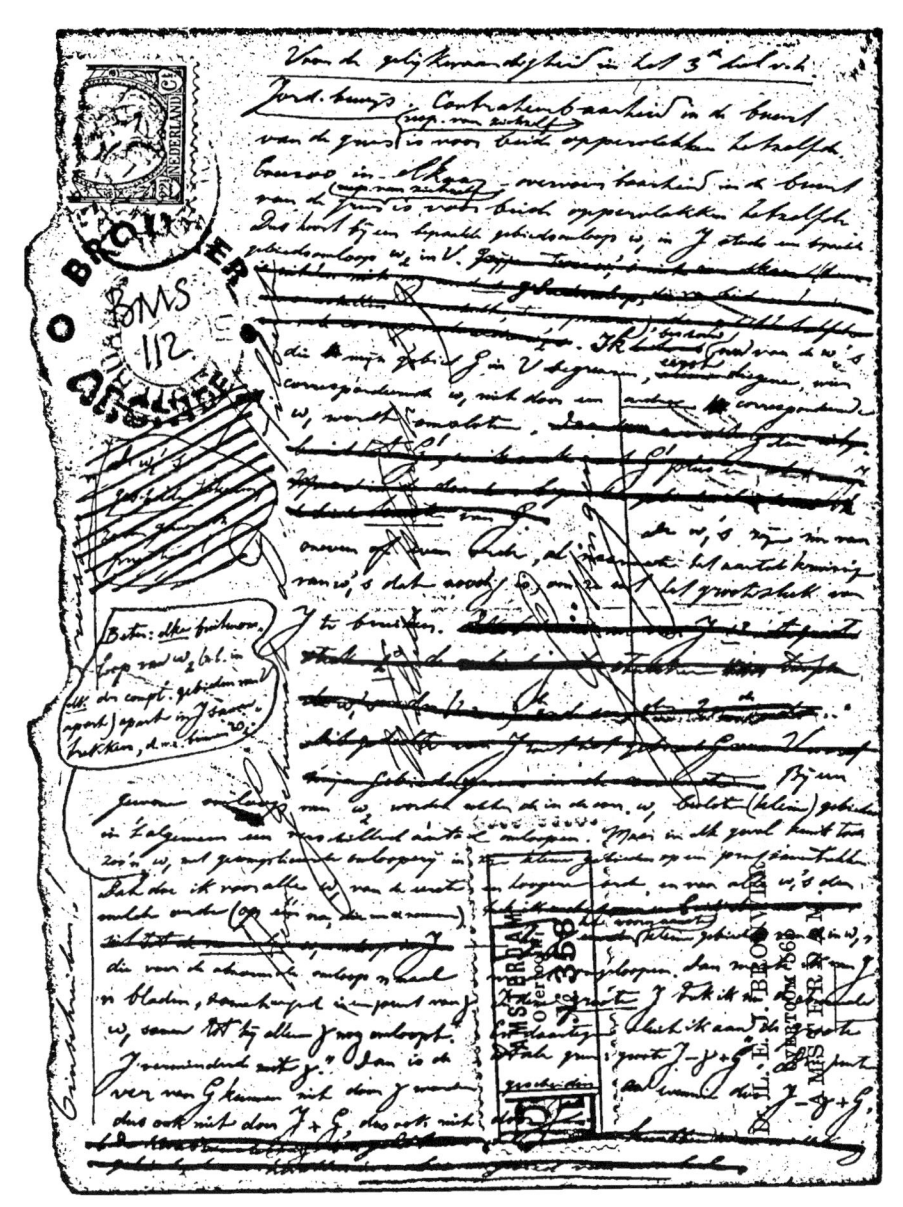

Topological notes on the cover of a letter.

The modern reader would prove the Phragmen–Brouwer theorem, like much of Brouwer's topological work, using homology.[79]

[79] Cf. [Dieudonné 1989], p. 207.

The revised version of the proof of the Jordan curve theorem was sent to Hilbert on the fifteenth of October 1909, with a covering letter that neatly summed up Brouwer's research intentions of the moment:

Sehr geehrter Herr Geheimrat,

Enclosed, I send you once more my proof of the curve theorem, and I hope very much that you find the presentation satisfactory.

Allow me to point out to you that we have no certainty about the measure of validity of the theorem *that the one–one continuous image of a closed curve again is a closed curve*,[80] and that there seems to be here a really difficult problem. (By a 'closed curve' we mean a point-set, which determines two domains in the plane, and which is identical with the boundary of both of these domains.)

There neither is a proof, that the one–one continuous image of a sphere determines two and only two domains in space.[81] Nobody, anyway, doubts the correctness of this claim.

Recently I found that, if one asks for those continuous transformations of the sphere into itself, which are univalent in one direction and *at most* two-fold in the other direction, such a transformation is always a one–one continuous image of the function $x' = x^2$. This gives me hope, that the whole Riemann theory of algebraic functions can be founded on Analysis Situs.[82]

This might perhaps even be the starting point for the theory of analytic functions, if one asks for such correspondences, which are built up from domainwise one–one and continuous correspondences.

Singular points, then, are those, for which no neighbourhood exists, in which the correspondence is one–one and continuous. Anyway, here a mass of problems belonging to Analysis Situs arises: what can the nature of the sets of singular points be, and what kind of singularity can there be in each point; which point sets, in which only the continuity is known, can one admit, so that the non-singularity of the points of these sets may be inferred from the continuity; and in the first place, whether one can correlate each of these correspondences to an analytic function.

Thus I return once more to the question you brought up in our discussion,[83] namely the behaviour of an analytic function in the neighbourhood of a nowhere dense perfect point set μ of exceptional points. Some of that I find treated in Pompéiu: 'On the continuity of functions with complex variables', *Annales de Toulouse* 1905. He proves among other things that, when the function is uniformly continuous, and the set μ has measure zero, continuity suffices to ensure analyticity, but that when the mea-

---

[80] Proved in [Brouwer 1912F, 1912L].
[81] Proved in [Brouwer 1911F].
[82] [Brouwer 1919G, 191N2, 1919S].
[83] Scheveningen, September 1909, cf. p. 128.

sure of μ is not zero, the function can very well be continuous in those points, but not analytic.

He also introduces a new notion, the 'length' of a set, and derives on the basis of this notion a second criterion.

Allow me to remind you of your promise, to have reprints of yours (on the Dirichlet principle, integral equations and the Obituary of Minkowski) and of Mr. Zermelo (Foundations of set theory) and Koebe (uniformization) sent to me? I express in advance many thanks.

My wife sends you warm greetings. In the hope of meeting again, with many greetings for *die liebe Frau Geheimrat*, whose recovery we would be very pleased to learn.

Yours truly

L.E.J. Brouwer.

Indeed, the coming years would show that Brouwer was as good as his word—topology would never be the same after the eventful years preceding World War I.

Brouwer investigated a number of topics in the Cantor –Schoenflies tradition, for example, of the type of the Cantor–Bendixson theorem (which was called 'Cantor's fundamental theorem' at the time). He generalized the theorem in various directions, in particular he proved a generalization by replacing the role of points in Cantor's theorem by closed components of a set.[84] In the 1911 paper Brouwer formulates and proves his *reduction theorem*, which he used to provide simple proofs of earlier theorems (including the theorem of Janiszewski and Mazurkiewics on irreducible continua). The topology papers contain many little gems that earned the author a well-deserved reputation as an original and ingenious topologist, slowly Brouwer's innovations have been absorbed in more general treatment, and the modern topologist would hardly be aware of their parentage. To mention just one case: in [Brouwer 1910B2] the first example of a topological group which was not a Lie group is given. The example is an early (perhaps the first) limit construction (in the categorical sense).

There is a hitherto unknown sequel to Brouwer's activity in the more traditional topology: in 1912 the mathematical world was sadly shocked by the death of Poincaré. After an operation, which was by itself successful, the great man had died from an embolism, only fifty nine years old. Shortly before his death his last paper appeared in the Rendicotti; on 9 December he had sent it to the editor with the comment that—contrary to his custom—he submitted an unfinished paper on a problem which he considered of the highest importance. This paper contained a statement, which has become known as 'Poincaré's last theorem':

> A measure preserving homeomorphism of a circular ring in the plane unto itself, so that the inner circle is rotated clockwise and the outer circle anticlockwise, has fixed points.[85]

[84][Brouwer 1910B2, 1911B2].
[85][Poincaré 1912].

Poincaré's theorem (or conjecture at this stage) would, of course, appeal to Brouwer, who had proved a number of fixed points on surfaces in the past years. Brouwer immediately set to work at the problem and convinced himself of the solution; there are a number of notes in his archive in which he attacked the proof. There are even a few sheets of a paper which he intended to present to the Academy. In a letter of 31 September 1912, however, he wrote Korteweg that 'It turns out that the solution of the Poincaré problem, before it is written in a matured form, it will take many more weeks. Before that time I would rather keep it under me; please don't discuss it with anybody. Only after I have finished the complete version, I will present an outline to the Academy.' There is no record of any publication of Brouwer on the topic. The many obligations that Brouwer had accumulated started to take their toll. In December 1913 he complained in a letter to Korteweg that he lived under a constant pressure on account of all kinds of obligations to foreign mathematicians, he had not been able to carry on his research which 'is resting already for one and a half year.' Brouwer had mastered the baffling problems of topology, but not the art of saying 'no'. In his capacity of the leading topologist his advice and judgement was all too often sought and given. The Poincaré problem was solved and published by Birkhoff in 1913. Brouwer reacted by sending Birkhoff a postcard with congratulations on the 'magnificent proof of Poincaré's geometric theorem', addressing the issue of a second fixed point according to Poincaré and Birkhoff.

5

THE NEW TOPOLOGY

5.1 Invariance of dimension

Although mathematics had been an exciting human enterprize at all times, the nineteenth century was particularly rich with excitement, often bordering on shock.[1] On numerous occasions old beliefs were shattered by the improved critical spirit of the century. The already heavily tried mathematical intuition underwent another traumatic experience in 1878, when a short paper appeared in which a professor at Halle proved an incredible result: there is a one-one mapping from the side of a square to the square itself (or from $\mathbb{R}$ to $\mathbb{R}^2$).

The author, Georg Cantor, had more surprises in store for the mathematical world, but this result, hidden under the nondescript title 'A contribution to set theory',[2] was a straightforward attack at common geometric intuition: 'one dimension is essentially different from two', or 'two independent variables cannot be reduced to one'. Given the ingrained intuitions and habits of traditional geometry there was enough reason for surprise at this new result.[3]

The result of Cantor seems to show that dimension is not the immutable rock it was always supposed to be. No mathematician prior to Cantor would have expected to fill up the square with a line, or to unwind the square as a line, like one unpicks a sweater. So was this the end of dimension as mathematics knew it? Dedekind tempered Cantor's excitement by pointing out that the bijection that Cantor had exhibited was highly discontinuous, and that the moral of Cantor's theorem should not be overestimated. Dimension would be in serious jeopardy if Cantor's mapping were continuous—if, roughly speaking, one could fill a square with a pencil without taking it from the paper and without passing through a point twice.

The mathematical community was rather inclined to keep dimension on its throne, and so it was generally conjectured that there could not be a bijection from the line to the plane, continuous in both directions.

Even stronger, Cantor's contemporaries would have been inclined to bet that one could not even fill up the square with a continuous curve that would be allowed to intersect itself (that is the pencil might pass through the same point more than once). The general feeling of mathematicians was that our ordinary lines, surfaces,

---

[1] This chapter makes essential use of Freudenthal's comments in Volume II of the Collected Works and of the paper *The Problem of the Invariance of Dimension in the Growth of Modern Topology I, II* of Dale Johnson.

[2] *Ein Beitrag zur Mannigfaltigkeitslehre*

[3] The history of this particular theorem is published by Emmy Noether and Jean Cavaillès, [Noether, E. 1937].

solids, etc., are continuous in nature, and hence a comparison as to their dimensions should take this property into account. So far this was an intuitive conviction; one of the tasks of the new discipline 'topology' was to make these ideas precise. The ultimate outcome was that Cantor's '*Gleichmächtig*' ('equivalent', 'of the same cardinality') was too general and arbitrary to reflect dimension faithfully.

Almost immediately after Cantor's paper appeared, efforts to save the invariance of dimension (and hence dimension itself) were made. Among the mathematicians who attacked the problem were Lüroth, Thomae, Jürgens, Netto and Cantor himself.

For an extensive discussion of the history of dimension the reader is referred to D.M. Johnson's magnificent papers of 1979 and 1981; for the present it suffices to record the achievements of Lüroth, who proved that one cannot give a topological mapping of a one, two or three dimensional space onto a higher dimensional one (1878, 1899). None of the other attempts succeeded. The proof for the three dimensional case was, however, so complicated, that everything seemed to point to the inevitable conclusion that, either the general problem was so hard that it ranked among the great unsolved problems, or that it required a totally new approach.

The majority of the practising mathematicians, however, was not disturbed by the state of affairs, and they were assured by no less a person than Schoenflies, in his chapter in the *Encyklopädie der Mathematischen Wissenschaften I*, 1898, that the problem had been settled by Netto and Cantor: a one-one correspondence between distinct R_n and R_m is never continuous. One wonders if Hilbert ignored the problem of the invariance of dimension on the authority of Schoenflies, or that he dismissed it as 'just another problem'; whatever his reasons were, the problem does not occur in his famous list of 1900. This may have been a matter of expediency, but it was certainly not the consequence of a lack of importance, as Brouwer's remark in his Rome talk shows (see below).

Another aspect of the same problem was discovered by Giuseppe Peano, the man who had created a symbolic language for mathematics, and who had proved the best existence theorem, so far, for differential equations. He had exhibited a continuous mapping from the segment $[0, 1]$ onto the square $[0, 1] \times [0, 1]$ in a three-page note, 'On a curve which fills a plane area'(1890). In plain words, one can fill in the whole square with a pencil without taking it from the paper.[4] Peano's example was followed by more, so-called, *space-filling curves*.[5] All these examples added to the urgent feeling of inadequacy of plain geometric intuition.

Brouwer was aware of the significance of the problem of the invariance of dimension at the latest in 1908, when he gave his address at the Rome conference; after defining Lie groups, he added in a footnote:[6]

---

[4]There is a fairly simple geometric representation of the above mapping, [Johnson 1979], p. 171 and [Young–Young 1906] p. 165, 291.

[5]Among others Hilbert and Moore, cf. [Young–Young 1906].

[6][Brouwer 1909B], p. 297.

Whether p [that is the number of parameters] is an invariant for each group, is an open problem as long as the 'non-applicability' of two spaces with distinct dimension numbers, is unproven.

There is a number of sources for Brouwer's research plans in this period, among them is the inaugural lecture *The nature of geometry*,[7] which Brouwer gave on 12 October 1909 at the Amsterdam University at the occasion of his appointment as a *privaat docent*. It contained his views on geometry, including topology. Brouwer's idea of geometry was closer to the Riemann tradition than to the Euclid-Hilbert tradition. The influence of Klein's Erlangen Program is evident:

> Geometry is concerned with the properties of spaces of one or more dimensions. In particular, it investigates and classifies sets, transformations and transformation groups in these spaces.[8]

He succinctly summarized the role of the transformation groups.

> Finally, it can often be shown that figures and operations with which we became acquainted in the smaller group, can be completely defined by properties invariant for the larger group; they can then be *more generally characterized*, though perhaps it may be useful for special questions afterwards to consider the smaller group. An example is supplied by the potential functions in two dimensions, which were at first characterized in Euclidean geometry. After a point at infinity had been added, and the conformal group had been introduced, it could be shown that such a function is completely defined by its invariants for the conformal group, and that the Euclidean group is inessential for it.[9]

In plain words, a larger group may solve some of the riddles that the small group leaves unexplained. The phenomenon was well-known through the Cayley–Klein approach to projective and non-Euclidean geometry.

It was in this vein that Brouwer studied geometry, and it also was his approach to group theory. Whereas the topic, properly speaking, is a part of algebra, Brouwer looked at groups with the eyes of a geometer.

The above considerations are general in nature, but they reflect Brouwer's own research activity. Of late he had been investigating groups larger than the traditionally considered groups, namely groups of topological transformations, that is of those one-one transformations that are continuous (preserve limits, in the intuitive approach) in both directions. As we have seen he strove for methods of logical simplicity, shunning extraneous concepts;[10] in particular he stripped away (or tried to do so) differentiability conditions whenever they were not essential, thus carrying

---

[7] *Het wezen der meetkunde*. The translation of 'wezen' by 'nature' is somewhat flat. *Wezen* expresses something more, like 'essence'.

[8] [Brouwer 1909A], CW I, p. 116.

[9] [Brouwer 1909A], CW I. p. 117.

[10] Not to be confused with *Methodenreinheit* (purity of methods) of older generations. Brouwer was quite prepared to use whatever means available.

Hilbert's wishes, as expressed in the fifth mathematical problem of 1900, to new areas.

The reader, who is used to modern topology, will have to keep in mind that the terminology took a long time in establishing itself. A lot of notions have gone through a long evolution, causing a good deal of confusion on the way. One such notion is 'connected'; as we will see later, Brouwer got into trouble over the interpretation of this notion. Likewise, he used 'one-one continuous' for what we now call a 'homeomorphism' or 'topological mapping'.[11]

The 1909 inaugural lecture is a most helpful document; it gives us a glimpse of Brouwer's private research program for topology. While presenting a survey of 'modern geometry', it lists a number of key problems:

- 'investigation of the general character of a system of several one-one continuous transformations [of a set] into itself'.
- 'a classification for analysis situs of the sets of points in a space from the viewpoint of analysis situs'.
- 'An immediately related problem is, to what extent spaces of different dimension are distinct for our group [of topological mappings]. Most probably this is always the case, but it seems extremely hard to prove, and probably will remain an unsolved problem for a long time to come.'
- 'in spite of many efforts no satisfactory proof has as yet been given for the seemingly very slight extension of Jordan's theorem, that the one-to-one continuous image of a closed curve is again a closed curve.'—'No more are we certain that a closed Jordan surface, that is a one-to-one continuous image of a spherical surface, splits the three-dimensional Cartesian space into two domains.'

A number of the above topics would soon be successfully attacked by Brouwer, although he apparently was at that time far from certain to succeed where others had failed. In particular he was rather aloof as regards the invariance of dimension. Nonetheless, within half a year the situation changed dramatically. Not only did Brouwer solve the problem of the invariance of dimension, but he enriched topology with (at least) two tools that enabled him to attack a whole range of problems.

The birth of the new topology has been recounted in Freudenthal's '*The cradle of Modern Topology, according to Brouwer's Inedita*'.[12] The first recorded evidence of the new ideas was found by Freudenthal in an exercise book of Brouwer, that had a label *Potentiaaltheorie en Vectoranalyse*, when he was editing the topological part of the Collected Works. The editing had already progressed to the final stage, when this notebook came into his hands; its title did not spell any surprising revelations, but when he opened it, two drafts of letters fell out. One was the draft of a letter to Hilbert, dated 1 January 1910, and the other of a letter to Hadamard, dated 4

---

[11] The first term was introduced by Poincaré (1885), the second one by Brouwer, [Brouwer 1919H].
[12] [Freudenthal 1975], cf. CW II p. 422 ff.

January 1910. The letter was written during the Christmas holidays,[13] spent with his brother Aldert in Paris. It shows that Brouwer had already developed two of his new methods that appeared in print only in 1911: the *degree of a mapping* and *simplicial approximation*. The letter to Hilbert ran as follows:

<blockquote>

Paris, New Year's morning 1910.

Dear *Herr Geheimrat*,

Best wishes for you and your dear wife for the New Year, for your health and for your scientific activity.

I am staying here for the Christmas holidays with my brother, the geologist; unfortunately my wife could not accompany me. In the middle of January my classes start again and I will return.

The harmony with Mr. Schoenflies has been re-established, and mainly through your intervention. I am enclosing the last two letters to him, which I have answered by saying that I am satisfied with his last version, and that I consider the matter closed.[14]

May I make a few comments on the univalent (not necessarily one-one) continuous mapping of a sphere κ onto a sphere λ? If one subjects it to the condition that it should be continuous both ways, then it is a one-one continuous image of a rational function of a complex variable [...] By this condition of the two-way continuity I mean that a closed Jordan curve around a point, L of λ, which converges to L, corresponds, for each point, K of κ, of which L is the image, to a closed Jordan curve around K that must converge to K.

Now, if we have two univalent mappings of a sphere (or a more general closed surface) K, satisfying these conditions, onto a sphere, L, and a sphere, M, then the question is raised what extra conditions should be satisfied so that the correspondence between L and M is a complex algebraic one in the sense of Analysis Situs.

If I return again to the general univalent and continuous correspondence between two spheres, then a finite number n can be indicated as its degree such that all correspondences of the same degree, and only these, can be transformed continuously into each other. In particular all correspondences of degree n can be continuously transformed into rational functions of a complex variable of degree n.

In order to define this degree, we introduce homogeneous co-ordinates x, y, z on κ and homogeneous co-ordinates ξ, η, ζ on λ, and consider first

</blockquote>

---

[13] CW II, p. 421, see also Brouwer to Scheltema 3 December 1909.
[14] Concerning the 'Analysis Situs'-paper, cf. p. 146.

the univalent mapping, which is determined for each domain by a correspondence

$$\xi : \eta : \zeta = f_1(x,y,z) : f_2(x,y,z) : f_3(x,y,z),$$

where f_1, f_2, f_3 are polynomials.
If we fix a positive orientation on both spheres, and choose in each point of κ this positive orientation, then each generic point of λ occurs p times with positive orientation and q times with negative orientation. One can show that $p - q$ is a constant for all generic points, which we call the degree of the mapping.
If the correspondence between x, y, z and ξ, η, ζ is not determined by polynomials, then one can approximate them by such polynomials, and one easily shows that those approximating correspondences have a constant degree, which we therefore also assign to the limiting correspondence. This degree is always a finite, positive or negative number. In particular the one-one continuous mapping of the sphere into itself has degree +1 if the orientation is not changed, −1 otherwise.

Now you know my theorem that every *one-one* continuous transformation of the sphere into itself which does *not* change the orientation, always has at least one fixed point. This theorem can be extended in such a way, that every *univalent* continuous mapping of the sphere into itself which does not have degree −1, always has at least one fixed point.

And I succeeded in extending the theorem to n-dimensional spheres. Then it reads: every univalent continuous transformation of the n-dimensional sphere into itself has at least one fixed point. The transformations of degree +1 form an exception for odd n, and those of degree −1 for even n. On the one hand for odd n and inverted orientation, on the other hand for even n and unchanged orientation, one-one transformations therefore necessarily have a fixed point Even more general is the result for univalent continuous mappings of the solid sphere into itself, for these have a fixed point anyway.

Once again best wishes for both of you,
as ever yours faithfully,

L.E.J. Brouwer.

This letter documents a spectacular breakthrough of Brouwer, soon after his exchange of letters with Hadamard. The work on vector fields, that preceded the discovery of the degree of a mapping, had brought Brouwer close to earlier work of Poincaré, and Hadamard had pointed out to him the relevance of Poincaré's investigations, including that of the indicatrix (introduced by Leopold Kronecker, 1869, and probably known to Gauss as early as 1840). The bold idea of stripping away the analytical notions, had, roughly speaking,

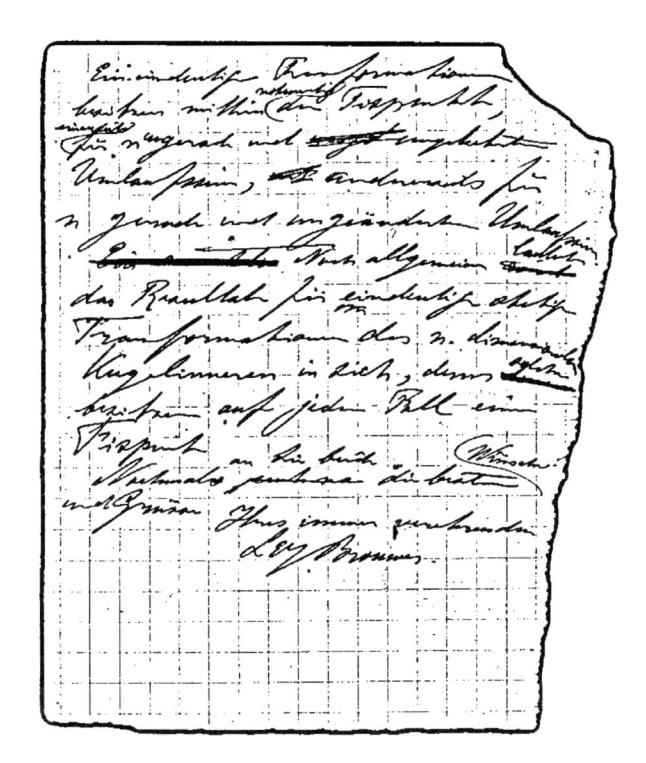

Draft of the letter to
Hilbert, 1 January 1919.

occurred in the short interval between his letter to Hadamard of 24 December and 1 January. There is a letter with a similar content to Hadamard (4 January 1910). Freudenthal, in his comments in the Collected Works II (p. 422, ff.) points out that the excercise book and the letters to Hilbert and Hadamard provide invaluable information on the birth of the mapping degree, and on the role of Hadamard.

> What has now also become clearer is Hadamard's role as midwife. Before I had never understood how Hadamard figured in this story and why Brouwer regarded him so highly in this context. Hadamard's work in this part of topology, Hadamard 1910, looks rather old-fashioned. Its strong dependence on analytic tools would have hampered rather than stimulated true topology. It was Brouwer's achievement to have shaken off the yoke of analysis from topology. In the genesis and maturation of his ideas, however, his intercourse with Hadamard must have meant more to Brouwer than can be expressed by mere citations and quotations.[15]

Hadamard's influence is acknowledged by Brouwer in his third paper on vector distributions. Hadamard, on his part, in his Appendix *On some applications of the Kronecker index*[16], gives the credit for the method used in this paper to Brouwer. As Freudenthal put it: According to [the draft] Y2 he learned the deeper reason for

---

[15] Freudenthal in CW II, p. 425.
[16] [Hadamard 1910].

it [the connection between fixed points on the sphere and singularities of vector distributions] in the correspondence with Hadamard, which does not mean that Hadamard revealed it to him'.

Freudenthal's discovery corrected the plausible but false impression that Brouwer first solved the problem of the invariance of dimension, and subsequently discovered the notion of 'mapping degree', possibly distilled from the invariance proof. The above letter provides convincing evidence that the mapping degree came first—the letter does not even mention the dimension problem as such! So, why did Brouwer publish his Invariance of Dimension-paper first? Whatever his reasons were, it was not a bad idea to start the overture to the new topology with a resounding trumpet blast. What better way is there to attract attention, than to solve a famous open problem?

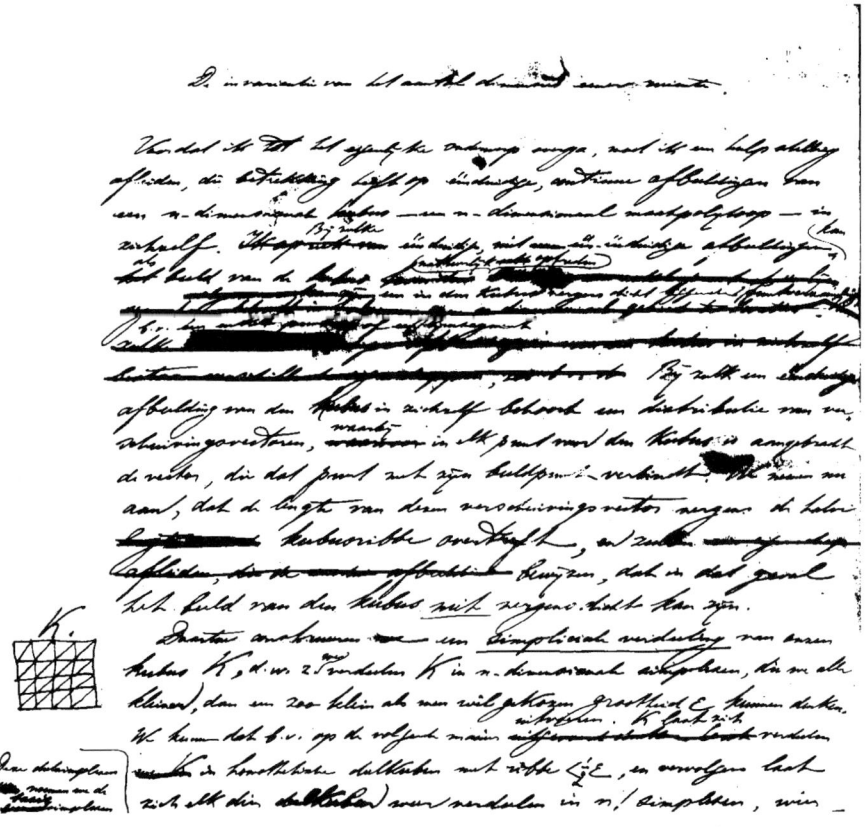

Page of Brouwer's lecture for the Dutch Mathematics Society on the invariance of dimension.

The solution of the invariance of dimension-problem, as a matter of fact, made use of the mapping degree, without introducing the concept by name. The paper was submitted before the summer vacation of 1910, probably before July, and it seems likely that Brouwer submitted it directly to Hilbert, who then passed the manuscript on to Blumenthal, the managing editor of the *Annalen* to process it for publication. A word of warning may, however, be in order. Brouwer's research was of a rich variety during the hectic years of 1909 and 1910. It is not certain when Brouwer started to investigate the invariance of dimension, but in a letter of 18 March 1910 he wrote to Hilbert that he was preparing a new paper for the *Mathematische Annalen* in which he had partly solved the invariance of dimension, in so far as he had shown that spaces of even and odd dimension are not homeomorphic. The paper was never published, and no drafts have survived, so we don't know what Brouwer's approach was in this particular proof. In all likelihood he found the complete solution before he had finished the manuscript of the partial one.

The manuscript of the invariance of dimension paper was sent to the *Mathematische Annalen* in June 1910, and the paper appeared in the issue that was dated 14 February 1911.

Although the material of the 'invariance of dimension', was rather out of the way for his Dutch fellow mathematicians, Brouwer presented a complete proof in a beautiful didactic form, at the October meeting of 1910 of the Dutch Mathematics Society. The manuscript, which has been preserved, shows us a clear exposition that could nowadays easily be used in an undergraduate course.[17]. The reactions to Brouwer's talk are not known, but it seems plausible that it conveniently raised the status of the young *privaat docent*, whose promotion to lecturer had just been turned down by the local authorities.[18]

Between the submission of the dimension-paper and its publication, something happened that added a tragic and dramatic note to the history of the invariance of dimension.[19]

Blumenthal had made a trip to Paris in the summer vacation of 1910, where he met Henri Lebesgue. The latter was a prestigious mathematician, who, among other things, created modern measure theory; the Lebesgue measure and integral have been household words in mathematics ever since their introduction. Lebesgue contributed to many areas in mathematics, in particular also to the young discipline of topology, which, as a matter of fact, touched on measure theory in quite a number of points. Blumenthal, aware of the importance of the topic, told Lebesgue about Brouwer's exploit. In a letter of 27 October 1910 Blumenthal informed Hilbert of his meeting with Lebesgue:

> We made a very nice trip to Paris during the summer holiday. Unfortunately, however, I did not see any mathematicians, they were all on

---

[17][Freudenthal 1979].

[18]Cf. p. 217.

[19]Freudenthal (CW II, p. 435 ff.) has given a thorough historic and mathematical analysis of the invariance of dimension episode. The reader is referred to Freudenthal's comments for more technical details.

holiday. That is to say, I have made acquaintance with Lebesgue, who happened to be in Paris. He is a very interesting man and told me that he possessed already for some time (*seit langer Zeit*)—not just one, but several proofs of the theorem of the invariance of the dimension number, which Brouwer has now proved in the *Annalen*. He has sent me one of those proofs for the *Annalen*, which looks very amusing. I have not examined it closely for the correctness of the underlying idea, for one can depend in the matter of details on such a shrewd man. If you want to check the paper in detail, it is at your disposal.

We don't know if Hilbert occupied himself with the matter, but when in February 1911 Brouwer's spectacular paper appeared, it was immediately followed by an extract from a letter of Lebesgue to Blumenthal, '*On the non-applicability of two domains belonging respectively to spaces of n and $n + p$ dimensions*'.[20] The tone of the message is somewhat patronizing. There is a ring of quiet amusement and superior insight in this note to the mathematical world, which had in all the years failed to find a proof:

Recently, when you talked to me about the proof of the impossibility of establishing a one–one continuous correspondence between the points of two spaces of n and $n + p$ dimensions, a proof by Mr. Brouwer that the *Mathematische Annalen* will publish, I have indicated to you the principle of several proofs of the same theorem. I will exhibit the simplest of these proofs; I will not occupy myself here with drawing from the argument other consequences than the theorem in question itself.

The principle which Lebesgue referred to was the marvellous *paving principle*[21] which reads:

If each point of a domain D of n dimensions belongs to at least one of the closed sets $E_1, E_2, \ldots, E_p$, finite in number, and if these are sufficiently small, then there are points belonging to at least $n + 1$ of these sets.

By 'domain', Lebesgue means closed, bounded, connected set. The principle is illustrated in the plane by the following example, Fig. 5.1, of the pattern of a pavement (which is the worst possible case: minimal overlap).

One immediately sees that there are points belonging to three tiles (we consider tiles with boundaries). The invariance of dimension is indeed an immediate corollary to the paving principle. Unfortunately, the paving principle itself is by no means trivial, at least not the way Lebesgue thought it was.

The records do not show whether Brouwer was informed of this letter before the publication, but it is highly unlikely. Brouwer would immediately have reacted! So we may assume that he was confronted with Lebesgue's letter when he opened

---

[20] *Sur la non-applicabilité de deux domaines appartenant respectivement à des espaces à n et $n + p$ dimensions (Extrait d'une lettre à M. O. Blumenthal)*, [Lebesgue 1911A].

[21] *Pflaster Satz*.

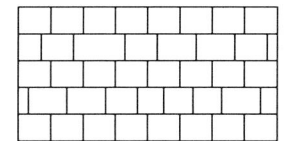

FIG. 5.1. Paving principle

his copy of the *Mathematische Annalen*—a nasty shock indeed! Imagine having settled an outstanding fundamental open problem of the day, only to be ridiculed by the 'simplest possible solution'. Brouwer indeed was furious; however, he quickly realized that Lebesgue's proof contained a serious gap. No matter how brilliant the paving principle was, as far as Lebesgue was concerned, it had the status of a conjecture. Nonetheless, Brouwer was upset; apparently few mathematicians could fathom the methods involved, and hence there was a serious possibility that—also in view of Lebesgue's barely hidden boasting—the credit would go to Lebesgue with, maybe, Brouwer as a fair second.

The reader has to bear in mind that the intricacies of topology were lost on the average mathematician. At best they would think of the geometrical figures and objects that were familiar from traditional geometry, but these did not display the pathologies that tricked even the best in the field. Now Brouwer's paper, which nowadays most readers would find clear and easy going, possessed a precision and a Spartan mode of argument that were unfamiliar to the mathematicians of the nineteenth century, who were accustomed to the pleasant narrative style of, say, Felix Klein. So a reader, presented with the papers of Brouwer and Lebesgue, would be tempted to agree with the Lebesgue paper, which was written in a beautiful convincing style, with a whiff of 'this is all very simple for those who would care to follow the indications given here', and skip Brouwer's paper as 'a difficult proof of the same thing'. Blumenthal was not the only one to fall for the temptation: in the *Fortschritte* the reviewer of Lebesgue's note—the same Engel who struggled with Brouwer's Lie group paper—virtuously cited the paving principle and accepted the paper on Lebesgue's authority. It is not unfair to say that the material was beyond him and most of his contemporaries, possibly even beyond Lebesgue. Even though the word 'priority' hardly crops up in the correspondence about the dimension problem—name not the rope in the house of the hanged man—it would not be human to claim that the following conflict was just about truth and correctness, and not about priority.

Henri Lebesgue. Courtesy l'Enseignement Mathématique.

More dispassionate scholars may point out that priority is not everything and that in the end history will see that justice is done. But unfortunately that is not always the case. There are enough examples where history has stuck the wrong label on theorems, methods, proofs, etc. Usually this concerns an item of local importance, and seldom a major result. It is unthinkable that, say, the proof of Fermat's last theorem would be attributed to the wrong person. In the period we are describing here, the dimension problem had a somewhat comparable status, it was one of the big challenges of modern mathematics. The fact that nowadays the theorem is part of a first course in algebraic topology is irrelevant, in Brouwer's days there were no tools available, and one had to go at it with bare hands. In the case of the dimension problem Brouwer had every right to be upset; it would already have been bad enough to find out that a well-known mathematician had a couple of solutions in his drawer, but it would really hurt if the priority would go to the wrong person on the basis of an unproven conjecture!

A key figure in the controversy that followed the publication of Brouwer's and Lebesgue's papers, was Blumenthal, a generally respected mathematician, who was active in complex function theory and applied mathematics.

He was born in Frankfurt a.M. in 1876, and studied mathematics in Göttingen with Klein, Sommerfeld and Hilbert. Blumenthal was the first Ph.D. student of Hilbert (1898) and he wrote his *Habilitationsschrift* in 1901.

He also studied for some time in Paris with Emile Borel and Camille Jordan. Since 1905 he had a chair in Aachen, not far from the Dutch border and close to Maastricht. Blumenthal and Brouwer became close friends and the latter used to visit Aachen regularly. Blumenthal was the managing editor of the *Mathematische Annalen*, the daily matters of the journal were completely in his hands, although he routinely informed Hilbert of the editorial matters. In effect he was a key figure of the *Mathematische Annalen* for the impressive period of 32 years. His involvement in the Brouwer–Lebesgue controversy was a natural consequence of his position. Either Brouwer had sent his manuscript right to Blumenthal, or Hilbert had passed it on. One would not blame Blumenthal for talking to Lebesgue about Brouwer's invariance proof, after all an editor may discuss papers with third parties. If Blumenthal can be blamed at all, then it is for not sending Lebesgue's note to a referee. That, however, was a bit difficult, because, apart from Brouwer, there was virtually no competent person to judge this kind of topology. His 'Analysis Situs'-paper had made that much abundantly clear. Anyway, it would have only have been fair to inform Brouwer about the Lebesgue note, since Lebesgue was also informed about Brouwer's paper.

Brouwer reacted immediately, he analysed Lebesgue's note and saw that the proof did not work. In early March, he submitted a short note to that effect to Hilbert, '*Remarks on the invariance proof of Mr. Lebesgue*'. The copy of this note bears 'Accepted Hilbert' in Hilbert's hand. Blumenthal, referring to this note in a letter of 14 March 1911 to Hilbert, could not match Hilbert's scientific composure. He was uncomfortably caught in the middle of a controversy that he could not quite fathom:

> I find the Brouwer–Lebesgue affair highly unpleasant, and in fact I am, on the whole, on the side of Lebesgue. That is to say: Lebesgue says explicitly that he assumes certain theorems; those have to do with certain linear equations and inequalities, and they will presumably be provable, in other words, the difficulty does not seem to be in that part, and the whole arrangement of the proof of Lebesgue is in my opinion, taking everything into consideration, a feasible and beautiful route to get to the dimension proof. Whoever reads Brouwer's note, does not altogether have this impression; the note has in my opinion an unfriendly and unpleasant ring.

Blumenthal advocated the withdrawal of the note, and he begged Hilbert to step in, should that be necessary. The matter was further complicated because

> On top of all that I, for one, like Lebesgue (according to an earlier communication) am not able to understand Brouwer's proof.

As a matter of fact, Brouwer's note was neither unfriendly nor impolite, at most somewhat condescending—possibly matching Lebesgue's tone. It only stated that Lebesgue's proof was defective, indicating the gap. Brouwer did not provide a counterexample, but confined himself to the remark that 'in any case considerable further elaborations are required'. Furthermore, Brouwer pointed out that Lebesgue's reference to Baire's work was off the mark, as 'the unproved theorems to which the problem is reduced,[22] are deeper than the problem itself'.

Lebesgue's reaction to Brouwer's note was brief, and showed that he did not yet fully grasp the difficulties:

> If I understand the remark of Mr. Brouwer correctly, it comes down to: I have announced that I will provide the facts that I qualified as evident, that does not replace a proof of these facts.
> I do agree with Mr. Brouwer on this point, I only add that I have not completely written out my proof, because I promised some time ago a paper on this topic to the secretary of the *Société Mathématique de France*.
> I do recognize that my formulation is very poor, since Mr. Brouwer has been able to believe that I had not seen the necessity of proving everything, and that till to-day it seems useful to him to point out this necessity to other readers.

Lebesgue's letter had convinced Blumenthal that Brouwer had nothing to fear. He explained in his letter of 25 March[23] that:

> ...whatever Lebesgue may stress, nobody doubts or disputes your priority for this fundamental proof [...] Lebesgue is, in his own opinion and that

---

[22] By Baire.
[23] Cf. [Johnson 1981], p. 191.

of the world, not your rival but your follower (*Nachfolger*).

As to Lebesgue's position, he wrote

> ...on the one hand Lebesgue's letter shows that he has a clear idea of the proofs of the tentatively accepted theorems, on the other hand he writes himself literally: Writing down the complete proof does not take long, and I am about to do so, but truly, it seems impossible to me to make my results look as ready-to-eat chunks (*bribe à bribe*) and I think that your readers, more generous than Mr. Brouwer, will give me credit until the appearance of my definitive memoir.

Blumenthal added a little sermon:

> From this communication it seems to follow, that it would not be right to publish your note before you have convinced yourself, that not just Lebesgue's note in the *Annalen*, but really *his whole line of proof* is defective. I am convinced that Lebesgue will put the manuscript that he had prepared, at your disposition for inspection. If necessary, I would be prepared to mediate in this respect. I would indeed like to draw your attention to the fact—and here I come to the heart of my impressions—that *your note has been formulated in a very rude form*, and that everybody must necessarily interpret it in such a way, that you consider the gaps stressed by you as irreparable, that is to say that you consider Lebesgue's proof as *false*; for false and incomplete is the same thing in this case. In my opinion, you can only accuse a man of Lebesgue's position of this, if you are completely certain of your ground.

Blumenthal and Brouwer

In the same letter Blumenthal announced that Lebesgue had withdrawn his reply to Brouwer's note, and written a new one. This document is not extant, so we cannot judge the contents, but it is not unthinkable that Lebesgue started to realize that the difficulties were not just minor formalities.

Brouwer immediately answered that, in view of Lebesgue's promise to provide a complete proof, he was happy to withdraw his note. In a letter of 31 March to Hilbert,[24] Brouwer expressed his gratification, he could now drop the matter. This was the more pleasing, he said, since letters of Lebesgue and Blumenthal showed that readers could misinterpret his action as a priority complaint, 'which it was not intended to be'. The statement may seem strange, but Lebesgue had satisfied Brouwer that there was no priority claim involved. Reconstructing the whole discussion, one is left with the impression that Brouwer was angered by the flamboyance, bordering on arrogance, of the older colleague, and by the apparent injustice of Lebesgue getting away with claiming a theorem without so much as a proof.

The appreciation of the mathematical world was, of course, another matter. Even though Lebesgue had made Brouwer's priority clear, there were others who did not have a clear grasp of the situation. For example, L. Zoretti, in a review of Schoenflies' survey of point set theory[25] stated that 'Chapter V contains an interesting study of the correspondence between two domains of n and $n + p$ dimensions, and the invariance of the notion of dimension. Very recent investigations of Messrs. Baire, Lebesgue and Brouwer have made a decisive step in the matter ...'.[26] Although the statement is so unspecific that it could hardly be wrong, it conveys to a reader the impression that somehow the mentioned persons had contributed in roughly the same proportion. As it happened, Brouwer had in the *Annales d'École Supérieure* pointed out a number of mistakes in a topological paper of Zoretti, which may have coloured the judgement of the latter, but it is equally well possible that Zorretti simply had failed to understand the matter. There was a similar misjudgement on the part of Emile Picard, when he presented in 1911 a report to the *Académie des Sciences* on the works of R. Baire:

> In 1907, in two notes entitled *On the non-applicability of two continua of n and $n + p$ dimensions*[27] Mr R. Baire has given a method to study the matter alluded to in the title. Some lacunas remained in the proofs which the author proposed to clear up. He was earlier than Brouwer and Lebesgue, who by different roads arrived at the theorem which was clearly anticipated by Baire, to wit that one cannot establish a bijective correspondence between a continuum of n dimensions and a continuum of $n + p$ dimensions.

One wonders to what extent Picard, an outstanding expert in the theory of functions, and familiar with mathematics at large, obeyed the academic tradition and polished up his report.

The *Fortschritte* soon realized that its reviewer had been a bit careless about the Lebesgue note; in its 1914 volume a brief correction was published (by Blaschke):

---

[24] CW II, p. 440.
[25] [Schoenflies 1908].
[26] [Zoretti 1911].
[27] *Sur la non-applicabilité de deux continus à n et $n + p$ dimensions.*

In the *Bericht* on p. 419 on the paper of Lebesgue it should be noted that the proof of Lebesgue is not correct. There is a gap, which the author so far has not filled. More can be found in a paper of L.E.J. Brouwer in *J. für Math.* 142, 151.[28]

Brouwer was under some apprehension, as he himself had not yet found a proof of the paving principle, but his private notes show that he was almost certain that Lebesgue misjudged the intricacies of this part of topology, and that he would not be able to give a proof. Lebesgue, as a matter of fact, sent Brouwer some more details, but declined to publish them in the *Mathematische Annalen*.

Lebesgue, who had evidently quite sound and effective intuitions on the matter of dimension, did not stop to provide a proof of the paving principle, but went on to present an alternative proof of the invariance of dimension to the *Académie des Sciences* (23 March 1911). The argument here is completely different, and it introduces the concept of 'linking varieties'. Again the underlying ideas are marvellous, but the note does not give satisfactory proofs or even proof sketches. Apparently he failed to grasp the value of the gem he had found.[29]

Brouwer, upon reading the above paper *On the invariance of the number of dimensions of a space and on the theorem of Mr. Jordan for closed varieties*,[30] quickly spotted its weak points, and prepared another critical note for the *Mathematische Annalen*.[31] In the accompanying letter of 9 May to Blumenthal, Brouwer motivated the submission of the new note to the *Mathematische Annalen* by referring to Lebesgue's curious behaviour. Not only had Lebesgue failed to observe common courtesy (again!) in the matter, he had also disassociated himself from an earlier promise to provide a complete version of the *Annalen* note, at the same time disowning all earlier proof attempts as hasty. These earlier attempts, said Brouwer 'are teeming with incorrect arguments, and are irreparably false'. And also the second *Comptes Rendus* note contained an 'incurably' false proof, to which Lebesgue stubbornly clung in spite of Brouwer's remonstrations. The long and short of it all was that Brouwer saw it as his unpleasant duty to publish the note on Lebesgue's invariance paper, but he said that he would withdraw the note if the editorial board would consider the note not to be in the general interest.

This note 'Comments on the invariance proofs of Mr. Lebesgue' (9 May 1911) and its successor (written in French for the benefit of Lebesgue) of 11 June 1911 were not published. The last one contained not only a critique of Lebesgue's papers, but also an account of the discussion so far. Brouwer complained that after he had withdrawn his note to the *Mathematische Annalen*, Lebesgue had not fulfilled his promise to provide a correct proof of the paving principle, but only sent inconclusive material:

---

[28] The dimension paper, cf. p. 177.

[29] [Alexander 1922] and [Alexandroff–Hopf 1935], Ch. XI elaborated the ideas involved.

[30] *Sur l'invariance du nombre de dimensions d'un espace et sur le théorème de M.Jordan relatif aux variété fermées*, [Lebesgue 1911B].

[31] [Johnson 1981], p. 198, 199.

...I studied his arguments repeatedly, but they remained obscure to me, and for the rest contained nothing that could not at a first glance be perceived by anyone.

Repeated questions were not honoured, so Brouwer concluded that it was impossible to continue the correspondence.

Blumenthal, to whom, most likely, those notes were addressed, was by that time convinced of Brouwer's viewpoint. He asked Brouwer[32] for an elaboration of his critique (in French) in order to confront Lebesgue.[33] There is a draft of Brouwer's answer; he complied with Blumenthal's request, but he insisted on providing the details for Blumenthal's private use only.

> For I have already so extensively and so often explained everything in writing, that I cannot tell him anything new.
>
> Exactly for this reason, the following explanation of his attitude has forced itself more and more upon me: that he directly after my first letter has recognized his lapsus, but that he was too vain to admit it, and that his further behaviour was determined by the hope of finding, perhaps later, a proof of the assumed statement, and by the necessity to gain time.

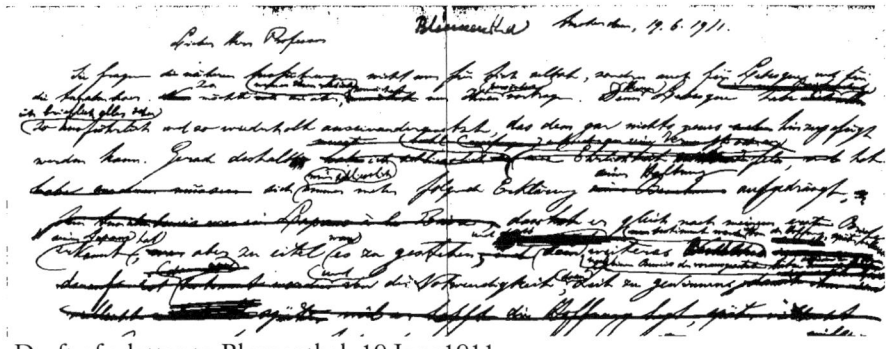

Draft of a letter to Blumenthal, 19 June 1911.

In the letter Brouwer suggested that Blumenthal should urgently press Lebesgue to provide a proof. This stratagem would force Lebesgue to furnish a proof, with the risk of committing new errors, or to plead failure. The letter ended with a request not to tell Lebesgue that Brouwer in the meantime had obtained a proof of the paving principle, because the latter considered it Lebesgue's obligation to clean up his own arguments first. A similar sentiment is expressed in the letter of 5 November 1911 to Baire, in which Brouwer told Baire that 'I have myself found a proof of the lemma of Lebesgue a couple of days after its publication, but I believe that I should not publish it and give Mr. Lebesgue the opportunity to perform his task himself.'

---

[32] Blumenthal to Brouwer 16.6.1911.
[33] [Johnson 1981], p. 203.

In the *Nachlass* of Brouwer there is a number of small scraps of paper, scribbled with Brouwer's fine detailed handwriting. Some of them are concerned with the Lebesgue affair; and the item Y4, dated 1912 by Freudenthal,[34] contains a proof of the paving principle.

In a letter of 2 July to Blumenthal, Brouwer wrote down (in French) the crucial step in the inductive argument for the paving principle. The note evidently was intended for publication, but it never appeared; finally its proof was published in the paper '*On the natural notion of dimension*'.[35]

On 8 July Brouwer drafted a letter to Blumenthal, in which he pointed out 'another gap in Lebesgue's so-called third proof' in the *Comptes Rendus* note.[36] Whether this letter was actually mailed is hard to say, since there is no Blumenthal Archive, but the fact that Brouwer kept the draft among his papers, suggests strongly that it was mailed. Blumenthal, in his answer of 14 July, admitted that his mediation attempts had failed, and that he therefore saw no objections to publishing Brouwer's Note on Lebesgue's proofs. He remained, however, cautious to the extent that he advised Brouwer to adopt the title 'On my and Lebesgue's papers on the invariance of the number of dimensions'. He declined however to publish Brouwer's proof of the paving principle—"I cannot do that before I have given Lebesgue a generous opportunity to speak for himself. For I intend to request from him his *Mémoire étendu*." Of course, he went on, you are free to publish your note elsewhere, for example, 'under supervision of the great Van der Waals in the *Verslagen*'.

Brouwer had some reason to feel that he was let down in order to save the face of his adversary. He drew his conclusions and published the proof and significant new material in the *Journal für die reine und angewandte Mathematik*[37]—a fateful choice as we will see later. Brouwer had made up his mind on the choice of journal in November, as appears from a letter of Hellinger, the assistant of the editor Hensel. This 'desertion' was quite a step for Brouwer, who had a very strong allegiance to the *Mathematische Annalen* and Hilbert!

The correspondence did, however, continue; in a letter of 19 August 1911, Brouwer told Blumenthal that he had reconsidered the first part of Lebesgue's *Comptes Rendus* note—and that he thought the theoretical basis of the linking of manifolds was a deep problem. His re-assessment of the concept of 'linking manifold' led him probably to the more positive view of Lebesgue's contributions in '*On the invariance of the number of dimensions of a space and on the theorem of Mr. Jordan for closed varieties*'.[38] The second part of the same letter contained the sketch of an elegant alternative proof of the paving principle that made no use of the mapping degree; it was never published.

---

[34] CW II, p. 440.

[35] *Über den natürlichen Dimensionsbegriff*, [Brouwer 1913A].

[36] [Lebesgue 1911b].

[37] [Brouwer 1913A].

[38] Sur l'invariance du nombre de dimensions d'un espace et sur le théorème de M. Jordan relatif aux variété fermées.[Lebesgue 1911b].

The name of a second prominent French mathematician has come up in the above story, that of René Baire. Baire had made a name for himself in the theory of real functions, where he had systematically investigated classes of discontinuous functions, now known as the *Baire classes*. Darboux had already initiated the study of discontinuous functions, but Baire had gone on to develop a systematic, abstract viewpoint. His work, together with that of Émile Borel, is the hard core of the so-called descriptive set theory.

In his note in the *Mathematische Annalen* Lebesgue had referred to notes of Baire in the *Comptes Rendus*,[39] suggesting that Baire's methods would yield a straightforward proof.

> Without doubt, Mr. Baire has not developed his proof; but it seems to me that, if one takes into account the hints given by Mr. Baire, there is nothing left to solve but difficulties of detail that are not very serious.[40]

Freudenthal pointed out that Baire and Lebesgue were not on speaking terms,[41] Lebesgue's reference to Baire could hence be seen as an attempt to belittle Brouwer and tease Baire at the same time.

Brouwer had, on seeing Lebesgue's note, sent reprints of his *Mathematische Annalen* paper to Baire, probably inquiring after Baire's alleged proof. Baire's first letter dates from 28 October, it contains no mathematics, but polite congratulations on the progress made in analysis situs. He declared that because of other commitments and a prolonged illness, he had not been able to pursue his research.

In his reply of 30 October, Brouwer diplomatically approached Baire concerning his proof of the invariance of dimension:

> Concerning your publication of 1907, I suppose that you yourself do not agree with the lines that Mr. Lebesgue devoted to you at the end of his false proof [...]
>
> The important theorems that you formulated in 1907 seem to me of a much more subtle character than the invariance, which is to me the most fundamental, but also the most crude property of the analysis situs of n dimensions.

He went on to say that he was unfortunately obliged to correct Lebesgue's statement, for it would, if correct, strip his own 1911 paper of all importance; this disagreeable task was not to be construed as a criticism of Baire. There is a short exchange of letters between Brouwer and Baire, but although Baire clearly resented being used by Lebesgue—their friendship of the days of the *École Normale* had gone stale when Lebesgue had gone in for career making[42] —he stuck to the view that, although he had not done so, one could without serious difficulties, but maybe at the cost of some lengthy exposition, prove the required statements on the basis of

---

[39] [Baire 1907a].
[40] [Lebesgue 1911a], p. 168.
[41] CW II, p. 439.
[42] Baire to Brouwer 5 December 1911.

his 1907 *Comptes Rendus* notes. 'On the other hand', he added, 'these statements don't form such a complete set as your statements 1, 2, 3 of p. 314'.[43]

As time went by, Brouwer lost sight of Lebesgue and the matter of the proof; and only when in 1923 Urysohn called his attention to Lebesgue's 1921 paper, Brouwer felt obliged to point out that this proof was basically his own 1913 proof, but complicated by Lebesgue's presentation.

The Brouwer–Lebesgue controversy more or less bogged down in a stalemate; Lebesgue clearly was not willing to concede the point, but he was equally unable to meet Brouwer's challenge. He eventually published a proof of the paving principle in 1921. In his *Notice sur les Travaux Scientifiques* of 1922 Lebesgue went so far as to state that

> On the occasion of the publication of the work of Mr. Brouwer, I indicated[44] the principle of an argument that also allows us to establish that theorem; I have recently[45] developed the proof.

The whole affair leaves a sad impression; on the one hand the older and established mathematician treated the topic and the newcomer as another occasion to demonstrate his superior intelligence, thus losing sight of common civility towards newcomers, but—and this is a more serious and purely mathematical shortcoming—also completely misjudging the complexity of the problem. On the other hand, the younger man could not distinguish between irritating vanity and intentional foul play. On top of that, Brouwer's somewhat inflated sense of justice demanded an unconditional surrender; the loss of face of his opponent did not worry him. There is a striking contrast between Brouwer's genuine mystic ideal of detachment from the world, of introspection and non-interference with fellow human beings, and the burning wrath that always was close under the surface, and that easily erupted at the least provocation. Brouwer bears a striking similarity to the character from Shaw's 'Androcles and the Lion', Ferrovius, with the strength of an bear and the temper of a mad bull, 'who has made such wonderful conversions in the northern cities'. Like Ferrovius, he could not resist the temptation of a battle against injustice and evil men—'When I hear a trumpet or a drum or the clash of steel or the hum of the catapult as the great stone flies, fire runs through my veins: I feel the blood surge up hot behind my eyes: I must charge; I must strike'.

This fighting spirit remained with Brouwer all his life—however, for all the agony it caused his adversaries, he certainly got his share in the suffering.

It is likely that, Lebesgue did not want to rob Brouwer of his priority, but he would not have minded changing the perspective of the dimension problem to his advantage by belittling Brouwer's efforts. One cannot escape the impression that the mathematical authorities of the day did little to see that justice was done.

---

[43] The n-dimensional Jordan theorem, [Brouwer 1911F].
[44] Cf. [Lebesgue 1911a].
[45] Cf. [Lebesgue 1921].

Blumenthal, as the managing editor of the *Mathematische Annalen*, did not display the decisiveness that was required of him in his function. His wish to keep on the good side of a recognized authority like Lebesgue, may have obscured his judgement. It is also remarkable that in a spectacular dispute like the one above, none of the older statesmen of the *Mathematische Annalen* stepped in. One might suppose that Hilbert would have seen the importance of Brouwer's methods, if not the theorem—also in view of the prior correspondence. But, alas, the whole matter was left to Blumenthal, who did not carry enough weight to address the problem. The episode also illustrates how little the mathematicians of the day understood the intricacies of topology. It is no exaggeration to say that only in the twenties a measure of understanding of Brouwer's methods started to spread. The influential book of Felix Hausdorff, *Grundzüge der Mengenlehre*[46] of 1914, that introduced large numbers of readers to the beauty of set theory and topology, remarked (p. 461) that 'The brevity of Brouwer's papers, which often forces the reader to fill in many details by himself, is most regrettable, in the absence of other impeccable and extensive expositions.' Brouwer had little patience with those complaints, he even felt rather offended at the suggestion (for example of Hausdorff) that his topological work should be redone. The full impact of Brouwer's topological innovations was not felt before the twenties. Although the delay may partly have been due to the First World War, the main reason was that it was generally viewed as difficult and inaccessible. Heinz Hopf reported that in 1917 Erhard Schmidt discussed Brouwer's topological methods in a course at the University of Breslau,[47] and when Schmidt moved to Berlin in 1920, topology became a regular part of the curriculum in Berlin. In America J.W. Alexander had grasped the new methods and started, from 1915 onward, adding his own contribution in the early twenties.

In spite of all the negative features of the conflict we described, there is at least one positive side effect: Lebesgue's obstinacy prompted Brouwer into feverish action; some of the papers of this period are reactions to Lebesgue's claims. In July 1910 Brouwer submitted his fundamental paper '*On the mapping of manifolds*'[48] to the *Mathematische Annalen*. It was the detailed elaboration of the facts mentioned in the letter of 1 January 1911 to Hilbert. The importance of this paper can hardly be overestimated, it contains virtually all the tools of the new topology. It was, so to speak, a short but exhaustive course in the topology of the future. Among the many concepts introduced here, we find that of *simplex*,[49] open and closed manifolds (based on simplexes), (n-dimensional-)*indicatrix*, *simplicial approximation*,[50] *mapping degree, homotopy of mappings* (under the name *continuous deformation*), *singularity index* (under the name *degree of the simplex*). The basic theorems, for example the preservation of the mapping degree under homotopy, are all proved by direct

---

[46] Basics of Set Theory.

[47] [Hopf 1966].

[48] *Über Abbildung von Mannigfaltigkeiten*, [Brouwer 1911D].

[49] Poincaré had already defined the notion in [Poincaré 1899]. Brouwer does not quote Poincaré, so presumably he was not aware of the paper.

[50] Note that in the letter to Hilbert of 1 January 1910, Brouwer still uses polynomial approximations.

geometric methods. The converse of this invariance theorem was established a year later in '*Continuous one-one transformations of surfaces into themselves, no. 5*'; this paper, in contrast to the earlier ones in the series employed the new tools and generalizes earlier results; it thus takes a special place in the long series.

5.2 The fixed point theorem and other surprises

The famous *fixed point theorem* of the (n-dimensional) ball is the quiet apotheosis of the 'mapping of manifolds'-paper. The theorem that made Brouwer's name a household word in mathematics and other disciplines was presented in one line, without any comments. Did Brouwer think that such a spectacular result, like good wine, did not need any recommendation, or was he showing off in modesty? We will not know.

Of all Brouwer's topological achievements, the fixed point theorem has appealed most to the imagination of the non-specialists. Had he been made a baron, then his coat of arms certainly should have contained a fixed point.[51] A number of popularization's have been put before the public, one of these illustrates the surprising omnipotence of topology: consider a cup of coffee which is stirred slowly in an arbitrary manner, then when the coffee has again assumed a state of rest at least one 'coffee point (particle)' has returned to its old position. The stirring instructions are necessary to prevent discontinuities, one does not want to have drops flying around.

The mapping degree was the prime instrument in Brouwer's topological investigations. It is surprising, and hard to explain why Brouwer, who had all the equipment he needed, never made use of, or developed, algebraic topology. In particular one wonders, why he never used homology theory, and all its marvellous conveniences. 'Instead', Dieudonné remarks,[52] 'he used his discovery to define rigorously the concept of *degree* of a continuous map, and then proceed, mostly by fantastically complicated constructions, relying exclusively on that notion, to prove the famous *Brouwer theorems*'. The trained topologist will, of course, see the implicit use of algebraic methods, but Brouwer did not make the important step to expand his insights into a systematic machinery for the algebraization of topology. He was and remained a geometer.

Part of the power of Brouwer's methods resided in his *simplicial approximations*; 'it may be said that Poincaré defined the *object* of that discipline,[53] but that it is Brouwer who invented *methods* by which theorems about these objects could be *proved*, something that Poincaré had been unable to do'.[54]

---

[51] The reader need not worry, nobility has not been created in the Netherlands for more then a century.
[52] [Dieudonné 1989], p. 161.
[53] Simplicial topology.
[54] [Dieudonné 1989], p. 168.

Although Brouwer's mapping degree was a new phenomenon in topology, it was foreshadowed on certain respects by earlier notions. There was, for example, the 'winding number' of a path around a point: walk around the point and count at the end how many revolutions you have made. The walk may be totally disorganized, going backwards and forwards in an unpredictable way.

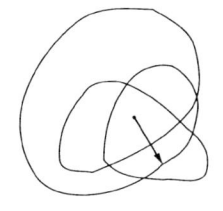

Note that if one looks from the point, a, and follows the moving point, the backward and forward neatly cancel out each other, so that the complicated figure above really comes to one revolution. The notion is captured by the following integral: $\frac{1}{2\pi i} \int_U \frac{f'(z)}{f(z)-a} dz$, which measures the angle swept by the vector following the movement of the point on the curve.

Winding number.

The idea of the mapping degree can be simulated by wrapping a rubber band around a cylinder. With a little bit of imagination, one can think of the band doubling backwards and forwards. What one gets in this case is a continuous mapping of the circle onto itself (looking in the direction of the cylinder). The mapping degree counts the number of times a point is covered, cancelling the 'backwards' against the 'forward'.

For two-dimensional pictures like the image of a circle, one has a reasonable geometric intuition, but for higher dimensions, it gets harder. Moreover, the phenomenon of the Peano curves had frightened people off. Mathematicians had come to distrust their geometric intuitions. Nonetheless, Brouwer succeeded in defining the correct notion; or, to be more precise, he proved that a higher-dimensional analogue of the winding number is a constant for a manifold (for example a surface) and then considered its properties in all kinds of situations. Using the mapping degree, he proved a wealth of new facts.[55]

A problem presented by Brouwer's topological research is that it is not immediately obvious what the underlying motivation is. Brouwer's published papers are no help in that respect. It looks as if he solves hard problems just because they present a challenge. This may partly have been responsible for the unfounded impression that Brouwer only practised topology to gain status and a secure job. There are a couple of clues, however. We have seen that his research into vector distributions may have been triggered under Poincaré's influence (Cf. p. 137); another clue is provided by a letter of Brouwer to Poincaré, related to the 'invariance of domain'. In this letter, undated, but probably written in the fall of 1911 (and only preserved in draft), Brouwer wrote "My proof of the invariance of the n-dimensional domain was inspired last year by the reading of your 'Method of Continuity' of volume 4 of the *Acta Mathematica*". The letter shows that Brouwer had familiarized himself with the geometric–analytic work of the grand master of European mathematics,

---

[55] After the ascent of homology theory, a much simpler definition of the degree of a mapping became available: let f be a continuous mapping from M to M', where M and M' are compact, connected, oriented (pseudo-) manifolds, then for $f^* : H_n(M; Z) \to H_n(M'; Z)$, we have $f^*([cM]) = c[M']$ where c is the mapping degree.

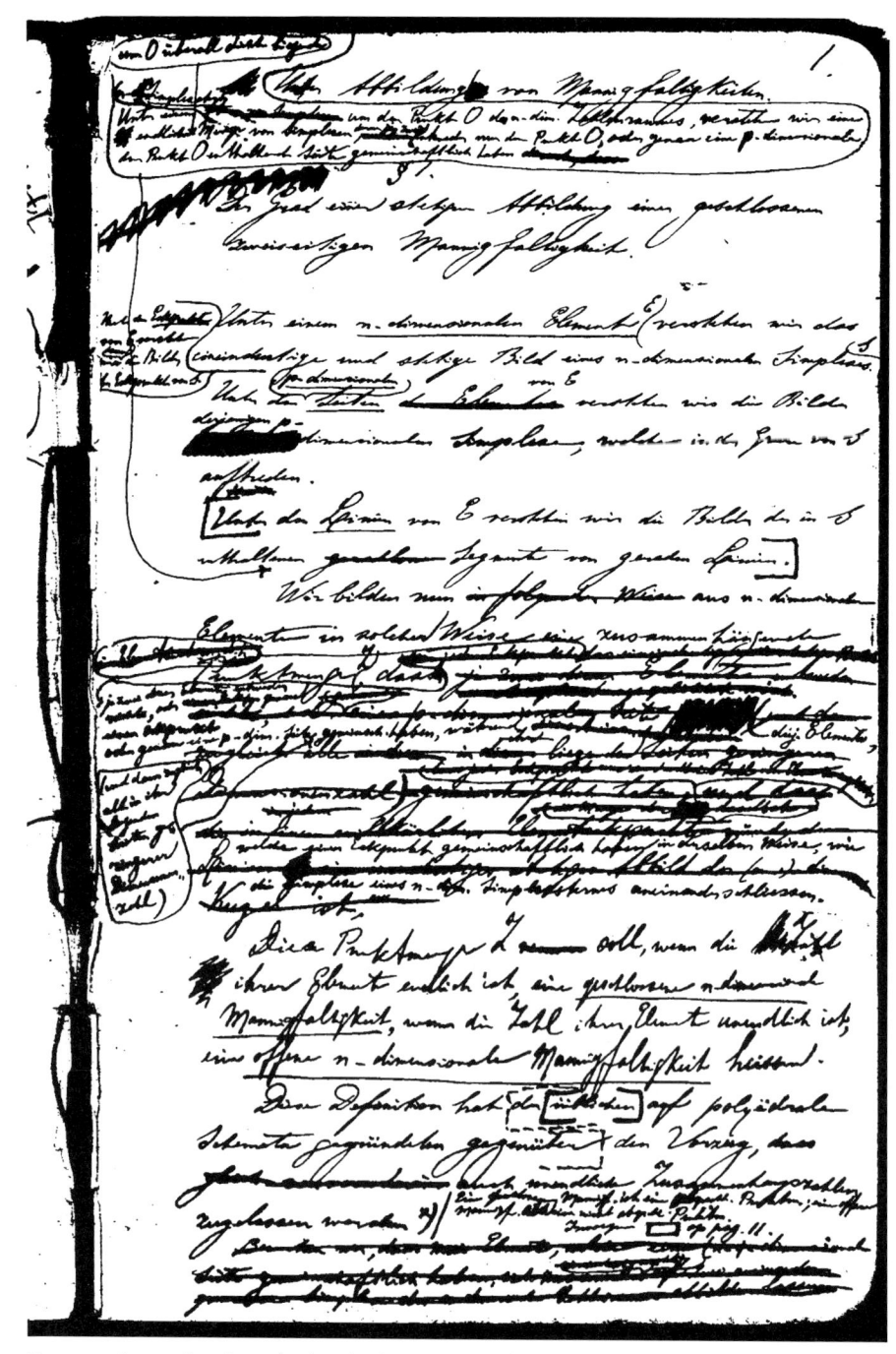

Brouwer's notebook with the draft of the 'mapping of manifolds'-paper.

and that in the middle of his general topological work he was looking for concrete problems areas. The 'Method of Continuity' is a tool from the domain of uniformization and automorphic functions (see below). It was devised to solve the basic problems in the field, but until Brouwer stepped in, progress was blocked by topological difficulties. Poincaré and Klein were actively looking for means to get the method to work,. They were joined in the beginning of the century by Paul Koebe. We note in passing that Koebe was one of the main speakers at the Rome conference; hence it is well thinkable that Brouwer had heard his talk on automorphic functions, and that he had as a consequence looked into the problems of uniformization. The method of Continuity was also mentioned in the draft of the letter to Baire (5 November 1911). We will return to this particular topic of uniformization later in this chapter.

In the paper *Proof of the invariance of domain*,[56] Brouwer considered the problem 'does a topological mapping map a domain in an n-dimensional space onto a domain?'

This paper contains the first public reaction to Lebesgue's dimension papers; in a footnote he characterizes Lebesgue's *Annalen* paper as 'insufficient' and the *Comptes Rendus* note as 'qua content identical with mine: –the deviations complicate the line of thought'.[57]

As Freudenthal pointed out, this is a strange formulation, 'Lebesgue's proof is not correct, and his method differs quite a bit from Brouwer's.[58] But Brouwer read Lebesgue's proof through the spectacles of his own knowledge, sympathetically filling in the gaps with ideas of his own, and then stated that it was his own proof, though needlessly complicated'.

The letter of 5 November 1911 to Baire discussed the invariance of domain, as claimed by Baire. Brouwer, however, could not see that Baire had solved the difficulties involved.

> As far as I can see, the indications, given by you in the *Comptes Rendus*, leave untouched the principal difficulty. For a long time I had searched for a proof; for $n = 3$ it is easy, for arbitrary n, I only succeeded last summer, by means of an argument that I afterwards have found again in the note of Lebesgue (*C.R.* 27 March 1911) where, by the way, it is in an almost unreadable form—and inexact—if one takes it literally.

The *invariance of domain* comes to the following: the image of a domain (a connected open set) under a topological mapping is again a domain. The theorem was proved by Brouwer in his *Mathematische Annalen* paper in 1911 by means of the *No Separation Theorem*, which in the simple case of the plane, tells that an arc of a Jordan curve (think of an arc of a circle) does not split the plane in two parts. The n-dimensional case is completely analogous.

---

[56] *Beweis der Invarianz des n-dimensionalen Gebiets.*[Brouwer 1911E].
[57] [Brouwer 1911C].
[58] CW II, p. 443.

Schoenflies had proved as early as 1899 the invariance of domain theorem for the plane, and Baire and Hadamard had shown that the theorem for arbitrary dimension was an immediate corollary of the n-dimensional Jordan theorem (which was still open at that moment). As we have seen Lebesgue had suggested that, up to some technical details, the work of Baire basically yielded the invariance of dimension.[59]

Brouwer had already discussed the matter with Baire in the above-mentioned correspondence, and in the paper *Proof of the invariance of the n-dimensional domain* he added in a footnote that:

> In so far as the developments of Mr. Baire, cited by Mr. Lebesgue, are concerned, the theorems to which the problem is reduced there [that is in the paper in the *Mathematische Annalen*] are deeper than the invariance of dimension.

Indeed, a fairly straightforward argument derives the invariance of dimension from the invariance of domain.

In the meantime Brouwer had realized that the theorem could also be derived from the second part of Lebesgue's *Comptes Rendus* note (cf. p. 166) and Baire's earlier work. However, in a two-page note *On the invariance of the n-dimensional domain*,[60] he gave a short direct proof, based on the degree of a mapping.

On 24 February 1912 Korteweg communicated to the KNAW a paper of Brouwer, that gave an exposition of his version of 'linking manifolds', which he had discovered independently.[61] This paper *On looping coefficients*[62] is, like all of Brouwer's topological work, geometric in nature. The proper way of dealing with this topic would be to use homological arguments, but those were only much later applied by Alexander. As it is, Brouwer's exposition is somewhat tortuous.

Already in a letter of 15.10.1909 to Hilbert, accompanying the manuscript of the Jordan curve theorem,[63] Brouwer had remarked to Hilbert, that the seemingly evident theorem 'the one-one continuous mapping of a closed curve is again a closed curve' was open, and that it seemed to present a 'genuinely hard problem'. To appreciate the problem one has to realize that the 'closed curve' was defined in the Schoenflies tradition. It was simply a bounded subset of the plane determining two domains of which it was the common boundary. Clearly, under this definition the problem is not trivial. It was settled by Brouwer in his paper *Proof of the invariance of the closed curve* in 1912 in the *Mathematische Annalen*, and summarized in the *Comptes Rendus*.[64] The technique of the proof is close to homotopy (no mapping

---

[59][Baire 1907a, Baire 1907b].

[60]*Zur Invarianz des n-dimensionalen Gebiets*, [Brouwer 1912C].

[61]See the letter to Baire of 5.11.1911, [Johnson 1981], p. 218.

[62][Brouwer 1912E2].

[63][Brouwer 1910E].

[64]*Sur l'invariance de la courbe fermée, Beweis der Invarianz der geschlossenen Kurve*, [Brouwer 1912F, 1912L].

degree, this time); Brouwer used in his paper the notion of *Zyklosis* (which probably goes back to Listing, 1847). The theorem of the paper is somewhat more general: the number of domains determined by a bounded closed connected planar point set is invariant.

The notion of mapping degree was also employed by Brouwer in the sequel of his series of papers 'Continuous one-one transformation of surfaces in themselves',[65] the fifth communication. The paper introduces the notion of *transformation class*, which we would now call homotopy class. This notion is used consistently in the statements of the paper. The content of the paper is that, for mappings of spheres into themselves, the mapping degree determines the homotopy class.

The last paper in the series, dealing with dimension, was published by Brouwer in 1913 in the *Journal für die reine und angewandte Mathematik* (see p. 168), a journal with a more restricted circulation than the *Mathematische Annalen*. It bore the title *On the natural notion of dimension*,[66] and it contained a fully fledged abstract definition of dimension. It will be discussed later in the context of the post-war developments of topology.

Brouwer briefly returned to topology after the First World War, but his main achievements belong in the period discussed here. One may well wonder why Brouwer, after his imaginative breakthrough failed to exploit the homological and homotopical features that were hidden in his work—and that had been envisaged by Poincaré! Indeed, in 1922 Alexander published a paper (the first draft was already finished in 1916) in which he used homological methods to prove his duality theorem, and which extended some of the Brouwer theorems.[67] Brouwer's reluctance to adopt the tools of homology theory may have been rather the result of a personal disposition, than an inability to do so. After all, a man of his calibre, could not very well miss the things that were staring him in the face.

On the whole, Brouwer was oriented towards the continuous and the geometrical. Even in his foundational work he eschewed the finitary, and attacked—broadly speaking—the problems of the continuum and Baire space (or Cantor space) rather than those of arithmetic or of a finitary nature. There are no publications in pure algebra among Brouwer's papers. It is clear that he knew his algebra, but he mainly applied it in geometrical context, that is to *support* the geometry, not to *supplant* it. Another explanation of his neglect of the new algebraic machinery may be Brouwer's inclination to find satisfaction almost exclusively in revolutionary innovations. His œuvre contains few if any routine, run-of-the-mill papers. After introducing some new idea, concept or tool, he experimented enough to be satisfied that it was worthwhile, but he was not the man to exploit his new ideas (let alone those of others) in an endless series of papers. That routine of mathematics just did not appeal to him. The grinding out of mathematical theorems was not his

---

[65] [Brouwer 1912K2].
[66] *Über den natürlichen Dimensionsbegriff.* [Brouwer 1913A].
[67] [Alexander 1922].

idea of mathematics. His concept of mathematics was rather that of the gentleman practitioner, or even better, the scientific artist, who would, in bursts of creativity, deliver high-quality products; but, as often as not, idle around.[68] His attitude is illustrated by a remark in his paper on the history of dimension theory:[69]

> I have restricted myself to the laying of the foundations of the theory of dimension, and refrained from further dimension theoretic developments, on the one hand because with the proof of the justification theorem[70] the intended purpose had been reached, on the other hand because an intuitionistic realization of the subsequent considerations (in the first place those which can be grouped around the 'sum theorem' and the 'decomposition theorem') was, in contrast to the justification theorem, not plausible.

It seems plausible to conclude that Brouwer was ready to leave the field to others after the basic facts had been established.

5.3 The Karlsruhe meeting and the Continuity Method

The next episode in Brouwer's career has to do with the so-called *automorphic functions* and *uniformization*. The topic is a traditional part of complex function theory and its subject matter is of the greatest importance for the theoretical understanding of analytic functions.

The uniformization problem[71] can be illustrated by a simple example: 'does $f(z) = z^2$ have an inverse function?' In the traditional highschool mathematics the answer is trivial: 'yes, take $g(z) = \sqrt{z}$'. Everybody will have experienced some qualms: why $\sqrt{z}$ and not $-\sqrt{z}$? This is, in a nutshell, the problem. So we could also have asked 'are there solutions of $w = z^2$ for z, how many are there, and what is their connection?' The solution to this problem goes back to Bernhard Riemann. The basic idea is to take the two-solutions idea seriously, and grant them each a limited autonomy. Take two copies of the domain, that is the complex plane, and give one to $\sqrt{z}$ and the other one to $-\sqrt{z}$, or in polar coordinates: $f_1(z) = \sqrt{r}e^{\frac{1}{2}i\varphi}$ and $f_2(z) = -\sqrt{r}e^{\frac{1}{2}i\varphi}$, where $z = r \cdot e^{i\varphi}$.

Each solution seems fine, until we discover that f gives to the same point two values, depending whether we consider it as $z = re^{i\varphi}$ or $z = re^{i\varphi + 2\pi i}$. Geometrically speaking, if we travel around the origin on a circle, we do not get the same output for f_1 after one full revolution, the same holds for f_2. The idea now is to combine both functions (and both domains): make one revolution on a circle in the

---

[68] Cf. p. 194 [Wiessing 1960], p. 143 ff.
[69] [Brouwer 1928C1].
[70] That is, Brouwer showed that $\mathbb{R}_n$ has dimension n.
[71] For a precise and general definition of uniformization the reader is referred to the literature, for example [Nevanlinna 1953], [Beardon 1984].

first copy of the z-plane and use f_1, then switch to the second plane, make another revolution and use f_2. Exactly where we switch is not so important, say at the negative imaginary axis. We now have a decent function defined for all z which acts in the inverse of z^2, and which also treats the two possibilities, f_1 and f_2, on an equal footing. The z-plane has been replaced by two sheets which are glued on along a line.

The geometric surface one gets by this cutting and gluing is called a *Riemann surface*, after its inventor. The process of extracting one decent function as the inverse of $w = z^2$ is called *uniformization*. Some of the greatest minds of the late nineteenth-early twentieth century devoted much of their energy and ingenuity to solve the general problem: how to uniformize general analytic functions of the form $f(z, w)$.

The algebraic functions are the traditional examples, for example how to get w as a function of z satisfying $w^3 + z^3 - wz = 0$.

The theory of uniformization became the meeting point of all the technical know-how of nineteenth century mathematics: Riemann surfaces, group theory, topology, function theory, algebra, potential theory, ... The two masters of the subject were Klein and Poincaré, who in playful competition tried to outwit each other.

The history of the theory of automorphic functions is covered by a veil of romance, drama and heroism, in particular through Klein's vivid description in his history '*Development of Mathematics in the 19th Century*', Vol. II.[72] Both Klein and Poincaré were looking for the final solution of the fundamental problem of automorphic functions from 1880 onwards, like a couple of King Arthur's knights searching for this Holy Grail—and bent on getting there first. In rapid succession Poincaré and Klein overtook each other, publishing while the ink was still wet. The stress took a heavy toll on Klein; in 1882 his health collapsed, and Poincaré had the field to himself.

A great many other mathematicians took part in the quest for the Holy Grail of uniformization: Schottky, Fuchs, Koebe, Fricke, Bieberbach, to mention the more prominent ones. Of those, Koebe and Fricke made it the main task of their lives to break the deadlock that the subject reached after the long series of successes of Klein and Poincaré. The efforts of those two masters of the *fin de siècle* were directed to the theory of *automorphic functions*. There is a close connection between automorphic functions and Riemann surfaces. It was expected that through the theory of automorphic functions one could solve the basic problems about Riemann surfaces; for Riemann surfaces were the key to the uniformization problem.

---

[72][Klein 1927]

Automorphic functions are complex functions which are invariant under a group of transformations of the complex plane. An example is the exponential function e^z, it is invariant under a vertical translation over the length 2π : $e^{z+2\pi i} = e^z$.

Now Riemann surfaces (at least the compact ones) can be viewed as spheres with handles, and the nineteenth century way to 'tame' them, was to cut them up, so that they could be flattened as a piece of the plane. For example, a torus (which can be topologically transformed into a sphere with one handle) can be cut twice, so that it becomes a flat piece (think of the inner tube of a bicycle tire). The correspondence between automorphic functions and Riemann surfaces was embodied in the fact that an automorphic function maps *fundamental domains* onto a cut-up Riemann surface. For a fundamental domain, one may think of the smallest part of the domain that is repeated (like a tiling) by the transformations. Think of e^z: if one knows the values for all z between the real axis and the line $y = 2\pi i$ (its fundamental domain) then one knows all values.

Felix Klein. Niedersächsische Staats- und Universitätsbibliothek Göttingen.

The question that remained to be solved at the end of the century was: 'does one get all Riemann surfaces in this way?' The process of passing from automorphic functions (and their groups) to Riemann surfaces was a continuous one, and Klein and Poincaré had hoped that exploiting this continuity would yield the desired answer. This route to a solution was called the '*continuity method*'; Klein tried to establish the result by the technique of counting parameters—some kind of dimension argument—one may think of the kind of use of dimension arguments in linear algebra. This counting method had however become highly suspect after Cantor's dimension lowering arguments, see p. 151.

In 1907 Poincaré and Koebe independently solved the uniformization problem along other lines. But such was the power of tradition that the automorphic function approach was still pursued after the hunting season was closed. Klein and Koebe remained active in attempts to save the continuity method. There is correspondence which shows that as late as 1910 the topic was still discussed. Whereas Poincaré had moved on to other areas after the successful wrapping up of the uniformization episode, Koebe remained active in the area. He was considered to be the outstanding expert in the field, and he certainly did not disagree on that point.

But even the indefatigable Koebe seemed to get enough of the topic around 1910, and to look for new challenges; Hermann Weyl reported in a letter to his Dutch friend Pieter Mulder 'Now Koebe has done with uniformization, and he is looking for another topic'.[73]

---

[73] Weyl to Mulder 29 July 1910.

Paul Koebe, the prince of the automorphic functions, was an ambitious function theorist, one year younger than Brouwer, born in Luckenwalde on 15 February 1882. He studied mainly in Berlin, where he wrote his dissertation under H.A. Schwarz. Subsequently he moved to Göttingen, where he was successively a *Privat Dozent* and later an extra-ordinary professor. In 1911 he was appointed full professor in Leipzig, where he stayed until his death on 6 August 1954.[74]

Koebe's mathematical oeuvre consisted almost solely of complex function theory, with emphasis on uniformization and automorphic functions. Between 1905 and 1941 he wrote 68 papers, among which there is a number of lengthy ones (eighty pages is no exception). When Brouwer was still one of the nameless academic crowd, Koebe had already earned fame by his solution of the uniformization problem, which put him in a bracket with Klein and Poincaré! During his life he got his share of recognition; a paper of his was awarded a prize by the King of Sweden, he got the Academic prize of the Royal Prussian Academy of Sciences (1914), and the Ackermann–Teubner prize (1922). Koebe was well aware of his importance and apparently did not try to hide it, which made him a natural target for some irreverent gossip.

Paul Koebe.
Niedersächsische Staats- und Universitätsbibliothek Göttingen.

Brouwer's sudden involvement in the defunct theory of automorphic functions at first seems rather surprising. Of course, Brouwer knew his complex function theory, and he had demonstrated his ability to handle the problems of analysis; yet, there were other things to consider for a budding topologist. The first hint of an interest in the problem area of uniformization may be found in his letter of 15 October 1909 to Hilbert (Cf. p. 143), in which he asked for some reprints including some of Koebe. Apparently Hilbert had discussed a number of topics with Brouwer during his vacation in 1909 in Scheveningen (cf. p. 130). It is not unlikely that Hilbert brought up the solution to the uniformization problem, which at that time was barely two years old. If so, he could very well have called Brouwer's attention to the continuity method. There are no documents that confirm Brouwer's interest in automorphic functions, but he must have considered the problem.

Light is shed on the activity of Brouwer by a letter from Brouwer to Poincaré, written after the Karlsruhe meeting (see below), most likely in the early days of December 1911. Not only did Brouwer state that his invariance of domain theorem was inspired by Poincaré's *Méthode de Continuité* in the *Acta Mathematica*—he also wrote that he had recognized that the continuity method was a consequence

---

[74]Cf. [Kühnau 1981].

of the Invariance of Domain, but after reading that Poincaré considered his 'exposition of the continuity method as perfectly rigorous and complete', he had started to fear that he had poorly understood Poincaré's memoirs. And so, he said, he had restricted himself to an oral communication at the Karlsruhe meeting of the German Mathematical Society, instead of adding it as an application to the Invariance of Domain-paper.

Thus Brouwer saw the solution late 1910 (or early 1911), but did not quite trust his judgement, and left the matter until the Karlsruhe meeting. Although he knew his complex function theory as well as the next man, he was no expert in automorphic functions. During a visit to Aachen he discussed the matter with Blumenthal, as appears from the last lines of the letter of 19 June 1911 that deals with Lebesgue's defective proof of the paving principle:

> ... and finally, I would like to learn from you which theorem of Analysis Situs that you mentioned to me in Aachen, as strictly necessary for the Continuity Proof of the existence of polymorphic functions on Riemann surfaces. From your latest letter I seem to conclude that it is not the Jordan theorem.

Blumenthal answered the question two months later, replying that he was no longer up to date in the theory of automorphic functions, but that he would be surprised if just the Jordan theorem, or even its converse, would do the trick.[75]

By that time Brouwer must already have solved the problem, for he was at the last minute added as an invited speaker to a symposium on Automorphic Functions at the annual meeting of the German Mathematical Society, DMV, from 27 to 29 September 1911. The first announcement listed Klein, Fricke and Koebe as speakers; Brouwer featured as an invited speaker in the second announcement. It is likely that he informed the organizers (in particular Klein) of his justification of the continuity method, and was immediately included in the list of speakers.

Brouwer's contribution to the automorphic function-session consisted of a redemption of Poincaré's original uniformization proof, by means of the filling of two gaps. The first one is of a technical topological nature, and the second one required the invariance of domain, which Brouwer had already proved, but not yet published, although Klein, of course, knew about it.[76]

Brouwer's talk was followed by a discussion, in which Koebe made a number of remarks. The reader who checks the printed report of the meeting[77] will be struck by the lack of coherence between Koebe's claim—namely that he had been able to prove the continuity property for *all* of Klein's fundamental theorems using Klein's general method—and Brouwer's rejoinder. The explanation is given in a letter from Brouwer to Hilbert of 24 February 1912:

---

[75]Blumenthal to Brouwer 26 August 1911.

[76]The Invariance of domain-paper was sent to the *Mathematische Annalen* on 14 June 1911, published November 1911.

[77]JDMV 1912.

Lieber Herr Geheimrat!

I am asking you for help and protection in a very unpleasant matter. [......] On 2 January I sent Koebe a copy of my letter, which I sent to Fricke in December and which has been submitted to the *Göttinger Gesellschaft der Wissenschaften*, and received roughly a week later the enclosed card. This card was followed on 14 February, not by the promised manuscript, but by the enclosed letter (together with my reply), in which I have underlined the statements to which my rejoinder refers.

Koebe can, however, not really mean the indicated claim, just as little as anybody who has heard my Karlsruhe talk. Therefore I can only perceive in Koebe's statement his objective to lend the matter in his next note the appearance, as if my letter to Fricke contained certain ideas, which I should have learnt from discussions with Koebe, whereas the real state of affairs in Karlsruhe was that I presented a complete continuity proof for the 'Grenzkreisfall', but that Koebe only contributed a vague inkling, that perhaps something could be done with his *Verzerrungssatz* [78] in the continuity method.[79] He said in the session of 27 September at the end of my talk the following: 'Since by my *Verzerrungssatz* nothing can happen under continuous modification of the modules, in my line of thought the efforts of Mr. Brouwer in the case of the difficulties of the invariance of domain and the absence of singularities of the module manifold are superfluous.' Whereupon I (emphatically) retorted: 'The *Verzerrungssatz* can only extend the result that Poincaré reached for the case of the *Grenzkreis* (boundary circle) and thus at the same time extend my continuity proof to the most general fundamental theorem; my contributions are, just as before, required in their full extent'. Thereupon Koebe pronounced the following nonsensical words: 'What Brouwer has done, I do with Poincaré series', and then Klein closed the discussion.

The report in the *Jahresberichte* contained a totally different version of Koebe's remarks. No mention was made of the *Verzerrungssatz*, but in Koebe's contribution to the discussion the claim was made that already before the meeting Koebe had carried out the continuity proof for *all* the fundamental theorems of Klein. Brouwer's reaction, according to the report of the discussion, consisted of the remark that the topological difficulties of the continuity proof came to the absence of singularities in the module manifold, and the invariance of domain, which were solved in Brouwer's talk for *all* of Klein's fundamental theorems. He added that these were neither mastered, nor avoided by Koebe's claim, but that a combination of Brouwer's method and that of Koebe were sufficient to cover the general case

---

[78]Cf. [Behnke, H, and Sommer, F. 1955].

[79]..., *Koebe aber nur eine gewisse Ahnung, dass sich etwas mit seinem Verzerrungssatz in der Kontinuitäts Methode lasse, mitbrachte.*

of Klein's fundamental theorem. The formulation suggests that both Koebe and Brouwer edited their parts in the discussion at a considerable later time.

It is all the more surprising therefore that Koebe in the *Fortschritte* of 1912 (published in 1915), reviewing the report in the *Jahresberichte* used the opportunity to modify his discussion contribution in the direction of Brouwer's letter to Hilbert: 'Koebe states that he can carry out the continuity proof for *all* fundamental theorems by means of the *Verzerrungssatz*'; Brouwer's remark is not mentioned at all. Koebe was having a field day anyway in this particular volume of the *Fortschritte*: he reviewed four of his own papers plus a paper of Plemeljs on uniformization. His own papers got lengthy extensive reports (at least in comparison to the traditional length of the reviews in the *Fortschritte*), and even Plemeljs' review was used for unabashed self promotion.[80]

Brouwer described, in his letter to Hilbert, his experiences during the Karlsruhe meeting:

> Only after longer private discussions, in which also Bieberbach, Bernstein and Rosenthal participated, Koebe learned after the talks, from 27 till 29 September, from me which partial result (by the way, formulated by Klein in the *Mathematische Annalen* 21, and at that time called the Weierstrass theorem) can be obtained via his 'Verzerrungssatz' and which part, to be settled by my contribution, still remains necessary. And, as the above mentioned gentlemen know precisely, in these discussions I have mentioned all the details of my present note.
>
> However, already at that time several warning voices said to me: 'All that, you have explained now to Koebe, you will only with the greatest effort be able to claim as your property, as soon as he has understood you', and indeed Koebe displayed some symptoms that seemed to bear out those voices. So when I was at home again I wanted to refrain from any publication on that particular topic, which is anyway rather far from my interest and with which I only casually concerned myself at the request of Klein, in order to avoid an unpleasant fight with Koebe.
>
> Only after Blumenthal prodded me, and after I had, moreover, heard that Klein would like to see a publication from my hand, the note of the thirteenth of January[81] came forth.

A considerable number of letters were exchanged in the Brouwer–Koebe conflict, but not all of them have been preserved; Brouwer wrote to Poincaré, and Koebe turned to Klein and Hilbert, and Brouwer and Koebe exchanged a number of letters.[82] As Brouwer explained in a letter to Hilbert on 31 May 1912, he had reluctantly intervened in a matter that he had rather wished to avoid:

---

[80]He quoted the author: 'The simplest and most natural of all these proofs [of the uniformization theorem] is the first one given by Koebe.' Modesty was not one of Koebe's defects.

[81][Brouwer 1912D].

[82]Of which only a few have survived.

For a better preparation of our coming discussion, two letters are enclosed, that explain to you the somewhat obvious question, why I had anything to do with Koebe at all. As a matter of fact my Karlsruhe contribution to the continuity proof consisted of two parts, of which the first one (the invariance of domain) had already been submitted for publication by itself in July,[83] whereas with respect to the second part (the extension of the group set to the set of automorphic functions with m poles), according to the enclosed letter of Bernstein (...) Koebe claimed priority. Since this second part did not seem very deep to me anyway, I of course hesitated to publish it, although Blumenthal pressed me to do so. Finally, in the beginning of November, I sent the manuscript of my talk to Fricke, with the question if he thought the contents were new, and deserved to be published. The information on Koebe (marked in pencil) that it contained, complicated the situation to the extent that I, when shortly thereafter, both Blumenthal and Fricke and also Klein (namely indirectly through Fricke) requested me to publish, I could not possibly do so, without contacting Koebe in order to get more certainty and clarity concerning his accomplishments, for otherwise I would be in danger that Koebe would accuse my publication of triviality and myself of plagiarism. In the correspondence with Koebe I then got only evasive answers to my specific question. The only thing I could get out of him was a mutual agreement to edit our notes on the continuity proof in [full] co-operation.[84] How he later broke his word and dragged out the matter, is known to you.

When Brouwer realized that his excursion to the kingdom of automorphic functions was going to be more than a quick trip, he wanted to make certain that he was not committing an error, so he wrote Poincaré,[85] asking him about details of his version of the continuity method. Although he was not personally acquainted with Poincaré, we may safely assume that the latter was aware of his work; Korteweg had written at some time in 1911 a letter to Poincaré, asking for a letter of recommendation for his student Brouwer. Since Korteweg's draft (which is the only surviving document from that correspondence) is undated, Poincaré may or may not have received the extra information on Brouwer before Brouwer wrote to him. It seems likely that Brouwer's solution to the dimension problem, coupled with the Lebesgue conflict had not escaped Poincaré's attention.

A number of letters was exchanged between Poincaré and Brouwer, but only a few have survived. One might wonder why Brouwer did not simply write to Klein, with whom he had already corresponded; one answer may be that Brouwer wanted to ask some specific questions relating to Poincaré's work, another might be that he trusted Poincaré's judgement in this matter better than Klein's.

The answer was dispatched on 10 December, and contained a brief indication

---

[83] [Brouwer 1912G].
[84] I.e. the notes for the *Göttinger Gesellschaft* submitted by Klein and Hilbert on 13 January 1912.
[85] Undated letter, mentioned above.

of Poincaré's solution.[86]

On 2 January 1912, Brouwer had indeed sent Koebe his manuscript of the '*Letter to Fricke*'. The latter played for time and did not send his manuscript in return, but instead wrote on 2 February 1912 a letter to Brouwer in which he listed his objections to certain passages of Brouwer's manuscript, even suggesting a formulation of a footnote that would do more justice to Koebe.

The manuscript of the 'Letter to Fricke' was already in the hands of Klein, who

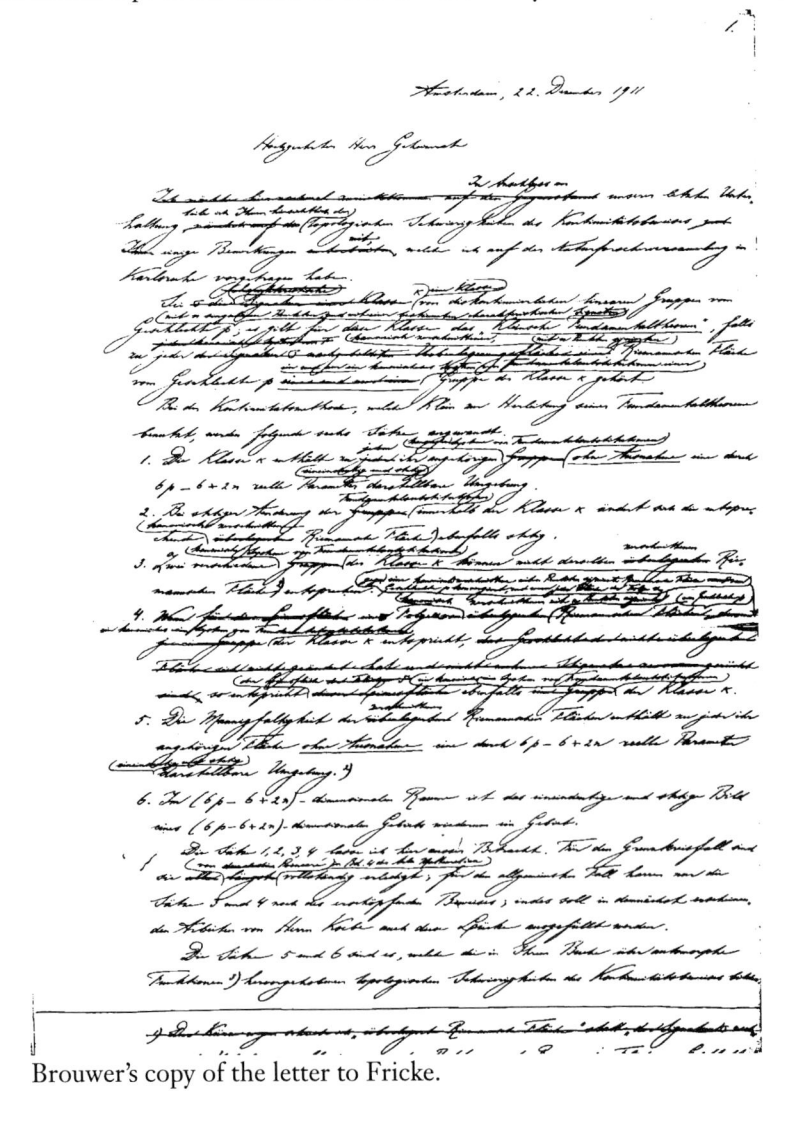

Brouwer's copy of the letter to Fricke.

---

[86] Cf. [Alexandrov 1972], [Zorin 1972].

submitted it on 13 January to the *Nachrichten der Göttinger Gesellschaft*. There is a draft of a letter from Brouwer to Klein, which shows that Klein had already gone over an earlier version (the real letter that Brouwer wrote to Fricke, most likely), and suggested some changes,[87] Brouwer must have met Klein's wishes.

It is not a coincidence that Hilbert presented at that same meeting a note of Koebe on the same topic 'The foundation of the continuity method in the area of conformal mapping and uniformizations'. The two notes were the ones that both parties agreed 'to edit in full consultation'. Brouwer had kept his part of the agreement, as we saw above, but Koebe simply ignored his obligation. Brouwer, not surprisingly (and probably quite correctly) concluded that he who refuses to send a promised proof, does not have one.

The 'Letter to Fricke' contained a summary of the theorems required for the continuity proof of 'the general fundamental theorem of Klein'. There are 6 theorems listed, the last two of which were singled out by Brouwer for further attention. One of them (no. 6) was literally the 'invariance of domain theorem', and the other one dealt with a technical aspect of Riemann surfaces, and Brouwer indicated how it could be avoided by a slight modification of the continuity proof. As to the remaining theorems, he remarked that Poincaré had already proved them in the case of the *Grenzkreis*, and that

> for the most general case only theorems 3 and 4 still resisted an exhaustive proof; in the meantime this gap too will be closed in forthcoming papers of Mr. Koebe.

Koebe objected to this particular passage; in the above mentioned letter of 12 February 1912 he suggested the following formulation to Brouwer:

> for the most general case, in particular for the theorems 3 and 4 and A,[88] the exact justification is still lacking, which however, Mr. Koebe, as a consequence of his tentative communication in the *Gött. Nachr.* (see in particular also the newest communication 'Foundation of the continuity method in the area of conformal mappings and uniformization' (1912)), has succeeded in solving completely, as he will shortly exhibit extensively in the *Mathematische Annalen*. The proofs, given by Mr. Koebe, also cover the only case considered by Poincaré, that of the *Grenzkreisuniformisierung*, and point—through the liberation from the idea of limit polygons and closed continua, introduced by Poincaré and borrowed from Klein–Fricke—to the decisive *life-giving*[89] progress, which at the same time means a return to Klein's old viewpoint of non-closed continua, which was vehemently attacked by Poincaré! Koebe's continuity method, by the way, yields another remarkable fundamental innovative distinction with respect to Klein, as Koebe in fact does not use theorem 4, although,

---

[87] Cf. Freudenthal's commentary; CW II p. 581 ff.
[88] One of the objections of Koebe to Brouwer's note.
[89] *Lebenspendend*.

> as Mr. Koebe told me, this theorem can indeed, with the help of the *Auswahlkonvergenzsatz*, be proved in the framework of Koebe's methods of proof.
>
> N.B. [This] can best be incorporated as a footnote, since it is not part of the letter.

Brouwer, quite sensibly, refused to comply, he sent a sharp reply to Koebe,[90] in which he answered some of Koebe's objections, for example Koebe's claim that Brouwer's method covered only the case of closed manifolds:

> Fortunately I am still in the possession of the abbreviated text of my Karlsruhe talk, which I enclose, so that you can no longer maintain that I used in Karlsruhe in the talk or in discussions the '<u>closed</u>' manifolds of Poincaré!
>
> That you could make such a statement, by the way, only proves that modern set theory must be absolutely unfamiliar to you. For the developments of Poincaré that work with the so-called 'closed' manifolds are pure nonsense and can only be excused by the fact that at the time of their conception there was not yet any set theory.

The letter closed with

> 'Why don't you send me a copy of your manuscript, as I did, and as you promised me?'

Ten days later Brouwer sent Hilbert the above cited letter, enclosing Koebe's letter. Brouwer complained that the promised manuscript did not come forth. The letter was accompanied by a copy of Brouwer's refutation of Koebe's points 'on which the rejoinder bears, which I have marked in blue (everything else is nonsense).'

Alas, Koebe again failed to send his manuscript, instead he wrote Brouwer (6 March 1912) that he was looking forward to the published version of the Karlsruhe-talk, but that he protested against the 'letter to Fricke', for: 'the arbitrator-like exposition given there does not give you permission to put the accomplishments of Poincaré and me in an unworthy and false light'. The promised exchange of manuscripts was not even mentioned! Thus Brouwer gave up hope of ever getting Koebe's manuscript. On 9 March he wrote to Hilbert:

> After sending you my latest letter, I got the enclosed card from Koebe. It neither brings the retraction of his false claims, demanded by me, nor the promised proofs of his note that he owes me. I now must give up hope of his return to common sense, and I therefore ask you to have my note for the *Göttinger Nachrichten* printed. All the same, it is important to me to answer here, for your information, Koebe's objections against my note. [...] Here Koebe is moving around in a circulus vitiosus; for on the one hand he demands me to praise his as yet unpublished work extensively, on the other hand he tries to prevent me from learning its contents.

---

[90] 14 January 1912, see CW II, p. 585.

After pointing out that Koebe was not aware of Fricke's *Würfelsatz* and its consequence for Brouwer's proof, and a fact about singularities, he closed the letter with:

> That the forthcoming note of Koebe does not contain falsehoods or insinuations is after all even more in Koebe's interest, than mine; for, in a possible refutation I could probably not avoid disgracing him irreparably.

Brouwer, nonetheless, was not unreasonable, he inserted on 20 May 1910 a small change in the first proof of the 'Letter to Fricke':

> ..., in the meantime Mr. Koebe has told me that in forthcoming papers he has completely filled the gap*)

---

*) There is already a tentative remark on theorem 3 in a note, submitted to the *Göttinger Nachrichten*

As Brouwer had written to Hilbert,[91] he did not object to quote Koebe in this particular detail, because

> I have myself verified in all detail that theorem 4 can be derived completely and generally from the *Verzerrungssatz*.

When this 'Letter to Fricke' appeared in the *Nachrichten*, Brouwer was confronted with an unpleasant surprise: the contested passage, quoted above, was changed into

> ... in the meantime Mr. Koebe has succeeded in filling this gap too. [2]

---

2) Cf. his tentative note of 13 January 1912 in the *Gött. Nachr.* 'The foundation of the continuity method in the area of conformal mapping and uniformization' and 'On the foundation of the continuity method' in the *Leipziger Berichte*.

The change is not spectacular, but it slightly strengthens the statement, and it suggests that Brouwer had seen, and agreed with the paper quoted. It does not, however, give credit, let alone priority, to Koebe for his Karlsruhe claims. Freudenthal described the episode in the *Collected Works*, vol. II (p. 575):

> Brouwer's note of 1 May 1913 makes an allusion to another incident. [Brouwer 1912D] had appeared with a change in the main text and in a footnote. The change implies a more positive acknowledgement of some claims of Koebe though not of those related to the continuity proof. This *unauthorized* change was signalled by Brouwer.
>
> Oral tradition tells a cloak-and-dagger story about this footnote: On some dark afternoon in March an unidentified person wearing a large hat, a turned up collar, and blue glasses called at *Dieterick'sche Univ.-Buchdruckerei W.Fr. Kaestner* in Göttingen, the printing office of *Göttinger Nachrichten*, and asked for the printer's proof of the next issue. He got it, and after a while he gave it back and left. The identity of this person has never been determined, nor is it known whether he made any change in Brouwer's reading proof, which of course disappeared after printing. I do not know how much of this story is true. To a trustworthy friend of

---

[91] Brouwer to Hilbert 9 March 1912.

mine who years later asked him about this incident, Koebe explained it as a trick somebody played on him. Though the revized edition of the footnote gives information which at that time was not publicly available, the hypothesis that it was a practical joke cannot at all be excluded in the Göttingen ambiance. Koebe was a picturesque character whose honesty and frankness forbade him to disguise his greatness as a mathematician; in order to escape embarrassing admiration he travelled incognito, and he often said that in his birthplace Luckenwalde the street boys shouted after him *Da geht der grosse Funktionentheoretiker*.[92]

Freudenthal told that when he was a student in Berlin, Bieberbach once heard that he came from Luckenwalde—'So you are one of the Luckenwalde street boys who run after Koebe to call 'there goes the famous function theorist', are you?' (which earned Koebe the nickname of *der grösste Luckenwalder Funktionentheoretiker*[93]). Koebe considered himself the rightful objective of unlimited admiration. He preferred not to register in hotels as 'Koebe', he travelled incognito because he could not stand waiters and housemaids asking him whether he was a relative of the famous function theorist,[94] and one of his celebrated sayings was *Europa spricht davon—Koebe versendet Separate*.[95]

Van der Waerden added another characteristic anecdote about Koebe: one time Landau gave a party for his Göttingen colleagues and Koebe was also present. Landau, who enjoyed jokes, at one point asked all his visitors to write on a piece of paper the name of the person who thought himself the most important mathematician. Everybody filled in 'Paul Koebe' and when it was Koebe's turn, he wrote 'Paul Koebe, and justly so' (*Paul Koebe und mit Recht*).[96]

It should perhaps not come as a surprise that Koebe, as a professor, had pronounced views on the relative importance of the various topics for the curriculum. For example, he told his students that they did not have to take Van der Waerden's classes. In Van der Waerden's eyes, this made him not just vain, but also rude.

> That such a character should be the target of a practical joke, in a place where practical jokes were not unusual, is not a farfetched hypothesis. In a letter to Klein of 25 May 1914 Koebe complained about gossip spread about him in Göttingen, related to the sticker in [Brouwer 1912H]. It seems incredible that for nearly two years Koebe had been unaware of the whole affair, but this does not prove hypocrisy. Gossip, if unjustified, goes a long way before it finally reaches the ears of the ones incriminated. From Brouwer's correspondence with Klein one can infer that Brouwer discovered the unauthorized change in [Brouwer 1912D] in the last week of June 1912. Koebe persevered in withholding the manuscripts of his

---

[92] There goes the great function theorist.
[93] The greatest function theorist of Luckenwalde.
[94] [Freudenthal 1984].
[95] All of Europe talks about it: Koebe is mailing reprints.
[96] Van de Waerden to Van Dalen 25 February 1992

publications even after they had been referred to in [Brouwer 1912D] without Brouwer's knowledge. It seems that even Klein did not succeed in changing Koebe's mind (Letter to Klein of 1 July 1912).[97]

The violation of his final proof sheets was a traumatic experience; Brouwer never again trusted the sanctity of the printer's shop again, it was not unusual for him to demand his proofs to be locked into the safe and the keys to be given to him.

Whoever pulled the trick on Brouwer, and by implication Koebe, must have been well informed. He changed exactly the part that Koebe had objected to. Koebe must somehow have been left out of the gossip circuit; in a letter of 25 May 1914, he wrote to Hilbert that he had been informed that Brouwer had put a sticker in his reprints, which contained a tacit improper reference to him.[98] 'As a result', he complained, 'all kinds of insinuations are circulating in Göttingen'.

We are running ahead of our history, however.

At the end of March Koebe wrote a soothing letter to Hilbert,[99] saying that he was about to send his proof sheets to Brouwer (but apparently he did not do so) together with Brouwer's own copy of his Karlsruhe lectures—but he had not been able to read that note more in detail because of lack of time. Anyway, he assured Hilbert, 'There is not the least reason to worry about a threatening priority conflict'.

Koebe's main tool was the *Verzerrungssatz*, and it enabled him to get round Poincaré's closure trick, but not in all cases. By the time he got to write his note for the report of the Karlsruhe meeting, he must have known that his initial belief, that the *Verzerrungssatz* could replace the invariance of domain, was unfounded. The aggressive tone of his letter of 12 February 1912 to Brouwer gradually made way for a more objective one, and in Koebe's paper 'On the theory of conformal mappings and uniformization'[100] Brouwer's view is done full justice.

When the report of the Karlsruhe meeting appeared, Klein committed a beautiful understatement in his introduction, by stating that the talks were followed by a 'stimulating discussion'. Brouwer was less than satisfied with the outcome, as may be clear from his sticker action. He was in particular incensed by the fact that Koebe had used the time between the receipt of Brouwer's manuscript and the final printing of the issue of the Jahresberichte to rewrite the history of the continuity method.

The brief but violent involvement of Brouwer into the domain of uniformization and automorphic functions ended formally with a paper *Über die Singularitätenfreiheit der Modulmannigfaltigkeit*,[101] in which he attacked the problem of the singularities in a direct way.

The commotion over the Karlsruhe talks soon died down. Fricke, in the Fore-

---

[97] Freudenthal in CWII, p. 575.
[98] See CW II, p. 571. The sticker summed up Brouwer's grievances mentioned above.
[99] Koebe to Hilbert 29 February 1912.
[100] [Koebe 1914].
[101] On the absence of singularities of the module manifold, [Brouwer 1912G].

word to Volume II of the 'Lectures on the theory of automorphic functions' (1912), lavishly praised Koebe (who indeed gave the impetus to the automorphic function saga, after the Klein-Poincaré episode) and the 'through his set theoretic papers renown[102] L.E.J. Brouwer'. The report on Brouwer's achievements was nonetheless somewhat inaccurate. Brouwer had already in the fall of 1910 informed Fricke and/or Klein of the progress involved in his invariance of dimension theorem, but the final result, the invariance of domain, was in the 'Lectures on automorphic functions, II' still announced as an open problem.

Brouwer did not forget the incident, and when he returned to the subject in 1918,[103] he inserted a footnote with a number of historical comments (partially coinciding with the sticker in the reprints of the Karlsruhe report): 'The quotation on page 604 of it [that is the letter to Fricke], [Brouwer 1912D], that refers to future publications of P. Koebe (who was at New Year 1912 in the possession of a copy of my letter to R. Fricke) has been inserted after the completion of the proofs by a person unknown to me,[104] without my knowledge or compliance; the footnotes in question became known to me only after their publication ... '.

In 1922 the Brouwer–Koebe conflict had a late revival connected with Klein's collected works.[105] Vermeil, an assistant of Klein, was preparing the third volume, which also dealt with the automorphic functions. He asked Brouwer permission to reproduce his part of the Karlsruhe proceedings. Brouwer answered, somewhat crossly, that the report of the genesis of the continuity proof of the general fundamental theorem of Klein was grossly misrepresented, and so permission could only be granted if Brouwer's own rectifications were incorporated. He added that Vermeil could, when in doubt, ask Bieberbach and Bernstein about the matter, both were present at the meeting. Bieberbach, said Brouwer, had made his view clear by referring to the Brouwer–Koebe continuity proof, whereas Koebe tended to advocate the terminology 'Koebe-proof' (or if it must be, 'Koebe-Brouwer proof').

Klein, in his comments in the *Gesammelte Mathematische Abhandlungen III*, on the continuity method, did not want to commit himself to either of the contestants (after more than 10 years!) and so when he came to the crucial point (p. 734) he appealed to the alphabetic convention, and wrote 'Brouwer–Koebe'. Klein's summing-up did Brouwer full justice, as it did Koebe. But one may guess that the style of Koebe was closer to Klein's heart, which hearkened back to the golden time of the eighteen-eighties. Even Koebe, first in the letter to Klein of 30 March 1912, and later in his publications gave Brouwer his due, albeit as an outsider who had the good luck to provide an essential tool. Koebe's wording of Brouwer's role in the theory of automorphic functions could easily be interpreted by a lesser man than Klein as if Brouwer more or less unwittingly had provided a key tool for the theory of automorphic functions, but that it took a man like Koebe to realize the

---

[102]*rühmlichst bekannt*.
[103][Brouwer 1919L2].
[104]*von einer mir unbekannten Hand*.
[105][Klein 1923].

significance of the 'invariance of domain theorem' for the continuity method. In his survey paper '*The essence of the continuity method*' of 1936[106] he referred to Brouwer in a way that did little to clarify the history of the continuity method:

> This, also in the following, for the final general foundation of the continuity method, important theorem [invariance of domain], was proved in fact for the first time by Brouwer.

The reader who consults Klein's History of the mathematics of the nineteenth century will find to his surprise that only Koebe is credited for saving the continuity method.

In *Zum Kontinuitätsbeweise des Fundamentaltheorems*,[107] however, Klein gave full credit to Brouwer. He even adopted Brouwer's presentation of the solution of the continuity method.

Brouwer's name has almost disappeared from the theory of uniformization, and even the textbooks that mention his name do not make clear what his contribution was. It certainly is the case that nowadays uniformization is carried out by more advanced methods, so the modest place of Brouwer in this respect is not so surprising after all. Even Koebe's name is no longer guaranteed a place in the treatises on the topic.

This chapter has told the story of the birth of the topology of the twentieth century. Brouwer's role in the process has undoubtedly been of the greatest importance; he gave topology new tools and showed how to use them; doing so he had a view of the promised land, but he did not enter it himself. With his knowledge of topology and algebra he could easily have developed algebraic topology, and thus reaped the harvest that was within his reach. That he did not do so is a surprise that is hard to explain. One possible explanation is that he was a geometer by nature, he liked to 'see' his mathematics rather than to embed it in a calculus. This is undoubtedly a valid point; nonetheless one can imagine that if Brouwer had been familiar with Poincaré's ideas about homology, he might have been carried away by the elegance and power of these tools. Who knows what power a beautiful theory may have over a man? On the other hand Brouwer did not have the common urge to exhaust an area of research, a career like Koebe's, consisting of a life of toil, polishing results, extending the theory a bit, trying an alternative approach, ..., was not his idea of mathematics. There was too much of the artist and the free spirit in him to be tied to a rigorous, ambitious research program. Finally, we must not forget that the foundations of mathematics were beckoning—there work was to be done as well.

There is a penetrating sketch of Brouwer's views in the book of Wiessing,[108]

---

[106] *Wesen der Kontinuitätsmethode*, 'after lectures held at the meetings of the German Mathematical Society in September 1913 in Vienna and September 1935 in Stuttgart', [Koebe 1936].

[107] On the continuity proof of the fundamental theorem, [Klein 1923].

[108] [Wiessing 1960] p. 143 ff.

who loosely interviewed his friend during one of Brouwer's rambles on the heath. It clearly shows Brouwer's aversion to treadmill of the mathematics industry.

> W: What kind of a figure are you in the world? What is your place in mathematics of the present and of the past? [....]
>
> B: I could only give you an impression in very global terms, but I am willing to try. You must imagine [...] that in the course of the centuries an increasing number of more or less mutually independent branches of mathematical thought have emerged, some of which, that up to now have found no material applications, may perhaps be used in connection with certain physical phenomena. [...] Such a future subservience to the physical sciences is in my opinion fairly unlikely for the branch of mathematical thought that I try to get accepted alongside the existing branches, and which distinguishes itself from those, among other things, by not treating a [particular] subject and by not recognizing axioms and postulates. [....] Basically my mathematical thinking is non-sensory internal architecture. You may compare these forms of thought to music or poetry. My first inklings of the possibility of such a mathematics emerged, I think, from discussions with my teachers at the time of my HBS and gymnasium study.[109] But only in my dissertation of 1907 I have started to give these thoughts a definite formulation. Since then this mathematics, nowadays called 'intuitionistic', has developed with interruptions. The recognition in professional circles of this work of mine came only in a rather slow tempo, with many ups and downs. It has by no means found general acceptation! Many view it, even now, as charletanism. There are also people, who say that it may be correct, but that it is totally uninteresting and not even new. If I had not, now and then, written about 'ordinary mathematics', I don't think a place at a Dutch University could have been found for me.
>
> W: But you have had, already in our early years, offers of professorships from abroad, I remember that all too well.
>
> B: Yes, that is right, but in thought and activity I feel far too much a Dutchman and in particular a Friesian Dutchman. And I would rather live here between Dutch friends to enjoy, and Dutch enemies to see through, than far away among strangers!
>
> W: What kind of mathematics do you teach now after all as a professor?
>
> B: Mainly ordinary mathematics, which I transpose here and there, depending on the degree in which topic and the receptivity of the audience allows this, into my own system of thought. Just now and then, when a particularly gifted and interested group of students presents an occasion, a course consisting exclusively of my own work evolves, and this course may extend over several years. On such an occasion I try to educate students,

[109]Cf. p. 4.

to whom I could eventually trust the further extension and dissemination of my theory.

Several former and present members of my audience (among them some foreigners, who have come to live here in Blaricum for that purpose) give me hope in that respect. And yes, if this will be fulfilled, that will give me great pleasure! Because I indeed really love mathematics for something other than mathematics. That is for the clear light it sheds sometimes on the general problems of life. And ultimately it is in the first place the problems of life that make my natural flow of thought find its way.

W: Could you formulate the special character of your mathematics in a way that is a bit clearer for laymen?

B: Well, let me try once more like this: although it will be again clumsy: consciousness gains access to free creation —which is my mathematics— as soon as it knows itself autonomous and immortal, ignoring objective knowledge and common sense. A condition, in my opinion, for all creation of truth and beauty.

W: I can imagine that you were occasionally called a charlatan.

B: Me too. All the more, as I said, since I don't like mathematics and it basically bores me.

W: [...] What would you, if it comes to that, rather have done than practise mathematics?

B: That is hard to say. Let me say: to have no subject and let my thoughts roam freely. Every attachment to a subject brings, as you will agree, that your realm of thought suffers a certain mutilation. And it is obvious that then one can only have pleasure in such a profession, if one is, as I sometimes observe with some people, supported and driven by ambition or conviction. But that has never been the case with me. Anyhow, life demands that you choose a profession. Well, then I think that science is for a man like me, who is by nature solitary, not such a bad sanctuary. One is less dependent on the public, and one can more easily preserve one's solitude, than if one takes up literature or the visual arts, not to mention music. For no matter how much pleasure and satisfaction art by itself may give to a person, society, I think, demands more violating concessions from artists than science does. But, I may have reached this conclusion also because I have no talent for practising art.

Even in his intuitionistic enterprise of the twenties, he could not bring himself to grind out the routine material that was required to give the program a good start in life.

The conflicts we have seen in this chapter, had a double role in Brouwer's scientific life. The positive influence was that they forced him to stick to the subject, and to use his ingenuity for the purpose of widening and deepening his ideas. The influence of Lebesgue has been (although Brouwer would have been loath to admit it) beneficial. Without the constant challenge of Lebesgue's bluffing, Brouwer could have left the field much earlier.

The same cannot be said for Koebe's role; Brouwer had little taste for the kind of function theory that Koebe practised, he would not have engaged in a competition with Koebe! As he said himself, he considered his contribution to the theory of uniformization a pleasing piece of fall-out, but no more. From the correspondence it appears that Brouwer would gladly have left the matter at his Karlsruhe talk. It was the incredible childishness of Koebe to claim the whole area of uniformization and automorphic functions as his personal domain, that triggered Brouwer's fighting spirit. The moral misbehaviour of the king of automorphic function theory angered him more than anything else.

These conflicts asked a heavy toll, Brouwer was ever prone to the return of his nervous breakdowns. Moreover, they tended to give Brouwer the reputation of a difficult man. He certainly did not belong to the meek of this world, but his conflicts always were the result of some instance of injustice, be it towards himself or towards others.

The first and intense topological period ended as suddenly as it began. After the dimension paper of 1913 Brouwer returned to his first love, the foundations of mathematics. Even here there are external as well as internal factors that influenced the course of his activities. If World War I had not isolated him from the international mathematical community, he might possibly remained active in the field under the pressure of fellow topologists, but when his ties to Göttingen were temporarily cut, no significant impulses influenced him.

6
MAKING A CAREER

6.1 Financial worries

Any graduate of a Dutch University, until long after World War II, was confronted with the pressing question how to reconcile earning a living and practising one's subject. For some disciplines this was less problematic, for example medicine and law, but in general a university degree did not guarantee a future in one's own subject. A master's *(doctoraal)* degree in most of the sciences and the languages could at best earn the recipient a position at one of the high schools or gymnasiums.

Brilliant physicists or chemists might be lucky and get a university position, but even they often had to serve a spell as a highschool teacher. Of course, there were industrial positions, but they were scarce. Even the famous Van der Waals and Lorentz had, before their academic career, to teach at high schools. Mathematics offered even less hope for a scientific career, there were no industrial jobs to speak of (apart from an incidental position with an insurance company) and the only way that led to an academic career ran via the teaching profession—a route followed by most academics. The successful doctorandus taught for a living, while studying and publishing in the spare hours, working for a Ph.D. degree, and hoping and waiting for some old professor to retire or die. The bright side of this practice was that the standard of teaching at high schools and gymnasiums was generally high; many a future professor taught Euclid or Homer at a high school or a gymnasium—and was fondly remembered for it.

Doctorandus Brouwer was not looking forward to such a career, for one thing, he was temperamentally ill-suited for the drill and the discipline of school life. But above all, he had drunk from the fountain of mathematical knowledge. He was in no doubt about his capacities and he had, as we have seen with *Life, Art and Mysticism*, an almost fanatic drive for preaching morals and mathematics.

The grant, that had enabled him to study in reasonable comfort, did not extend beyond the doctorate; on 14 June 1907, the St. Jobs foundation transferred a final sum of DFl 224.10, and closed the books on Brouwer. This did not mean that Brouwer was entirely without financial means: on 1 September 1905 he had bought his mother-in-law's pharmacy, and so, one would think, his wife and he could expect a reasonable livelihood. Appearances were, as so often, misleading, as we shall see.

The pharmacy at De Overtoom in Amsterdam.

The pharmacy had a long history; the house itself dated back to about 1600. It was originally a well-sized tavern, called 'The Angler',[1] at that time it belonged to the town of Nieuwer Amstel (the predecessor of the present Amstelveen). From the last part of the eighteenth century on it was mainly occupied by surgeons. The father of Lize, J. de Holl set himself up in 1873 at the Overtoom as a family doctor with a surgery and pharmacy. The large building was not only a home for a rapidly expanding family, but it also housed those assistants and servants who were interned. In 1880 Jan de Holl died, 46 years old, leaving behind a widow and seven children (cf. p. 53). The widow, having no training or qualifications in the medical profession, had to close the medical practice, but she continued the pharmacy with the help of a so-called 'provisor', that is a professional pharmacist who carried the responsibility for the pharmaceutical side of the shop. This was required by law, it guaranteed the professional expertise. It was, however, a heavy financial burden on the widow De Holl-Sasse. Some of these provisors lodged with Mrs. De Holl, a circumstance that was grudgingly accepted by the family De Holl. The pharmacy had to provide a livelihood for Mrs. De Holl and whatever children were still dependent on her. Lize and her daughter Louise had, after the divorce from doctor Peijpers, moved in with Mrs. de Holl. Louise was effectively raised by her grandmother while Lize studied pharmacy. When Brouwer bought the pharmacy, he had to accept a substantial financial responsibility towards his mother-in-law and her dependents.

---

[1] *De Hengelaar*, formerly *De Boerendans*. We gratefully acknowledge the generous help of Mr. and Mrs. J.A.L. van Lakwijk-Najoan, who provided valuable information on the pharmacy of Brouwer.

The acquisition of the pharmacy by Brouwer was therefore certainly not the solution to his financial problems, on the contrary! The salary of the provisor still had to be paid, and Brouwer had accepted the contractual obligation of an annuity of DFl 1100.- for Mrs. de Holl-Sasse. His friend Carel Adama van Scheltema, among others, had to stand surety. The transaction confronted Brouwer with problems that his academic training had not prepared him for—at the last minute Brouwer discovered that the transaction carried a notary fee of DFl 800.-, a considerable amount, which was due within a week of the transaction.[2] It was again Scheltema, who had to rescue his younger friend, he lent the amount to Bertus on security of his house in Blaricum.

> I am terribly sorry that I have to scrounge like this: a friend that you hardly ever see, and then to get all sorts of horrible chores from him, must be painful and disturbing for you with your soul that is halfway that of a grocer. But your free poet's soul is indeed above such a thing, and 'generous' means 'royal'.[3] I just wanted to say that I know that such sacrifices bother you, and that I value your sacrifices. But you know that if the occasion arises you can count on me as firmly as you would on yourself. Well, bye now. The money must be there, and at once.
> Now, just do this calmly for me—with such a receipt with a pledge you truly put nothing at risk—and then go on and live your regular cautious artist life, that I, in fact, fear to disturb.
> Bertus.

Scheltema complied with a heavy heart; he could, in all fairness, not see, why he should be a security to guarantee Brouwer's mother-in-law a lifelong annuity of DFl 1100.-. We note in passing, that Mrs. de Holl lived until a ripe old age, so that, in as far as annuities are a kind of a gamble, Brouwer certainly was not a winner. His relationship with his mother-in-law always was strained, to say the least. The matter of the periodic payments definitely did not endear her to him. She occasionally suffered from mental breakdowns, and spent some time in mental institutions. On the whole she was a good and loving mother and grandmother, but the relationship with Brouwer remained uneasy, to put it mildly. Keeping in mind that Brouwer did not even get along with his own father, it would be too much to expect him to develop the obligatory filial love for his mother-in-law.

Although one should not speak of poverty, life was definitely not without its financial problems for the young couple. Matters improved when Lize, after successively passing her doctoral exam in September 1907 and her special pharmacists exam in December 1907, established herself in the pharmacy at the Overtoom, and ran the shop as a fully licensed pharmacist without the help of an expensive provisor. Lize's doctoral diploma caused no stir in the sober household; Brouwer reacted no more enthusiastically than if buying a new pair of shoes; Louise gave the day a

---

[2] Brouwer to Scheltema 23 August 1905.

[3] The pun is lost in translation: '*royaal*' stands for 'generous', 'handsome' and 'koninklijk'. Recall that Scheltema and Brouwer considered themselves kings of spiritual realms, cf. p. 37.

festive touch by buying a small bottle of wine and by cooking some floury potatoes which her mother liked so much.[4]

Passport photos of Lize and Bertus.

In spite of the resulting improvement of the financial basis of the pharmacy, the pecuniary worries persisted. One should keep in mind that in the first part of the century pharmacies were in general not the goldmines they became later. Moreover, even at the best of times, Lize was a poor hand at management. The correspondence between Brouwer and Scheltema, time and again, mentions late repayments, new loans, etc. Scheltema himself lived modestly but well on the money his father had left him, but after an American investment of his had plummeted (1907) he had to avoid extravagances.

By the time Brouwer's dissertation approached its final stages, he had already lost the unencumbered freedom of the student, he had published respectable mathematics, and he started to view the scientific community with different eyes. The transition from student to breadwinner did not leave him cold; society started to exact its dues:

> Life is a magic garden. With wondrous, softly shining flowers, but among the flowers the gnomes are walking, and I am so afraid of them. They stand on their head and the worst is that they call out to me that I must also stand on my head; once in a while I try to do so and die with shame; but then sometimes the gnomes cry that I do it very well, and that I am indeed a real gnome too. But on no account I will fall for that.[5]

Indeed, Brouwer was getting his share of recognition in Dutch mathematics; his brush with Jahnke had done him no harm, on the contrary—Korteweg's opinion of

---
[4] Oral communication Mrs. Peijpers.
[5] Brouwer to Scheltema 7 September 1906.

his student was confirmed, and the mathematicians P.H. Schoute (the author of a successful book on higher-dimensional geometry, professor in Groningen) and W. Kapteyn (professor in Utrecht, the brother of the famous astronomer) started to notice Brouwer's achievements.

After finishing his dissertation, Brouwer published a kind of philosophical appendix *The unreliability of the logical principles* (Cf.p.108). This paper contained the revolutionary rejection of the general validity of the principle of the excluded third—it remained virtually unknown until it was translated in the collected works in 1975. While continuing his study of modern mathematics, Brouwer prepared two papers for the International Mathematics Congress in Rome; both were offshoots of his dissertation, one on Lie groups and one on Foundations.

Most of the information on the period 1907–1908 is to be found in the correspondence with Scheltema and with Korteweg; the letters exchanged between Brouwer and Scheltema tell a tale of friends growing up into the world, and of dreaded but unavoidable separation. The friends were no strangers to the frailties of mind and body—almost verging on hypochondria:

> That what we have in common also starts to whither, if the soil, our common youth, dries up. Must we now, resign ourselves to that, and each for himself let things go in the coarse mansion of society, and light its chandeliers, and grace its door-posts?[6]

Even Scheltema, the artist, was caught up in the general malaise of defeated youth and past:

> I have given up the struggle for the 'superman'—I could reach him now and then, but not lastingly [...]. You wrote in a recent letter that we drift apart—perhaps, but there is such a deep difference in principle between us:—the other day I read Nietzsche's 'Birth of Tragedy' and then I thought of you and I felt things clearer; read it some day if you do not know it. You are *Dionysus* and I am *Apollinius*, and the world we live in is *Alexandrinius*.[7]

The characterization may not have been so far off the mark, a good deal of Dionysus could probably be found in Brouwer!

Just before Lize's final exams, the Brouwers made in August 1907 a walking tour through Belgium. Brouwer was a lifelong devotee of long walks. The tour to Italy has already been mentioned, but also in Holland he used to ramble through the landscape, sometimes accompanied by his youngest brother Aldert. There are reports that they made long hikes together through North- and South-Holland.[8]

Physical exercises were always a favourite pastime of Brouwer, he would gladly undertake some exhausting extravaganza. In the winter of 1908, for example, he skated, in a fierce wind and on poor quality ice, all the way from Amsterdam to

---

[6] Brouwer to Scheltema 8 July 1907.
[7] Scheltema to Brouwer 6 August 1907.
[8] Oral communication C. MacGillavry.

Rotterdam and back.⁹ The exertion proved too much, three days of high fever followed, and he asked for a notary to make his last will. The fever, however, disappeared as suddenly as it had come.

When Brouwer was not abroad, he spent his days in the hut in Blaricum, or in Amsterdam in the apartment over the pharmacy, where he often occupied himself with the administration of the pharmacy. Lize regularly stayed at the Overtoom, where she supervised the pharmacy. There is a well-known (and well confirmed) story that at times, when Lize was in Amsterdam and Bertus in Blaricum, Lize would take a pan with food to the point of departure of the *Gooische tram*, where it was put next to the driver's seat. At the other end Bertus would then pick it up at the stop almost in front of his house. The Gooische tram was the normal means of transport between 't Gooi and Amsterdam and Brouwer was a regular passenger.

This tram had acquired a certain notoriety as a result of the not infrequent accidents; it received the nickname of *Gooische moordenaar* (murderer of 't Gooi). The fearful potential of the tram was brought home to Brouwer, when Lily van der Spil, the fiancée of his brother Aldert, was run over by it, and lost both legs.[10]

Lily had studied *bouwkunde* (architecture) in Delft, where she met Aldert. The two married on 2 June 1909; they got three children, the eldest of them was named after Bertus.[11]

In the flurry of all the new mathematics that he absorbed, Brouwer worked wherever he could, but he definitely preferred the peaceful surroundings of Blaricum; he loved to bask in the sun in his garden with a wide brimmed straw hat, working on an old-fashioned draft board and, depending on the weather, in a state of partial undress.

It is surprising to read in a letter of Scheltema, that this man with the sharpest mind in the country, did not escape the lures of the irrational. The poet, who was in fact a far more sober-minded person than his friend the mathematician, scolded Bertus when Lize wrote him that Bertus was upset on account of a fortune-teller's prediction: 'shame on you for that silly nonsense'.[12] Even in the time of his budding productivity, Brouwer found time to keep up the correspondence with his old friend. The friends went on to deplore the steady detoriation of their original innocent state and to provide each other with words of solace and advice. Brouwer, in passing, mentioned his successes in mathematics, and Scheltema kept Brouwer informed about his literary achievements. There is one, somewhat enigmatic, theme that keeps cropping up in their exchange of letters, Scheltema repeatedly reminded Brouwer of some promise, described in vague terms. Before leaving for Italy, where he intended to spend a prolonged visit, he urged Bertus: 'If at all possible, fulfil your promise before I leave again'.[13] And, indeed, in May of the next year, Brouwer re-

---

[9] Altogether a distance of more than 130 km.

[10] Brouwer to Scheltema 2 June 1909.

[11] Like his father, he studied geology. He ended his brilliant career as the Chairman of the Board of the Royal Shell.

[12] Scheltema to Brouwer 11 October 1908.

[13] Scheltema to Brouwer 4 September 1908

ports that the promised object is ready, and that he had prepared one for himself as well. In his reply the next day from Florence (sic) Scheltema dwelled on the promised object and spoke of a talisman, that embodied a 'possibility' that can give rest, 'How many abysses of the soul go by, that alarm less with a talisman'
In a later letter he talked of 'Charon's pennies' and 'non plus ultra's'. One gets the strong impression that Scheltema, who feared and abhorred the tortures and indignities of a pitiless disease (his father had died of a brain tumour—this had left an indelible impression on Carel!) had asked Brouwer for two suicide pills from the pharmacy, one for him and one for his wife.

After some pressing reminders, Brouwer prepared the coveted objects, 'The promised object is ready for you. I have at the same time made one for myself too.'[14] And on 7 July 1909 he wrote Scheltema that 'The talisman and taliswoman (forgive me the lugubrious joke)–good for half a century–are quietly waiting. Perhaps that certainty will already put you at ease!'

6.2 First international contacts

In April 1908 Brouwer travelled to Rome to attend the international conference of mathematicians; there is a short report of his experiences and impressions in a letter to Korteweg.[15] The letter is well worth reading, as it paints the emotional impression of a young man who suddenly finds himself in the presence of the great ones.

> I am very glad that I made this journey. In the first place because it made me feel more healthy and vigorous; but even more because of the admiration and respect that I felt for mathematics under the impression of this congress, seeing and hearing the heroes of abstraction, and by the aura of 500 honest thinkers acting on me. Poincaré was a revelation to me, and also Darboux and Picard made strong impressions. In general I recognized in the impressive heads practically all those, for whose work I had got the highest regard. But furthermore, the sight of the persons generously provides hints as to the choice of authors that I will read later on. For example, I will never turn to Mittag-Leffler for instruction after seeing his superficial pompous face; but certainly to Darboux, of whose work I know as yet nothing.
>
> In general it seemed to me that the French are mostly the leaders, mostly command the central parts; Hadamard and Borel as well as the three mentioned above. They, rather than the Germans, seem to me to possess an instinct for what is truly important, a kind of aristocracy in the choice of their topics. To reach a point from where you can, like Poincaré, give an address on 'The future of mathematics' (*L'avenir des*

---

[14]Brouwer to Scheltema 12 May 1909
[15]Brouwer to Korteweg 20 April 1908.

mathématiques),[16] of which everybody feels the reliability as a guidance in his work, seems to me the highest ideal a mathematician can strive for. My respect for the Italians and Americans has not increased, and I am convinced more than before, that they cultivate unimportant parts, without much feeling for the direction in which the main body is moving.

I did not get much pleasure out of my talks; the one on continuous groups was the last one of the morning, and the five preceding speakers were so boring, that after each talk a part of the audience disappeared, and there were only thirteen left at my talk. Nonetheless there was one, who, as Versluys[17] said, followed [the talk] attentively from beginning to end, and whose manner of applauding—again according to Versluys— clearly showed that he had fully understood the purport. But when, after a few brief words with the chairman, I left the podium, the man had disappeared, and I have never seen his face, or found out who he was. The other talk on cardinalities had a much larger audience, but due to lack of time I was allowed to talk for only 10 minutes, so that it was so vigorously condensed that it was not done justice. But on account of the mass of other impressions, I did not get much out of my own talks.[18]

On the way back Brouwer stopped in Milano, presumably to see the city and the art collection. He had an extensive knowledge of Italian art and architecture, having already visited Italy before, and later in life he was a regular visitor. He loved to talk about his Italian impressions, and even in his old age he retained very sharp memories of Italy, he could give instructions of a surprising precision to his friends: 'when you come to the cathedral square in (say) Lucca, turn right at the little barbershop at the corner, enter the small alley opposite the post office and then ...'

When Scheltema spent a year in Italy, Brouwer advised him:

Listen Carel, I feel obliged to give you a gentle warning concerning Rome. It could be that you will not fare like me, but I was so fed up after a stay of 3 weeks in Rome, that I must advise you to make your arrangements for a stay of a year with great reservations.

I always had a hard time leaving Florence; there I would leap for joy, my whole life long; in Rome I would get on the train with a sigh of relieve; it is a suffocating, an oppressive, an evil place on our beautiful earth [...].

---

[16]As a matter of fact, Poincaré fell ill during the conference, and someone else had to read his lecture. It is likely that Brouwer had nonetheless seen, or perhaps even met, Poincaré earlier during the conference.

[17]Professor in Delft, the only other registered Dutch mathematician.

[18]Looking at the list of the participants of the congress, [Castelnuovo 1909], one gets the impression that Brouwer could have met almost everybody that was going to be of interest for his later career. Emile Borel, Felix Bernstein, Blumenthal, Carathéodory, Cartan, De Donder, Dehn, Hahn, Hardy, Levi-Civita, Koebe, Hadamard, Hilbert, Poincaré, Zermelo. Even Jahnke was there. But even though the congress had not reached the size of the modern mega-meetings, it is hard to conclude from the list of participants who met whom.

> Don't leave Florence before visiting Siena and Lucca. In the first there are a few admirable Sodoma's, in Lucca the most beautiful Breughel in the world. And in particular, the landscape around Siena with its cool red angular mountains strikes one as miraculous, after the singing blue and white of Florence.[19]

And half a year later he wrote:

> Have you already been in Paestum? That really is a place, where the belief in the reality of beauty created by man can be absorbed. I would rather live in Naples than in Rome.[20]

Scheltema and his wife spent the year 1909 in Italy, and a charming book of impressions was published in 1914.

6.3 Climbing the ladder

Back in Amsterdam, Brouwer resumed his freelance activities, he tried to get a foothold in the few mathematical enterprises that were open to a nameless beginner. He started, for example, to do review work for the editors of the *Revues Semestrielles*, a Dutch review journal, and he hinted to Korteweg that he would be prepared to take over some of the regular review duties,[21] should Schoute or Kluyver wish to stop ('although, of course, I do not ask for it').

At the same time he put out his feelers in Groningen, where P.H. Schoute, who was soon to retire, held a chair. When he expressed his awe for the high status of a professor to Korteweg, the latter reassured him: the scientific virtues and vision were the prime criteria, but he warned his student that routine duties would certainly conflict with his present studies:

> There is only one aspect of your consideration, that I think I have to react to. It concerns the 'loftiness' of the position of a professor. In my opinion the position of a professor is neither higher nor lower than any other one. As with any other position, one has to look for the man who will best carry out the duties. That here the scientific virtues and the scientific insight come in the first place is clear. But not *only* those. If we had one big university instead of four, it would be different. Then one could afford the luxury of professorships, such that each scientifically prominent man—for there are not so many of those—could find a place.[22]

Korteweg pointed out that the position of professor brought many teaching duties and other obligations that would probably conflict with Brouwer's research activities. But he made it clear that he would respect Brouwer's decision, no matter how it would turn out. Korteweg had a reason to be concerned: Brouwer was fully

---
[19] Brouwer to Scheltema 3 December 1909.
[20] Brouwer to Scheltema 11 June 1910.
[21] Brouwer to Korteweg 1908, undated.
[22] Korteweg to Brouwer 8 November 1908.

emerged in his topological research. In 1909 he published his first paper on Lie groups in the *Mathematische Annalen*, and at the same time his series of papers on transformations on surfaces and on vector distributions started to appear in the proceedings of the Academy. At this moment in life Brouwer needed all the time he could find to fulfil the promises of his topological genius.

As it happened, Brouwer was spared the difficult choice, since The Groningen University could only offer him a one year teaching job until the chair could be filled by a more senior man.[23] He turned the offer down and thanked God that he had the courage to avoid scientific suicide and to cut short his career. As he remarked to Scheltema;[24]

> It was not at all that attractive: when the term was over, some professor from Delft would be appointed in Groningen, and I could have moved to Delft, where a professor is something like a supervisor at drawing classes.

With the example of the physics department in mind (where Zeeman's succession of Van der Waals had been successfully negotiated), Brouwer sounded Korteweg on the possibilities of a position as a lecturer or an extraordinary professor in Amsterdam. The answer was disappointing, no extension of the mathematical staff (consisting of Korteweg and Hk. de Vries[25] was considered in the near future.[26]

In view of Brouwer's unusual qualities, Korteweg had no problem to get, in April 1909, the young Brouwer admitted as a *privaat docent*. Even this modest step in his career caused Brouwer a great deal of anxiety, when the formalities had to be completed in June.

In a letter to Korteweg of 8 June he ventilated his many doubts. He was afraid of squandering his energy on matters of secondary importance, whereas he had decided to dedicate himself fully to mathematics, in order to attain the best he could offer. Korteweg must have tried to arrange a position for Brouwer that allowed him time to concentrate on his research and to teach a few advanced courses of his choice. But that was not what Brouwer had in mind. He considered *privaat docenten* in extra-curricular subjects to be social climbers, and he hoped, he said, that Korteweg knew him well enough by now to understand that social status did not interest him. He was willing to be a *privaat docent*, but only if he was of any use to Korteweg or the faculty:

> As far as my personal wishes are concerned, I hope, at least as I am still immature as a mathematician, like I am now, to remain excused from teaching; and even later if I am not needed. But in any case, I will never be able to teach classes, where I am at the mercy of the *bon plaisir* of the au-

---

[23] In 1909 Frederik Schuh was appointed. He was also a student of Korteweg, six years older than Brouwer.

[24] Brouwer to Scheltema 1 March 1909.

[25] Another student of Korteweg. He had been a professor in Delft for one year before he was appointed in Amsterdam (10 October 1906)

[26] Korteweg to Brouwer 16 February 1909.

dience. I have seen that successively with Van der Waals, Jr, Mannoury and Kohnstamm, and I have felt more and more that I am not capable of something like that.[27]

Brouwer viewed his appointment as *privaat docent* with some mistrust; it was another step on the road to his assimilation into the system, but, as he wrote to Scheltema;[28]

> I finally succumbed to the pressure to become a privaat docent here in Amsterdam, in some subjects that have my sympathy. If it suits me, and I suit them, then it will probably result in a lecturer's position or an extraordinary professorship with a very restricted duty, and without official tasks. Well, that is, I believe, on the long run fairly acceptable to me.
> Moreover, when they held out their hand to me after I turned down Groningen, I could not very well turn my back on them any longer. Yet, it made me melancholic, when I read my appointment to *privaat docent* in the newspaper; and I almost started to cry for feeling sorry for myself. It is, after all, a 'Sic transit', but my youth passes, and I work more and with more pleasure than before.

And so, at the age of 28, Brouwer made his first step at the academic ladder. Compared to newcomers at the German Universities he was already fairly old, and although he had displayed sufficient mathematical talent and revolutionary ideas in his dissertation, no great feats had put his name on the mathematical map. Compared to, for example, Koebe, who became famous at the age of 25 for solving the uniformization problem, and Von Neumann, who was to revolutionize set theory at the age of 22, Brouwer was a slow starter.

At the time of his appointment to *privaat docent*, he was entering the field of topology, and gradually mathematics became his true love. He could not muster the same feelings for society:

> I did not live for long in Amsterdam; being a respectable man oppresses me, and after my classes I flee to 't Gooi and gloriously pace the heath as a tramp. That works out wonderfully, as long as I don't have the misfortune to run into a student here in my vagabond outfit. In class I always look spick and span. I am getting ever fonder of my subject, but I detest society more and more.

The relationship between Korteweg and Brouwer was apparently of the sort that allowed for the inclusion of extra-mathematical activities; when Korteweg, who was mildly involved in politics, discussed the coming parliamentary elections, Brouwer—surprisingly—offered his support in campaigning for the Liberal Party.[29] Whether he actually took part in any campaigning is unknown. Anyway, it left its traces in

---

[27] Brouwer to Korteweg 8 June 1909.
[28] Brouwer to Scheltema 12 May 1909.
[29] Cf. p. 35, Brouwer to Korteweg 22 June 1909.

the correspondence in the form of a political positioning in the Dutch party system.

Brouwer expressed his gratitude to Korteweg by presenting him the collection of all the papers that Korteweg had communicated so far to the Academy, bound in a small volume,[30] with the inscription

> To prof. D.J. Korteweg in grateful memory of the submission to the Academy of these essays of his student.
>
> L.E.J. Brouwer   September 1909

The period following the dissertation seems to belie Brouwer's fervently claimed principles. There is no calm resignation to fate; even though he claimed in his letter of 12 May 1909 to Scheltema that a minor academic position of limited visibility would do for him. No sooner had the position of *privaat docent* become his, then he started to worry, complain and look for better positions. Even in the case of a highly strung, well principled young man like Brouwer, the flesh turned out to be weak—even a practising mystic apparently wants his comfort and recognition!

Job hunting kept Brouwer's mind continually occupied, and in July he had found another career perspective: the Teyler Museum in Haarlem was looking for a candidate to fill the position of director of Teyler's physical cabinet and editor of the *Archives du Musée Teyler*. Brouwer judged this to be an excellent place for a person like himself with no urge to teach and with an extensive research program. He expected that a director could devote almost all his time to research.[31]

One can imagine his interest in the position—having lived in Haarlem, he knew the oldest museum in Holland well, and it was, as it is to-day, a delightful institution. It was the first science museum built as such, a monument from the days of the Enlightenment, founded by a rich merchant couple, Teyler Van der Hulst, in 1778. It has a marvellous collection of scientific instruments, for example the electrical machine of Van Marum (1784) and an substantial art collection, containing paintings, prints and drawings. An important part came from the collection of the 17th century Queen Christina of Sweden.

The position of director would certainly have allowed ample time for personal research. However, in spite of a modest measure of lobbying, the application was not successful.

6.4 The shortcomings of Schoenflies' *Bericht*

Mathematically speaking, all was well with Brouwer, he was rapidly finding his way into the impenetrable fortress of topology. He was conducting investigations in the topology of surfaces, vector distributions and in general topology. The research in

---

[30] Kept in the University Library at Amsterdam.

[31] Brouwer's idea was not as eccentric as one might think: Lorentz filled the position of director from 1912–1920 after he had given up his full professorship at Leiden.

general topology was the direct consequence of his work on Lie groups, as we have seen (p. 143), Schoenflies' monograph on point set topology was far from reliable, so that Brouwer had to redo the material for himself.[32] The result was the famous 'Analysis Situs'-paper. Brouwer had submitted it to the *Mathematische Annalen*, and this led to an exchange of letters with David Hilbert, the leading editor. Brouwer was undeniably flattered by Hilbert's attention.

> I received the enclosed postcard from Hilbert, from which I conclude with somewhat mixed feelings of satisfaction, that my latest Academy communications[33] have drawn his attention.
> Somewhat mixed, as I say, because I value that publication much less than my other work: anybody whose thoughts would wander in that direction would have found the results. The value of a mathematical composition lies, like that of every work of art, in its penetration, and not in some surprising and popularly comprehensible result, no more than that the value of a Dutch painting lies in the little windmill.[34]

The 'Analysis Situs'-paper, in the meantime, held the promise of some tangible recognition. Brouwer had sent a copy of the manuscript to Schoenflies, who returned an elaborate letter—in Brouwer's words: 'probably because I pressed him hard, ...'. The paper made a clean sweep of set theoretical topology, as presented by Schoenflies in a prestigious monograph, '*The Development of the theory of point sets, II*'. 'Nonetheless', wrote Brouwer to Korteweg, 'I am glad to have finally a bite, and to receive more than a kind postcard at my request'.[35]

Brouwer had indeed scored; Schoenflies was one of the leading set theorists, in spirit the successor of Cantor. He had mainly carried on the Cantor tradition in point set topology. Schoenflies was an acknowledged expositor; he was the author of a large number of papers on point set topology and abstract set theory. His work culminated in two influential monographs on set theory, which had the ill-fortune to become obsolete soon after their publication. Schoenflies was a close friend of Hilbert, which partly explains the role of the latter in the relationship between Brouwer and Schoenflies.

Schoenflies, clearly, was taken unawares by Brouwer's manuscript. His letter to Hilbert, written almost directly after the receipt of Brouwer's paper,[36] shows that he suspected that after all these years of preparations the control over the field of topology was slipping from his fingers. He was in the position of a general who at his victory banquet is informed by a common soldier that the battle is not won but lost:

---

[32] For the mathematical aspects of the *Bericht* episode, see section 4.6

[33] [Brouwer 1909F2, 1909G2] the first paper of the series on continuous maps of surfaces into themselves and the first one of the series on vector distributions.

[34] Brouwer to Korteweg 22 May 1909.

[35] Brouwer to Korteweg 18 June 1909.

[36] Schoenflies to Hilbert 22 May 1909

> The paper of Brouwer has made me half glad, half sad. Glad—because I see that the younger generation has started to study the *Bericht*; sad, because among the abundance of geometrical forms, a possibility has escaped me. For this is the essential content of Brouwer's paper.

He sadly acknowledged the shortcomings of the *Bericht*. In a moving passage, he reminded Hilbert of the adverse circumstances under which both *Berichte* [37] were produced.

> You have by now also experienced that one can reach a point where the mind starts to fail. Alas, I have not had the good sense to pause in such a case, but I devoted all my energy to reach the goal.

It appears that Schoenflies had corrected the proofs of the last *Bericht* during a summer-trip, missing some of the finer points in the atmosphere of vacationing.[38] The letter closed with an expression of annoyance at the formulation of the opening of Brouwer's paper; in Schoenflies' opinion it rather suggested that the author considered the *Bericht* as largely incorrect. Schoenflies went on to say that this impression was, however, contradicted by the accompanying letter of Brouwer (which he enclosed).

Notwithstanding Brouwer's lethal criticism, he still hoped and expected that minor revisions would suffice. In that optimistic spirit he was already considering the revision of the *Bericht*. He clearly underestimated the damage discovered by Brouwer.

Whereas Schoenflies' letter to Hilbert had the character of a lament to a trusted old friend, the first reaction to Brouwer was (understandably) more guarded. Taking his time to think the matter over, Schoenflies had concluded that the situation was not hopeless. He suggested some minor adaptations, but he could not make sense of the figures of the manuscript, partly because the letter did not match the text, and also he did not see the point of Brouwer's criticism of his notion of 'curve' and 'accessibility'. Brouwer's suggestion to publish his criticism together with a reaction of Schoenflies, was gladly accepted. [39]

The latter had in so far recovered that in the same letter he undertook a defence of his earlier results against 'admonitions which seem to me partly unjustified, partly exaggerated'. It clearly distressed him to be taken to task by a much younger man, who did not bother to hide his disapproval, and who had repeatedly used the term *ungenügend* (inadequate) in connection with Schoenflies' topology. In the years before the war there has been an extensive correspondence between Schoenflies and Brouwer. The first group of letters dealt with the Analysis Situs paper, and the second group concerned the new editor of the *Bericht*. Unfortunately the correspondence is far from complete, most of Brouwer's letters are missing.

---

[37] of 1900 and 1908.
[38] Brouwer to Hilbert 24 June 1909.
[39] Schoenflies to Brouwer 17 June 1909.

In a prolonged exchange of letters, Brouwer tried to convince Schoenflies of the shortcomings of his proofs, and the latter continually grasped at straws to defend the remnants of his *Bericht*. He did not realize that something was basically wrong, but attributed the gaps in the proofs to the nature of the *Bericht*, which he saw as a quick, convenient and, above all, readable introduction of the reader to the problems and results of topology. In particular he assumed that the attentive reader could provide further details when necessary. Here, evidently, was a clash between two scholars with different notions of exactness. Schoenflies belonged to the older tradition in which the foundations of topology were taken for granted, and where geometrical intuition often took the place of mathematical argument. Indeed, Brouwer's (counter-) examples showed that the sophistication of topology required more than the conviction of 'first impressions'.

Artur Schoenflies. Niedersächsische Staats- und Universitätsbibliothek Göttingen.

Reading the correspondence, one gets the impression of a growing exasperation of the older man. When all of his arguments bounced off Brouwer's granite wall of exactness, he must have felt that the younger man was not susceptible to reason— 'Should it then really be impossible to convince you in writing', and 'I can now only wait and see if you will stick to your arguments concerning Ch. IV, §12. Nonetheless, I hope to have finally convinced you that here the error is on your side'.[40]

But Brouwer did not give up, he kept trying to set Schoenflies right. Schoenflies felt that his dignity was at issue, rather than his topological expertise.

Brouwer may have been under the impression that Schoenflies tried to save his face by making light of the inadequacies of his work. Here he misjudged Schoenflies, who protested: 'Anyway, I don't think anyone could think that I try to belittle your paper or to diminish my mistakes. In my opinion the opposite is the case.'

Schoenflies had turned to Hilbert for protection from this young Turk, and Brouwer had likewise appealed to the undisputed Master of Mathematics, who as editor in chief of the *Mathematische Annalen* had a vested interest in the matter. He informed Hilbert that Schoenflies did not fully realize the defects of his presentation, and that, should he stick to his views in the forthcoming paper, some extra space would be required to enlarge upon the details of the defective parts. Brouwer strongly stressed the importance of the topics, so the corrections in his opinion were really necessary. He wrote to Hilbert:

> I am awfully sorry that Mr. Schoenflies feels insulted by something that rather resulted from my appreciation, but I don't want [to run the risk]

[40] Concerning decomposition theorems, Schoenflies to Brouwer 14 August 1909.

> that someone could reproach me with the slightest semblance of justification, that I publish scientific trivialities, especially in the *Annalen*; nor do I want to allow that the supplement of Mr. Schoenflies contains even the slightest innuendo in this sense, and I kindly beg you to make this explicitly clear to Mr. Schoenflies, and to allow him extra space only on the ground of the objective content of my remarks.[41]

Schoenflies' excuse that he did his proof-reading in a hurry during his vacation, was interpreted by Brouwer as an implicit complaint that Brouwer had resorted to nit-picking—the idea being that between gentlemen one does not mention the smaller mistakes.

> With respect to isolated inaccuracies, which do not impede the general flow of the theory, Mr. Schoenflies, if I understand him correctly, calls attention to the fact that he did the proof-reading during a summer trip, that this explains the minor slips, and that it is a petty thing of me to locate them. I thought, however, that even these little corrections were worthwhile, in view of the great importance of the subject.

and he continued:

> In order to rule out that Mr. Schoenflies could feel the least bit hurt, I just would like to insert at the end of my introduction the following note (by the way, with my warmest intentions): 'I explicitly stress that this paper does not seek to diminish in any way the great value of Schoenflies' discoveries, but rather that it tries to emphasize the value more clearly.'

In the paper itself this note was modified to:

> 'I expressly stress that this paper does not seek to diminish in any way the great value of Schoenflies' discoveries. It is just that the considerable consequences have given me reason for this criticism, which by the way, does not essentially concern the largest part, namely the theory of simple curves.' [42]

In the bargaining that followed, Brouwer softened parts of his argument and further sugared the introduction, in the hope that this would pacify Schoenflies.

The affair went on for the better part of a year, and eventually Hilbert got fed up with Schoenflies' defence tactics; at least, in December Schoenflies complained about the rather unfriendly tone of Hilbert's letter. And in reaction to Hilbert's exclamation, that the readers had already lost interest in the matter, he retorted that they had not even seen the papers, and that he would, when necessary, revise his attitude with respect to the *Annalen*. It appears that Hilbert felt no compunction in letting scientific interest prevail over old friendship. By the end of December, both parties had reached an agreement, and the *Annalen* could publish the papers.

The paper sent shock waves through the mathematical community. Hardly anybody

---

[41] Brouwer to Hilbert 24 June 1909
[42] [Brouwer 1910C].

had guessed that the road of topology contained so many pitfalls. For Brouwer, the Analysis Situs affair had after all a pleasing consequence, the paper contributed substantially to his status in the mathematical community.

Brouwer's exploits had already drawn the attention of the reigning king of mathematics, David Hilbert. Brouwer's work on Lie groups and his subsequent research in topology had satisfied Hilbert that he was dealing with a clever young man, the talks in the dunes at Scheveningen (cf. p. 128) doubtlessly had reinforced this impression. To be noted by the Göttingen group of mathematicians was already a feat by itself, but Brouwer's luck did not stop there. He found himself a niche in the community, if not the hearts, of the Göttingen mathematicians in a remarkably short period. The correspondence with Hilbert soon led to plans to visit Göttingen in person. Already in 1910 he made plans for a visit in the summer vacation, but the trip was cancelled because of family matters. A year later he made his entree in Göttingen; he wrote Hilbert that 'I will be in the Harz for a couple of weeks and on the way I'll stay for a few days in Göttingen. I am very much looking forward to get to know the people and the situation over there'. The next year Brouwer again visited Göttingen, this time accompanied by Lize: 'Next Sunday I will come to Göttingen via Löhne, Hameln, Elze.'

David Hilbert. Niedersächsische Staats- und Universitätsbibliothek Göttingen.

Again he combined his visit with a stay in the Harz mountains; the *Brocken* was his destination. In this short period Brouwer managed to make many friends for life, Felix Bernstein and Hermann Weyl, to mention two of them. He became one of the 'extra-territorial' members of the Göttingen group. Brouwer immediately made himself popular with the company, he was an expert conversationalist with an inexhaustible fund of stories and a sharp insight in most branches of mathematics. Göttingen benefitted by acquiring the support of the leading topologist, but Brouwer paid the price of a considerable increase in workload, consisting mostly of referee jobs and general advice.

6.5 Privaat Docent

With the first term of the academic year 1909–10, Brouwer's duties as a *privaat docent* had started. Traditionally, *privaat docenten* (like lecturers and professors) gave an inaugural lecture in order to present their professional views on their subject. This was a formal address delivered in the auditorium (aula) of the university, in the presence of colleagues, friends and future students. Brouwer gave his lecture on

12 October 1909 with the title 'On the nature of geometry'[43] a beautiful programmatic exposition of the latest in geometry, including the space-time geometry under the Lorentz transformations, complete with moving systems and clocks. Curiously enough, there is no reference to Einstein, and only Lorentz with his 'relativity postulate' and transformations is mentioned. Brouwer drew from this geometrical-physical theory the conclusion that the *a priority* of space and time, and hence of Euclidean space and time, had become untenable. Later, in the collection *Mathematics, Truth and Reality*[44], Brouwer added 'This lecture, were it given at present, would undoubtedly bring up in its first part the 'general theory of relativity', introduced after 1909, which would not have influenced the epistemological conclusions'.

The last part was devoted to Analysis Situs (topology), and contained a list of prominent problems, see p.154. We note in passing that the better part of those problems was to be solved by Brouwer within a few years!

The writing of the inaugural lecture had to be squeezed in between many other activities in a letter of September 1909, Brouwer complained to Korteweg that he was in the middle of the quarterly administration of the pharmacy and that the correspondence with Schoenflies had entered a new phase—at that moment only two weeks were left for composing his text.

The inaugural lecture ends with a credo, that strikingly reflects Brouwer's personal taste, and that had a prophetic ring. After discussing the foundation of geometry in the light of Analysis Situs he stated that:

> Thus, one does not have to ban co-ordinates and formulas entirely from other theories if one succeeds in basing them on analysis situs, but the formula-free, the 'geometric' treatment will be the point of departure, the analytic one becomes a dispensable aid.
> It is the possibility and desirability of this priority of the geometric treatment, also in parts of mathematics where it does not yet exist, that I primarily wanted to point out in the preceding pages.[45]

Two days later his classes began: projective geometry and analytic geometry. These courses remained his regular contribution to the mathematics curriculum during the first years; a short (and no doubt polished) description of Brouwer's teaching can be found in the student's almanac, which provided among other things reports on the courses of the year before. The almanac of 1911 describes the contents of Brouwer's courses in projective geometry (for first- and second-year students in physics) and the 'suppletion' course in analytic geometry. The projective geometry course contained a fair amount of material: for example an axiomatic treatment, the 'numerical-model', the introduction of cross ratios and the derivation of the Euclidean and non-Euclidean metrics. The reporter remarked that:

---

[43] *Over het Wezen van de Meetkunde*, Cf. p. 153
[44] *Wiskunde, Waarheid en Werkelijkheid*, 1919.
[45] [Brouwer 1909A], p. 23.

> The modest size of the audience was an indication that the course may have been too much for freshmen, perhaps also because the high-speed teaching of Dr. Brouwer made it difficult to acquire a good insight in this topic, that was completely new to us. But those who have taken the trouble to work through it till the end, will certainly agree with our great admiration for this course of Dr. Brouwer.

The report of the next year ends on the same note:

> It seems to us that the audience would have benefitted more from Dr. Brouwer's lectures, if those would have been less fast.

The gist of these reports is confirmed by most of Brouwer's students; his courses were models of elegance and precision, but patience with the less gifted student did not belong to Brouwer's virtues. A former student[46] reported a comical event in the late thirties. The audience was left far behind by Brouwer during a certain course, so the students took courage and sent a delegation to Brouwer with the request to reduce his tempo of lecturing. After listening politely to the delegation, Brouwer answered: 'Allright, I will talk as slowly as the gentlemen think'.

In October 1909, Brouwer had a mild difference of opinion with Korteweg; he had found a considerable simplification in a derivation in a paper of Van Uven, the mathematics professor at Wageningen, and asked Korteweg to present a short note to the Academy. Korteweg resolutely refused: 'An alternative derivation of a formula does not belong there'. In spite of Brouwer's insistence, Korteweg stuck to his guns; the note was not published.

As we have seen in the preceding chapter, Brouwer was fully emersed in his research in topology, he had already obtained his results on fixed-points on the sphere, and he had opened the cleaning-up operation in elementary (plane) topology. And now on the instigation of Hadamard he had started to read the *Mémoires* of Poincaré, and this was bearing fruit in the form of simpler proofs of some of his earlier results:

> the result appears here as a surprise, whereas in the original proof I gradually constructed the transformation, thus forcing myself step by step to admit the invariant point.[47]

In spite of Brouwer's growing recognition among mathematicians, there was no comparable academic recognition at home. He made himself useful in Amsterdam by his teaching, and by taking care of the collection of mathematical books and journals in the University Library. In those days there was no mathematical institute and no independent mathematics library; the central building of the university in the Oudemanhuispoort contained lecture rooms and 'faculty rooms', the latter served mostly as a meeting place for faculty members and as a cloakroom for those

---

[46] F. Kuiper
[47] Brouwer to Korteweg 24 December 1909. The letter was accompanied by a copy of Brouwer's letter to Hadamard.

who had to lecture. Those professors who had no laboratories or clinics dropped in to give their courses and went home again to conduct their research. Brouwer was no exception to this pattern; the pharmacy was his pied-a-terre in Amsterdam, but his heart drew him to Blaricum.[48]

The academic routine dragged on, and Brouwer, who was totally absorbed by his topological research, grudgingly fulfilled his tasks.

Korteweg, in the meantime, tried to improve the prospects of his brilliant student. He saw that the average teaching load in Amsterdam surpassed the level that was generally thought acceptable. By appointing Brouwer, one could kill two birds with one stone, the teaching load would be reduced, and the mathematical genius of the young man would be rescued from real and imaginary dangers.

In order to underpin his arguments for an extension of the number of mathematicians at the University of Amsterdam, Korteweg conducted a small investigation into the general situation in mathematics in Holland. The results are interesting, because they shed light on academic mathematics in the Netherlands.

We reproduce here Korteweg's list of the mathematics departments:

University of Amsterdam[49]
    Professors: J.D. Korteweg and Hk. de Vries
    teaching duties: mathematics, mechanics (that is mathematical physics, mechanics), astronomy.

University at Leiden
    Professors: J.C. Kluyver and P. Zeeman[50]
    teaching duties: mathematics, mechanics

University at Utrecht
    Professors: J. de Vries and W. Kapteyn
    teaching duties: mathematics.

University at Groningen
    Professors: P.H. Schoute and F. Schuh.
    teaching duties: mathematics.

The load varied from 8 hours a week (Leiden) to 10 hours a week (Amsterdam).

There were more mathematics professors in the Netherlands than the ones just mentioned, to be specific, in Delft and Wageningen, but Korteweg did not want to compare those (technological and agricultural) institutes with the universities.

The number of graduates in mathematics, physics and astronomy in Amsterdam was also steadily rising:

---

[48]see p. 207.
[49]The Amsterdam University was a municipal university. The other ones were state universities.
[50]Not the famous physicist.

	candidaats degrees	doctoraal degrees
1880 – –1884	4	2
1885 – –1889	5	4
1890 – –1894	1	4
1895 – –1899	11	3
1900 – –1904	10	5
1904 – –1905	9	9

After ample consultation, Korteweg and De Vries addressed the City Council of Amsterdam with a request for an additional lecturer. The arguments were not unreasonable, and they could be (and are) repeated at any time and any place: too much teaching, supervision of Ph.D. theses and research. Moreover, Korteweg argued, this exceptional man, Brouwer, 'who can be considered as the equal of the best mathematician of our time', should be attached to our university.[51]

The authorities, however, did not share Korteweg's views; even though Korteweg had pointed out that the university could have Brouwer for one thousand, say fifteen hundred, guilders, they did not see the justification of an extra lecturer.

The rejection hurt Brouwer, he had just been passed over in Delft, where 'a younger and lesser man' was appointed.

Already in June he had written to Scheltema that he was fed up with the job, and that he wanted to quit after the summer vacation. He had expected an extraordinary chair, but since nothing of the sort was considered, he felt no obligation to go on with teaching for a marginal fee. The old dream of a quiet hermit's life was returning:[52]

> ... do you remember Watt's principle? It says that the vapour-pressure in a container is determined by the lowest temperature occurring in it. Likewise, the intensity of our sensing and thinking is determined by the most impure of the activities that we are involved in, no matter how little. He who teaches two courses a week, withers like every professor, even though he is for the rest surrounded by the fragrance of flowers, and shone upon by the sun.
>
> Thus I now ponder the question, how I can decently, without quarrelling, get rid of my job. For the rest I am turning more and more into a home-lover, and I hardly leave any more my beloved terrain, where I dig, weed, prune and philosophize. Yesterday and the day before I have tarred, in the blazing sun, my faithful hut in a scraggy reddish brown.
>
> Maybe this is caused by the advent of old age, but again and again all my desires subside, and make place for cheerful self-satisfaction, and the world that interests me contracts more and more. Last year I was still en-

---

[51] Letter to the city council, 6 October 1910.
[52] Brouwer to Scheltema 11 June 1910.

thusiastic to break the bread with the star of the mathematicians,[53] but now I find that idiotic too, and I no longer admire my colleagues.

In spite of his aversion, Brouwer did not give up his position, marginal as it was. With varying hope he muddled through the coming years, no doubt keeping his morale up by his research activities and the international recognition that started to come his way.

6.6 Korteweg's campaign for Brouwer

In 1910 Korteweg undertook another action for his student. If, so he reasoned, the City Fathers do not see the importance of furthering the career of the gifted Brouwer, and, worse, if they do realize the importance of securing the services of such an exceptional scholar, then they may be helped in getting the right perspective by an act of recognition from an unsuspected side. And so he started a campaign to get Brouwer elected to the Academy of Sciences; some skilful lobbying got him the support of his colleagues in the Academy and of the Dutch mathematicians. Moreover, he approached some of the leading international mathematicians, Hilbert and Poincaré, pointing out to them that, although the attempt to find a position for Brouwer had failed, the situation was not hopeless, and that a membership of the Academy was a powerful recommendation, that could not so easily be ignored. Both Hilbert and Poincaré reacted positively, Hilbert's answer is preserved in the form of a draft written on Korteweg's letter.[54] Apparently Poincaré was not fully aware of Brouwer's latest feats, but he was sufficiently impressed to write a flattering recommendation.[55] Hilbert on the other hand had seen almost all of Brouwer's work in topology, and he had a sharp eye for outstanding mathematics. His letter is so much to the point that it would be a pity not to reproduce it here:

> I am very pleased that you will try to get Mr. Brouwer a chair at the University. I wish that you may succeed not only in the interest of Brouwer, but also of science. For, I consider Brouwer a scientist of unusual talent, of the richest and most extensive knowledge and of rare ingenuity. The area in which Brouwer has been particularly active is that of the theory of point sets, a theory, that, as you know, interacts in an essential way with almost all disciplines of mathematics, and therefore its development is one of the most important tasks. Moreover, it is Brouwer's characteristic not to contend himself with easy successes that are offered by researches of a general nature, but he attacks each time a special, difficult, deep problem, and he leaves off only when he succeeds in obtaining the fully satisfactory solution. I am thinking here in particular of his solution to the problem of the finite continuous groups and his marvellous proof of the Jordan curve

---

[53] David Hilbert, cf. p. 128.
[54] Korteweg to Hilbert 6 February 1911.
[55] The letter of Poincaré has not been found.

theorem.[56]

That your small country repeatedly produces in the most varied areas outstanding scientists, is a pleasing phenomenon that may fill you with pride. But this phenomenon imposes on the government and the leading authorities the duty of special efforts. I hope therefore that your efforts on behalf of Brouwer will be successful.

The letter beautifully illustrates the confidence of the old style academic in dealing with the authorities: one simply has to tell them their duties to science!

But even Hilbert's authority did not have the magical power that one would nowadays assume. The matter was further complicated by the circumstance that the Academy had already a fair share of mathematical members, and Brouwer's appointment would either increase the number of members, or it would be at the cost of some other discipline.

In the meeting of 25 March 1911 Brouwer was proposed for membership by a most impressive array of scholars: Korteweg, W. Kapteyn, Cardinaal, Hendrik de Vries, Schoute, Jan de Vries, Kluyver, Lorentz, H.G. van de Sande Bakhuyzen, Zeeman, Van der Waals, J.C. Kapteyn, Hugo de Vries, Winkler en Hubrecht.

At the next special meeting, 28 April 1911, the voting took an unexpected course. In the first round the largest number of votes was cast for Brouwer (21), but still less than required to get elected. And so a next round of voting took place. This time another candidate, Kuenen, got the required statutory 24 votes and was elected. In the final round Brouwer was only two votes short of the required number, and thus only Kuenen collected enough votes to join the Academy.

Korteweg reported (not quite truthfully) in October to Klein that only one vote had failed to get Brouwer into the Academy.

The parallel struggle for recognition by the curators of the university was not easily decided either; for one thing, it involved money! In September 1911 Brouwer, irritated by the earlier rebuff from the City Council, informed Korteweg that he would gladly continue his lectures on analytic geometry, since that was a personal favour to Korteweg, but that he refused to lecture on projective geometry, since that should be a favour to the authorities.[57] Apparently, Korteweg succeeded in calming down his hot-headed student—the course on projective geometry was given after all.

However prominent Brouwer had become through his majestic topological papers, in the privacy of his home, he was again beset by thoughts of the futility of it all. In a letter of 7 November 1911, he bared his soul to the only friend who had seen glimpses of the real Brouwer, the by now famous poet Carel Adama van Scheltema:

... Although I am nowadays fairly fertile, and have gradually gathered some international fame and envy, you must not get an overly serious

---

[56][Brouwer1910E].
[57]Brouwer to Korteweg 10 September 1911.

> impression of my work. For I have, as ever, the intimate certainty that mathematical talent is of the same sort as an abnormal growth of the nail of the big toe.
> Yet, at congresses I play for the popes of science the role of an enthusiastic ensign, but, when I paint *in flammender Begeisterung* (with burning enthusiasm) conversations rich with thoughts, with the prospects that inspire my work, my apparently absorbed gaze quenches itself with the monomania of their features, and sees in some of them inconsolable, imprisoned heroes, in others poisoning kobolds, and in the latter the unprecedented hangmen of the first. And while I am physically permeated with the sensation of being in hell, my eyes beam with sadistic lust of sympathy.
> My productivity therefore will never bring forth a grand creation, because it is exclusively fertilized by a mocking dissection of what exists.[58]
> None of my colleagues will, however, fathom this, although some of them eventually get ill at ease in my presence; they go round and speak evil.

The above passage provides a revealing glimpse of Brouwer's insight in his own motives, but it also shows the traditional tendency of the high-minded to flog themselves mercilessly, to confess evil thoughts. Whereas Brouwer certainly was not then or later in life, a dull-but-respectable citizen, he was not as contrary as he wanted Scheltema to believe. Indeed, he could never bring himself to take science deadly seriously, but he definitely had come to love his subject, and he had rather strict ideals concerning it. Even his bleak views on the scientific community are artistically exaggerated. No doubt, his rather unusual behaviour, choice of words, of topics, would surprise, or even shock, people, and make him an easy target for gossip or worse, but his relations with his colleagues were not disastrous; with some of them he had lifelong relationships of friendship and in general he recognized honesty and ability.

Brouwer's mildly cynical view of his learned colleagues, as described in the above letter, sometimes manifested itself when his natural inclination for mocking ran out of control under the influence of blatant pompousness, often combined with insignificance. One such instance was recorded by an eyewitness: at a conference, during the lecture of some pretentious mathematician, Brouwer had taken the precaution to stand at the back of the hall, so that he could unobstrusively leave if things got unbearable. When, however, the speaker ventilated a particular shocking enormity, Brouwer could no longer control himself and managed to fall flat on his back, as if struck by thunder.[59]

If anything, it was his honesty and his high standards that, combined with his emotional character, lent him a reputation of 'difficult'. We have already seen his conflicts with highly respected members of the mathematical community, many more are to follow.

In Dutch mathematical circles Brouwer was by now fully recognized. He had

---

[58] A surprisingly modest position for a man who has just solved one of the famous problems of Cantor!
[59] Oral communication E. Hölder.

presented two talks in 1908 and 1910 at the meeting of the Dutch Mathematical Society,[60] and in 1911 he was the speaker at the November meeting with his talk *The theorem of Jordan for n dimensions*.[61] If his position in Dutch mathematics can be measured by the role he played in the professional organization, the *Wiskundig Genootschap*, we note that he was appointed as a member of the Committee for prize essays, together with Ornstein, the theoretical physicist from Utrecht at the meeting of 26 April 1913. Shortly afterwards, at the April meeting of 1914, Brouwer was elected President of the Mathematical Society for a two-year term.

Korteweg, in spite of the earlier rebuffs, refused to accept the verdict of the city council of Amsterdam as final. He set out to organize another campaign for Brouwer's appointment, this time he not only argued the practical need for more teaching support, but also the exceptional qualities of the candidate. As in the case of Academy campaign, he collected some weighty references. This time he solicited recommendations from Hilbert, Poincaré and Emil Borel and Felix Klein. Borel answered somewhat guardedly, 'The papers that he has published so far are interesting and sometimes deep, they allow one to hope that he will one day arrive at important discoveries....' [62] Borel's judgement is certainly surprising, Brouwer had already changed topology beyond recognition—what else could one want? Perhaps Borel was aware of Brouwer's foundational ideas (after all, Borel was at the Rome conference, and he could have attended Brouwer's lecture, or read the paper) and expected spectacular contributions. Borel also wanted to know if the position that was envisaged for Brouwer was a newly created one, or if there was a competition involved; the wording of the recommendation, he said, would depend on the nature of the position. Korteweg patiently explained that he was thinking of a new position. The final recommendations have apparently not been preserved.

In the meantime the lecturer Van Laar,[63] a contemporary of Brouwer, had resigned and Korteweg saw an opportunity to propose Brouwer as his successor. In the faculty meeting of 24 April, Korteweg vigorously defended the view of the mathematicians that a replacement for Van Laar was necessary, the teaching for chemistry and biology students required a separate course and an extra teacher. He proposed Brouwer 'who in a few years time had been able to become one of the leading mathematicians of our time'. The faculty, no doubt conveniently impressed by the letters of recommendation of Hilbert, Klein, Poincaré and Borel, unanimously supported Brouwer for a position of extraordinary professor. The switch from 'lecturer' to 'extraordinary professor' rather suited Brouwer; he was definitely

---

[60] In the October meeting of 1908 he gave a talk *On plane curves and plane domains* (see p. 142, and in the October meeting of 1910 he gave another talk, this time *The invariance of the number of dimensions*.
[61] 25 November 1911, published in [Brouwer 1911F].
[62] Borel to Korteweg 2 January 1912.
[63] The faculty was divided on Van Laar. Some found him useful, but others considered him to suffer from delusions of grandeur, for example, he insisted that an honorary doctorate was due to him. There was no support to keep him on. In fact, he was a man with a fine reputation in thermodynamics. He indeed got an honorary degree in Groningen in 1914. For more information see [Emmerik 1991].

not looking forward to the treadmill of teaching mathematics for applications.

Not only had Korteweg to embark on a long march along the various dignitaries of the City Council; he also had to convince the faculty that mathematics required and deserved reinforcement. The operation 'a chair for Brouwer' was carried out parallel to a new initiative for Brouwer's membership of the Academy.

This time Brouwer' candidacy was proposed at the meeting of the physics section of the Academy on 30 March 1912, by Korteweg, Schoute, W. Kapteyn, Jan de Vries, Kluyver, Cardinaal en Hendrik de Vries. At the special meeting of 26 April 1912, Brouwer was this time elected in the first round. He again got the largest number of votes, this time 37, hence he was elected straight away. The meeting went through 7 rounds of voting before the last member passed the prescribed threshold of 24 votes.

The Minister of Internal Affairs informed the Academy on 15 May that Her Majesty the Queen had approved the appointments of the gentlemen de Sitter, L.E.J. Brouwer, Boeke, J.C. Schoute, van Hemert en Wertheim Salomonson.

Early 1912 things started to move in the university; the curators, in particular their president, the Mayor of Amsterdam, began to see the request of the mathematicians with more sympathetic eyes. Korteweg and De Vries had visited the Mayor,[64] and convinced him that Brouwer was 'extremely competent' and that the connection between the University of Amsterdam and Brouwer should be given a firmer basis. In May the faculty was asked to submit a formal request for Brouwer's extraordinary chair. From then on everything went smoothly. In July 1912 the appointment was made official. Of course the decision had been favourably influenced by the consideration that the final proposal only mentioned an extraordinary chair, with a yearly salary somewhere between 2000 and 2500 Dutch guilders, which—as Korteweg pointed out—was less than the salary of a highschool teacher (HBS).

After some more deliberations, Brouwer was indeed appointed. He expressed his joy in a letter to Klein,[65] thanking him for writing a supporting testimonial.

At the time of his appointment Brouwer had already obtained more fame and recognition in mathematics then all his Dutch colleagues together. Nonetheless the title of 'professor' acted as a strong booster of his self confidence.

On 14 October 1912 he gave his inaugural lecture[66] in the aula of the university. This time the topic was the foundations of mathematics. The title *Intuitionism and Formalism* is significant for Brouwer's views on the topic; according to him the prime contenders for the foundations of mathematics were the formalists and the intuitionists—no mention of logicism, in particular Russell, is made. The lecture is interesting as a review of the situation in the foundations, but apart from some

---

[64] cf. Korteweg to the Mayor 24 March 1912. The letter was accompanied by the recommendations of Hilbert, Klein, Poincaré and Borel.

[65] Brouwer to Klein 21 June 1912.

[66] published in 1912, reprinted in [Brouwer 1919B], [Dalen, D. van 2001] and translated in [Brouwer 1913C].

refinements of certain views, it is basically a reformulation of the dissertation.[67] It owed its influence to an English translation, published by Arnold Dresden in the Bulletin of the American Mathematical Society. This lecture made the name '*intuitionism*' a household word for a particular mathematical–philosophical tradition that goes back to (at least) Kant. Brouwer reserved the name for the French tradition of Poincaré, Borel and others. For his own new brand he coined the term '*neo-intuitionism*'. He had in fact used the term one year earlier in a Dutch review of a book of Mannoury.[68]

The lecture can be seen as the apotheosis of Brouwer's first foundational program. There is relatively little new material; it consisted mainly of a survey of the various foundational views and of an exposition of the intuitionistic criticism.

One interesting point worth mentioning is that Brouwer briefly compared the real numbers of the intuitionist and the formalist; the first, he said, recognize only decimal expansions given by laws, whereas the formalists allow expansions 'determined by elementary series of freely selected digits'.[69] Although one might conclude that this was a wholesale condemnation of arbitrary sequences (and the admissibility of 'choice') one should keep in mind that Brouwer is talking about the continuum. Here a undetermined sequence would not yield a well-determined, individualized real number. This would not rule out arbitrary sequences as legitimate objects. We will see that he later reconsidered this view, cf. p. 239. Already in the 1914 review of Schoenflies' revised *Bericht*, he accepted arbitrary choices as legitimate objects for intuitionists.[70]

Brouwer in the main building of the University in *De Oudemanhuispoort*, posing for the newspaper picture. 1912.

Considering the number of projects (with accompanying conflicts) Brouwer was involved in, one is inclined to think that job hunting would have been the least of his

---

[67] Cf. [Dalen, D. van 1999]
[68] [Brouwer1911A]. This was pointed out to me by Dennis Hesseling.
[69] [Brouwer 1913C], p. 92.
[70] [Brouwer 1914], p. 79.

problems. But one should keep in mind that his mathematical activities, including an almost incessant travelling, heavily burdened his financial reserves: furthermore the annuity of Brouwer's mother-in-law (the substantial sum of DFl 1100.-) heavily burdened the budget of the Brouwer couple. It may be remarked in passing that Brouwer made himself a perfect nuisance in the matter of the annual payments. He often was late and had to be reminded of his obligations by the family of Mrs. de Holl. The pharmacy was far from a gold mine, and the fee of a *privaat docent* was not sufficient for the lifestyle of a leading, travelling and letter-writing mathematician; the extra income of a professorship, was most welcome. In fact, his brother Aldert, who worked as a geologist in the Dutch Indies, financially supported Bertus. To put not too fine a point on it, Bertus was slowly accumulating a debt!

One should also not underrate the psychological effects of the lack of recognition at home: his contemporaries (who were not in the same mathematical league!) were getting chairs, and he was left out. There are no specific complaints about persons, but there is no doubt that Brouwer thought himself passed over by the establishment. This was partly due to his personal preference for the geographical location. Not only was he strongly attached to Amsterdam and to het Gooi as a perfect environment, but he was also subjected to a boundary condition that nowadays is generally recognized, but that was rather uncommon in his days: his wife had a job. Not only just a job, but one that could not so easily be transferred to another town—one has to keep in mind that the pharmacy had a tradition, and that a pharmacy, say in Groningen, would not have the same emotional value.

So when Korteweg in April 1913 warmly recommended Brouwer for a mathematics chair in Groningen, which had became vacant upon the sudden death of P.H. Schoute, Brouwer was faced with a difficult decision.

Korteweg had reached years ago the conclusion that his student should, if possible at all, be kept in Amsterdam; moreover, he was ready to act on his own if the authorities did not share that view. Since the negotiations with the curators did not yield palpable results, he generously proposed that he and Brouwer should switch chairs. At the time Korteweg was 65 years old,[71] and he had found the recognition that was due to his solid mathematical production. He felt that at this point it was of tantamount importance to put Brouwer's career beyond risk. The thought may seem strange from our present viewpoint, sooner or later a university, be it in Holland or abroad, would make him a fair offer. But one has to keep in mind that there was always the possibility that Brouwer could bid farewell to mathematics and academia at any time; the mystic in him was still powerful enough to make him give up the worldly affairs of science and mathematics.

Korteweg presented his plans to the curators of the university, who invited him and De Vries to their meeting of 28 June for further explanations. The curators expressed their concern for Korteweg's pension rights, and when that was settled

---

[71] the retiring age was 70.

to everyone's satisfaction, they wondered if the faculty would after all, even with Brouwer and De Vries as full professors and Korteweg as an extraordinary professor, require an extra mathematics professor. Korteweg confessed that he had not even consulted the faculty on his present proposal, but that the faculty would doubtlessly react quickly to a request of the curators for support. He made it clear that under the present circumstances his proposal was motivated by the offer of a Groningen chair to Brouwer. The curators were apparently convinced of the importance to keep Brouwer in Amsterdam, and so they sent the proposal with their official backing to the city council.

Lize, in a letter to Brouwer's brother Aldert and his wife Lily in the Dutch Indies, soberly summed up the situation:[72]

> Last week Friday Professor Schoute in Groningen died. Bertus can be appointed there as an ordinarius; of course we cannot accept it, but it may lead to a rise in salary or an appointment to ordinarius.

The choice between Amsterdam and Groningen caused the Brouwers a good deal of unrest, the weighing of all factors kept them busy. Lize had made up her mind,

> I am not in favour [of Groningen]. It would cause great changes, Blaricum would no longer be useful and everything should be sold. The pharmacy would also be over. Groningen is in the north, but Bertus would again wish to live in the country in Anlo or Gieten. The main advantage would be that we should at last have a regular household. This way of life is not easy. Bertus finds the travelling back and forth rather a welcome change. Yet I have an idea that a regular household will not suit Bertus. *Enfin*, for the time being we can only wait.

So here was Brouwer, confronted with a difficult dilemma: go to Groningen and get a substantial raise, or stay in Amsterdam. While conducting highschool examinations in The Hague (recall that university professors spent part of their summer vacation touring the country, conducting oral examinations at gymnasiums), he asked Hilbert for advice.[73] Should he accept the offer from Groningen, which left him totally free to teach and conduct his research, but

> where I will find in the petty provincial town fewer sympathetic colleagues than in *Amsterdam*, i.e. in a lively big city, which has always been closely connected with my life, where I have my faithful home in Blaricum, and where the dunes are close by.

Amsterdam had in the meantime adopted Korteweg's scenario, Brouwer was offered Korteweg's chair. With the chair went, however, the obligation to teach 'mechanics'; Brouwer feared that his research would suffer under the demands of this particular piece of applied mathematics. From the letter to Hilbert, one gets

---

[72] Lize to Aldert and Lily 4 May 1913.
[73] Brouwer to Hilbert 16 June 1913.

the impression that Brouwer had already made up his mind, he was horrified of an exile in (what was usually considered) the northern wilderness—'The small-town contacts with the conventional pressure must be terrible in Groningen, and there is no nature around at all', he wrote to Hilbert. But above all, his heart was in Amsterdam and Het Gooi.

A month later the matter was determined by 'force'; the University at Groningen issued an ultimatum, and Brouwer opted for Amsterdam. Korteweg had deftly manoeuvred Brouwer's chair through the faculty, on 26 June he told the faculty that the University of Groningen demanded Brouwer's decision before next Saturday. The board of the university did not object to the arrangement, so he told. And thus, the faculty unanimously accepted the proposal. Now Brouwer's toil of years of research was rewarded by a full professorship at the University of Amsterdam, a young university, but one which had already gathered a considerable amount of prestige, and which had added now a precious pearl to its crown. Brouwer thus became a full professor with all the prestige it carried in those days, in addition it brought him a salary of DFl 4000 (+ DFl 500 lecturing fees).

While the Universities of Groningen and Amsterdam were considering to offer positions to Brouwer, there was another university interested in the same young scholar. The University of Göttingen was looking for a successor to the eminent mathematician Felix Klein, who was about to retire. The mathematics department wished to fill the vacancy with a mathematician with a geometric approach to mathematics, and not just in the narrow sense, but in intimate connection with all the areas of mathematics. The faculty listed three candidates, who had mastered the present prominent problem, which emerged from the connection of set theory and function theory: 'What are the most general ideas one has to form about the continuum and the structures which are contained in it?' The three candidates were Carathéodory, followed at a distance by Hermann Weyl and Egbertus Brouwer (*ex aequo*). The choice of the faculty clearly shows that Hilbert, as the most influential man, was sufficiently impressed with Brouwer, to think him good enough for Göttingen. The available material strongly suggests that Brouwer was not aware of the Göttingen proposal (else Korteweg would have used it as another strong argument to convince his curators).

One should not get the impression that Brouwer was after the comfortable life of a research professor, he certainly had his problems combining a flourishing scientific practice with the aspects of teaching, but on the whole he, then as well as later, took his teaching duties quite seriously. Apart from the routine calculus courses for the science faculty, he gave courses on topics closer to his research interests. In 1913 he taught a course on the theory of functions, and he introduced a novelty—undoubtedly inspired by the German example: namely a seminar in pure mathematics for the older students. The seminar consisted of talks by students on recent advances in mathematics, and it was obligatory for the *doctoraal* exams.

Furthermore, Brouwer was—against his wishes—sucked into the committee-circuit; he had to attend faculty meetings, official exams, academy meetings, etc. On top of that he was rather occupied with writing reviews for the *Revues Semestrielles*, a Dutch review journal. Brouwer and his colleague Hendrik de Vries drew some comfort from a certain degree of innocuous tomfoolery that they practised at exams and meetings. De Vries excelled in providing comments under his breath, that heavily taxed Brouwer's facial muscles.

Brouwer's teaching consisted at that time of a number of topics, a considerable part of his time was taken up by the standard course Differential- and Integral Calculus for chemistry students. As Lize put it,

> ... it has to be taught fairly superficially, this is very dull for Bertus. He requires the students to make exercises, and is of course very cross if this is done poorly. Last week he had a young lady, who in this respect gave little reason for being content; he made her work at the blackboard for half an hour, so that she almost fainted. As a contrast he called somebody to the blackboard, whom he knew would solve the problem in 5 minutes'.[74]

The sudden elevation from beginner to expert, and from *privaat docent* to professor, seemed to have taken Brouwer by surprise. As so often happens, the newcomer was loaded with tasks that soon cumulated to more than he could handle. A number of these tasks were undertaken in the expectation of a better position, but some were indeed almost forced upon him. The signs of the pressure on Brouwer began soon to tell; on new year's eve he more or less desperately wrote to Korteweg, reporting on the prize essays of the *Wiskundig Genootschap* (Dutch Mathematical Society), that

> With the continuous pressure, under which I live as a consequence of all kinds of obligations with respect to foreigners, for the fulfilment of which time is lacking me, and as a consequence of the standstill of my mathematical researches (to which my opinion, the task of a professor and member of the Academy forbids me to resign), I am forced to refrain from all unnecessary work.[75]

The situation, however, did not improve, and Brouwer kept complaining. And so it is hardly surprising that Korteweg deemed it necessary to set his gifted student right.[76]

> Amice, De Vries already told me how much Göttingen[77] occupies your time and I quite understand that at this moment you do not want to take on a lecture.[78] Anyway, my request was only a consequence of my endeavour to raise the level of these talks as much as possible, and I half and half expected that you would excuse yourself this time.

---

[74] Lize to Aldert and Lily 4 May 1913.

[75] Brouwer to Korteweg 31 December 1913.

[76] Korteweg to Brouwer 4 June 1914.

[77] Brouwer's research contacts were mainly with Göttingen; moreover, he was involved in refereeing for the *Annalen* even before he became a member of the editorial board.

[78] At a meeting of the Math. Soc.

Less did I expect your divulgence. That your professorship suits you so little, distresses me very much. I view this, however, as a subjective phenomenon, indeed connected with your great talents, like everything in a particular human being is more or less connected, but I don't consider it inseparable from such talents.

In my opinion our physicists, who are members of international Academies, and for a prolonged period had no fewer official encumbrances than you (Van der Waals, Lorentz, who took over [Kamerlingh-] Onnes' course for the medical students) prove that. It is thus difficult to accept that six courses a week, partly elementary in nature, a few examinations a month (with an ample four months of nearly undisturbed vacation (sic!)) would prevent anybody from expecting scientific work, also of the highest level.

If this is the case with you, then there is for you really nothing but to accept a German chair, and that occasion will not fail to come up, although I expect that also there, 'hampering' influences will arise, if you are so sensitive to them. It is another question if you could not, if you come, with me, to the conclusion that the difficulty is to be found *in yourself*, do something or other to lessen the conflict.

For example, prepare your courses in the vacation, so that you are always well ahead, and that you have to prepare yourself each time only briefly. That would take away much of the nervousness and agitation that otherwise is inherent to the teaching of new material for the first time.

The reaction of Brouwer to this fatherly advice is not known, but a month later Korteweg wrote another letter in the same vain. The direct cause was Brouwer's wish to withdraw from the prize-essay committee of the Mathematical Society. His success in the world of mathematics had its consequences: the demands from academic and mathematical organizations, started to absorb all his time and energy. And in an act of desperation he had blamed the stagnation of his mathematical research on the professorship. Korteweg put things into a more realistic perspective, when he wrote[79]

> Whereas under different circumstances I would be most pleased with your honourable appointment to editor of the first among the mathematical journals of the world,[80] this is now not whole-heartedly, and for more than one reason.
>
> *In the first place* I view the work with which you are swamped by the Göttingen people as a very serious and *lasting* hindrance for the continuation of your own work, and yet you will be in the long run, also in Germany, be judged *by that*.

---

[79] Korteweg to Brouwer 13 July 1914.
[80] Brouwer was appointed editor of the *Mathematische Annalen*, cf. Brouwer to Klein 10 July 1914.

In the second place I foresee that you will absent yourself more and more from Dutch mathematical life, whereas one expects, in my opinion justly so, the opposite attitude of a Dutch professor [...]

In the third place I fear that you will look for the cause of your diminished fertility there where it is not, or only for a small part: in your professorship, and this will thus appear more and more as a pure nuisance. [...]

I must concede one thing, in order not to become unfair with respect to you. Namely this, that elementary courses seem to present you with great difficulties, because they arouse your impatience, and seem to make you temporarily unfit for other work.

Korteweg's diagnosis was perfectly to the point, and, in view of later developments, prophetic. Indeed, the lure from Göttingen was an unparalleled temptation, —to be elected to the mathematical elite! The price was dear: Brouwer spent a great deal of time refereeing papers for the *Annalen*. And, necessary and honourable as it might be, it was not sensible to exchange research for supervision.

Korteweg, who had given up his chair for his student, had good reasons to be bitter, but his main worry seemed the fate of this student. He pointed out that, contrary to Brouwer's complaints, the Netherlands and its Universities were on the whole a congenial environment for the young scientist, and that Brouwer had indeed profited from the liberal possibilities.

The reaction of Brouwer has not been preserved, but we may assume that he accepted the consequences of his position; he did not give up his chair, nor did he severe his fresh links with the *Annalen*. The outbreak of the war solved the problem of the claims of Göttingen by brute force: the demands of the *Annalen* were temporarily reduced to more moderate proportions.

6.7 Schoenflies again

So far Brouwer had led the life, not only of a fundamental researcher, but also of a free-lance philosopher and a moralist. Lately another burden, that had nothing to do with his administrative or teaching duties, had been added to already overcrowded schedule: the cooperation with Schoenflies.

The comprehensive survey of Schoenflies, commissioned by the German Mathematical Society, had presented a state of the art in set theory and topology. The two volumes appeared in 1900 and 1908. The volume on topology had attracted Brouwer's attention; he had studied it closely and used it as the background material for his own research. As we have seen, the monograph was not proof against Brouwer's scrutiny, it had revealed a number of serious lapses. The 'Analysis Situs'-paper had resulted from this study.

In view of the importance of the topic, and also of the so recently discovered mistakes in the topological part of the *Bericht*, Schoenflies started to prepare a new edition.

Brouwer, as the recognized expert on topology, had been receiving from various quarters the request to write a book on set theory, since the existing books and encyclopaedia contributions were too superficial and undependable. When Brouwer was visiting Göttingen in the summer of 1911, he was confronted with new attempts to get him to write such a book; he suggested that a satisfactory solution could be reached with little loss of time 'if he were given the opportunity to supervise the book of Schoenflies during the preparation, and when necessary to improve and supplement it'.[81]

At the time Brouwer was fully engaged in research, and the writing of a book would seriously interrupt it. When that same summer Brouwer visited Fricke in the Harz, the latter offered to mediate between Brouwer and Schoenflies. This, at first sight, fortunate solution proved disastrous; Schoenflies wanted to restrict Brouwer's role to the correction of mistakes, and Brouwer, naturally, wanted to improve the book in depth and to update it.

The ensuing bickering upset both parties, but Brouwer suffered most; his quick and emotional mind could not cope with the slow pace of the older man, who fought a war of entrenchment against Brouwer's innovations, and who often failed to grasp Brouwer's corrections and improvements. He would for example, ask Brouwer to prepare part of a chapter, only to tell him later that the text should be incorporated in another chapter; Brouwer reacted,

> That is not how Schoenflies can make use of my time, which has its value, just as much as his time. I am neither his assistant, nor I am helping him out of personal friendship, but because I find it important that a good book on set theory should be produced, and because I am at the moment the most competent person in this area.'[82]

Again and again, Schoenflies would try to escape the iron control of Brouwer. He apparently did not come up to the standard of exactness that was set by Brouwer in topology, and by Zermelo in set theory. Even Hilbert's influence was not sufficient to keep Schoenflies on the right track. He regularly abbreviated, reformulated, or even cancelled Brouwer's corrections, 'He starts to abridge industriously much of what, after an endless correspondence, finally was formulated correctly. He must be *very* overworked, for he makes mistakes, for which any student should be ashamed.'[83]

Schoenflies kept rehashing Brouwer's contributions, and relapsing into his old mistakes. In reaction, Brouwer would repeatedly appeal in desperation to Hilbert, begging him to bring Schoenflies to reason. Schoenflies, also turned to Hilbert with the urgent plea to rescue him from the hands of this merciless Dutchman, just as he done at the occasion of the 'Analysis Situs'-paper.

Unfortunately the correspondence concerning the revised version of Schoenflies' book is deplorably incomplete, but enough is left to understand the mutual

---
[81] Brouwer to Hilbert 16 April 1913.
[82] ibid.
[83] Brouwer to Hilbert 16 June 1913.

irritation of both men.

Schoenflies enjoyed a reputation as an excellent expositor, and he understandably reserved the right of the final wording of the manuscript. Brouwer's censorship was far from complimentary for a man of his age and status; in fact he repeatedly tried to rebel. Once he sent in despair Hilbert a proof sheet, with a sentence underlined in red, that, according to Brouwer, contained a logical mistake, challenging Hilbert to find it—'If you find it, I will gladly give you a present'.[84] More than four letters were exchanged between Brouwer and Schoenflies to discuss this particular sentence!

The whole affair was heavily taxing Brouwer's patience, some of the letters are therefore rather harshly formulated; at one occasion Schoenflies had, for example, inserted a statement to the effect that a certain theorem could be proved using Brouwer's methods, whereas Brouwer had given a counterexample to the theorem. Utterly astounded Brouwer demanded to know if he was to be quoted as the intellectual father of the proof of a falsity. Given Brouwer's emotional sensitiveness, one can easily understand Mrs. Brouwer, when she added a post script to a letter from Brouwer to Hilbert; 'If my husband becomes insane because of Schoenflies, he has to thank you for it.'[85]

A page from the proofs with Brouwer's corrections.

The new edition of the *Entwickelung der Mengenlehre und ihrer Anwendungen* appeared in 1913.[86] The book must have been a sad disappointment to the experts in the field, all the developments of the last few years were ignored. Almost none of Brouwer's achievements in the 'new topology', or even in Cantor–Schoenflies topology, were incorporated. It was as if the shock of the Analysis Situs paper had paralyzed Schoenflies. He left out all the material concerning curves, dimension, etc. Brouwer's influence apparently did not go far enough to add material, only to correct or to omit.

In a sense the book was already obsolete by the time it appeared. It was almost immediately superseded by Hausdorff's *Grundzüge der Mengenlehre* (1914). Also, as far as Brouwer's wishes were concerned, the book was a disappointment— no account of his recent advances in topology was given, and the demand for a coherent exposition remained. His stratagem, to incorporate the newer material into Schoenflies' book had utterly failed. Although Brouwer is mentioned here and there, there are hardly any recognizable traces of his mastery of set theory and

---

[84] Schoenflies to Hilbert 8 July 1913.

[85] Mrs. Brouwer to Hilbert 11 September 1913.

[86] Although the title page says 'published jointly with Hans Hahn', Schoenflies is the only author. Hahn acted as a critic and adviser (albeit not so prominently as Brouwer), he was to write a second volume on real functions which never appeared. In 1921 Hahn published his own monograph *Theorie der reelle Funktionen I*, [Hahn 1921].

topology.

In the introduction Schoenflies profusely thanked Brouwer for his 'unselfish and abundant' support, pointing out that not only he, but 'the collective mathematical world' was in debt to Brouwer for the exact and useful form of the book.

Brouwer had lost his last struggle with Schoenflies, when he tried to force Schoenflies to insert a statement to the effect that the cooperation of Brouwer did not mean that he had given up his constructive convictions in favour of the set theoretic ones; Schoenflies flatly refused to do so. It seems plausible that the suppression of Brouwer's intensive involvement in the preparation of the volume was the result of a compromise.

Some of the objections of Brouwer can, however, be found in his review of 1914, in which he not only objects to some parts that become meaningless, when seen from an intuitionistic viewpoint, but also pin-points some weak spots in the 'classical' treatment.

With this item the first act of Brouwer's mathematical career ends, a number of topics will reappear in the sequel.

There is no doubt that the cooperation with Schoenflies heavily taxed Brouwer's resources; the material in itself was not a serious obstacle, but the endless bickering, pointing out mistakes, giving new proofs, only to find that Schoenflies did not grasp the point and preferred his own defective arguments—it all wore out Brouwer. A more equanimous man would have borne his cross quietly, but Brouwer—with his fierce convictions of right and wrong—was easily tempted to see insults, and hence to repay in force. Conflicts in general affected him strongly, to the extent of physical afflictions. So it is not unthinkable, that Schoenflies' hard headedness had worn out Brouwer's resilience to the extent of the drying up of the fountain of topological productivity. The complaints of Lize, who knew her husband well enough, are telling. She complained in the above quoted letter to Aldert and Lily, that the Schoenflies project had cost Bertus the priority for Poincaré's last theorem.

The ultimate recognition, which generously redressed all the hardships of the past years, had come in July 1914, when he was co-opted by the *Mathematische Annalen* as an editor. In a letter of 10 July to Felix Klein, Brouwer expressed his appreciation for this 'high honour'. The appointment was indeed the crown on his scientific work; the journal carried a prestige that surpassed that of scientific societies and academies. Under Klein's regime the *Annalen* had become the leading journal in mathematics, and the editorial board was more than just a collection of editors, in Carathéodory's words it had been moulded into

> a kind of Academy, ... That was in my opinion the main reason why the *Annalen* could claim to be the first mathematics journal in the world.[87]

As we have learned from Korteweg's letters, Brouwer was already working for the *Annalen* before his elevation to the rank of editor. Brouwer told Schoenflies in 1921 that he was an informal editor (*Mitarbeiter*) from 1911 to 1914.[88]

---

[87] Carathéodory to Courant 19 December 1928.
[88] Brouwer to Schoenflies 17 January 1921.

At the age of 33 Brouwer had now reached the highest position in the republic of mathematics, ten years after his first papers, and in spite of his attacks on the prevailing views on the foundation of mathematics.

7
THE WAR YEARS

The Great War that brought tragedy and hardship to the warring nations, passed by the borders of the Netherlands, a country that took its neutrality for granted. The Dutch mobilized on the first of August 1914, all political parties supported the government in its energetic policy of neutrality. The average citizen had only vague ideas concerning the reality of war, after all, since Napoleon's defeat, the nation had only known the short skirmish of the Belgian uprising, and the far and exotic military exploits in the colonies. The European conflict was generally considered to be 'none of our business', and sympathy for the Entente and the Central Powers was fairly evenly distributed.

Brouwer, who had visited in July cathedrals in France with his stepdaughter Louise, had returned home before the outbreak of hostilities, firmly resolved to stay home. But soon his curiosity got the better of him, and early August he travelled to the war scene, where he reportedly saw quite a bit during a stay of a week.[1]

The immediate consequence of the war for Brouwer was a disruption of his international contacts. The last few years he had come to consider Göttingen as his second scientific home; ties of admiration and friendship bound him to Hilbert, Klein, Blumenthal and others. And, although it was not impossible for a subject of a neutral nation to travel in the territory of the belligerent nations, it was far from simple to obtain the required visa and permits. So, in effect, his contacts were limited to correspondence.

Whether the isolation was the main cause, or perhaps the lure of the foundations, the war years saw a return to the problems that were raised in the dissertation, and that had been left unsettled. War or no war, the university went about its business, and Brouwer had to carry his teaching load like his fellow professors. Apart from the standard courses (which rather bored him—but which he always took seriously) he taught a number of courses on subjects of his own choice. In 1912, still a *privaat docent*, he had started a course on algebraic functions, and after his appointment to extraordinary professor he introduced courses on Theory of Functions (*Functierekening*), which were actually courses on point set theory. The university catalogue announced the 1912 course as *Functierekening*, but Brouwer listed the course in his own notes as *Point sets*. In addition he gave in his first professorial year another course: *General Set Theory*.

From 1913 onwards he taught also the courses on Mechanics, inherited from Korteweg. Through the years he offered courses in *Higher Mechanics* (in two parts,

---
[1] Brouwer to Scheltema, 15 August 1914.

I and II) and the *Theory of Oscillations*. Furthermore he regularly taught courses on geometry, described in the catalogue as follows:

> *Projective Geometry I* (1913–14)—Projective coordinates, Projective and reciprocal transformations, invariants, Conics and sheaves of conics, higher algebraic curves, multiplicity of intersections, connection between equations in point and line coordinates, multiplicity of main points of a tangent, Plücker relations, n-th degree dependence of point systems, genus of algebraic curves, theory of unicursal curves.
>
> *Projective Geometry II* (1914–15)—Continuous groups, non-Euclidean Geometry, Foundations of Geometry.
>
> *Set Theory* (1915–16)—Finite and infinite cardinal- and ordinal numbers, the fundamental operations with well-ordered ordinals of the first, second and third domain, general theory of well-ordered sets, intuitionistic generation of point sets, solution and internal decomposition of point sets.

In particular the kind of set theory that had been initiated by people like Baire and Emile Borel (what we now would call the theory of real functions, or descriptive set theory) was close to Brouwer's personal interests. This course was at first intuitionistic in the spirit of the dissertation, only after 1915–16, when Brouwer knew how to handle his new intuitionism, based on choice sequences, it became fully intuitionistic in the modern sense.

There is not much material that helps us to form an idea of the courses during the first years of Professor Brouwer; the first notable piece of evidence is a set of handwritten course notes of his 1912–13 lectures on set theory. It sheds light on the evolution that Brouwer's foundational views underwent in the years of introspection and search.

7.1 Sets and sequences—law or choice?

In order to appreciate the solitary quest of Brouwer for the truly constructive mathematics one has to go back a little in history and view the early approaches to constructiveness as advocated mainly by Borel. Virtually all mathematicians will recognize the numbers $1, 2, 3, 4, \ldots$ as effectively given. The German mathematician Leopold Kronecker went even so far as to say that 'the natural numbers are made by God, and everything else is the work of man', which was the poetic expression of his belief that the only legitimate mathematical objects (on the side of number systems, geometry was not included) were those that could be reduced to natural numbers. We know lots of those objects: integers (as pairs of natural numbers, for example, $(5, 2)$ stands for 3 and $(2, 5)$ for -3), rational numbers (as pairs (n, m) with $m \neq 0$, for example, $(5, 2)$ stands for $\frac{5}{2}$).

Irrational numbers can also be given by means of natural numbers, for example, if they are roots of polynomial equations, but in general they do not have such simple representation. Therefore, Kronecker banned the irrationals that were not decently representable from mathematics. A feat that did not contribute to his popularity among his contemporaries. Fortunately he was one of the giants of algebra and number theory, so his authority did not depend on his foundational views. Kronecker was an outspoken man, and he did not hesitate to attack well-respected and -established colleagues. His attacks on Cantor and Weierstrass were notorious. He was the hornet of the end of the nineteenth century, and he did not hesitate to call the work of certain contemporaries totally devoid of meaning. In the eyes of most mathematicians Kronecker was an eccentric, albeit brilliant, man, but his rejection of that what generally was accepted as progress, gave him the reputation of a reactionary and kill-joy. Kronecker stressed over and over that only mathematical objects could be admitted that were constructible in finitely many steps. It was not unusual to contrast in Kronecker's days the 'algebraic' and 'logical' methods, the latter were abstract, such as Cantor or Dedekind used, the first were finite and effective.

After Kronecker, the French mathematicians, led by Poincaré and Borel, started to cultivate certain constructive arguments. Poincaré, as we have seen, was close to Brouwer in certain respects. The French constructivists stressed in particular the 'definability in finitely many words'. Their view, which was generally shared by mathematicians of a constructive leaning, was that 'infinite objects', i.e. objects that could not be built in finitely many steps from natural numbers, should be presented by a finite definition, so that there was a finite guarantee (building instruction) for their existence.[2] Remarks to that effect can be found in the literature, Otto Hölder, for example, in his book *The mathematical Method* explicitly demands that if one wants to give a set with infinitely many points, one has to prescribe a law.[3] Thus, for instance, the irrational number $\sqrt{2}$ is effectively determined because there is a specific procedure to calculate all decimals. A real number whose decimals were successively determined by flipping a coin (e.g. 0.0011101010111110...) was thus not be admitted.

Brouwer seemed to share this view still in 1912, when he gave his inaugural lecture 'Intuitionism and Formalism':

> Let us consider the concept: 'real number between 0 and 1'. For the formalist this concept is equivalent to 'elementary series of digits after the decimal point', for the intuitionist it means 'law for the construction of an elementary series of digits after the decimal point, built up by means of a finite number of operations'. And when the formalist creates the 'set of all real numbers between 0 and 1', these words are without meaning for the intuitionist, even whether one thinks of the real numbers of the formalist, determined by elementary series of freely selected digits, or of the real numbers of the intuitionist, determined by finite laws of construction.[4]

---

[2] An analysis of the ideas of the French constructivists can be found in the excellent and instructive survey[Bockstaele 1949].

[3] *Die Mathematische Methode*, [Hölder 1924], p. 98. He refers to an earlier statement of the same tenor in 1892.

[4] [Brouwer 1913C], p.91.

Before we conclude from this quotation that Brouwer rejected sequences of 'freely selected' numbers, let us recall that the above refers to the points of the continuum. These points must be given by a law in order to turn them into distinguishable, individual points; a freely selected decimal expansion would (viewed in that stage of Brouwer's program) never determine a specific point. Hence there seems to be no convincing grounds for ruling out freely selected points altogether.

The real numbers of the inaugural lecture are given by decimal expansions. Such an approach is more restrictive than the treatment in his dissertation. It is not implausible that Brouwer chose the 'decimal' approach on didactic grounds. After all, the lecture was for an audience of non-specialists. It took Brouwer until 1921 before he publicly renounced the adequacy of decimal expansions for representing the full continuum.

It should be pointed out that the terminology of *Intuitionism and Formalism* is that of 1912, that is to say, before Brouwer had started his own modern brand of constructivism. So, 'intuitionist' is a loose term, but refers at this time mostly to the French school. It is not simple to compare Brouwer's first intuitionism with the earlier and contemporary forms of constructivism or intuitionism. Borel, for example, insisted on the 'finite definability' condition for mathematical objects, but this notion hardly plays a role in Brouwer's writings. With Brouwer 'constructiveness' and 'algorithm' are more or less immediate—one recognizes a law when one sees one. That is to say, Brouwer's algorithms basically are the result of his natural number construction (the two-ity intuition) but he did not bother to investigate the possible algorithms. Brouwer agreed with Poincaré on the topic of mathematical induction, but he vigorously disagreed with Poincaré on the identification of 'existence' and 'non-contradiction'. With respect to the continuum Brouwer differed from his fellow constructivists, in that he recognized a special 'continuum-intuition'. Something like that can be found in Borel's work, who recognizes a 'geometric continuum' as immediately given [Borel 1898] edition 1928, p. 16.

However, Brouwer goes essentially further than his contemporaries by rejecting full classical logic. Before World War I, all constructivists, including Brouwer, agreed on the point of individual objects: they had to be given by a law. Similarly, Brouwer accepted only those sets that could be obtained in specific ways, as indicated in his dissertation (Cf. p. 115).

From this point of view it is natural that Brouwer rejected, along with the majority of his colleagues, sequences generated by choices. The latter did indeed occur in the literature mainly as pedagogical examples that illustrated objections against free choices. The earliest examples are to be found in the book on function theory of Paul Du Bois-Reymond[5] in the discussion between the *Idealist* and *Empiricist*, the first explicitly considers decimals determined by throwing a die.

---

[5] *Die allgemeine Functionentheorie*, 1882.

The topic of 'choice' became popular after Zermelo's epochal proof of the well-ordering theorem based on axiom of choice;[6] this paper triggered a heated debate on the legitimacy of the notion of choice in mathematics, a discussion that reverberated long after set theory had returned to 'business as usual'. Borel, in particular, considered the status of series determined by choice; in his eyes an uncountable number of choices made no sense at all, but countable choices were allowed if there is a procedure that guarantees the effectuation of each choice after a finite amount of time.[7] In 1912 he even considered choices made consecutively by a large number of people, but only for probabilistic and pedagogical purposes.

On the whole, one may conclude that constructivists were not inclined to accept choice elements, and classical mathematicians, say set theorists, were not interested in them.

Brouwer was worried all the same by this state of affairs, and for a quite specific reason: at best the continuum of the semi-intuitionists[8] is denumerably unfinished, and hence must have measure 0. This, obviously, was totally unacceptable for him. As late as 1952 Brouwer advanced this as the prime argument against the semi-intuitionists. Furthermore, the foundational scene was haunted for years and years by the paradox of Richard.[9] This paradox, which used only the most simple tools, highlighted the inherent dangers of the notion of definability. Consider real numbers that are defined by finitely many words, these definitions can be put into a linear sequence $D_0, D_1, D_2, D_3, \ldots$ (roughly in the same way the words in a dictionary are put into the alphabetical order) then we can define a new number as follows:

> the n-th decimal of r is a 0 if the number defined by D_n has an n-th decimal which is not 0, and it is 1 if this decimal is 0.

Here, evidently, a real number is defined, which is distinct from all the numbers which were defined by $D_0, D_1, D_2, \ldots$ Hence the list cannot contain all definitions: Contradiction!

This paradox seriously worried the mathematical world; Poincaré, Russell, and others, went to great lengths to avoid the phenomenon. Only much later, when the notion of definability was better understood, did the fear of Richard's paradox subside.

Poincaré banned Richard-like phenomena from mathematics by restricting its methods to so-called predicative ones. The peculiarity of the above defined number is that it is a definable number which uses in its definition all definable numbers, hence also itself. Poincaré saw such notions, which he called *impredicative*, as the bane of mathematics. Russell, too, tried to safeguard mathematics by imposing

---

[6] [Zermelo 1904].

[7] [Borel 1908], [Troelstra 1982].

[8] We will from now on call the French intuitionists, Borel, Lebesgue, Baire, Poincaré, Hadamard, and their associates *semi-intuitionists* in accordance with present usage. Brouwer used in his later publications the name *old-intuitionists* for his predecessors.

[9] [Richard 1905].

his *vicious circle principle*: 'no object shall be defined in terms of a collection that contains it'. This particular issue 'predicative vs. impredicative' is completely absent from Brouwer's writings, but he was no doubt familiar with it, since Richard's paradox was treated in the courses of his friend and teacher Mannoury. Although Richard's paradox may not have been uppermost in his mind, it is not improbable that it added to his dissatisfaction with the state of intuitionism.

Slowly, Brouwer's attitude towards infinite sequences started to change; in his review of Schoenflies' *Bericht* on set theory[10] he explicitly mentioned sequences of choice as legitimate objects.

The course on set theory in 1912–13 was still fairly conventional; however strong Brouwer's philosophical convictions were, he had not yet discovered how to do justice to them in mathematics. Brouwer's private notes were kept in a black exercise book; he conscientiously wrote down the course material on the right-hand page and recorded emendations, corrections, remarks on the left-hand page. He always used a fine steel pen, and produced an incredibly fine but still legible writing, often meticulously crossed out in the pattern of a fine grid. Extra lines were inserted in an even finer handwriting, and sometimes pencilled remarks (unfortunately less readable) were added. Notebooks such as this one present us with a kind of 'history' of the development of his thoughts. This particular notebook starts at 5 November 1912 (oh golden days of leisurely teaching!). Its first line defines the cardinal number of a set, closely following the pattern set by Cantor. On the whole the choice of material is fairly traditional; cardinals, ordinals, orderings, basic point set topology, and measure theory are treated. The constructive view point is observed, but rather as a limitation of traditional methods, than as a new approach to mathematics. Repeatedly Brouwer corrected himself later by inserting criticism of parts of the course, e.g. after showing that finite sets yield the same number under all countings (the fundamental property of finite sets), he went on to show that for infinite sets 'exactly the opposite holds'. But afterwards he noted on facing page 'the weak point, into which we will not enter ... is the tacit assumption that each set is finite or infinite', and on the same page he spotted another weak point, namely why is a set with a denumerable set of points removed again a set? This remark, by the way, shows that in 1911 Brouwer still stuck to the notion that only sets generated in a specific way, as indicated in the dissertation, make sense. The well-ordered sets are not explicitly defined in the notes (presumably Brouwer used the notes partly as a summary of the lectures, so that a few notes would often be enough to keep track of the course) but knowing Brouwer's partiality to Schoenflies' book, we may assume that he used the classical standard notion; in his remarks after the lecture he questioned this notion. One remark runs 'According to my Schoenflies-criticism there can, by the way, be no other ordinals than countable ordinals' and a few lines later he suggested 'Maybe this is correct after all [that is the argument that all countable ordinals belong to the second number class]; for we can demand for well-orderings that they are brought about by the two generation principles; but then the theorem

[10][Brouwer 1914], p.79.

is self-evident, a tautology, that is perhaps the cleanest view.' Here one sees the genesis of Brouwer's theory of ordinals.

Brouwer did not use the Cantor–Bernstein theorem, which according to him 'must be regarded as an open problem, ... but the proof of this theorem is considered inconclusive by many mathematicians', hence he defined 'as large as' by 'A can be injected into B and B into A'.

The real breakthrough came in 1916. In the academic year 1915–1916 Brouwer taught an advanced course, the theory of point sets (listed in the university catalogue as 'Set Theory') which started in January. According to the catalogue it was continued the next year, but Brouwer's private notes suggest that he started the course all over again. The university catalogue gave the following description of the 1916–17 course:

> Point sets and internal sets. Deductions and derivations of point sets. Inducible properties of point sets. Covered and measurable point sets. Measurable properties which do not define point sets. Connection between measurable properties and internal limiting sets.

Brouwer's handwritten notes are clearly those of the 1915–1916 course, but with emendations added in the margins and on the backs of the pages. It is of course possible that Brouwer made these notes for purely private purposes, but it seems far more likely that he gave an improved course in 1916–1917 and started again from scratch, but this time using new insights.

The 1915–1916 course was basically an improved version of the 1912–1913 course. However, during the course, and most likely before the new course, Brouwer must have gotten a revolutionary insight that changed the whole perspective of the course.

The private course notes of Brouwer witness the revelation that shaped modern intuitionism: the original text was written in the characteristic hand with a steel pen, but added in pencil in the margin of the very first page, there was the casual formulation of a revolutionary breakthrough, taking place in a quiet class room in Amsterdam.

After announcing that he was going to do mathematics from an intuitionistic point of view—admitting that he had himself often applied the *principium tertii exclusi* in his work, 'which then probably has not yielded correct, but only non-contradictory results'—the notes in the margin go on

> First say something about sets defined *mathematically* and by *comprehension*.
> A mathematical thing is either an element of a previously constructed fundamental sequence F (governed by induction, like the sequence ρ) or a fundamental sequence f (which is not finished and not governed by induction) of arbitrarily chosen elements from F or a finite set. With such a sequence one can work very well, if one has always *at each phase* for the

finite thing d or the fundamental sequence r that is derived from it, to work with a suitable initial segment of f (r is then in general *also* never finished) [...] A set is now a law by means of which a d or an r is derived from f; this r can then, for example, contain also relation symbols (e.g. ordering ones), so that the law can, for example, lead to well-ordered sets or other ordered sets, or to functions (one can indeed not get the set *of* ordered sets or the set of well-ordered sets). In addition one can accept *pseudo-sets* defined by comprehension, better called *species*, and one can call one species of a higher cardinality than another one, or two species equivalent.[11]

The definition reappears in full prolixity in the first *Begründungs*-paper of 1918 (cf. p. 314). It is helpful to compare both formulations to see that finally the choice-feature had gained a firm foothold.

The significance of this concept of set and species is that Brouwer had found another way to deal with the genesis of sets. Already in the dissertation particular methods for creating new sets were given, but here is a new method which makes use of the notion of arbitrary sequence.[12] The basic distinction between sets (later called spreads) and species is introduced in the 1916–17 notes. The mere sanctioning of choice sequences is by itself not yet remarkable, for Du Bois–Reymond and Borel had already considered sequences of choices, before him. In a crucial respect Brouwer goes beyond those authors, he indicated how one can use choice sequences in actual mathematics, showing that it was not a mere curiosity without any uses.

Brouwer had in the above note formulated the key to the success of the choice sequences: if you don't know how a sequence is going to be continued, use only the first values that are given so far. This already demonstrates a piece of common sense that goes a long way in everyday mathematics: suppose a real number is given by a converging sequence of rational numbers $a_0, a_1, a_2, \ldots$ (a so-called Cauchy sequence) where the a_i are chosen arbitrarily, that is to say the choices are not restricted as long as one takes care that the sequence converges. Here is an example: first choose a natural number n, and after that choose from $n-1, n, n+1$, say a_1, next choose from $a_1 - \frac{1}{2}, a_1, a_1 + \frac{1}{2}$, say a_2, next choose from $a_2 - \frac{1}{4}, a_2, a_2 + \frac{1}{4}$, etc.

Now we want to multiply the resulting number by 2. This is most easily accomplished by multiplying all the approximating rationals a_i by 2. The moral is that for the n-th approximation of the resulting number, one only needs the n-th approximation of the original.

So far, so good: it allows one to use choice sequences in practice, but somehow it is a cautious approach: if you can reduce an operation to an operation on initial segments then you are home. But can it always be done? This basic question was answered by Brouwer only a few pages later:

---

[11] ρ is the sequence of natural numbers; 'governed by induction' is an old fashioned way to express 'given by a law'.

[12] Brouwer was not quite consistent in his use of 'fundamental sequence'. Usually (and certainly in the early foundational papers) this was a lawlike sequence, but from time to time Brouwer also speaks of fundamental sequences of arbitrarily chosen elements. One has to be careful in reading his papers, usually the meaning is clear from the context, but not always!

	0	1	2	3	4	5	6	7	8	9	10	11	12
f_0	1/2	47	18	0	1	0	3	3	7	10	.	.	.
f_1	0	12/13	7	2	4	1	0	89	4	7	.	.	.
f_2	0	5	6/7	3	7	1	1	1	0	7	.	.	.
f_3	5	8	3	19/20	2	5	4	7	2	1	.	.	.
f_4	3	9	45	2	3/4	7	1	0	4	6	.	.	.
f_5	0	0	45	23	76	1/2	0	76	31	95	.	.	.
f_6	64	0	1	1	4	63	3/4	6	9	1	.	.	.
f_7	0	6	9	4	45	5	1	7/8	9	0	.	.	.
f_8	.	.	.	.	.	.	.	.	.	.			
f_9	.	.	.	.	.	.	.	.	.	.			

FIG. 7.1. Cantor's diagonal method.

The impossibility of mapping all elements of f_1 to distinct elements of r follows from the fact that the choice of the elements of ρ would have to take place at a certain point of the (forever unfinished) choice sequence, and in this way all continuations of such a *finite* choice-branch determining the element of ρ will obtain the *same image in ρ*.[13]

This remark is the essence of Brouwer's Second Proof of the non-denumerability of the set of all natural number sequences, which has found a modest place in the margin of the note book. Thus Brouwer gave a beautiful and bold argument that replaced the famous diagonal argument of George Cantor. The latter argument is one of the most powerful inventions of the nineteenth century, and it is surprisingly simple.

Consider the set of all infinite sequences f of natural numbers, that is to say, each f is a sequence of natural numbers. Suppose that this set is countable, that is that it can be written as $f_0, f_1, f_2, f_3, \ldots$. Then one can put the sequences and their values in a sort of infinite chess board, see Fig. 7.1.

Now consider the values in the diagonal: $f_0(0), f_1(1), f_2(2), f_3(3), \ldots$; we change these values as follows: add 1 to each of them, thus we obtain a sequence d with $d(i) = f_i(i)+1$. This sequence does not occur in the list $f_0(0), f_1(1), f_2(2), f_3(3), \ldots$, for if $f_m = d$ for some m, then $f_m(m) = d(m) = f_m(m) + 1$, which is impossible! The result can be summed up as 'the set of all infinite number sequences is not denumerable'.[14]

Now Brouwer had in the 1912–1913 and 1915–1916 courses faithfully reproduced Cantor's diagonal argument,[15] but after the recognition of choice sequences as sound mathematical objects, he realized that the nature of choice sequences forbade an enumeration of all choice sequences for a totally different reason. Indeed, this would require the assignment of natural numbers to choice sequences, such that distinct choice sequences would correspond to distinct natural numbers. Now

---

[13] f_1 and ρ stand for the sets of all choice sequences of natural numbers and the set of the natural numbers.

[14] [Cantor 1892].

[15] Brouwer had seen that the diagonal method is perfectly acceptable to a constructivist; this is implicit in the inaugural lecture of 1912; it is explicit in the set theory lectures, which were constructive (in the spirit of the dissertation); Brouwer provided the non-constructive arguments with the *caveat* 'this is not constructive'.

let 273 be assigned to the choice sequence f, then the assignment is effected after, say $f(0), f(1), f(2), \ldots, f(2001)$ have been given. But then any two sequences that start with $f(0), f(1), f(2), f(3), \ldots, f(2001)$ are assigned the same natural number. Hence one can never satisfy the requirement that distinct choice sequences get distinct natural numbers under the assignment. In simple mathematical terms, Brouwer's argument says 'The functions from $\mathbb{N}^{\mathbb{N}}$ to $\mathbb{N}$ are continuous, and hence there is no bijection from $\mathbb{N}^{\mathbb{N}}$ to $\mathbb{N}$'.

The above 'fact' that Brouwer invoked in the quotation above has become known under the name *Brouwer's continuity principle*.[16] The insight that the mathematical universe not only contains infinite sequences given by laws, but also arbitrarily chosen sequences, was a significant step towards a viable constructive mathematics. The 'poor but honest' doctrine of Kronecker and the semi-intuitionists, was very upright and respectable, but it could without some extra foundational insight never progress beyond a cautious fragment of traditional mathematics.[17] One can imagine that such a practice did not appeal to an imaginative person like Brouwer; but the main objection to a life in the shadow of classical mathematics was that it did not make use in a proper way of reflection on the power of human thinking.

Between the discovery of the virtues of choice sequences and their appearance in print some time lapsed. The new approach was however, albeit in a rather cryptic way, already announced in a short note with corrections and addenda to the dissertation.[18]

> However, it has lately become clear to me as I hope to explain in a paper that will appear before long, that the limits of set theory can be extended.

Brouwer's 1916–1917 course on set theory can indeed be considered as the watershed between the, fairly negative, intuitionism of the dissertation and intuitionism as we know it today.

7.2 The International Academy for Philosophy

At the same time that Brouwer was discovering a new basis for intuitionism, he was drawn into a philosophical enterprise that has gone down in history as 'Significs'. It brought together a remarkable and mixed company. Significs did not come out of the blue, but it had its roots in the last years of the preceding century. The main actors in this ambitious and idealistic project were Frederik van Eeden—a well-known Dutch author, Gerrit Mannoury—the self-made mathematician and

---

[16] 'Functions from Baire space to the natural numbers are continuous'. In traditional mathematics this is, of course, not the case.

[17] Note, however, that Bishop managed to regain a good deal of mathematics in the Kronecker–Borel tradition. The price to be paid was a general strengthening of assumption.

[18] [Brouwer 1917A1].

philosopher, Jacob Israël de Haan[19] —a man of letters and of law, Henri Borel[20] —a sinologist and journalist, and Brouwer. Each of these persons had a colourful life, and the significs group still stands out as the most imaginative, and indeed the only, collective philosophical scheme in the Dutch tradition. The intellectual drive behind significs was to be provided by Mannoury, but the pioneer of this particular brand of philosophy of mind and language was Frederik van Eeden.

Van Eeden was born in Haarlem in 1860, he came from a well-to-do family, his father had taken over the family's bulb trade.[21] Frederik studied medicine in Amsterdam, where he was a rather successful student; he became Rector of the Student Corps, the highest honour in the world of students.

Ever since his school years, Van Eeden had felt literary ambitions, and when still a student, he wrote a masterpiece, which was the main source of his literary fame: *De kleine Johannes*[22] (1887), translated in numerous languages, it appeared in English under the title 'The Quest'. Dutch literature at that period was in a phase of transition from the lofty tradition of the nineteenth century Dutch letters to a more impressionist style. The old guard had founded an authoritative literary journal *De Gids* (The Guide), and the new generation expressed its sentiments in a counter journal *De Nieuwe Gids*, perhaps not the *summum* of originality but sufficiently adequate as a symbol of protest. Van Eeden's talents did not escape the young poets and authors, and so it was more or less natural that he was invited to join the select group of founders of *De Nieuwe Gids*.

Frederik van Eeden.
Courtesy Frederik van Eeden Archive

As a medical doctor, Van Eeden was equipped with a good dose of curiosity and initiative; indeed, he opened in Amsterdam the first practice for psychiatry together with a colleague (1887). He fostered a keen interest in almost all phenomena that had to do with the manifestations of the mind, he was equally interested in hypnosis, spiritualism and occultism. As if his medical practice and his literary activities were

[19] Among friends he was called 'Jaap' or 'Joop'.

[20] The Dutch sinologist and author. This Henri Borel will frequently be mentioned in the history of Significs. The reader should not confuse him with the French mathematician.

[21] There is a two volume biography of Van Eeden, [Fontijn 1990, Fontijn 1996].

[22] Little Johannes.

not yet enough, he undertook an experiment of practical social reform by founding an utopian colony, and engaged in philosophical studies. His early occupation with philosophy produced a neat monograph *Rhetorical foundation of understanding*,[23] a treatise containing Van Eeden's view on ratio, communication, mind, mathematics, mysticism, etc. Although Van Eeden was not a trained philosopher, he was sufficiently well-informed to understand the developments of his day. In particular his views on mysticism and on mathematics are on the surface vaguely similar to Brouwer's views; he fully realized that one did not exclude the other.

Van Eeden was strongly influenced by Victoria Lady Welby, a sometime Maid of Honour to Queen Victoria (who was also her godmother).[24] When Van Eeden participated in 1896 at a conference for experimental psychology in London, as a delegate for the *Société d'Hypnologie et de Psychologie*, he met Lady Welby, who had presented a paper '*The use of the 'Inner' and 'Outer' in Psychology: Does the Metaphor Help or Hinder?*' She invited Van Eeden to Denton Manor, where they could discuss topics of mutual interest. Lady Welby's research centred around the notion of 'meaning', the functioning of signs in ordinary communication. Gradually a friendship developed between Van Eeden and Lady Welby that made him stand out among her numerous correspondents, such as Bertrand Russell, Ferdinand Tönnies and C.S. Peirce; their friendship was also Van Eeden's ticket to the higher strata of English society. This did not, however, dull his social conscience. Far from it, he gradually conceived the plan to found a commune, or, in the usage of the day a *colony*. Under influence of, among others, the leading expert of Dutch flora and geology, Jac. P. Thijsse, he had come to a strong appreciation of a way of life in balance with nature.

After reading Thoreau's 'Walden' his mind was made up, he was going to found a colony in which he expected to interest people of a like mind. In 1898 Van Eeden bought some land in Bussum, a small town in Het Gooi,[25] not far from Laren and Blaricum, and soon a few cottages were erected: the colony 'Walden' was a fact. The colony attracted a number of people, for example, Nico van Suchtelen, the author of a number of novels and later director of the *Wereldbibliotheek*, an idealistic publishing company, and also Brouwer's friend, Rudolf Mauve, the architect who designed Brouwer's cottage. The colony became the best known institution of its sort in Holland, no doubt through the fame of its founder. It was, in contrast to the colony of the International Brotherhood in Blaricum, not based on religious principles, but a purely social-humanitarian enterprise. Poor management and the unavoidable bickering, aggravated by financial problems, soon put an end to the undertaking. Van Eeden indeed lost a sizeable part of his mother's fortune on his pet project. Although the experiment could not be seen to support the colony-idea, the name 'Walden' has ever since been cherished under a romantic veil. Van Eeden

---

[23] *Redekunstige grondslag van verstandhouding*, published in *Studies* 1–3, 1897.

[24] Victoria Alexandrina, the Hon. Lady Welby, wife of Sir William Welby-Gregory, 4th Baronet (1837–1912). I am indebted to Adrian Mathias for the information on Lady Welby.

[25] Cf. p. 59.

could not bring himself to settle down for the management of his social experiment, various other causes claimed his attention, and so the idealistic experiment foundered, not least because of negligent leadership.

In spite of the failure of the Walden colony, Van Eeden saw a special role for himself in a world in spiritual confusion, and slowly the idea of a ruling cultural elite pressed itself onto him. In 1910 his path crossed that of some like-minded individuals, Poul C. Bjerre, a Swedish psycho-therapist and author, and Erich Gutkind.[26] the author of a remarkable book, *Sideric birth—Seraphic wanderings from the death of the world to the baptism of the deed*.[27] After reading a book of van Eeden, Erich Gutkind had sent his own book, in which 'The essence of man was set free from all naturalist determinedness and constituted above all nature and physics in the transcendent essence',[28] to Van Eeden. Van Eeden was enthralled by this work, in which he recognized a spirit in tune with his own. After meeting the author in person (at first Van Eeden ascribed the 'Sideric birth' to Buber), he wrote enthusiastically to Lady Welby

> I am here now in daily conversation with Volker, the author of *Sideric Birth*. There is your man, the man the world is looking for, or rather waiting for, without looking or knowing.[29]

Van Eeden and Gutkind set themselves to get together a small band of noble minds, they talked to Gustav Landauer, an anti-marxist anarchist, author and philosopher, and Martin Buber, the Jewish religious philosopher. They prepared a 'Call', an appeal to all the best minds in the world, to come to an understanding. 'We will dub it: A call to all kingly men, for the transcendent conquest of the world, by deeds of royal love ... Our motto will not be 'proletarians of all nations unite!', but *noblest* of all nations unite!'[30] It was Van Eeden's firm conviction that significs, the study of signs, communication and meaning, could and should play a key role in the council of the 'Men of a Royal Mind' (*Koninklijken van Geest*), that he and his new companions envisaged. Van Eeden had not realized however, that Gutkind, Buber and Landauer did not share his views on language at all. On the contrary, Gutkind stressed the central place of 'the breaking down of the fetters of language', appealing to Mauthner's 'Critique of Language'.[31]

The promised call was published by Gutkind and Van Eeden in 1911, '*Conquest of the world through heroic love*'.[32] Each author had written his own part, and, as to be expected, Gutkind reproduced the ideas of his 'Sideric Birth', while Van Eeden stressed the importance of significs. In spite of the lack of positive reactions, Van Eeden pressed on, and in 1914 he again travelled to Berlin in order to further the

---

[26] Pseudonym *Volker*.
[27] *Siderische Geburt—Seraphische Wanderungen vom Tode der Welt zu Taufe der Tat*, [Gutkind) 1910].
[28] [Gutkind 1930].
[29] Van Eeden to Lady Welby 7 October 1910.
[30] Van Eeden to Lady Welby 16 November 1910.
[31] [Mauthner 1906].
[32] *Welt-Eroberung durch Helden-Liebe*, [van Eeden 1911].

realization of a circle of noble minds. In consultation with Gutkind, Rang, Buber, Walther Rathenau and Max Scheler, the founding of the so-called *Forte Kreis* was scheduled in the fall of that year. The meeting place was to be *Forte dei Marmi*, a small place north of Pisa, and the list of (tentative) participants was impressive: Barlach, Bjerre, Henri Borel, Buber, Däubler, Dehmel, Van Eeden, Gutkind, Landauer, Mombert, Florens Rang, Rathenau, Romain Rolland, Stehr, Susman.
The war superseded all plans, and soon the high-minded feelings gave way to more nationalistic ones. Thus the Circle came to its end even before it was founded. Rang and Gutkind fell prey to the virus of militant nationalism and the members of the belligerents readily forsook their supra-national ideals. The hostilities effectively prevented any further action in the planned direction, but Van Eeden, not willing to betray his ideals, did not quite give up hope. He directed his attention for the time being towards his native country. And so we find suddenly in Van Eeden's diary an entry (22 October 1915):

> Beautiful weather. Yesterday I dined with Borel at Professor Brouwer's, the mathematician. A sympathetic man, with the charming manners of a genius. It is curious that so often mathematical genius goes hand in hand with freedom of judgement and noble character. A refined, sharp, spiritualized head, clean-shaven, with young wrinkles. He is only 34 years old. Dressed in a white linen suit. In a conversation he at times, as immersed in thought, sits down on the floor. He had already conceived the ideal of the *Forte-Kreis* for himself. His wife is a fine type, with a high forehead, and brown eyes, like a model of the primitives, Van der Weyden or Metsijs.

Van Eeden was immediately taken with Brouwer, he had recognized a kindred spirit who might be won for significs and for a possible substitute for the Forte-Kreis. It is interesting that Van Eeden's impression of Lize Brouwer confirms to Brouwer's, who fondly spoke of her Memlinck face.

Brouwer, on his part, appreciated Van Eeden and reciprocated his friendship. The relationship between the two men lasted for the better part of the rest of Van Eeden's life, but although the author surpassed Brouwer's friend Scheltema in fame and influence, this friendship was not to reach the same intensity and purity. The circumstances were, of course, totally different. Brouwer had become a famous man, albeit not in the eyes of the average Dutchman. He had reached a stage in life, where recognition was no longer as urgent as it had been for the beginner of the period after his dissertation. After all, what could a man with an international reputation, who was accepted as an equal by men like Hilbert, Klein and Poincaré, gain from a connection with a Dutch author? The answer is probably that the popular fame of a scientist can never match that of an artist (of whatever kind)—at least during his life. There was a certain mystique and notoriety surrounding the charismatic Van Eeden, that, added to a certain similarity in philosophical and social-cultural outlook, made feel Brouwer flattered by the attention of the great man. Van Eeden, definitely, was a man recognized by his fellow country men as an outstanding character—although the opinions ranged from 'superior author and playwright' to 'social amateur' and 'poseur'.

The Great War was seen in Holland from the detached viewpoint of a neutral spectator, the Dutch were somehow convinced that neutrality was, since the Napoleonic wars, their rightful rôle in history, and the horrors of war reached them only through the reports and misery of the bands of refugees from Belgium. For Brouwer the war was a challenge of a moral and ethical nature, his first brush with the facts and fiction of the events of 1914 was in connection with the so-called *Declaration of the 93*. After the first military actions of the Germans at the western front, a number of accusations of violations of the code of war started to circulate in the allied press. For quite a number of otherwise quite reasonable German scholars, this imputation seemed a wilful act of slander, designed to denigrate a nation with recognized cultural standards. In reaction to the allied reports, a number of scientists published a pamphlet, 'Appeal to the World of Culture', (4 October 1914).[33]

The 'Appeal' protested against accusations that laid the blame for the war at Germany's doorstep, and that painted the Germans as wilful destroyers of the Belgian cultural heritage (for example, the sack of Leuven[34]). In a six fold *Es ist nicht wahr* (it is not true) the 93 scholars vehemently denied the charges. The list of the signatories contained—a cynical coincidence!—one of the former candidates of the Forte-Kreis! Interestingly, the list contained only one mathematician: Felix Klein had become the victim of temporarily enhanced national feelings and misleading information, cf. p. 337.

The recent events even found a modest echo in the scholarly confines of the Dutch Royal Academy; the minutes of the meeting of November 1914 reveal that Brouwer, under the heading of 'other business', called the attention of the meeting to 'the phenomenon that scientists of renown of various nationalities had published manifestos in their quality of scientist, creating the impression that the leaders of science had reached, *on the basis of their scientific reflections*, the conclusion that a military and fiscal separation between the nations is necessary to protect the higher interests of culture'. He asked if the Academy could not prod the International Association of Academies into making a statement on the issue. After some discussion the chairman acknowledged the importance of the issue, but 'answered Mr Brouwer that perhaps individual persons could better than an official body, such as the Academy, do something to repair this impression with the public.'[35] The manifesto of the 93 was to haunt the German scientists for years to come, it tainted them with the image of servants of military imperialism. A good many of them soon regretted signing this declaration without access to reliable information.

Of course, the manifesto of the 93 was not the only, and not even the first, excursion of scientists into the uncertain domain of politics. To Brouwer, actions of this patriotic sort were anathema, he had, in line with his views of *Life, Art and Mysticism*,

---

[33] *Aufruf an die Kulturwelt*, [Kellermann 1915].
[34] August 1914
[35] *Verslag van de gewone vergadering der wis- en natuurkundige afdeling XXIII (1e gedeelte)* p. 828.

developed a strong disapproval for demagogic arguments and notions, such as 'fatherland' (*vaderland*). He considered the moral-emotional content of such terms as totally misleading and harmful, a cloak under which interests of certain groups were pursued. In short, he had become a thorough internationalist. His views are spelled out in a review of two books, *In the light of the flame of war*[36] by Frederik van Eeden, and *Patriotism, Philanthropy and Education*[37] by Carry van Bruggen, published in *de Nieuwe Amsterdammer* of 3 February 1917 under the title *Anti-nationalistic Literature*. Brouwer upbraided both authors for being spineless, because they still considered nationalistic manifestations as a product of an autonomous life-tendency of the individual. According to Van Eeden these were a desire for self-denial and absorption in a larger community, and according to Van Bruggen they constituted a quest for support of one's own shaky pride and self-confidence:

> The authors thus maintain, rather nebulously, the insight that the state is the institute of regulated parasitism; in the first place of the community of citizens on the spontaneous intellect-less life within the confines of the state, in the second place of the prosperous surface of that community on its needy core, and that the durability of this institution rests solely on the fact that anyone who shares in its management, gets a stake in its continuation.

Brouwer rejected the solutions of Van Eeden and Van Bruggen, to wit the cultivation of the common intuitive sense of justice and an anti-national drill at the elementary schools, and pointed out that 'the perpetuation of the fatherlands by the interested parties can only seriously be impeded after eradicating the word-of-power 'fatherland' itself'. Brouwer claimed that the abolishing of 'words-of-power' is the most effective way of fighting social injustice.

> The suppression of one's fellow man by means of private property and trade would be impossible without words-of-power, such as *entrepreneur, interest rate, profit*. And the morphinizing of the conscience, required for the perpetuation of social injustice would be impossible without words-of-power like *happiness, religion, art, civilisation, genius, duty, reliability*. These *means of defence of injustice*, among which the word *fatherland* is one of the most powerful, can only be destroyed by ending the present *anarchy in the formation of words*. And thus there emerges the primary social task, imposed on us by our conscience: the founding of an *institute for language-reflection* that first will have to brand the *words-of-power of injustice* in extensive analytical manifestos, and that subsequently will have to occupy itself with *the forming of words for the realization of the common intuitive sense of justice*, as soon as the eugenetics for this sense of justice will have created a possibility for development.

---

[36] *Bij 't licht van de oorlogsvlam*.
[37] *Vaderlandsliefde, menschenliefde en opvoeding*.

From this review Brouwer emerges as stern a judge of the arrangements of human society, as befits the erstwhile author of *Life, Art and Mysticism*. Comparing, however, the above exhortations with the message of *Life, Art and Mysticism*, one observes a major distinction. The new element is the role of language, and even a language institute is advocated—this foreshadows the International Institute for Philosophy. The negative role of words-of-power must rather be viewed through the eyes of the mystic than those of a socialist, pacifist etc. The objectionable terms, in Brouwer's perspective, are nothing but tools of the established powers in the struggle for domination of mankind, a domination which is in direct conflict with the non-interference principles of the mystic. Brouwer's proposal is simple enough: remove the terms, and then the oppressor will be robbed of his efficient means of indoctrination. There is, of course, no promise that the immoral effects will disappear, but at least their perpetuation will be made harder. Brouwer was quite consistent in this particular view; after the Second World War the same advice is repeated.

The earlier tendency of Brouwer to blame the evils of human society on its unlimited growth is repeated forcefully in the above review. He judges Van Eeden's appeal to 'common intuitive sense of justice' a sad illusion:

> So long as the health of men languishes so sadly, that repulsion is the dominating intuitive feeling they instil in each other, and, as a consequence, everybody's instinct for self-preservation compels him to find a place where he does not have to get closer to his disgusting fellow citizens than he chooses—for which only a parasitic constitution offers chances. And there is only a prospect of an improvement of human health after an end has been put to the anarchy of *procreation* and the population density has been reduced to such an extent, that the atmosphere of free breathing of single individuals gets a bearable scope.

This passage may elicit a smile from the modern reader, after all Brouwer lived in a country with fewer than 6 million inhabitants, against the estimated 16 million of today. Generally speaking, life in his days in Holland was, with respect to privacy and ecology, a paradise compared with the present situation. So what was there to complain of, would one be inclined to ask? Probably this is an instance where an exceptional person with prophetic gifts is able to read the signs of the times better than politicians and administrators.

The above is quite characteristic for Brouwer's views on the inadequacies of mankind with respect to the world and life on it.

7.3 Family life

In the early years of the war Brouwer's little household underwent some change, a new housemate joined the family.

His stepdaughter Louise had been a source of anxiety to Lize and Bertus, she was not an easy child to get along with, and the intellectual prospects of her education interested her preciously little. Not being cut out for an education in a gymnasium or HBS[38], it was decided that she should enrol in a so-called domestic science school.[39] This particular school on the Zandpad in Amsterdam had a splendid reputation, it attracted girls from various social strata, including the daughters from the better families. The girls were taught all they should know about running a household. Various careers were open to the graduates of this school; Lize and Bertus had decided that Louise should select her courses so that she could become a teacher in laundry techniques.

Louise joined the school as an intern student. Little is known about her progress, but eventually she left the school without a certificate. Brouwer, who did not like to leave things to chance, had decided that it would be wise to keep an eye on Louise's friends, he therefore asked for the list of students. He studied this list carefully and found two suitable girls on the list; one of them was the youngest daughter of a notary. Brouwer let Louise know that she could bring this girl home to Blaricum. The young lady, Cor (or Corrie) Jongejan, was more or less the opposite in most respects of Louise. She was fun-loving and flippant; she preferred the tennis court over classes, at school she was cheeky. In contrast to Louise, she had finished highschool (HBS) before entering the present school. Like Louise, she did not make much progress in the domestic sciences, much to the chagrin of her parents, who despaired of Cor's future. Under these circumstances, Brouwer's interest in the girl appeared to them something of a godsend. It occurred to them that perhaps this stern professor could knock some order and respect into the pretty head of their daughter. Brouwer was approached by them with the request to take the supervision of Cor in hand and somewhere in 1914 Cor joined the Brouwer household as a kind of general secretary-factotum.

Bertus and Lize.

Louise could not stand Cor, and the new arrangement was not to her liking.

---

[38]Cf. p. 4.
[39]Huishoudschool.

Most sources confirm that Louise's opinion was mostly dictated by (an understandable) jealousy. Cor was an engaging person, who could make pleasant conversation and who easily made friends. The somewhat sullen, sombre Louise had to content herself with a modest place in Cor's shadow.

Cor was about twenty-one years old when she became Brouwer's assistant. She performed all kinds of secretarial duties, she copied, for example, the manuscripts of Brouwer, at first by hand and later on a typewriter. In addition she copied (most of) the manuscripts that Brouwer handled for the *Mathematische Annalen*. In the beginning she was rather awed by the impressive scientist, but eventually her jolly nature won from her timidity, and she started to treat her employer on a less formal footing. In fact Cor became an intimate friend of Brouwer, she often accompanied him on his travels, and Lize came to accept her as a normal part of family life.

The family of Cor was less pleased, the ties with Cor were cut. Nonetheless, when her father became terminally ill in 1918, Cor went back to him and helped to nurse him in his last months of his life.

Cor inherited the tidy sum of Dfl. 60.000,- from her father. Back in Blaricum she handed the money over to Brouwer, with the words 'I don't know what to do with it'. Brouwer invested the money in various ways for her, for example, after the war he bought a house for her in the Harz (Harzburg), which served as a *pied à terre* for visits to the area, which was close to Göttingen and to Brouwer's usual health clinic.[40]

The avowed international traveller Brouwer found the limitations, brought by the war, hard to bear. He desparately longed for an occasion to cross the borders again. Although travel abroad was not easy, he did manage to make a short trip to Germany in the spring of 1915. For a person like Brouwer, who loved travelling, and who felt at home in the most diverse countries; being cooped up in Holland was one of the worst aspects of the war. In March he wrote Schoenflies a postcard, asking what the travelling conditions were; he had read in the Dutch newspapers 'that travelling in Germany for non-Germans was not quite safe, and that one runs the risk of suddenly being locked up for several days, for a closer inspection of one's identity, even if one has a passport.' Schoenflies, with whom he had excellent relations once the editing of the Set Theory *Bericht* was out of the way, must have reassured him, for he visited Schoenflies in Frankfurt a.M. during the Easter vacation.[41]

7.4 An offer from Leiden

In the meantime a vacancy had come up at the University of Leiden and H.A. Lorentz, the leading Dutch physicist, tried to win Brouwer for Leiden. He personally advocated Brouwer's candidacy for the mathematics chair:

---

[40]The house is still there; an old man who had lived in Harzburg his whole life, vividly remembered Professor Brouwer *mit seiner Kusine*. He recalled that Brouwer offered his father to invest the family capital of only 18 Marks into Dutch guilders, which was strictly forbidden by the currency laws in this period of galloping inflation after World War I, just to help them out. The father refused the offer.

[41]Brouwer to Schoenflies 28 March 1915, 10 June 1915.

> nothing would please me more than that you could decide to change Amsterdam for Leiden, ...
> we all think it of the greatest importance for the flourishing of the faculty. In particular Kluyver, de Sitter, Ehrenfest and I would appreciate collaborating with you. You may be assured that you will be received warmly and with open arms.[42]

The two knew each other well enough, both being members of the Royal Academy, of which Lorentz was President. When Brouwer was still an unknown *privaat docent* Lorentz had reacted to the inaugural lecture of Brouwer in 1909, and this led to an exchange of views on the theory of relativity.

The offer from Leiden had its attraction for Brouwer. The Leiden faculty was apparently convinced that the addition of this prestigious mathematician to its staff was worth a few sacrifices. It told Brouwer[43] that they would not object if Brouwer were

> to put into practice the view that the discharging of the duty of a professor rather consists of the pursuit of his own scientific work and of being available for students who work on their own, who seek supervision and advice, than of regular courses in stereotyped theories, which have already for years found clear expositions in books.
>
> An oral clarification of the offer was summarized by my colleague Ehrenfest with the words 'Thus materially nothing will be required of you other than your presence'.[44]

This generous offer contrasted sharply with the practice in Amsterdam, where according to Brouwer, the teaching load was not taken so lightly. He complained that 'since the acceptance of my duties I have been handicapped in my research activities in a discouraging manner'. The offer was probably not really considered by Brouwer, but it set him thinking about his own situation in Amsterdam.

Without taking Ehrenfest's statement too literally, he would at least like to get some freedom from the board of the Amsterdam University to organize his activities after his own insights on the basis of the Leiden model. That is to say, no undergraduate teaching, and mostly supervision and consultation where advanced students were concerned. The Leiden offer gave him some bargaining power, which he used to ask the Board of the Amsterdam University for two specific commitments:

> 1. To raise the strength of the teaching staff in mathematics to a level comparable to that of the state universities, at the retirement of Korteweg in three years time (N.B. Amsterdam had 2 full chairs and 1 extra-ordinary chair, but the teaching duties included mathematical astronomy and theoretical mechanics).

---

[42] Lorentz to Brouwer, 11 June 1915.
[43] As reported in a letter from Brouwer to Zeeman, 19 June 1915.
[44] *Also materiell wird von Ihnen nichts weiteres verlangt als dass Sie da sind.*

2. On account of his weak health, Brouwer could not live in a city, but according to the rules he had to live in Amsterdam. Thus he felt obliged to keep a town house and a house in the country. He therefore requested permission to take up his residence beyond the limits of Amsterdam.

P.H. Zeeman, A. Einstein, P. Ehrenfest. Courtesy Rijksarchief Haarlem.

Brouwer had already taken the initiative to discuss the matter with the municipal authorities, and on 18 June he had an audience with the Mayor of Amsterdam. The Mayor was, he wrote to Zeeman,[45] favourably disposed, but he could not possibly go so far as to promise an extra lecturer immediately.

The Mayor was as good as his word, Brouwer's case was discussed in the meeting of the curators of 25 June. No objection was raised to Brouwer's request to be released of the obligation of living in Amsterdam. The remaining requests were another matter: the curators flatly refused to discuss the distribution of chairs after Korteweg's retirement. The board also devoted their attention to the mode of teaching proposed by Brouwer. In the absence of facts it was hard to form an opinion, they agreed, but nonetheless they questioned the idea of a professor not teaching regular courses, but making himself available to students at certain office hours. One easily recognizes here an echo of Ehrenfest's ideas. The curators were

[45] Brouwer to Zeeman, 19 June 1915.

not particularly happy with such a radical innovation, 'For very gifted and diligent students this may be possible, but for the great majority this certainly is not sufficient'. The curators decided that they could not approve of a manner of teaching where the professor would only be available for consultation, but they did see that some consideration was necessary to enable Brouwer to continue his researches. Finally they had no problem granting the mathematicians an extra DFl 500 a year for a seminar library.

Thus in the end Brouwer obtained permission to live in Blaricum, but in most other matters he probably had to content himself with promises of a sympathetic reception of later proposals. The fact that, after all, he preferred Amsterdam with its teaching load over the generous offer from Leiden, strongly suggests that the combination 'Amsterdam–pharmacy–'t Gooi' outweighed almost any other situation.

The faculty was more than pleased that Brouwer had turned the offer from Leiden down; the chairman, at the meeting of 6 February 1916 expressed his pleasure that Brouwer 'had withstood the siren call of Leiden, and thus was saved for Amsterdam'.

The recognition that Brouwer had already got in the form of an editorial post at the *Mathematische Annalen* was further confirmed when Blaschke asked him in November 1915 to write a book on the new developments in topology for the publisher Teubner.[46] The invitation suited Brouwer's plans, and he answered that he would gladly write a book on Analysis Situs.[47] No drafts or manuscripts of this book exist, it is doubtful if he ever started to work on it. For at the same time he was rebuilding mathematics from a constructive viewpoint, and soon most of his active research time was taken up with his new intuitionism.

7.5 Significs and Jacob Israël de Haan

Next to mathematical research, the significs project of Van Eeden occupied Brouwer most. The initially vague plans to create a substitute for the Forte-Kreis, began more and more to take form. Van Eeden had gathered a small circle of intellectual supporters around him, Henri Borel, Jacob Israël de Haan, Gerrit Mannoury, and Brouwer. Each of these men had considerable intellectual capacities, although it is fair to say that on the level of pure intellect none of them was a match for Brouwer. The most colourful and remarkable among them was De Haan. Born in the same year as Brouwer, he had to find his place in life the hard way; he was the son of an orthodox Jew, and initially aimed at a rather modest goal. He got a teacher's education in Haarlem (1896–1900) and was a schoolmaster for some time. All the time he carried on a literary production; in 1904 he created a shock in Holland with the publication of a homo-erotic book *Pijpelijntjes*. De Haan had already in 1899 dedicated a number of fairly tearful poems dedicated

---

[46] Blaschke to Brouwer 4 November 1915.
[47] Brouwer to Blaschke, draft 19 November 1915, Cf. CW II p. 410.

to Van Eeden and sought contact with the man whom he viewed as a literary and spiritual father figure. After an anti-religious period he returned to the orthodox faith and became an ardent Zionist. In 1903 he started his law studies in Amsterdam, which were concluded with the *doctoraal examen* in 1909, although he could not attend the courses, since he had to earn his living. After finishing the formal study he became a popular *repetitor*, a free-lance teacher who drilled law students for their exams. Not content with this routine, he decided to embark on research for a dissertation. The subject he selected was 'Liability, Responsibility and Accountability'. Through Van Eeden he became acquainted with significs, met Lady Welby, and started to study the language of law and to subject it to signific analysis.

In 1916 De Haan defended his dissertation[48] in Amsterdam; in the meantime Van Eeden had informed Brouwer about De Haan's projects. This proved to be of great importance for De Haan, the law faculty was rather in doubt as to the scientific content of the dissertation, *Legal Significs and its applications to the notions: Liable, Responsible, Accountable*. Van Eeden noted in his diary that one of the law professors

Jacob Israel de Haan. Courtesy University Library Amsterdam.

> had found the whole dissertation unscientific, and he would never have accepted it, let alone *cum laude*. He considered persons like Lady Welby and Clifford Albut as not belonging to science and not to be taken seriously. Only through Brouwer's defence the *cum* had come through.[49]

The combination of De Haan's unusual personality, Brouwer in the role of examiner, and the role of Van Eeden as a *paranimf*[50] lent an air of eccentricity and glamour to the public defence.

Brouwer's relation to De Haan was ambivalent, on the one hand he recognized

---

[48][Haan 1916].
[49][Eeden 1971], diary 8 February 1916.
[50]Traditional formal helper of the candidate during the defence of the dissertation, cf. p. 119.

the scientific importance of his work and ideas, on the other hand he initially fostered an antipathy towards him. The antipathy is not surprising, De Haan was aggressive, and he had a sharp tongue and pen. Brouwer himself was a man to be handled with care, but De Haan surpassed him in this respect. Van Eeden patiently laboured to bring Brouwer and De Haan closer together, and, indeed, Brouwer conquered his 'physical antipathies', so that the two could get to grips with the signific problem.

Brouwer wrote two critical reviews of De Haan's dissertation, dissecting the central notions and assumptions.[51] He used the occasion to expound once more his views on language and communication, already laid down in *Life, Art and Mysticism*, but applied here to the context of law:

> Individuals with a parallel orientation of will, resting on voluntary or compulsory acceptance of a common determination of values, maintain and differentiate this parallelism by *words*, that is signs, for each individual associated to a moment of life, and hence providing an impulse to the will of the receiver, or influencing his disposition of will. The fact, that a large part of the will-parallelism that supports the action of words, has durably turned into the belief in an objective 'Anschauungs' -world, common to all individuals, does not alter the fact that word-utterances are more or less developed verbal imperatives, and that therefore *addressing* always comes to ordering or threatening, and *understanding* to obeying; and that the 'education of the masses' comes to a further differentiation of its parallelism of will, that is to say to preparing it for more complicated and varied obedience. The meaning of a word is never completely understood, not only because it is nothing but an influence expected by the sender on the acts of the receiver, and is barred from a complete realization of expectations, but also because each individual belongs to more groups, each based on a private parallelism of will, maintained by a private language and at the alternating use of these languages necessarily falls into confusing inaccuracy.[52]

The above passages are, so to speak, logical consequences of Brouwer's philosophical point of departure; the interesting thing is that he goes a step beyond his earlier work, in analysing the underlying mechanism of social and legal order, the interaction of more individuals. De Haan's goal, the systematization and objectivization of legal language, is in Brouwer's opinion unattainable, because

> practical life requires many more primary notions, than language can offer in primary words and compositions of words.[53]

The primary notions are the ones close to the 'self', the egoic ones, conceived (or experienced) by the individual. In particular, 'unaccountability' is such a primary

[51] [Brouwer 1916C,1916D].
[52] [Brouwer 1916C].
[53] Ibid.

word, and its incorporation in legal language is according to Brouwer, problematic (if not impossible) because

> Every judgement of the acts of voluntarily switching off the intellect, presupposes a recognition of the spiritual values of life, in which morality can demand the subordination of the intellect, but for which the ordinary language has no words. The incorporation of the notion 'unaccountability' in the understanding of legal order is already therefore restricted to the few occasional instances of total stupefaction of the intellect (in which cases at best psychiatrists may be competent).[54]

De Haan accepted Brouwer's criticism, and started to read Brouwer's earlier work; the influence can be seen in De Haan's inaugural lecture as a *privaat docent* on 31 October 1916. Soon after this address attempts were made to create a chair for De Haan in the law faculty, a chair devoted to (legal) significs. Van Eeden, Brouwer and Mannoury considered this chair a unique possibility to give significs a firm footing, and they warmly supported De Haan's efforts.

The ideal shared by the small group was expressed by De Haan in his inaugural lecture:

> However, we consider a higher education incomplete, as long as significs is missing from the curriculum. We consider it necessary that the significs of each special science will be taught, and general significs to boot. This will lead us to the Understanding of the Sign and the Sign of the Understanding.

The attempts to establish De Haan failed, although his name occurred on the list of candidates for a vacancy in 1917. Brouwer wrote a warm recommendation for him, saying that

> I consider the rebuilding of the language of law, by him, the most noble work, to which one can devote one's efforts, in this time of crisis of social evolution, because there is, in my eyes, no other effective means, that can guarantee moral tendencies a lasting influence on the social relations.

Brouwer expected that De Haan would 'convey to the students in addition to the required quantum knowledge, a powerful and wholesome orientation on the foundations and the social significance of their discipline'. None of the efforts led to anything and De Haan was left a disillusioned man.

7.6 Van Eeden and the International Academy

When Van Eeden and Borel met Brouwer in September 1915, the two were already involved in a project that envisaged nothing less than a philosophical academy. The war had made it quite clear that philosophical considerations had given way to the nationalistic ones, and that the unity of the Noble of Spirit had proved illusory.

Independently of Van Eeden a small group had proposed a meeting in the summer of 1915 that should combine people of all denominations to express religious

---
[54] Ibid.

feelings across the borders of the various creeds. Borel and Van Eeden made contact with the leading personalities, Countess Van Randwijk-de Jonge and a broker in tea, J.D. Reiman. The initiative was taken to found an Academy of a philosophical nature, and a committee was installed, consisting of Van Asbeck, Van den Berg van Eysinga, Blok, Brouwer, Van Eeden and Reiman. Brouwer probably met Van Eeden at the first meeting of this committee (21 September 1915). Since Van Eeden soon saw in Brouwer a supporter of the Forte-Kreis idea, Borel and Van Eeden started to visit him in order to get better acquainted. Brouwer was made chairman of the committee, and Reiman secretary.[55] Van Eeden reported in his diary:

> Brouwer presided [24 October 1915]. He was sometimes exceedingly naive in his plans. But it was good all the same. This is already an elaboration of the Circle.

The committee chose *Oud Leusden* as the site for summer schools, with room for cabins for the students. Oud Leusden is a village just south of the town of Amersfoort, and, to run ahead of our story, it still is the location of the 'International School of Philosophy'. The founding of a philosophical institute put all those involved in a state of pleasant excitement; meetings were held, and plans and rules were proposed. The project, however, did not work out the way it should have. In January 1916 a conflict arose; Brouwer's way of chairing the committee had met with incomprehension and disapproval. His proposal of an ascetic lifestyle for the school added considerably to the confusion. The relation with the philosophical establishment in Holland was also a moot point. The Van Eeden group wished to exclude the Bollandists from the school, something that antagonized the company at large. Bolland, the philosophical giant from Leyden, had refused to take part in the activities, since he considered a committee with a poet (Van Eeden) in its midst, 'strange company'.

Henri Borel, the sinologist. Courtesy University Library Amsterdam.

The conflict erupted at the meeting of 10 January 1916; Brouwer clashed with Reiman, who matched Brouwer in authoritarian behaviour. Brouwer wanted to get the action on its way by consolidating a core of the five active members—Henri Borel, Bloemers, Van Eeden, Reiman and himself. The remaining members would then be encouraged to join the group if they accepted its principles. Reiman did not agree to this procedure, and Brouwer was forced by the meeting to give up the

---

[55] The history of the founding of the philosophical institutions, mentioned here, can be found in [van Everdingen 1976], [Schmitz 1990], [Heijerman–van der Hoeven 1986], [Fontijn 1996]. Furthermore, Van Eeden's diary is an invaluable source of facts and of running commentary, [van Eeden 1971].

chair; thereupon Borel, Brouwer and Van Eeden left the meeting. The rift could not be bridged, a reconciliation attempt on 27 February failed, the parties clashed again, and each went its way.

Reiman and his group went on to found on 13 February 1916 the 'International School of Philosophy' at Amersfoort. The school had a building erected at the Doodenweg in Leusden; the architect was K.P.C. de Bazel, who was one of the leading architects of the period in Holland. In June 1917 the chairman Reiman opened the main building.

The same building is still standing today, and the school still functions more or less as intended. The International School of Philosophy has been the scene of some distinguished meetings, and the list of speakers boasts a impressive cross-section of the philosophical community. The following list is a small selection: Adler, Bernays, Bloch, Buber, Cassirer, Church, Dürckheim, Heidegger, Koyré, Landgrebe, Naess, Quine, Reichenbach, Schweitzer, Tagore, Tarski, Tillich.

The loss of the initiative to Reiman irked Van Eeden, he noted in his diary that 'Reiman is a dwarf compared to the giant Brouwer'[56] and after the final clash he described Reiman with the words:

> The man is the dupe of his all too great ideas. He cannot bear them. He feels himself a chosen one, a tool of God, and hence becomes a fanatic and upsets everything.

Brouwer, with his over-sensitive nature, suffered intensely—in Van Eeden's words:

> Now it became clear to me how extremely sensitive he is. The person of Reiman makes him sick. It gives him visions, and he sometimes shouts 'Reiman!' to get rid of him.

The group around Van Eeden went its own way; Van Eeden tried to revive the Forte-Kreis formula in the form of an International Academy for Philosophy. It was the small but select group that had left the fold: Borel, Brouwer, Mannoury and Van Eeden. The organization of flourishing societies was not their strong point, they rather formed an exquisite band of individualists. We have already met Mannoury, he was in a sense, in an extracurricular way, the mentor of Brouwer. His philosophical interests made him stand out in Holland, but in addition he was a prominent leftist and writer–pamphleteer. Together with Brouwer, he represented the philosophical expertise of the group. Henri Borel was a man of many trades, albeit mainly literary, he was a (free-lance) sinologist and earned a living as a journalist and author. His literary fame rested on two early novels *Het Jongetje* (The little boy) and *Het zusje* (The little sister), (1899, 1900), and numerous books and brochures on China, Asia and its religion, art and philosophy. Borel was an expert gossip, he freely mingled with the leading men of letters and the artists of the day, there is many a volume of correspondence of Borel and one of his artistic or intellectual relations. His contacts with Van Eeden went back to 1889, he lived at Walden for some time in 1905. After meeting Brouwer in 1915, he grew very much

---

[56][van Eeden 1971] Diary 11 January 1916.

attached to him.

The next steps soon followed; on 6 March Van Eeden, Bloemers and Brouwer discussed the principles of the Academy, and at the end of the month a circular letter was sent out to prospective members, explaining the reason for parting from the Amersfoort-group, and announcing their plan to found an Academy of a different nature. This academy should consist of a limited number of members, to be elected by co-option, and a board of trustees 'made up of local that is Dutch persons who are well-known and sympathize with the plan'. The circular ended with a call for members of this board. The result must have disappointed the initiators: only Mannoury sent in a positive reply; that did, however, apparently not stop them.

The acquaintance with Van Eeden had considerably widened Brouwer's circle in the society of 't Gooi. He was, of course, already part of the local scene. In the small towns with their artists and intellectuals, he was recognized as a special character, the tall, lean professor in mathematics—a very clever person no doubt, but hard to follow. After joining the circle of friends of Van Eeden, Brouwer's contacts multiplied. He met the main characters of the Walden episode and also a number of members of the fringe. Life was full of little events, tea's, birthday parties, friendly visits, discussions and the like. There is an enormous difference in Brouwer's lifestyle before and after the First World War; one might guess that the international isolation combined with the the somewhat Bohemian atmosphere of het Gooi gave Brouwer a taste of a rather un-Dutch 'dolce vita'.

The School of Philosophy might be a serious matter, but it did stop Brouwer nor Van Eeden from enjoying the pleasures of the microsociety of het Gooi. Van Eeden's fame made him a natural centre of a small court of admirers, and the curious Brouwer, with his appetite for personal relationships (which seldom touched his inner soul) enthusiastically took part in the meetings. Although he was on Van Eeden's turf, he saw no reason to share Van Eeden's preferences or dislikes.

Van Eeden was no stranger to conflicts,[57] and at that time he was in the final stages of a conflict with his secretary, Holdert. Brouwer, not influenced by Van Eeden's aversion for Holdert, developed friendly relations with Holdert and his wife Gerda.

The famous Van Eeden attracted a rich variety of visitors and followers, ranging from devote admirers to curious loafers. He was in particular surrounded by artists and quasi-intellectuals with a penchant for 'deep thoughts'. Brouwer watched the circle around Van Eeden with quiet amusement, and from time to time took part in the exchange of wit and profundities, often sitting with crossed legs on the floor or reclining in a picturesque pose. Some members of Van Eeden's entourage became regular visitors or friends of Brouwer, for example, the aforementioned Holdert and Borel.

At Van Eeden's place there was also a couple by the name of Langhout, which lived in 't Gooi not far from Van Eeden. Langhout apparently had tried to impress

---

[57] cf. [Fontijn 1990]

the company with 'deep' conversation, he had defended the view that artists should suffer in poverty in order to create good works of art. Van Eeden recorded in his diary that

> Langhout did this, hyper-idealistically, with the exaggeration of a superficial person. And Jaap [de Haan] could not bear to hear this from somebody whom he had heard described as a sluggard by Brouwer himself.

Tine, Mrs Langhout-Vermey, was eventually to become a close friend of the Brouwer family. Brouwer already knew Tine before her marriage; he enjoyed her conversation and her company. He thought she made a colossal mistake marrying Langhout, whose only virtue was that he was well-to-do; indeed, on the day of the marriage, Brouwer went all the way to the town hall to talk her out of it.

In fact, Brouwer and Tine became more than just 'good friends'. She occasionally accompanied him on foreign trips, thus taking the place of Cor Jongejan. Lize was perfectly aware of Bertus' attachments, but far from being jealous, she was often relieved that 'Bertus was taken care of'.

Brouwer and Van Eeden were for a time fairly close, they met regularly and discussed all and sundry. Van Eeden, for example, visited Brouwer at the Amsterdam Academy on 28 February 1916 to discuss a couple of matters. He recorded in his diary that Brouwer told him about Lorentz' talk:

> ...who now has found the connection between gravitation and electromagnetic phenomena. An event as important as the laws of Keppler ... I talked with him until half past six. And again and again, tears come to my eyes, by the feeling of gratefulness for his understanding, for the space he gives me ...[58]

Brouwer, curious as ever, also accompanied Van Eeden to spiritistic seances. The latter suffered from guilt feelings after leaving his first wife. The subsequent death of his son Paul drove him in desperation to frequent the sessions of a spiritistic clairvoyant. Van Eeden was strongly influenced by the clairvoyant, a young girl. She played an important role in Van Eeden's later conversion to the Catholic Church. In this exalted company Brouwer was probably rather out of place, his curiosity and the complete absence of the required awe was not appreciated. When he attended on 30 August 1918, a day of light, as the seances were called, he was told to hold his tongue, because 'the days of light were no common seances'!

At the same time Van Eeden's diary tells us that remorse did not work lasting changes in a man like Van Eeden; it appears that Brouwer and Van Eeden shared a lively interest in the other sex, for example, they attended the concerts of the pianist Henriette Roll together, and were both under her spell. Van Eeden, it seems, did not appreciate the presence of a younger, and virile, competitor for the lady's attention.

The already disturbed relationship between Borel and Van Eeden did not help to set Van Eeden's mind at ease. He must have complained in a letter to Lize about

---

[58]Ibid. 28 February 1916.

the influence of Borel on Brouwer, for in a letter of 18 December 1917 she tried to reassure Van Eeden. 'Bertus is on a cycle tour', she wrote, 'so I have to use the occasion to take away your distress'. According to her everything was part of a malicious plan of Borel; Bertus had broken off his relations with Henriette, and any talk of a new amorous relation with an unnamed lady (probably one of Van Eeden's female admirers) was totally unfounded. 'Bertus always tells me everything, and I am not aware of anything'. This little episode illustrates the tensions that existed even among 'men of a royal mind'!

In the meantime the relationship between Van Eeden and Borel had detoriated; each went out of his way to point out the inferiority of the other. There is a passage in Van Eeden's diary in which he reflected on the proper attitude with respect to the fair sex, apparently inspired by Borel's behaviour:

> In connection with the case Borel [...] I will in future demand two things from those who wish to count themselves among my friends: (1.) that they avoid needless talk, keep themselves under control and mind what they say; (2.) that they maintain their self-control in sexual matters. These two things are demands of discipline and dignity. He who does not meet these demands and does not make all efforts to do so, not only harms himself but also his friends. His place is not in the company of the Good and Royal. Women acquire power over the man who is above them through their sexuality. If the man tolerates this power, he sinks and loses his sense of self-respect [...]. For the masses it doesn't do harm, there the values are the same. But in the company of the good, the woman should also be good. There are women whose only power lies in their sexuality. We should avoid those. What is innocent among common people, is dangerous among the Good and Royal. Among them the greatest discretion and chastity must be the rule. [...] and this is not only meant for Borel, but also for Brouwer.[59]

Clearly, Van Eeden was worried about Brouwer, whom he deeply loved and admired. Brouwer could joke and flirt like the next man, albeit with refinement and style, and Van Eeden apparently was inclined to view this as demeaning. Van Eeden's concern about sexual matters did not just spring up overnight. He enjoyed female company and was an easy prey of his amorous urges; being often surrounded by female admirers, he did not lack occasion to practise his romantic inclinations. Van Eeden left his first wife after more than twenty years of marriage. The stern demands on others may very well have reflected the realization of his own frailty.

At this time Van Eeden started to read Brouwer's Delft Lectures, *Life, Art and Mysticism*; since the book was not likely to be on the counter at just any book shop, one may well presume that the author had drawn Van Eeden's attention to it, or—more likely—given him a copy. Van Eeden was thunderstruck by the lectures, which

---
[59] Diary Van Eeden, 11 August 1916.

were rather close to his own thoughts and writings. He immediately set himself to write a belated review of the book, which appeared as a series of essays, which were printed in the weekly, *De Nieuwe Amsterdammer*,[60] under the title 'A Potent Brew'.[61] In Van Eeden's words one can still hear the resonance of the shock that must have been felt by other readers of Brouwer's book as well:

> These one hundred pages of Dutch prose are indeed the most powerful, but also the most terrible, in my opinion, that have been published in this century. They are beautiful and deep and full of truth. But they are fiercely revolutionary, completely hostile to our whole society. They conflict directly with the order, the religion and law of the people. In this manner they are similar to many prophetic words—and it would be an almost ridiculous inconsistency, an unforgivable inanity, of mankind that poisoned Socrates, stoned prophets, crucified Jesus and burned Bruno, to let this formidable doom-brewer walk free without hanging him or at least interning him behind barbed wire. But look—the man is a Professor at the University of Amsterdam and a member of the Royal Academy...

Van Eeden's life was a perpetual series of social and philosophical actions and positions, at all occasions presented and defended with a strong emotional force. His discovery of Brouwer and his mystic-philosophical views came at a time when he was under the spell of his ideas of an intellectual elite; *Life, Art and Mysticism* fitted his views of the moment perfectly, and Brouwer's conversation must have contributed considerably to Van Eeden's admiration for ideas which were so close to his own. The review series reflect Van Eeden's admiration for his new friend, the only point where he refuses to follow Brouwer is in his stern judgements on females.

It was the second time that fate brought Brouwer into contact with a poet–author; compared to the strict privacy of his friendship with Adama van Scheltema, the relation with Van Eeden had another dimension. It could not possibly have escaped Brouwer that Van Eeden was intellectually no match for him, but it was no small matter to be a close associate of the most controversial poet and social reformer of the period. It did, however, not compare to the intimacy of the friendship of his youth. All the same, the relation with Van Eeden lasted and played an important role in this phase of his life. The two went swimming, exchanged visits, went on a cycle tour—and founded an academy. Indeed the thirty-six-year-old mathematician and his friend, who was twenty-one years his senior, made a long cycle tour all the way to Limburg, some 400 km in all. Van Eeden was rather depressed and the ride with a younger and more energetic friend heavily taxed him; there is a charming description in his diary, where he honestly noted that 'I enjoyed it when Brouwer had a flat tyre. So little pleasure did the trip give me'.

The preparations for the 'Academy' were set in this atmosphere of informality and friendship, with its small gossip and jealousies. The serious rift between

---
[60](Cf. p. 15).
[61]*Een Machtig Brouwsel*. 'brouwer' is Dutch for 'brewer'

Borel and Van Eeden did not break up the company, but was characteristic of the tension between the participants, who all saw themselves as superior minds.

In view of the modest size of the founding group, new members were more than welcome. The first addition was the theoretical physicist Ornstein, professor at Utrecht, introduced in 1917 by Brouwer. Ornstein attended a few meetings, but soon gave up signifilcs.

The statutes of the Academy, together with an invitation to join, were sent at the end of December 1916 to a number of prominent foreigners; the list of invitees contained among others the names of the candidates for the ill-fated *Forte-Kreis*, Walther Rathenau, Martin Buber, Gustav Landauer, Erich Gutkind, Ernst Norlind, and some more names mainly suggested by Brouwer. The mathematicians Schoenflies, Peano, Birkhoff, and Mittag-Leffler subscribed to the ideas of the Institute. In addition, Rathenau and the authors George Davis Heron and Allen Upward reacted positively. This rather meagre response did not put off the initiators, but did not spur them into action either. This situation changed when Brouwer got Mannoury to join the group. It is fair to say that Mannoury took the organization in hand and saw to the realization of the high-minded intentions.

Brouwer on his bicycle.

In his book *De Hollandse Significa*,[62] Walther H. Schmitz has analysed Brouwer's conceptions that led to the original 'Academy'-approach. He has shown that Brouwer's proposal was based on an earlier treatise of the philosopher–sociologist Ferdinand Tönnies, 'Philosophical Terminology',[63] in which an international academy of scientists is advocated. This academy of Tönnies was to be assigned a very special task: a wholesale upkeep and creation of terminology.[64]

Brouwer's idea clearly had a good deal in common with Tönnies' Academy. The founding declaration of the academy that was envisaged by Van Eeden bears the stamp of Brouwer. Content and formulation of the plan are reminiscent of *Life*,

---

[62][Schmitz 1990].
[63][Tönnies 1899, Tönnies 1900, Tönnies 1906].
[64]Cf. [Schmitz 1985], [Tönnies 1906].

Art and Mysticism. And, like its predecessor, the Forte-Kreis, the Academy was to be based on merit and expertise.

The documents relating to the enterprise give ample information on the goal and the means of the academy, which eventually was baptised *International Institute for Philosophy*.[65] Its goal was 'the renewal of the valuation of the elements of life of the individual and society'.

The task of the academy was summed up as follows:[66]

1. To create words of spiritual value for the languages of the Western nations, and thus to give a place to spiritual values in the mutual understanding of the people of the West (therefore a 'declaration of spiritual values of human life').

2. To indicate in the present legal order and in the commercial production developing under its protection, the elements that most strongly suppress or stun the spiritual tendencies. And to propose restrictions which are desirable on that account, on the sphere of influence of legal order and technology.

3. To mark in the principal languages the words that suggest spiritual values for notions that ultimately root in the pursuit of personal security and comfort. And, for that reason, to purify and make precise the goals of democracy in the direction of a world-state with exclusive administrative competence.

It is plausible to see Brouwer's hand in the above points; the influence of Van Eeden and Borel was probably restricted to minor details.[67]

Brouwer's interest in the academy project can be seen as a continuation of the line of thought of *Art, Life and Mysticism*. The mystic, apparently, was willing to suspend pure introspection for the sake of improving his fellow humans. In spite of his declared distrust of the power of language and communication, the part played by Brouwer in the signific enterprise, shows that far from renouncing the use of language, he had made up his mind to 'make the best of it'!

The group had set itself a quite ambitious goal; it wanted to avoid the mistakes of earlier philosophers who had attempted to carry out similar tasks single-handed, thus restricting the desired reform of language and thought to the author and a handful of readers. The social impact of these projects had therefore been negligible. The present group saw its common activity as superior in that respect, since 'the group of independent thinkers would take on the same task *jointly*', thereby the 'insights *formed in mutual understanding* will have obtained automatically a linguistic accompaniment, *suitable to find a place in the mutual understanding of the masses.*' One could say that the members considered their exchange of ideas and formulations as a useful experimental testing ground for the use of their linguistic proposals in society at large.

On the last day of 1917 Brouwer and Mannoury crowned the months of discussion and planning at the notary's office by officially registering the *International*

---

[65] *Internationaal Instituut voor Wijsbegeerte* .
[66] Manifesto of March–April 1916, cf. [Schmitz 1990].
[67] Cf. [Schmitz 1990], p. 220 ff.

Institute for Philosophy at Amsterdam, on behalf of Bloemers, Borel, Van Eeden, De Haan, and Ornstein. Thus the Institute was born; it was a small but select band. Unfortunately the international character did not extend beyond its name; the war may have been to blame for that. There was a number of sympathisers, but the actual deliberations took place in Holland, and so we may take it that the international aspect was for the time being more theory than actual practice.

The International Institute for Philosophy was, in legal terms, a foundation (*stichting*). It had set itself a quite specific goal: the above mentioned *renewal of the valuation of the elements of life of the individual and society*. The means for realizing this goal were:

1. The founding and maintaining of an International Academy for practical Philosophy and Sociology.
2. The founding and maintaining of a school for the dissemination of the notions and notion-relations[68] formed by the Academy.

The prospectus of the institute was sent to several thinkers, among whom were Martin Buber, Gustav Landauer, Erich Gutkind, Ernst Norlind and Walther Rathenau. Most of them were unaware of the history of the manifesto, in particular they could not be aware of the role that Brouwer's ideas played. Those who were familiar with Brouwer's earlier philosophical publications would have grasped the underlying intentions, but even Mannoury did not understand Brouwer's aims. He wrote much later[69] 'What Brouwer exactly meant by those 'words of spiritual value' of point 1, and how he planned to put those words in the service of the spiritual reformation of the world [...] never became quite clear to me.' Thus it is not surprising that the other invitees were uncertain about the enterprise. Both Buber and Gutkind reacted critically. Buber, who was already acquainted with Henri Borel, directed his comments to the latter. Borel had enthusiastically described the personality of his new friend, 'He is a professor in mathematics, and it is indeed this mathematics that has made him a mystic'.[70] When Buber sent his comments on the manifesto, Borel passed the letter on to Brouwer.[71] The objections of Buber were directed against Brouwer's idea of 'word-creation'. He questioned the possibility of the creation of words by a collective:

> The creation of a word is for me one of the most mysterious processes of spiritual life, indeed I admit, that for my perception there is no difference in essence between what I call here creation of a word, and that what one may call the emerging of the Logos. The genesis of a word is a mystery, that takes place in the kindled, opened soul of the world-dreaming, world-thinking, world-discovering man. Only such a word, generated in the spirit, can be generated in man. Therefore it can, in my opinion, not be the task of a community to make it. It rather seems to me that a corpo-

---
[68] begrippen en begripsverhoudingen.
[69] Mannoury *Nu en morgen. Signifische varia.* (unpublished) 1939. Cf. [Schmitz 1990], p. 226.
[70] Borel to Buber 7 December 1916.
[71] Buber to Brouwer 1918, [Gutkind 1919].

ration like the one planned by your friends, should and must only allow itself the cleansing the word as a goal. The abuse of the great old words is to be fought, not the use of new ones learnt.[72]

Buber's idea of 'word' is separated by oceans from Brouwer's; for Buber it is a mythical object, directly connected with the knowing subject and the designations. For Brouwer words are utterances intended to influence fellow men:[73]

> The word of the Occident does indeed in various cases have in addition to its material, a spiritual value, but the latter is always subjugated to the first, and where the first one has acquired a more certain and permanent orienting influence on the activity of society, in the sense that it induces isolated individuals, in their pursuit of physical security and material comfort, to hinder each other as little as possible and wherever possible to support each other; the latter lacks any influence on the legal state of affairs (except possibly when it is used itself for the devious realization of injustice); therefore its influences are weak, fleeting and local. Words, which have exclusively a spiritual value and which are suitable for the in- and exhaling of the *Weltgeist* and for the orientation on the observance of Tao, are non-existent in Occidental languages; should they exist, then their influence would be paralysed by the mutual physical hatred, rooted in mutual distrust of the purity of their birth, of people who live too closely together, and which hampers the pursuit of material comfort by the single individual only mildly, but which considerably hampers the in- and exhalation of the *Weltgeist*. The introduction of the first word of exclusively spiritual value into the general human understanding will, as a phenomenon, be inseparable connected with the understanding of the intolerability of this physical hatred, and thus will be the immediate cause a legal regulation of human procreation. Only then a possibility for this introduction will be provided, if the 'Mystery of the genesis' of the word concerned, has not been taken place in the single individual, but in the mutual understanding of a society of clear feeling and sharp thinking people, who are, by the way, not too close in the material sense.
> Yours truly,
> Prof. Dr. L.E.J. Brouwer

Clearly, Brouwer and Buber missed each other's point; this certainly is surprising, as Brouwer was a patent mystic, but he drew the line quite a bit earlier than Buber. Whereas Brouwer considered language as a rather external phenomenon that only fleetingly touched the inner ego of the individual, Buber was inclined to include language in the spiritual domain; thus an understanding, or even a compromise was unlikely.

Gutkind's objections were of another kind, he doubted if European languages

---

[72] Buber to Borel 17 March 1917.
[73] Brouwer to Buber 4 February 1918

could simply incorporate or assimilate words of a high spiritual value. 'In any case there should be available [...] a higher form of society, the simple research society does not qualify.'[74] Brouwer replied:

> The word cannot wait for the higher form of society, because the higher form of society waits for the word. For on the one hand the timely creation of new words, fitting the new values, participates to a great extent in each liberation process, in each revolution, on the other hand it is the lack of language-critical reflection, which is the cause that the self-revolutionizing of society still proceeds discontinuous and with devastating crises, and which yields with respect to the necessities a continuous, enormous backlog, which keeps mankind in lasting tension.

It may be remarked here that Brouwer was not one of those scientists who preach the overriding importance of the word, and then go on expressing themselves with a minimum of regard for precision. His personal use of language was of a legendary refinement. Mrs. Vuysje, the daughter of Mannoury, described Brouwer's conversation as an almost artistic process: Brouwer could in the middle of a sentence pause to savour a particular expression, and to replace it after consideration by a more suitable one. She remembered him, standing in the garden stretching out his hand, tasting a particular word and contemplating fitting equivalents. The result was not a faltering flow of words; Brouwer had mastered the technique of refinement to such a degree that he managed to produce gentle, flowing sentences of incredible length without discernible interruptions. Of course, he could also loose his way in one of those excessively intricate sentences which he loved so dearly. Max Euwe recalled with admiration the beautiful, long sentences of Brouwer. Till the end of his life Brouwer had this precious gift of formulation.

The Institute from its inaugural meeting on 12 September 1917 onward met monthly at Walden where Van Eeden lived, at Mannoury's, or at Brouwer's, be it in the pharmacy in Amsterdam or his house in Blaricum. The attempt to gather an international group had utterly failed, the whole organization was in the hands of the small group that had been involved from almost the beginning of the project. The driving force was from now on Mannoury, who eventually came to dominate significs; he had become an extraordinary professor in Amsterdam in 1917, probably with the active help of Brouwer and apparently against the will of Korteweg. The minutes of the faculty meetings and the available correspondence suggest that a rift had appeared between Brouwer and his former promotor.

After long and earnest preparations and discussions, the International Institute for Philosophy finally got on its way in 1918, observing all the necessary formalities to a nicety. Van Eeden, the most influential member on the national level, published a number of essays on significs in the weekly *De Groene*, and gave a talk on *Intuitive Significs* at the University of Amsterdam (13 March 1918). Brouwer announced this talk in *Propria Cures* with a brief introduction to the subject, the message of it was by and large identical with his comments on Buber:

---

[74][Gutkind 1919].

The means of production and the legal system of contemporary society accept human individuals as mutually completely separate centres of fear and desire, fighting each other or cooperating in the fight against third parties. Therefore nobody can be happy in the *activity of this unholy society*. And those who search for happiness or holiness elsewhere, find in the words of contemporary languages, which are after all nothing but the command signals of the social labour rules, no impulses for energetic thoughts, at best they find in their sound and rhythm sources of moods, poor in consequences. If, now and then, in restricted groups better means of production and legal systems were created, which allowed for a holier way of life of the individuals, then these could not be lasting, because their maintenance and enforcement had to rely on the *common language* as a means of communication, so that these reformers were forced to composes the signals of their new society out of that of the rejected society, and thus remained subordinate to the suggestions of the latter. *Intuitive signifcs* is concerned with the creation of *new words*, which form a code of elementary means of communication for the systematic activity of a new and holier society.[75]

In August 1918 Brouwer, Van Eeden, De Haan and Mannoury appointed Paul Carus, Eugen Ehrlich, Gustav Landauer, Fritz Mauthner, Giuseppe Peano and Rabindranath Tagore as members of the associated Academy. The first practical activity of the Institute was the compilation of dictionaries for the legal and socio-economic aspects of society; the topic turns up regularly in various unofficial documents, and it was discussed at numerous meetings, but concrete advances were not achieved. After the First World War new initiatives were taken.

7.7 Faculty politics

In Amsterdam Brouwer had also become involved in faculty politics. In 1916 a vacancy came up in the chemistry department and the chemistry professor in charge displayed some tactics that rather put off Brouwer. The man went so far as to declare at one point that candidate X had better rights to the position than candidate Y, but that he nonetheless preferred Y as 'it was in *his* interest to promote Y, who would assist *him*'.

At roughly the same time Brouwer had again approached the curators of the university with his desiderata for the increase of the number of professors at the mathematics department. Since Korteweg was to retire in 1918, plans for his succession were being made, but Brouwer had independently presented his wishes to the board. Brouwer's plans were apparently overly ambitious; at least the curators, at their January meeting in 1917, observed that Brouwer had admitted in conversation that he had realized that he was asking too much. Nonetheless Brouwer

---

[75] *Propria Cures* 9 September 1918.

made it clear that he would appreciate it if the university were to allow the mathematicians to appoint a lecturer. After ample discussions with De Vries and after a (more perfunctory) consultation of Korteweg, Brouwer suggested the appointment of Mannoury. Brouwer had already scored a minor victory in 1916: the university had granted him an assistant (for DFl 600.– a year) beginning January 1917.[76] But Brouwer wanted more, he had a clear goal: a superior mathematics department. Should the University of Amsterdam prove reluctant, he could easily move somewhere else. It was no secret that the state universities aimed each at two chairs in mathematics, and the minister seemed favourably disposed to grant Utrecht two chairs; Leiden then would soon follow. The Curators of the Amsterdam University seriously considered the possibility that 'Brouwer and Mannoury would then go to Utrecht, and Amsterdam would not be able to get such first class mathematicians as replacement'. This might eventually lead to the departure of Prof. Hendrik de Vries. A proposal to ask the city council to prevent the departure of Brouwer, and to grant the immediate appointment of a lecturer, who could be promoted to full professor after Korteweg's retirement, was, however, not accepted, as the curators were of the opinion that nothing should be done to upset Korteweg; they well realized the obligation of the university to this father of Amsterdam mathematics. So it was decided that the matter would be arranged in agreement with Brouwer's wishes after Korteweg's retirement. At the January meeting the chairman informed the curators that the City Council had no objections to Brouwer's desiderata.

The lecturer position being out of the question, the faculty sent a warm recommendation for Mannoury's appointment as an extra-ordinary professor to the curators. The faculty lavishly praised Mannoury's geometrical work, 'of which one does not know what to admire more, the penetration of his intuition, or the powerful deductive ability'. In particular his 'surprising simplifications in the proof of Dedekind of the fundamental theorem of formalist arithmetic' was mentioned with approval. The faculty even went so far as to refer to Mannoury's verbal contributions to science: 'In particular it has been noted at the meetings of the Dutch Mathematical Society, that Mannoury if he is present, always takes part in a masterly manner in the discussion, and subjects of the most diverse character are equally perfectly commanded'. The faculty did not mention that Mannoury, in spite of his recognized ability had so far published only seven papers in mathematics, one book and an inaugural lecture as a 'privaat docent',[77] and that after 1901 he had not published a single research paper in mathematics.

The second candidate was J. Wolff, Brouwer's fellow student, who passed his final exams a year after Brouwer. The style and content of the recommendation betray that Brouwer had a hand in its formulation.[78] The reader will recognize Brouwer's formulation, cf. p. 45, of the address at Mannoury's honorary doctorate.

---

[76] Brouwer to Zeeman 8 May 1916.
[77] On the significance of mathematical logic for philosophy, 1903
[78] Faculty to Curators, 9 January 1917.

The proposal did not meet with serious difficulties, and on 24 April 1917 Mannoury was appointed. On 10 October he gave his inaugural lecture, *On the social significance of the mathematical form of thinking*,[79] in the presence of the Brouwers, Joop de Haan and his wife, and Frederik van Eeden.

Brouwer had in the past years seen enough of faculty politics to advocate more stringent arrangements. On 3 April 1917 the faculty, following a proposal of Brouwer, took two important decisions. It split the faculty into sections: (a) mathematics, physics, astronomy, (b) chemistry, ..., (c) biology, ... Furthermore a committee for the expansion of the faculty was installed, with Brouwer and Zeeman as members of section (a). Brouwer devoted a considerable amount of time to the reorganization, but as it dealt mostly with the expansion of the chemistry department, we will not go into the activities of the committee.

In the final years of the war the relationship between Brouwer and Korteweg seemed to have deteriorated, and both the negotiation for the Defence committee (see below) and the procedure for Mannoury's appointment led to suspicions on Brouwer's part that Korteweg had turned against him. How far this was the case, cannot be ascertained. It is not unthinkable that Korteweg did not share Brouwer's enthusiasm for Mannoury. Although it was not hard to appreciate Mannoury's originality and spiritual conversation, it was almost exclusively directed towards politics and significs, and it could not have escaped a sharp observer like Korteweg that Mannoury's attention was no longer directed at mathematics. The cause of the estrangement cannot easily be pinpointed. In 1915 there was no cloud on the horizon, Brouwer and Korteweg had lively discussions on mathematical topics. Brouwer was at that time studying Korteweg's paper 'On the stability of periodic plane orbits',[80] and he wondered if Korteweg had missed a case. Korteweg truthfully answered that he had difficulty in recalling his considerations of 29 years ago, but he immediately returned to the problem of 1886 and started thinking of solutions to Brouwer's questions. A day later Brouwer confirmed Korteweg's idea, writing 'I just found that your conjecture is correct, and that the possibility of a system of perturbed orbits, which I had in mind, is ruled out'.[81]

A little later Brouwer and Korteweg discussed a problem of triple tangents to algebraic curves. In rapid succession letters were exchanged,[82] and in the last one Brouwer wrote:

> I have been able to fix now the matter of the general presence of double and the general absence of triple tangents, even in a fairly simple way, so that it surprises me that I have not hit on the same idea at any of the three occasions upon which I 'discussed it in class'.

During this exchange some insensitive remarks must have been made by Brouwer,

---

[79] *Over de Sociale Betekenis van de Wiskundige Denkvorm.*

[80] *Ueber Stabilität periodischer ebener Bahnen*, Wiener Berichte 1886. Brouwer to Korteweg 15 October 1915.

[81] Korteweg to Brouwer 17 October 1915, 18 October 1915, Brouwer to Korteweg 19 October 1915.

[82] Brouwer to Korteweg 29 November 1915, 1 December 1915, 2 December 1915.

for on 11 December Korteweg referred to Brouwer's card of 2 December: 'A small remark about your card of 2 December. That we should henceforth never talk about scientific topics seems to me a rough remedy. It would, as I feel it, cause a distorted relation between us. The word 'talk' is not intended by me in contrast to a 'serious discussion', but to 'writing'. Even if I talk about a scientific topic, I am really 'serious'. Even though what I say it is not always correct, I strive for its correctness'.[83]

So much is certain that Korteweg did not bow to the whims of Brouwer and did not hesitate to correct him in the faculty meeting, if the occasion arose. In 1918, at the time that the new chairs in mathematics were arranged, the relation had already deteriorated to an intolerable extent. Brouwer complained in a letter to Mannoury about Korteweg's behaviour, 'K. has since two or three years no longer any cooperation from me, since I saw that almost all his efforts serve to cripple Dutch mathematics, efforts which he will doubtlessly continue to his dying day.' The correspondence suggests that Mannoury's appointment raised questions that could not just be brushed away.

One year later Mannoury became the successor of Korteweg as professor in Analytic and Descriptive Geometry, Mechanics and Philosophy of Mathematics. His course on philosophy was regularly devoted to significs, and so the topic gained a modest but secure foothold in Holland.

Gerrit Mannoury in his Vlissingen years. Courtesy J. Mannoury.

Mannoury was a kind and understanding teacher, revered by his students, who taught and published in a very personal style. There is a large number of publications from his hand of a philosophical–social nature; his inaugural address, with its title '*On the social significance of the mathematical form of thinking*' set the tone for his activities in the university and in society. After a long career he bade his farewell to the university with a valedictory lecture, entitled '*The beauty of mathematics as a signific problem*'.[84] After his retirement in 1937 he devoted himself to the writing of a comprehensive text on significs;[85] he remained unusually active. In spite of all good intentions, the signific movement was to remain an almost exclusively Dutch affair, and after Mannoury's death it gradually withered away.

Although Brouwer had already more than enough duties to keep him occupied, he felt that he had to keep an eye on the quality of the staff at the mathematics

---

[83] The reference to Brouwer's card is erroneous. There must have been a verbal exchange.
[84] [Mannoury 1937].
[85] [Mannoury 1947, Mannoury 1948].

departments in Holland. And so he was naturally interested in the local appointments of mathematics professors. Strictly speaking, the appointments were a local affair, and external advice was usually purely a matter of courtesy or prudence of the local faculty. Brouwer, however, considered the choice of mathematics professors, no matter at which university, of national interest. As a consequence he was always willing to give advice, or, as some would say, to meddle.

When in 1916 the mathematics chair at Utrecht became vacant through the death of Kapteyn, the physicist Ornstein wanted to appoint Brouwer. Apparently Brouwer was vaguely interested; he was willing to consider the matter if a third mathematics chair would be forthcoming. He clearly was aiming at a mathematical centre in Holland that could stand comparison with existing international centres. The faculty committee nominated two respectable, but somewhat unexciting candidates: Van Uven (who had a chair at the agricultural institute (*hogeschool*) and Rutgers. Ornstein opposed this choice, and put down his own list: Valiron, Denjoy, Rosenthal and F. Bernstein.

Brouwer, who was not a member of the committee, suggested a list of Dutchmen: Mannoury, Wijthoff, Droste, Rutgers, Boomstra and Bockwinkel. After ample discussions *en petit comité*, it was decided that a short list, consisting of 1. Fréchet, 2. Bernstein, would be proposed to the faculty, and, should the faculty turn this down, a second short list: 1. Droste, 2. Boomstra, as a kind of reserve. In March 1917 Brouwer again interfered; he now recommended to the committee the following candidates: 1. Denjoy, 2. Valiron, 3. Bernstein. Brouwer's number one was a young French function theorist with a wide interest. His œuvre contained papers on topology, set theory, function theory, measure theory, ... In topological matters he was close to Brouwer's interest. His fame was in particular based on his process of 'totalization' and for the Denjoy integral.

The faculty adopted Brouwer's suggestion, and Denjoy was approached. He made two conditions: a salary of DFl 6000,- and a contract, renewable annually. The faculty was far from happy, because they feared, rightly so, that Denjoy did not want to give up his French position, and would only use Utrecht as a temporary base. However, the faculty appreciated Denjoy's reputation as a competent analyst of the French school and the appointment was duly made on 19 July 1917. He indeed combined a French and a Dutch academic position, which must have made life pretty uncomfortable for him. During the war he regularly made the trip between Holland and France by sea, going via England. The dangers of this routine were far from imaginary, on one occasion his boat to England was torpedoed; the faculty recorded in its minutes with some relief that Denjoy had safely reached Harwich.

Brouwer and Denjoy worked in perfect harmony; they exchanged papers and manuscripts and Brouwer presented Denjoy's work to the Royal Academy, and nothing seemed to hint at the eventual violent disruption of their friendly relations.

In the period following his topological research Brouwer had, voluntarily or not, become involved in a number of activities outside the academic sphere. We have seen some of this above; some of Brouwer's actions may have been the unconscious fall-out of his recent contacts with artistic and political circles, but underlying all his projects and plans there was this strong conviction of right and wrong, his aversion to injustice, and a deep passion for the underdog. His rationality saved him from the fate of Cervantes' hero, but nonetheless he became the champion of a number of causes where a more sober and calculating man would have remained aloof. The period of the First World War was perhaps richer than usual of challenges to a man with the moral and social conscience of Brouwer. The following episode is another example of his involvement in politics.

7.8 The Flemish cause

The direct cause for his activity in this case is somewhat unclear, Brouwer may have had some information from Belgian refugees, but it is more likely that his attention to the position of Flemish higher education was called by third parties. The correspondence with Schoenflies seems to indicate that certain German circles were trying to get support for the German interference in civil matters in the occupied part of Belgium. On 1 July 1916, Brouwer wrote that he wanted to talk to some Belgians and to collect some data before he could answer Schoenflies' question. In his next letter, Brouwer said that he had talked to numerous Flemings, and that he had diligently studied the Belgian Constitution and the Law of Higher Education. The problem for which Schoenflies, or somebody else at Frankfurt, had tried to enlist Brouwer's assistance, was that of the position of higher education in Flanders.

In 1914 there was still no university for the inhabitants of Flanders with Flemish as an official language. There had been harmonious efforts by the Flemish community to redress this wrong, but the German invasion had brought the cooperation to an untimely end. Some Flemings suspended all actions and remained neutral, and others used the German intervention to get a Flemish University realized.

After reading the relevant legal documents Brouwer soon concluded that 'to my vivid disappointment I have *not* been able to convince myself that the German occupying authorities have the right to make the Gent University Flemish against the wishes of the Belgian Government.'[86] And thus he claimed the 'in dubiis abstine' for himself as a politically neutral person. It is not clear what exactly was asked from Brouwer, it could be that efforts were made to find professors for the Flemish University at Gent. He added, strictly confidentially, that he could possibly find Dutch colleagues, who were '*also politically*' and in particular in the matter of the Belgian matter, completely on the German side, and who would probably be ready for the desired efforts.' The letter was followed by another one[87] in which he explained that the The Hague Convention did not allow the German occupation authorities to change the legal status, as long as the public order did not force them to do so, and thus turning the University of Gent Flemish (*Flamandisierung*) could not be justified. He refused therefore to cooperate in this matter, 'no matter how unsympathetic the Belgian Government is to me, in that it has refused the Flemings this

---

[86] Brouwer to Schoenflies 5 August 1916.
[87] Brouwer to Schoenflies 27 August 1916.

right [of their own university] for 80 years'.

Although he did not wish to interfere with the internal affairs of the Belgian state, Brouwer did publish an open letter to the Belgian Government in *De Groene Amsterdammer*, which contained a summary of the injustice done to the Flemish community. He concluded that the inclination of many of the Flemish to feel a quiet sympathy for the, unjustified, German efforts to turn the University at Gent into a Flemish university, was excusable in the face of the suspicion that the Belgian government 'would after the war, forgetting that four-fifth of the Belgian defensive army consisted of Flemish men, violate the Flemish rights as before'. Therefore he invited the Belgian Government to declare that it would recognize Flemish rights after the war by adopting an earlier proposal along those lines.[88] He closed the letter with the rather provoking words

> In this way it [the Belgian government] would provide to everyone's satisfaction the proof that it has the moral courage to refuse unconditionally, not only to hand over the total Belgian population to German imperialism, but also half of the Belgian population to French imperialism,

In the same year Brouwer considered an action against the German government on the issue of the use of Belgian forced labour. From the correspondence it appears that he could not get convincing corroboration; some rumours said that Belgian unemployed workers went voluntarily to Germany because of the high wages. Not only in the above cases, but in general, Brouwer was extremely careful to investigate the matter at hand before undertaking any action, and so in the latter case he probably did not want to proceed without sufficient grounds.

In 1917 Brouwer organized another protest action, this time in Holland against the treatment of a Dutchman, J.C. Schröder, who had been sentenced to three months in prison for making his political views public. Brouwer asked his colleagues at the Academy to sign an address against the judicial action, which 'can endanger in this way our complete constitutional freedom of the press, one of the main guarantees of our personal freedom'.

7.9 Air photography and National Defense

In the middle of the war a totally unexpected thing happened: Brouwer showed national feelings. For a mystic with avowed internationalist views this certainly is surprising. The review article (p. 249) with its scathing remarks on the word 'fatherland' would have almost ruled out a sudden rallying to the defense of the nation. Nonetheless, there is a curious episode in which Brouwer had a brief infatuation with defence matters.

When Brouwer travelled in Germany at the occasion of his visit to Schoenflies, cf. p. 252 (which, as far as we know, was his only wartime visit to Germany), he learned that the Germans, and presumably the other belligerents, employed numerous young mathematicians for the transformation and measurements of air reconnaissance photographs. Furthermore, he heard that many Academy members

---

[88] De Nieuwe Amsterdammer 19 June 1916.

were enlisted as scientific advisers of the general staff.

Wishing to kill two birds with one stone, that is to say, to stay out of the regular army and to serve his country, he had decided to start his own investigations in photogrammetry. He was completely serious about the importance of a systematic use of mathematical methods as a means to extract exact information from air reconnaissance photographs, which usually were taken under varying angles, and showed all kinds of distortions. As early as September 1915 he sent a memorandum to the minister of defence, drawing his attention to the importance of photogrammetry and offering his services.[89] The letter was answered by the chief of staff, General Snijders.[90] The general said that he was much indebted to Brouwer for his suggestions, but the army had at its disposal excellent topographic maps (1 : 25.000), and that it was very well possible to indicate the precise position of an object, observed from the air, on such a map. Here the matter ended as far as the army was concerned.

That Brouwer was not just day-dreaming may be inferred from the fact that he seriously studied the available literature, and prepared a number of papers, which were published in aviation journals.[91] After the rejection of his proposals by the army, there was little Brouwer could do. Another chance to promote his idea came, however, in 1917, when Professor Jaeger, a physicist from Groningen, agreed with Brouwer that it would be profitable for the nation if proper use were made of the intellectual capacities of its academic specialists. He wrote to Brouwer, asking him to get Zeeman interested, so that a small group of academicians, including Brouwer, could consult Lorentz on the matter. Apparently there was considerable discontent in academic circles with the firm intentions of army headquarters to ignore the possible impact of science on warfare and military management. As Jaeger expressed it: 'The military high command has had for three years the possibility of improving the army by more efficient use of the intellect that has been enlisted. The result was totally nil, because there is no initiative at all [...]. Thus, nothing can be expected from the common sense and initiative of the military command, *a fortiori* not in times of panic. And thus it has to come from us'.

Formally, the Academy had every right to approach the government, as the rules of the Academy specifically allowed it to advise the government on matters of national importance. Zeeman supported the idea, and sounded Lorentz on it. When Lorentz also agreed, action was taken.

It may be remarked in passing that Lorentz wanted to go further and proposed that all professors would be excused from military (or para-military) service,[92] in order to be available for special duties as advisers. An attempt to get a dispensation for members of the Academy foundered, however. In December 1917, the Minister

---

[89] Brouwer to Minister of Defence 18 September 1915.

[90] Snijders to Brouwer 12 October 1915.

[91] [Brouwer1916F, 19117C, 1919R].

[92] In 1915 the so-called *landstorm* was created, an organization comparable to the National Guard. It was intended for volunteers who had completed their first military training, but who were not called up in the mobilisation.

made it known that he saw no reason to put Academy members on a par with, members of the judiciary, civil servants of the social service, aldermen, and the like.[93]

Soon after the October meeting of the Academy a meeting with the responsible minister, Lelie, was arranged. The Academy was represented at this meeting by Brouwer, Jaeger and Schoute (the meteorologist). The government reacted positively on the proposals, and what would seem more natural than that the temporary committee, in particular the originators, Jaeger and Brouwer, would join the executive committee? The authorities decided otherwise, however, and neither Brouwer nor Jaeger were included.

Brouwer was despondent; he was back at the dark depressions that ruined his student years. He started to draft letters to Lorentz, whom he held for a large part responsible for the unfortunate outcome. He was also convinced that his old teacher, Korteweg, opposed his plans; he may have had a point there.

H.A. Lorentz. Courtesy Rijksarchief Haarlem.

Turning things over in his mind, Brouwer got more and more excited. The horror of another period in the army arose in his mind's eye, and he frankly declared that this was more than he could stand.[94] He sat down and summed up all his grievances in a first draft, three days later he finished the letter itself, which was much more composed in nature; the draft reveals some of his strong emotions. He felt insulted because 'his unwritten rights as an initiator' were violated, without even a warning or an apology. In particular as the president of the Academy, Lorentz had the duty to guard the rights of the members. At this moment it seemed that Holland would remain neutral, and that the risk of being drafted had vanished. That was, in Brouwer's opinion, no excuse for the army to drag its feet in the matter of photogrammetry. For if at any moment Holland should get involved in an armed conflict, it was obviously too late to start build up a photogrammetric service.

The extent to which Brouwer's mental stability was threatened may be illustrated by a draft of the following letter from Mrs. Brouwer to Lorentz (possibly not sent, there is no copy in the Lorentz' archive).

> Dear Professor Lorentz, I am worried about my husband; he is so hurt that he is passed over as a member of the Commission for Scientific Research for National Defence, that I am at a loss what to do with him. Since

---

[93] Brouwer to Zeeman 17 December 1917.

[94] Cf. p. 51. The passage occurs in the draft but not in the final letter (Brouwer to Lorentz 16 January 1918).

he has heard this, he cannot do anything, and not so much because of the matter itself, but rather because of the hostility of his elderly colleague and teacher, whom he surmises is behind it. As you know he initially took on this new study in order to get, in case of a war, the opportunity to offer the country his services in a more scientific manner than otherwise his rank of corporal would require. Gradually he has pursued the study with an ever increasing interest, and prepared the instalment of the above mentioned committee. When the funds for the practical elaboration of his plans were made available, he was very glad, and quietly awaited, without suspecting any harm, the outcome. He thought to be able to get to work soon, and was full of plans for the future laboratory. And now this. Each day he wants to write to you, but gets so enraged each time, that he cannot find the right words. What will be the end? He will not resign himself to it, which seems even to me impossible. How everything turned out that way, can be guessed, but not ascertained. But an explanation will have to be given, why the *auctor* of the whole conception has been removed from the arena. How often do I deplore, that the hospitable and safe Leiden has beckoned him in vain, as a quiet hiding place for unrest and conflict. It would have been the fulfilling of my dearest wishes in the interest of my dear husband. And yet it were not the personal reasons or matters of health, also mentioned in your presence, but commitments of an indebtedness to those who had him appointed in Amsterdam, and who allowed him everything that the other side offered in advantages, [...].

This letter confirms the nature of Brouwer's depression. He felt deprived and insulted. Lize deplored in the letter the obligations that had effectively prevented Brouwer from accepting the chair in 'the hospitable and safe Leiden'. The letter also hints at 'remaining years of struggle with another colleague', this is most likely a reference to Korteweg, who could be quite stern with Brouwer, and who deplored what he considered to be the *prima donna* airs of his student. Brouwer for his part grew dissatisfied with the limited scientific exchange with his old teacher. A degree of disillusionment on both sides cannot be denied, and Brouwer may very well have shown a lack of filial respect for the older man. But to what extent Korteweg's opposition was imaginary cannot be ascertained.

On 26 January Lorentz talked to Brouwer about the choice of the members of the committee and it appeared that Brouwer was prepared to accept the inevitable. Lorentz must have been struck by the importance Brouwer attached to the membership, so that he must have pulled some strings. On 15 February Lorentz wrote to Brouwer, informing him about the discussions in the planned committee, and asked if he was willing to join the project as an extraordinary member of a subcommittee, the one 'for the terrain photographs made by aircrafts'. Lorentz confessed that he felt rather out of place in the committee itself, 'surrounded by experimenters and technicians', but that the minister had insisted that the committee should contain some members of the academy.[95] Brouwer accepted Lorentz' offer, and so he found himself in due time an extraordinary member of no less than two subcommittees.

The Royal Decree of 20 February 1918 appointed Lorentz (chairman), Zeeman (secretary), S. Hoogewerff, C.A. Pekelharing and F.A.F.C. Went to the board of the 'Scientific Committee of advice and research in the interest of public welfare and defence' (*Wetenschappelijke Commissie van advies en onderzoek in het belang van Volkswelvaart en Weerbaarheid*).

Brouwer figured on the membership list of the various subcommittees of the Scientific Committee as an extraordinary member of the subcommittees for X-rays, and of the one for Photogrammetry. Lorentz himself was the chairman of the latter one.

On 2 May 1918 Lorentz wrote to Zeeman that he had discussed the photogrammetry problems with the other members of the subcommittee, Brouwer and Schoute (See above). The three had agreed that it was desirable to add a military expert to the subcommittee, the name of Major H. Walaardt Sacré, commander of the military aviation section at Soesterberg,[96] was mentioned.

Furthermore the members of the sub-committee thought it a good idea to have also a physicist on the sub-committee.[97] They suggested F. Zernike, who had already, as an assistant of Kapteyn, occupied himself with measurements of photograms. It fell to Brouwer to sound out Zernike; this apparently was unproblematic, since Zernike was duly added to the sub-committee.

Brouwer, Schoute and Zernike did not lose much time. On 20 June, Brouwer reported to Lorentz, that they had visited Soesterberg. They had found that the photographic service mostly took vertical shots, but the staff was readily convinced that in a time of war one had often to work with photographs taken from various angles, and that it was desirable to be able to draw conclusions from these photographs too. It was agreed that Lieutenant Meltzer, would make a number of photographs of the city of Utrecht from various angles. It would then fall to Brouwer to transfer the information to the maps. The plans were, however, thwarted because there was no gasoline available for the flight!

Again in July the three visited Soesterberg, and this time inspected the cameras and the equipment. As a fair number of the cameras came from aircraft that had been interned on Dutch soil, they were far from uniform, differing in focus, means of orientation, etc. The previous arrangement was repeated, but before any results became available, the sad news arrived that Meltzer had died.[98]

In order to secure the investigations, Brouwer proposed to Lorentz to add two physicists from Zeeman's laboratory, Van der Harst and Bosch, as reserve officers

---

[95] It is a fact that Lorentz could not say 'no' to requests from the government. Another instance is his involvement in the 'Zuiderzee project', which took up much of his valuable time.

[96] A military airfield of long standing. After the second world war an American squadron was based at Soesterberg.

[97] As already suggested in Brouwer to Lorentz 19 February 1918.

[98] Brouwer to Lorentz 8 September 1918.

to the Soesterberg staff.[99]

Brouwer's own investigations led him to inquire with our national aeroplane builder, Anthony Fokker, into the state of photogrammetry. Fokker sent copies of papers on the importance of air photography for surveys and on the Cranz–Hugershoff method. He told Brouwer that 'all experiences which were gained in Germany during the war, in particular also a number of excellent experts, who belong to the top in the area, are at my disposal'.[100] In the same letter he offered his cartographic services for future projects in air photography, in particular in the colonies.

The more the work advanced, the less place there was for a gifted amateur, such as Brouwer. He slowly withdrew from the area, but not without at least taking part in some of the photography runs. This early experience with actual aviation remained one of his treasured memories.[101] In 1920 he became a member of a committee that had to report to the ministry of war on 'air photography' and 'air cartography'. On 13 July a letter was sent to the minister, which, Brouwer remarked, would probably be the final burial of the subcommittee.

During the last year of the war Brouwer had gathered enough insight into the intuitionism new style, to start the preparation of a series of expositions on the subject. The first one appeared in 1918 and the following issues in 1919 and 1923; they were all published as individual pamphlets in the *Verhandelingen* (Acts) of the Royal Academy, two of them under the title *Begründung der Mengenlehre unabhängig vom logischen Satz vom ausgeschlossenen Dritten* (The founding of set theory independent of the logical proposition of the excluded third) and one on function theory with a similar title. The choice of his title was curious, and it invited misunderstanding. An outsider could easily get the impression that this was a logical project with a weakened axiomatic basis. As we have seen, Brouwer's constructive mathematics incorporated genuine *mathematical* principles (involving continuity and transfinite phenomena), so the choice of name could hardly be called felicitous. Maybe Brouwer had not yet made up his mind as to a proper name for his project, but it is also possible that he considered the name 'intuitionism' a bit grand for a one-man business. The term 'intuitionism' occurred frequently in his early foundational papers, but there it still refers to a vague kind of constructivism *à la* the French school, a restriction of classical practice, rather than an autonomous, legitimate mathematics. Only in 1920 Brouwer started to refer in writing to his mathematics as 'intuitionism', cf. [Brouwer 1920J].

One has to keep in mind that Brouwer was not after the founding of a school of constructivism in a small corner of the building of mathematics, but indeed after a complete revision of all of mathematics. Hence a label, no matter how tasteful, would always reduce his efforts to another local specialisation. The emergence of

---

[99] First having obtained Zeeman's advice. Brouwer to Zeeman 12 July 1918.
[100] Fokker to Brouwer 24 March 1919.
[101] Oral communication C. Emmer, Brouwer's friend and family doctor.

the term 'intuitionistic mathematics' may therefore have been a by-product of the foundational conflict of the twenties.

The war years, although they kept Brouwer in Holland, did not bring him much spare time. His contacts with the group around Van Eeden, his teaching, research, his professional contacts in the University and in the Academy, kept him busy enough. Furthermore he cultivated a large circle of, if not friends, at least close acquaintances: another time consuming matter!

His contacts with his closest friend, Scheltema, remained rare. On one occasion the Scheltema family called in Brouwer's help to knock some common sense into the head of Carel's brother Frits, who had developed a virulent pro-German attitude. If and how Brouwer reacted has not been recorded.

The contacts between the two were mainly kept up through correspondence, Brouwer sadly sighed, 'I would be so happy to talk to you sometime, but Blaricum and Bergen are so far apart, and my spare days get so scarce.' Indeed, meetings were becoming a thing of the past. Even the birthdays were no longer occasions for visits. Brouwer reported in a letter of 25 February 1916, that he could not visit Scheltema on his birthday; he was not actually ill but

> if I stay out of bed for more than one hour, my heart starts to behave in a funny way, so that I stayed in bed for a whole week and read all the volumes of the adventures of Arsène Lupin.[102]

This minor detail shows that Brouwer, in spite of his acquired status, could still enjoy the pleasures of the adventure novels in the shade of recognized literature.

Scheltema had by this time become the pre-eminent socialist poet, with an immense readership, his poems were to be found in the homes of the working class, but not there alone. They were set to music and sung in the homes all over the Netherlands and by choirs of all parts of the population. In the years 1917–1918 he wrote a number of theatre plays, *The Naked Model, Great days in Knollenbroek*[103] and *The guilded Pretzel*.[104] The Brouwers got free tickets for the premiere of the *Great Days in Knollenbroek* and they intensely enjoyed the farce. Brouwer complimented his friend, 'Altogether I am grateful to you for your play, and I believe that you have shown yourself up to this genre.' A year later the Brouwers attended the performance of *The Guilded Pretzel*.[105]

The war had both positive and negative effects on Brouwer. It had given him a chance to concentrate in isolation on the tough problems of a revision of mathematics in the sense of his program, but it had kept him from his natural audience: the Göttingen mathematicians. It also had introduced him into the circle of artists and intellectuals around Van Eeden. Mathematicians will probably deplore this as

---

[102] The gentleman-burglar Arsène Lupin, was the hero of detective novels of M. Leblanc.
[103] A fictitious village.
[104] *Het Naakt Model, Grote dagen in Knollenbroek , De vergulde Krakeling.*
[105] Brouwer to Scheltema 27 September 1917 and 8 September 1918.

a waste of time, but it satisfied a need of Brouwer to take part in social-cultural life. After all he had strong convictions on a rich variety of issues. His moral position might have become more balanced, but it certainly had not evaporated.

So at the end of the war Brouwer had re-thought his position, both in the esoteric domain of mathematics and in the world of his fellow human beings.

8
MATHEMATICS AFTER THE WAR

The end of the First World War was greeted by Brouwer with relief. He immediately set out to re-establish his international contacts. During the war years he had lost contact with his foreign colleagues, and whether induced by his isolation, or as a consequence of his changed interests, he had almost totally dropped his research in topology. There was also a marked drop in international correspondence, among the remaining letters (which, of course, may totally misrepresent the actual number) there are only a few to and from Blaschke, Birkhoff, Carathéodory, Mittag-Leffler, Scholz, Kerékjártó, Buber, Study, Denjoy and Schoenflies. The correspondence with the last mentioned had miraculously sprung from their rather acid exchanges at the time of the Analysis Situs paper and the new edition of the *Bericht*. The two men had become good friends, and although Brouwer did not take Schoenflies quite seriously in his topological capacity, he had come to appreciate his older correspondent as a good friend and colleague. During the Easter holidays of 1915 Brouwer even managed to overcome the bureaucratic obstacles to visit Schoenflies in Frankfurt a.M.

We should not be surprised that Brouwer was keen to revive his old contacts. The Göttingen mathematicians were however one step ahead of him; he was elected to the *Gesellschaft der Wissenschaften zu Göttingen*, and in a letter of 28 August 1919 he warmly thanked Hilbert, whom he guessed to be the instigator of this honour. This was his first official recognition by a learned society over the borders of his native Holland. As a matter of fact, Brouwer had already been put forward to the *Göttingen Gesellschaft* in 1917 by Hilbert, Landau and Klein. Brouwer was recommended for his topological work only:

> To Brouwer are due, in the first place, important contributions to set theory, the general solution of the problem of the invariance of dimension, the generalization of the Jordan curve theorem to space and n dimensions; he has moreover applied his set theoretic methods with success to existence theorems from the theory of automorphic functions'.

The proposal was for some reason withdrawn and submitted the next year, when it was crowned with success.

Brouwer was keenly interested in the developments in Europe after the war. After the armistice between the belligerents was signed (11 November 1918), he wrote identical letters to Klein and Hilbert, expressing his hopes for the future:

May the healthy heart of your fatherland overcome the present crisis; and may the German lands soon prosper to unknown bloom in a world of justice.[1]
    This is wished to you by
        your Brouwer.

As a matter of fact, Brouwer was seriously worried by the turn things were taking. Like most Dutchmen, he had abhorred the war, and having many friends in the camps of both the Central Powers and the Entente, he was keenly aware of the misery on both sides. At the end of the war one had not to be a trained politician to recognize the mood in the world; the victims of the German attack had made up their mind to exact war damages from the losers, and to make certain that Germany's role as a middle European power would be over.

Already in May of 1918 Brouwer and Van Eeden had approached the American consul in the Hague with a plea for a just peace. Brouwer had at the occasion proposed a conference of the scholars of the belligerent nations. Two months later another meeting was organized with a representative of the American legation.[2] Needless to say that their good intentions did not accomplish much. The event illustrates that Brouwer was willing to put his ideals into practice, irrespective of their chance of success. However unlikely it was that the two men were to succeed where even President Wilson failed to temper the feelings of revenge of the victors, one should keep in mind that the couple, Van Eeden–Brouwer, had no doubts about their duty. The spirit of the foundered Forte Kreis was still very much alive in their minds, and they took their obligations as spiritual leaders in a torn world extremely seriously.

The return to mathematics did not prove easy. After the hectic years of the great topological breakthrough, Brouwer had, partly because of the interruption of his international contacts and partly because his mind looked for other challenges, left the mainstream of the new mathematics that he had helped to create himself. We have seen his involvement in philosophical projects, and his return to the foundations of mathematics. Also, his academic duties and his social-political conscience laid to a considerable extent claims to his attention. So when he could return to the beckoning haven of topology, a multitude of other interests and obligations stood in the way. Politics, albeit not along any party-lines, ever tempted Brouwer to actions and views for a better world. In addition to that kind of activity Brouwer had developed an interest in academic affairs. It could not escape anybody that he was by far the strongest man in the mathematical wing of the faculty. Korteweg used to have a certain natural authority over him, but with advancing years Brouwer started to resent the ideas of his old teacher, cf. p. 272. It is not difficult to see that gradually Brouwer became more and more interested, and bogged down, in academic

---

[1] Brouwer to Klein and Hilbert, 25 November 1918. '*Möge das gesunde Herz Ihres Vaterlandes die heutige Krise überwinden, und mögen die deutsche Landen alsbald zu ungeahntes Blüte gedeihen in einer Welt der Gerechtigkeit! Das wünscht Ihnen Ihr Brouwer*'.
[2] Diary of Van Eeden 22 May 1918, 24 July 1918.

organizational matters, appointment committees, and the like.

Although the war had not touched the Netherlands, the country did not escape its by-products; a general scarcity had made itself felt, food was rationed and the inevitable war-profiteering reared its head here and there. When the guns fell silent at all fronts, a wave of revolution and social unrest swept through Europe, scourging in particular Germany; it even reached Holland.

The most prominent socialist policitian at that time was Pieter Jelles Troelstra. He was the leader of the *Sociaal Democratische Arbeiders Partij*. In this position he had to steer a middle course between the radical proletarians and the fashionable intellectual leftists. The latter were not quite taking Troelstra seriously; from the pinacles of theoretical purity, they critically followed his exploits. Leading personalities like Henriette Roland Holst called him disparagingly, 'Troeleman'.[3]

On the same day that the armistice was signed in Compiègne, 11 November 1918, Troelstra, called on the workers to take over the state. But neither the party nor the workers were prepared to take the required drastic actions. The Dutch revolution misfired, and Troelstra admitted his miscalculation two days later in the parliamentary debates. The more cautious and conservative part of the nation had shown its distaste for revolutionary adventures by a mass demonstration in support of the Royal Family on 18 November. As one can imagine, there was an understandable fear among the progressive intellectuals, that Troelstra's rather rash imitation of the historical events elsewhere in Europe would result in a freezing of social progress in Holland. Among the reactions, we note an open letter to the Dutch citizenship, in *De Groene Amsterdammer* of 30 November, bearing the signatures of a number of prominent citizens, including Brouwer's, warning that the events of November were not isolated phenomena, and that it was an illusion to hope and expect that modest concessions would rid society of a discontent proletariat. The letter ended with an eloquent appeal to the press, to provide a counterbalance to the conservative press:

> I think, that we still are at the eve of great events. It will partially depend on the position and insight of the bourgeoisie, along which roads the coming changes will be guided. And in that process, the information, given by the press will be of great influence. Let the inclination of a part of that press to make an effort to keep the material interests of the bourgeoisie intact as long as possible with fake concessions or with force, be somewhat compensated by statements from the other side. Imponderabilia carry weight too.

This letter called attention to a number of problems that were apt to be overlooked in the euphoria after the failed revolution. It was published in the weekly of Brouwer's friend from his student days, Henri Wiessing. Wiessing had already become a prominent man in Holland; his political preferences were clearly on the far left. As a student, he had been active in the social democratic scene, and later

---

[3][Wiessing 1960], p. 201.

he, like so many of his generation, became a communist, although *not* a party member. After a short career as a correspondent for the *Algemeen Handelsblad*,[4] Wiessing became, at the age of twenty-nine, editor-in-chief of *De Amsterdammer*.

After a conflict of a political nature, Wiessing left *De Amsterdammer* and founded his own weekly, *De Nieuwe Amsterdammer*, popularly known as *De Groene (Amsterdammer)*.[5] Wiessing succeeded in attracting writers of good standing, for example Van Eeden. Brouwer too wrote from time to time in *De Groene*; his choice of this particular magazine was probably more a consequence of his friendship with its chief editor, than of its political colour. It would go too far to see in Brouwer's choice of *De Groene* a commitment to a particular political view, or even to the views of Wiessing; it was probably a matter of personal loyalty and partly a matter of convenience. As today, it was not always easy to find an outlet for one's ideas (in particular when they tended to be eccentric and radical, as Brouwer's views often were). When *De Groene* ended its short life in 1920 in a cloud of financial problems, Brouwer wrote a summary obituary in the last issue:[6]

> You have dared to wish to cultivate a magazine that would serve ideals instead of interests, principles instead of party leaders. Such a thing conflicts with the laws of the dynamics of society.

Brouwer had already published a review in *De Groene Amsterdammer* in 1915; it is rather curious that Brouwer had done so, since the subject was a mediocre pamphlet with the impressive title *Geometry and Mysticism*. It was written by a mathematics teacher, H.A. Naber. In spite of the trifling occasion, Brouwer could not let an opportunity go by to ventilate his insights.

> As the making and observing of mathematical forms in the *Anschauungs-world* is a preparation for, and a consequence of, the *intellectual* self-preservation of man, and since theoretical mathematics can only be defined as the activity of the *intellect in isolation*, and since furthermore, mystical vision only begins after the intellect has gone to sleep, practical nor theoretical geometry can have anything to do with mysticism.
>
> Therefore, already the above title betrays in the author, the lack of the first principles of epistomological insights that have the last twenty years become widely accepted among the practitioners of mathematics and physics, in particular thanks to Poincaré, and in which our time, no matter how coarsened otherwise, possesses a wisdom, that all historical periods of civilization have lacked.
>
> The text of the book is for the greater part no better than the title makes us fear; for example, the following three facts are put forward in one of the three lectures: that a nice two-pronged fork can be drawn with

---

[4] For a long time there were two quality newspapers that stood out in the Dutch newspaper world, the *Algemeen Handelsblad*, based at Amsterdam, and the *Nieuwe Rotterdammer Courant*. The two have merged in the nineteen seventies.

[5] De '*Mosgroene*'.

[6] The *Groene* was later properly resurrected; it is still flourishing today in its own modest way.

the help of a pentagram; that in the old days by the word 'cross' sometimes a fork was meant; that the Pythagoreans revered the letter Y. Conclusion, hold your breath: the Holy Grail is really a divining rod.

In addition to its repugnant title and text, the book contains, however, a collection of very beautiful illustrations; details of the cathedral of Amiens and other buildings; furthermore striking form rhythms from the animal and vegetable kingdom. The geometric analyses of some of the pictures made by the author are not uninteresting, and they suggest the conviction that what causes the aesthetic sensation in architecture is the impression that a large wealth of forms is governed by a few surprisingly simple geometric relations.

The last paragraph of the review is interesting as it hints at Brouwer's aesthetic views:

> ...so the emotion of beauty is a purely intellectual sensation, and it is only the physical disturbance in the central nervous system caused by the above that wakes up in our diseased bodies the frozen consciousness of God; a condition that can indeed also be produced by other physical means, such as isolation, fasting and the adopting of various Eastern attitudes.

It may be pointed out here, that for the average mathematician or academic, Brouwer's mystical convictions were all but invisible. The most strikingly philosophical passages had been removed from the dissertation, and the only philosophical paper, so far, had appeared in a rather obscure Dutch philosophical journal. The early book, *Life, Art and Mysticism*, had a rather limited circulation, and no more than a handful of Dutchmen were aware of its existence. Only in the late twenties Brouwer started to publish his philosophical convictions. In the company of the significists, Brouwer freely discussed intuitionistic and philosophical matters, but little of it reached the general public.

8.1 How to appoint professors

In July 1918 Brouwer stirred up the otherwise quiet pond of the Dutch universities by launching a revolutionary proposal; in the *De Nieuwe Amsterdammer* of 27 July he published a letter to the editor, which discussed the procedure for professor-appointments. The Dutch universities and institutions of higher education had only at the end of the nineteenth century joined the company of prominent scientific institutions in neighbouring states. Around the turn of the century, the corps of academic teachers contained a number of excellent scientists, but at the same time there was a group of rather invisible scholars that may have excelled in teaching a somewhat cautious curriculum. This rather inhomogeneous company obtained its new members via co-option. In practice, if a vacancy occurred, the faculty got together, discussed the possible candidates, and after ample consideration proposed a candidate to the Board of Curators. There was no obligation to consult outside experts, neither was there a generally accepted check-list of required qualities. In fact this system basically survived until the nineteen seventies. One did not apply for chair; one was asked. As we have seen in the case of the appointment of Denjoy, Brouwer did not hesitate to interfere in matters of other faculties if he considered it necessary (and if he could make himself heard). He was evidently dissatisfied with the general procedures for the appointment of professors, in particular the stress on teaching. And so he formulated his views in the 'Groene':

THE ARRANGEMENT OF APPOINTMENTS OF PROFESSORS
Since in each country there ought to be institutes for scientific education, and that thus the principal character of the universities should not be abandoned in favour of their secondary task of vocational training, at least the full professors of the universities should exclusively be appointed on the grounds of their scientific merits. All other personal properties are, for the proper fulfilment of their position, of infinitely less, and moreover of a significance, which can only be establishable *a posteriori*.

Now, how can guarantees be obtained, that a comprehensive view of the scientific importance of the appointee has indeed been the decisive factor? By motivating the recommendation with an analysis of the scientific work of all eligible persons, carried out by qualified committee, under the obligation of publication in a scientific journal that is read by the international colleagues, to whom the committee thus becomes answerable. [...]

The editor in chief, Brouwer's friend Wiessing, had not taken the risk of promoting a discussion that would fall flat, so he had sent letters to all the members of the Academy, requesting their views on the appointment issue.[7] He put it to them that 'Where it is for fellow countrymen of an importance which hardly could be overrated, that the appointment of the scientific leaders in the country is made by means of a system, which gives the greatest possible certainty that the most competent will be appointed, it seemed to us the responsibility of our weekly, not to be content with the publication of professor Brouwer, but to ask the opinion of his confrères of the Royal Academy'.

The next issue brought two positive reactions. One of them, Ariëns Kappers,[8] gave his unqualified support to Brouwer's proposal, pointing out that 'students not only learn to practice a branch of science, but chiefly to further this branch of science, that they learn to work at the frontier of the field, and that their teacher, encourages them and leads them in this respect.'

The other writer observed that the existing old-boy network was a serious obstacle for the proper functioning of the universities, and that thus Brouwer's proposal was a painful infringement on the academic etiquette.

De Groene of 25 January 1919 contained a selection of the replies to Brouwer's challenging proposal. Only five out of more than forty respondents were supporting Brouwer; the objections mainly concerned the public features of the proposed proceedings. The reactions ranged from 'good idea, but it won't work' to 'nonsense'. Some of the respondents used the occasion to point an accusing finger at

---
[7] Wiessing to members KNAW 21 October 1918.
[8] A former fellow member of CLIO, cf. p. 14.

the City Council of Amsterdam, which indeed handled appointments of professors in a manner that differed from the procedures of the state universities. There had been occasions where the Council had not hesitated to ignore the proposals made by the professionals, be it by changing the order of recommendation, or by an appointment that had little to do with the advice of faculties or trustees.

A particularly negative reaction to Brouwer's proposals read:

> The preparation of professor's appointments, as desired by Prof. Brouwer, will, in my opinion, result in a big fuss. With that item we are already well-stocked.

Zeeman observed, that under Brouwer's proposal, gifted young candidates with few publications but with a bright perspective would not stand a chance. It is in this context worth noting that Brouwer himself was in such a position at the time that Korteweg tried to secure a position as a lecturer for him.

Brouwer replied in *De Groene* of 1 February, refuting the objections one by one. He referred, in reaction to the claim that 'In our country there is hardly a case known that a scientist of more than standard importance, not has been appreciated', to the embarrassing history of Stieltjes:

> T.J. Stieltjes, undoubtedly the first Dutch mathematician of the nineteenth century and perhaps of all times, was here regularly passed over at the occasion of professor-vacancies and had as a consequence to work under extraordinarily difficult circumstances, until France took pity on him and offered him a chair at Toulouse, where he died young. Had there been a system of appointments, as proposed by me, Stieltjes would have been saved for his fatherland and been spared many difficulties; not improbably he would have lived longer, and accomplished more. This gain, colleague..., would not have been bought too dearly with a little more fuss.

Brouwer also observed in his reaction that the 'Groningen system' of asking the advice of sister faculties in the case of vacancies, was put forward as an argument that the old system worked perfectly well. He did not see the weight of this remark, as few faculties followed the Groningen example. Moreover, he added, 'I know a faculty, in which a member, who brought up the desirability of the Groningen system, had little success, and got the remark "What is the use, everybody only recommends his own students" '.

And here the matter rested. Nothing was done and professor's appointments remained largely the concern of small groups in the local faculties.

8.2 The return to topology

Although Brouwer had spent the war years mainly investigating foundational matters, and had more or less given up topology, topology had not given up Brouwer, and, the war being over, it was natural that mathematicians should turn to him for criticism and advice.

His own postwar activities in topology were largely 'unfinished business' from the first topological period. Apart from a number of corrections or additions to earlier papers, Brouwer published an elementary proof that a function, continuous on a closed subset of $\mathbb{R}_n$ can be extended to the whole space. This proof had the characteristic features of Brouwer's approach; whereas De la Vallée Poussin and Harald Bohr had proved the same result by means of series or integrals, Brouwer's proof contained just the bare essentials. The theorem had, however, already been proved by roughly the same methods by Tietze in 1914; it is now known as the Tietze-extension theorem. Brouwer had missed the paper, which, by a curious coincidence, was published in the same journal that published Brouwer's last pre-war topology paper! Brouwer acknowledged Tietze's priority in a note a year later.[9]

Another paper that appeared in the *Mathematische Annalen* dealt with the topological aspects of the Lebesgue measure; it showed that measurable sets in the plane lacked topological invariance.[10] By means of an intricate construction Brouwer gave examples of closed, nowhere-dense subsets of the unit square that could be topologically transformed into sets with any measure between 0 and 1. Similar results had already been established for the real line, for example, by Bohl and Carathéodory, and by Brouwer himself. He had communicated this fact to Blumenthal in a letter in 1913. The style and technique of the measure theory is that of the Schoenflies *Bericht* from 1913, which is not surprising as Brouwer was in a certain way the godfather of that book.

Brouwer's main efforts in the postwar-period were directed towards the topology of surfaces. The first postwar paper in the series, 'On one-one continuous transformation of surfaces in themselves'(sixth communication),[11] presented to the Academy on 30 November 1918, seemed the continuation of the old series, but it was in fact wholly different in style and method. The paper was presented as an elaboration of a footnote in the letter to Fricke of 1912.[12] It offered a purely topological proof, and an extension, of an analytical theorem of A. Hurwitz[13] on birational transformations of Riemann surfaces of genus 1. The letter to Fricke was part of the Koebe affair. In the above paper Brouwer used the opportunity to set the record on the matter straight—in particular the incident of the faked note, cf. p. 189.

The topology of 1918 was no longer the same discipline it had been before the war. The germs of the early years had developed into an independent subject, the significance of which for mathematics at large was fully realized. When mathematics resumed its normal life after the war, there were a good number of specialists in the subject; without pretending completeness, we mention M. Dehn, P. Heegaard, H. Weyl, F. Hausdorff, A. Schoenflies, M. Fréchet, W.J. Alexander,

---

[9] [Brouwer 1918C, 1919E], [Tietze 1914].
[10] [Brouwer 1918D].
[11] [Brouwer 1919L2].
[12] [Brouwer 1912D].
[13] [Brouwer 1919L2], [Hurwitz 1892].

H. Tietze, G.D. Birkhoff, W. Dyck, A. Denjoy, O. Veblen.

More and more mathematicians realized the importance of the subject, and incorporated topological methods and terminology, in their papers and books. At the beginning of the twenties there were already some influential books and monographs dealing with the young discipline: A. Schoenflies, *Bericht über die Mengenlehre I* (1900, 1913), II (1908), M. Dehn and P. Heegaard, *Analysis Situs, Enz. d. math. Wiss. III* (1907), F. Hausdorff, *Gründzüge der Mengenlehre* (1914), H. Weyl, *Die Idee der Riemannschen Flächen* (1913), O. Veblen, *Analysis Situs* (1916), W.H. Young and G.C. Young, *The theory of Sets of Points* (1906).

The landscape of topology after the First World War became, thanks to the seminal work of pioneers, such as Poincaré and Brouwer, a bustling scene. Brouwer's new methods were catching on, although it took some time before they were fully understood. A new generation of young topologists joined the field, and among the first to turn to Brouwer was a young Hungarian, Bela von Kerékjártó, who had already given a talk in Göttingen. Kerékjártó had submitted a paper to the *Mathematische Annalen*, and it was handled by Brouwer as the editor mainly responsible for topology. Probably the topic stirred some of the old enthusiasm in Brouwer—he published in the years 1919–1921 some 14 papers on the topology of surfaces. The first one, in the *Mathematische Annalen*, immediately followed Kerékjártó's paper.[14] It opened with the words:

> 'The results of the preceding paper of Mr. v. Kerékjártó were known to me for several years; I told Mr. Bernstein in 1911, among other things, the following two theorems:
>
>   1. Every periodic one–one and continuous transformation of the sphere with invariant indicatrix is topologically equivalent to a Euclidean rotation.
>
>   2. Every involutary [i.e. periodic] one–one and continuous transformation of the sphere with inverting indicatrix is topologically equivalent to a euclidean reflection in the centre, or in a plane through the centre.

Brouwer pointed out that his old proof was not simpler than Kerékjártó's, but that it was easier to generalize. He listed a number of generalizations in the paper, which combines the techniques of Riemann surfaces and combinatorial topology in a manner of quiet routine, but it does not introduce new unexpected insights. The topic was pursued further in a paper presented to the Academy in Amsterdam.[15] In a following series of papers Brouwer carried out the classification (enumeration) of continuous or topological mappings of various surfaces, for example the torus and the projective plane.

Brouwer was sufficiently taken with Kerékjártó to try to find him a position in Holland. In 1921 he asked Ehrenfest if there was a possibility to get him a temporary job in Leiden; the answer, regrettably, was negative.

---

[14] [Brouwer 1919S], [Hurwitz 1892].
[15] Submitted 29 March 1919. [Brouwer 1919N2].

That Brouwer was not blind for the mathematical shortcomings of the young man appears from the covering letter of a recommendation for Kerékjártó, written to F. Riesz.[16] He pointed out that Kerékjártó was '(both in human and scientific respect) restless and rash', and that he was not certain that he would ever 'conquer these traits sufficiently to become a dependable researcher and a useful teacher'. Being the responsible editor for most topological contributions to the *Annalen*, Brouwer had seen ample evidence of Kerékjártó's performance as a researcher: 'Of all his submissions so far, only a small part was, after thorough revision, fit for publication: the greater part had to be rejected without any hope of redemption'.

Brouwer's interest in topology got a second impulse from the young Danish mathematician Jakob Nielsen, who had learned his topology from Dehn in Kiel. After serving in the war in Belgium and Turkey (in the German army, he became a Danish subject only after the war) Nielsen visited Göttingen in the summer term of 1919. There he submitted a paper to Klein for the *Mathematische Annalen*, and Klein duly asked Brouwer to have a look at it and sent him the manuscript.[17]

Brouwer answered a month later, that the contents were valuable, but that the paper had to be straightened out both in presentation and precision. Nielsen soon contacted Brouwer, and a correspondence ensued.

Apparently Brouwer did a thorough job analysing the paper, for Nielsen replied that, already in 1912, he was well aware of the weaknesses in the dissertation, but that under the pressure of the time schedule he could not afford to elaborate the matter.[18] Brouwer was truly impressed by Nielsen, and he did not hesitate to express his appreciation.

Jacob Nielsen. Courtesy Vagn Lundsgard Hansen.

The short episode of the Nielsen paper is instructive because it illustrates how Brouwer handled papers for the *Mathematische Annalen*. On the twenty-first of October he reported to Klein that he had started to check the paper, adding that he

---

[16] Brouwer to Riesz 22 November 1921.
[17] Klein to Brouwer 12 September 1919.
[18] Nielsen to Brouwer 18 October 1919.

was certain that correct and direct proofs would emerge, provided he could be sure that the author would not be allowed to discuss the printing of the paper simultaneously with the managing editor. This seems to suggest that editors and authors were sometimes confronted with conflicting instructions; Klein, it appears, was sometimes easier to convince than his editorial staff. One can easily imagine the possibilities for conflicts!

Brouwer added that

> Only because Carathéodory strictly maintained this principle with respect to Kerékjártó, I have been able to get something good out of the young Hungarian and only because Blumenthal was too lenient with Juel, a lot of confused nonsense could be published.[19]

Brouwer did not leave it there. He even went so far as to say that he would welcome a guarantee that a paper, sent to him to handle, could only be accepted by him—especially in this case where the authors had contacts with Göttingen. Klein's answer (if there was any) is not known. For Nielsen, Brouwer's editorial activity did work out all right; on 9 November the paper was accepted.[20] The paper dealt with fixed points of topological transformations of the torus. Nielsen's work on fixed points matured in 1927, he established a lower bound for the number fixed points of continuous maps for certain surfaces. His work has been generalized to wide classes of topological spaces, and the minimal number of fixed points has become known as the *Nielsen number*.

Nielsen's work had re-awakened Brouwer's interest in the topic. He wrote a paper, that immediately followed after Nielsen's paper in the *Mathematische Annalen*,[21] extending the results to continuous mappings of the torus and the Klein bottle. Both authors showed that the minimal number of fixed points is determined by the homotopy class.

The last paper of the postwar series of Brouwer's papers on the topology of surfaces dealt with the characterization of continuous mappings of finitely connected surfaces.[22] In March Brouwer presented a paper on transformations of the projective plane to the Royal Academy; this paper also dealt with the minimal number of fixed points.[23]

Although Brouwer had gradually turned his mind almost exclusively towards intuitionism, topology was to re-enter his life dramatically in 1923, when suddenly a young topological genius, Paul Urysohn, emerged and continued the research where Brouwer had left off. This story will have a prominent place in volume 2.

---

[19] Brouwer to Klein 21 October 1919.
[20] [Nielsen 1920].
[21] Nielsen's paper was submitted 15 January 1920 and Brouwer's 20 January 1920.
[22] [Brouwer 1921D].
[23] [Brouwer 1920B2].

8.3 Brouwer, Schouten and the *Mathematische Annalen*

A man like Brouwer, with his sometimes overly strict views on quality, correctness and presentation, would be a jewel on any editorial board, and the *Mathematische Annalen* definitely made a wise choice by enlisting his services as an expert in the field of topology and related areas. We have already seen that Brouwer took his duties extremely seriously, to the extent that Korteweg had to speak words of warning. It is beyond doubt that certain authors profited from Brouwer's careful judgements, for example Nielsen and Kerékjártó. There is equally little doubt that many an author was bitterly disappointed by Brouwer's stern attitude. In the course of his editorial career at the *Mathematische Annalen* Brouwer ran into a few cases where the author did not wish to acquiesce in the rejection of his paper. Given Brouwer's uncompromising views, this sometimes led to unpleasant exchanges, which even disturbed the peace of mind of the chief authority of the *Mathematische Annalen*, Felix Klein.

One such incident took place in 1919. It announced itself innocently enough in a letter from Klein to Brouwer.[24] Carathéodory, one of the editors of the *Mathematische Annalen*, had written to Klein from The Hague, that Brouwer had doubts concerning a manuscript of J.A. Schouten. 'Fortunately', replied Klein, 'I have not undertaken anything definite in this respect.' He had gone no further than a innocuous private remark to Schouten, to the effect that in his opinion, papers connected with those of Einstein and Lorentz, would find their proper place in the *Mathematische Annalen*.

The author to whom Carathéodory refererred in his letter, was a Dutchman, two years younger than Brouwer, who had studied electrical engineering in Delft, at the Institute of Technology (*Technische Hogeschool*). In 1914, after completing his dissertation, he was appointed a professor in Delft. Schouten soon became world famous for his contributions to differential geometry. His name is forever connected with the Tensor Calculus, a discipline that furnished geometry with a formalism of great intricacy, and appealing to those with a taste for ingenious calculations. At the end of the First World War, Schouten was well established, a number of his papers had been published in the *Mathematische Annalen* and the *Mathematische Zeitschrift*.

Brouwer had handled the above mentioned manuscript for the *Mathematische Annalen* and, after consulting E. Study, a recognized authority in (among other things) differential geometry, rejected it. Study, a man with a reputation for caustic wit, did not mince words. He summarily judged that 'from an objective discussion with such a confused head, I do not expect any benefit for him'.[25] This paper, which consisted of an application of Schouten's 'direct analysis' to the theory of relativity, did not qualify for the *Mathematische Annalen*, said Brouwer, 'in the first place because the author does not understand the art of presentation, and in the second place (what is more important) because his achievements consist, briefly said, of the

---

[24] Klein to Brouwer 8 September 1919.
[25] *Von einer sachlichen Auseinandersetzung mit einem so unklarer Kopf verspreche ich mir keinen Gewinn für ihn.*

veiling of the results, found earlier by inventive authors, in a new (but thick and opague) robe'.²⁶ But, he added, 'What Mr. Schouten is lacking, is not talent, but erudition and moderation, so that I don't exclude in any way the possibility that he will make a good mathematical author in the future'.

Weitzenböck, the Austrian expert in differential geometry and the theory of invariants, when asked by Brouwer for his opinion, was not impressed by Schouten's formalistic virtues; he had referred to the book 'Foundations of Vector- and Affinor Analysis' as 'the terrible book that he has committed'.²⁷ It cannot be said that Brouwer had acted out of personal animosity or subjective motives, for he had himself introduced Schouten in the *Mathematische Annalen*:

> As a matter of fact I am more or less guilty of prematurely drawing the attention of the editors of the *Annalen* to Schouten by recommending to Blumenthal, in the summer of 1913, the paper *On the classification of associative number-systems* for publication (since then published in vol. 7b). This was done after a prior inspection of the prominent point of its contents, that is the 'Principle of the continuing self-isomorphism' with respect to its novelty. For the value of the paper stands or falls with this novelty (which I could not judge myself). As I believe, Blumenthal subsequently sent the manuscript to Hölder, who definitely accepted the paper, and only later it turned out that the above principle had already been presented by Cartan in a much more perspicuous way.

The problem of the rejected paper dragged on for many years. Apparently, the matter went back to 1917; there is a confused correspondence about a manuscript, submitted in the summer of 1917. The issue is not quite clear, it appears that Schouten had sent Brouwer his manuscript bound in two parts, and that at some later time he had, for whatever purpose, asked Brouwer to return it temporarily to him. At a later time he had again sent it back to Brouwer, this time bound as one volume. Brouwer thereupon refused to accept the manuscript in its new form; in those days the making of a simple copy was nothing less than the complete retyping or the copying by hand of the original, and without such a copy it could not be ascertained in how far two versions were identical. Although one may be inclined to think Brouwer overly fussy in the matter, one must keep in mind that he had a long and bitter experience in the publishing business. Both the Lebesgue and the Koebe affair had taught him that in matters of claims to results, priorities, etc., one had to exercise extreme care. He certainly did not want to relax his standards in his role of editor of the *Mathematische Annalen*. It cannot be said that he did or did not suspect attempts at doctoring the manuscript; he merely took the position that the identity of the manuscripts that he sent and received could not be established beyond doubt.

Schouten on his part failed to see why a manuscript, after a bit of rebinding,

---

[26] Brouwer to Klein 19 September 1919.

[27] "*das schreckliche Buch das er verbrochen hat*". *Grundlagen der Vektor- und Affinoren Analysis* [Schouten 1914].

should be refused by Brouwer.[28] Schouten firmly denied any obligations on his part regarding the form of the manuscript:

> 'In answer to your letter of the twentieth of this month, I inform you that the promise, made in my letter of 20 November 1919, of which a copy is enclosed, does not admit any other interpretation than the putting at your disposal of one of the manuscripts in the state in which this manuscript had been submitted to you in the summer of 1917. No promise at all has been made by me concerning the state of the *volume* and the *manner of binding*.'

Two years before Brouwer had already tried to sever his contacts with Schouten in a biting letter.[29]

> Dear Sir, I have informed you some time ago of my opinion, that the mentalities of you and of me do not lend themselves for mutual understanding. At that occasion I asked just for a message whether you wanted, on legal grounds, to have your duplicate manuscript back from my archive.
>
> Your letter, which I received thereupon, would have been opened in this Christmas holiday, were it not that in the meantime I heard from my friend Ornstein (with whom you already at an earlier occasion have tried to find out, as I assume in accord with your morals, but according to mine in a most improper manner, my more intimate feelings) that you had again approached him in this matter, 'in order to avoid bickering'(!). The opening of your letter has thereupon not taken place, and neither will the missive, received today, be read by me. As I have ascertained in the meantime that my position with respect to the keeping of your duplicate manuscript is legally not strong, I will at my own expense have a copy made, and then will send back the copy that belongs to you. And now I request you urgently, to leave me alone. In my capacity of member of the Academy, member of the Board of the Mathematical Society and editor of the *Annalen*, I have always considered myself obliged to reserve a large part of my time on behalf of coming young mathematicians, and you have so far profited to a considerable extent. In exchange I do not require expressions of gratitude or excuses for the trouble (although words to this effect were never totally omitted by others, if they took up my time in the same way you did), but indeed the strictest respecting of this procedure, which I consider as the right one in fulfilling this sacrificing task. And after your deficiency in this respect—also after the given hint—you have

---

[28] Schouten to Brouwer 25 March 1920, erroneously dated 1919. In Schouten's letter of 19 April 1929 this is repeated, but probably on the grounds of the dating of Schouten's carbon copy. In the letter it is mentioned that Brouwer had no right to the manuscript, so it is not unlikely that Schouten had withdrawn the paper. The question then remains why Brouwer should wish to see the manuscript, as stated in the letter.

[29] Brouwer to Schouten 9 January 1918.

automatically brought an end to the availability of my time on your behalf (including the reading of your, for me incomprehensible, letters).

The above is a vivid illustration of both Brouwer's handling of manuscripts for the *Mathematische Annalen*, and for the uncompromising stand on matters which he deemed essential in mathematical affairs. He had an almost exaggerated sense of the scientific responsibilities of editors; in particular he found it his duty to keep a precise record of the various versions of manuscripts. To that end he had copies made of most papers, a job mostly performed by Cor Jongejan. The discussion with Schouten remains vague on account of the incomplete correspondence, but, probably Schouten had in one way or another offended Brouwer's standard of propriety and this, added to Brouwer's aversion to cumbersome formalisms, had wakened in him the almost puritanical ferocity against the obnoxious (but probably unaware) sinner.

It is likely that Schouten did not accept the rejection of his paper without an appeal to the higher authority of Klein, and so the matter dragged on for a while. In August 1920, Brouwer again turned to Klein to complain about the time consuming verification of a paper of Schouten (possibly a version of the old one). The letter gives the impression of an editor who is slowly reaching his boiling point. Brouwer stated that the evaluation of Schouten's paper consisted in effect in the stripping away of all the tiresome and worthless symbolism and 'to trace among the great mass of trivialities, the few essential theorems, and finally to find out at which places, not cited by the author, those theorems, in so far as they are correct, have appeared earlier in the literature.' Since a thorough checking of the manuscript would take more of Brouwer's time than he found justifiable (after, as he said, protecting the *Annalen* a couple of times for the affront of the publication of a paper of Schouten) he declined to take any more responsibility for Schouten's products.

Indeed, the Schouten paper was no longer mentioned in the correspondence between Klein and Brouwer. Klein, who wavered between the hope that something valuable might after all be hidden in Schouten's paper and his abhorrence for the 'unbearable way' in which the subject matter was put in a particular symbolism, had asked Hermann Weyl to give his opinion on the Schouten paper.[30] Klein suspected a personal feud between the gentlemen, who accused each other of lust for power, respectively absence of any accomplishment. Weyl answered a month later that he would be happy to give an opinion, but that he was not quite without prejudice:

> Brouwer is a person I love with all my heart. I have now visited him in his home in Holland, and the simple, beautiful, pure life in which I took part there for a few days, completely and totally confirmed the impression I had made of him. He is certainly not domineering; the violence of his behaviour towards Koebe as well ,at present, towards Schouten, I believe, is mainly based on the circumstance that his being revolts against the impurity that it discerns here. [...]

---

[30] Klein to Weyl 6 October 1920.

> I don't know enough about the facts to play the judge here; but I think that Brouwer's reproach is not wholly without grounds. I learned to know Schouten first from his paper on hyper complex numbers in the Mathematische Annalen, and I was enthusiastic. However, at that time I did not know the subject at all; afterwards I saw that almost everything that I, on the basis of his presentation, thought to be Schouten's property, was already known for a long time, and that he added only a few small items, that partly made it worse instead of better,[31] and partly even contained mistakes. His big paper on the *Analysis zur Relativitätstheorie*, contains, I believe, nothing new. (I must admit that I could not work my way through the symbolism of Schouten). The way in which he mentions there the notion of parallel shift (mentioning Levi-Civita only casually, whose paper of 1917 he claims to have read only after finishing the manuscript, although he gets the Rendicotti sent from Palermo, and this paper is made use of in other papers which he quotes) does not quite make a reasonable impression.

Weyl, on the whole shared Brouwer's negative opinion of Schouten's symbolic practice, and his general judgement of the relativity paper was just as negative as Brouwer's. He remarked, however, that he had seen in the summer two manuscripts of Schouten, one without interest on Riemannian geometry, and one with a beautiful result, 'a real accomplishment'.

Klein must have felt uncomfortable in his position between Brouwer and Schouten; he would have appreciated a friendly solution to the conflict. In this vein he contacted Weyl, who conveyed the idea to Brouwer. But the latter was not to be placated:

> It is not clear to me what Klein means by a 'reconciliation' between Schouten and me. For Klein there can only exist the relation between Schouten and me, that I have rejected a manuscript of Schouten that I had received for refereeing; it is known to Klein that this has been done on the ground of its plagiarizing character, and not on personal grounds.
>
> Indeed, I have of late no longer taken notice of Schouten's publications; but as nothing would please me more than to revise an unfavourable opinion on an author, as soon there is an occasion to do so, I would like to ask you to indicate to me the place in the journal of the 'positive achievement' you mentioned, [...]; should I find there a new (and not somehow copied, for example, from Cartan, Bull. Soc. Math. 45, 57–81) theorem on the mapping problem concerned, I will be glad to see and acknowledge such.[32]

The reader may well wonder why this conflict has been given so much attention; one reason is that it illustrates the absolute refusal of Brouwer to water down his

---

[31] *waren Verballhornungen*.
[32] Brouwer to Weyl, 1 January 1921.

mathematical standards, another reason is that the relation between Schouten and Brouwer remained hostile for a prolonged period. Only in 1929 both parties decided to make peace; it took the mediation of Weitzenböck and Van der Woude to bring the parties together. On 9 April 1929 Schouten wrote a conciliatory letter to Brouwer, in which he accepted the responsibility for the conflict concerning the above mentioned manuscript and its bindings in Brouwer's files, and on 24 April Brouwer informed his mathematical colleagues that he no longer felt a barrier between himself and Schouten.[33] The enmities were, however, to be resumed—fiercer than ever—after 1945.

8.4 The offers from Göttingen and Berlin

The above mentioned visit of Weyl to Brouwer was not a mere coincidence. Brouwer was playing a subtle game; the post-war activities had brought certain changes at the German mathematics department. In Göttingen a vacancy had arisen through the departure of Hecke, who had moved to Hamburg, and in October 1919 the faculty had offered Brouwer a chair.[34] Without exaggeration this could be called the apex of Brouwer's career: at the age of 39 he was offered a tenure position at the world's topmost mathematics department. The offer, by the way, once more underlined the absolute harmony between Brouwer and Hilbert. Where Klein, might have had his doubts as to the desirability of the gifted, but obnoxious hothead, Hilbert was not plagued by any such doubts. We may safely assume that he was only vaguely aware of Brouwer's foundational program. Apart from the translation of his Inaugural address and the Rome paper, only a few foundational remarks of Brouwer had appeared in English or German.[35] To Hilbert he was the topologist with a side interest in foundations. Brouwer's reaction to this offer of a chair in Göttingen has not been preserved; this is rather surprising, after all, the event certainly must be considered the ultimate recognition of his status. The absence of documents could possibly be explained by the fires that have destroyed part of Brouwer's private archive.

There were more parties interested in the Dutch topologist; it never rains but it pours: the philosophical faculty of the University at Berlin had also decided to make Brouwer an offer; on 19 December it proposed to the Minister of Science, Art and Education to fill the vacancy, left by Carathéodory's departure,[36] by Brouwer, Herglotz or Weyl (in this order). Brouwer was recommended warmly for his work in topology,

> His papers deal mainly with Analysis Situs, for which he has provided the firm foundation, long sought in vain. Among the big results of his research

[33] Brouwer to Jan de Vries, J.C. Kluyver, Hk. de Vries, Van der Woude, Weitzenböck, 24 April 1929.
[34] Akten der Phil. Fak. Az. II PH/36–d, Dozenten Generalia), 30 October 1919.
[35] German: [Brouwer 1908A, 1914, 1918B], English: [Brouwer1913C], Dutch: [Brouwer 1907, 1908B, 1908C, 1918B].
[36] Carathódory left Göttingen in 1918 for Berlin. In 1920 he accepted a call from Athens, and subsequently went to Smyrna in order to organize the founding of a mathematics department in the new university.

we only mention the proof of the invariance of dimension and the decomposition of space by means of closed surfaces, the theorem of invariance of domain, the theorem of the existence of fixed points of continuous deformations of the ball, and the far-reaching group theoretic researches on continuous transformations. These theorems, of which Brouwer had already made the most fertile applications for the proof of the basic uniformization theorems in function theory, promise also undreamed-of consequences for the modern theory of physical and astronomical differential equations. Brouwer is equalled in the originality of his methods by none of the mathematicians of the younger generation. Brouwer has a most extensive and deep knowledge at his disposal, not only of pure, but also of applied mathematics. He speaks German fluently, and has given excellent talks in the German language for academic societies. His political persuasion is on the whole sympathetic towards Germany (*Deutsch freundlich*).[37]

Given the choice between Berlin and Göttingen, the former may have appealed more to Brouwer than the latter, and indeed for the same reason that Amsterdam appealed to him. The cultural atmosphere of Berlin probably suited him better than that of the (after all) provincial town of Göttingen. Berlin and Göttingen both had a reputation for mathematical research and for top mathematicians, but Göttingen had Hilbert, and Hilbert was the universally accepted icon of mathematics. Hilbert's influence was due to a number of factors; his mathematics was superior not only in the sense of quality, but also of elegance, choice of topics and innovative character. As an organizer and teacher, he was second to no one, and a worthy master in the tradition set by to Klein. Mathematically speaking, Göttingen would thus have been the natural place for anybody offered the choice. Brouwer's apparent but undocumented preference for Berlin, therefore, had probably something to do with non-mathematical attractions.

Hilbert probably summed up the matter quite aptly in a letter to Weyl, in which he declared himself unable to see why Weyl should choose Berlin:

> Which I can quite well understand in the case of Brouwer and Landau— Brouwer only wanted temporarily to stay in Berlin, and getting to know Berlin, just as the nimbus of getting an offer in the capital, were his motives...[38]

As it happened, Brouwer was visiting Berlin late December 1919 in order to reestablish, on behalf of the newly established International Academy for Philosophy, the contacts with the people from the Forte Kreis. He was wished *bon voyage* by his friend Henri Borel, with the words: 'Please give Gutkind and Rang, a gem of a

---

[37] Bestand Philosophische Fakultät, Humboldt Universität, Nr.1468, Bl.313. The recommendation was signed by Carathéodory, Erhardt Schmidt, Planck, H.A. Schwarz, Nernst, Cohn. The qualification 'Deutsch freundlich' was taken quite seriously after the war. The German nation felt itself misunderstood and surrounded by hostile nations.

[38] Hilbert to Weyl 16 May 1920. ETH Hs.91: 606.

man, a firm handshake, and a respectful kiss from me to Mrs. Gutkind.'[39]

Brouwer's friends and his colleagues were really worried that the honour of a chair at one of the great Universities of Germany (and of the world), would be more than Brouwer could resist. Jacob Israël de Haan wrote to Van Eeden in January 1920, 'I hear that Brouwer is going to Göttingen, quite a blow.' De Haan was misinformed about the place, but not about the threat that was posed to Dutch mathematics, and, what was closer to his interests, to significs. Lorentz, too, showed his concern about Holland's possible loss of Brouwer. He wrote Brouwer that he hoped that Brouwer would stay in Holland.[40]

Brouwer was indeed using this trip to Berlin with the intention to engage in some bargaining. He was offered enough incentive to start a new life in Berlin, but he would perfectly be willing to remain in Amsterdam, provided the City Fathers could make him a similar offer. He did visit the Gutkinds, as intended, we will return to that part of the visit in a later section.

In a flurry of negotiations Brouwer had to hurry back to Amsterdam to attend the faculty meeting on the twenty-first. Five days later Brouwer had a personal interview with the Mayor of Amsterdam, Mr. J.W.C. Tellegen. The results must have been satisfactory, for Brouwer felt that he had a reasonable prospect of obtaining a suitable compensation from the City of Amsterdam in exchange for turning down the Berlin offer. He left for Berlin almost immediately, much to the chagrin of his friend Frederik van Eeden, who wrote in his diary 'I heard that Brouwer was in the country and left again for Berlin. He had wounded himself with some glass, when he kicked in a window in his house, because it was closed. And that without telling me. He loves me after all in another way than I love him.'[41]

In Berlin considerable pressure was put on Brouwer to get him to join the faculty. Von Mises, writing from Dresden about mathematical business, expressed his hope that Brouwer would accept, or maybe would already have accepted, the Berlin offer.[42] It was generally seen as a matter of considerable prestige to get Brouwer attached to the university.

The Board of Curators of the University met on 2 February and discussed the threatening loss of Brouwer for Amsterdam. As the Mayor was by City rule the President of the Board of Curators, he could give the information to his fellow curators 'straight from the horse's mouth'. He listed the conditions that Brouwer had formulated, and that were submitted by the faculty: an extra credit of DFl 10.000 for books and journals for the mathematical seminary, two lecturers and a raise of Brouwer's salary to the legal maximum. The Curators decided to ask the City Council for one lecturer and to consult the faculty on the two other items.

---

[39] Borel to Brouwer 20 December 1919.

[40] Lorentz to Brouwer 25 January 1920.

[41] Van Eeden's diary 27 January 1920. Brouwer was indeed short tempered where such obstructions as closed doors or the like were concerned. Brouwer used to drop in occasionally to see a journalist of the *Handelsblad*, living in Blaricum, with whom he was on good terms. Once, when he did not find the man in, Brouwer got so angry that he kicked in a window (oral communication Mr. Crèvecoeur).

[42] Von Mises to Brouwer 28 January 1920.

Brouwer, knowing the date of the meeting of the Curators, lost no time to inquire after the results of the meeting. He wrote from Berlin to the Mayor[43] that he had been received in such a friendly way and with such offers and concessions, that, should he have to disappoint the University of Berlin, he would insist on breaking the news in person, rather than to resort to a written statement. The letter also contained a business proposal with respect to the request for a mathematics library. Would it not be possible, he asked, to convince the Board of the City to provide him already now with the money he had asked for, so that he could get the books and journals for a much better price than when ordered from Holland? Even though the gigantic sum of DFl 10.000 would buy the university a nice collection, he could buy 2 or 3 times the quantity of books at the spot by direct negotiations.[44] He planned to order them in person in Leipzig (where the publisher Teubner had its home base). We have already seen that Brouwer judged an independent mathematics library of the greatest importance, cf. p. 255. He wished to follow the Göttingen example of a library where mathematics staff and students could have direct access to books and journals, without the involvement of an extra organization that had to get the books out of the storage rooms, to register the books, etc. In short, he wanted a collection that was optimally geared to the needs of the researcher and the student. In his letter to the Mayor he called this collection the '*handbibliotheek*' (reference library).

The Mayor answered a week later, he was happy to inform Brouwer that things had gone well for him.[45] The City Council had shown no objections to grant Brouwer's wishes, provided Brouwer would stay in Amsterdam. There were, however, a few minor details to be settled. In the first place the Mayor could not just put the promised DFl 10.000 at Brouwer's disposal. At the very least Brouwer would have to confirm his commitment to stay in Amsterdam, in the second place the transfer of such sums to a foreign country required special caution. These matters could possibly be settled later. As to the lecturers' positions, the Mayor had hinted at the meeting of the City Council that some money might be saved by offering the lecturers' positions in combination with a part-time teaching position at one of the municipal gymnasiums or high schools.

A week later Brouwer was back in Holland; he had definitely chosen to stay in Amsterdam. The German colleagues still had high hopes of winning over Brouwer, as a letter of Schoenflies showed. He had already asked Brouwer if it was true that he had accepted the chair in Berlin, but Brouwer answered him that he would stay in Amsterdam—he had received a really attractive offer.[46] The rumours about Brouwer's Berlin chair circulated nonetheless through Germany. The *Jahresberichte* reported that Brouwer had accepted the chair in its 1919 volume: *Professor Dr. L.E.J. Brouwer an der Universität Amsterdam wurde zum o. Professor der Mathematik an der*

[43] Brouwer to Tellegen 4 February 1920.
[44] Although the galloping inflation was still two years away, the German *Reichsmark* was not as strong as it used to be, and certainly weaker than the guilder.
[45] Mayor to Brouwer 12 February 1920.
[46] Schoenflies to Brouwer 16 February 1920, Brouwer to Schoenflies 22 February 1920.

Universität Berlin ernannt. (JBDMV, 1919. p. 61 Italics). Van Eeden could set his mind at ease; the signific enterprise would not founder on account of Brouwer's departure. He wrote in his diary that Brouwer had dropped in to see him on the nineteenth, and that on his return home Brouwer had found two laundry baskets full of mail. Brouwer settled with relief in his beloved Het Gooi. Van Eeden visited him on the twenty-fourth at home and joined the Brouwers three days later for Brouwer's birthday party. Brouwer was approaching forty and had reached the top position in Dutch mathematics, the maximum salary, and the largest department in the country. His demands had been fulfilled, and he could now build himself a department comparable to that of Göttingen.

The official aspect of Brouwer's agreement was efficiently taken care of. The Curators were told by De Vries that two lecturers were really necessary, and so the faculty obtained the promise of the appointment of two extra lecturers. This eventually proved, as we will see below, the cause of some embarrassment.

The lecturer positions remained vacant for the time being. Brouwer had, as we will see, made new plans.

The City Council held its part of the bargain and voted Brouwer the maximum salary, DFl 10.000. The official decision was conveyed to Brouwer on 27 April.

8.5 The Academy—how Denjoy was elected

In 1919 new members of the KNAW were to be elected, and some lobbying had already been going on. Already in January, Hk. de Vries had asked Korteweg to support the candidacy of his former Ph.D. student F. Schuh, and to write the required letter of recommendation. De Vries was doing so on behalf of a number of fellow Academy members, including Brouwer. Frederik Schuh was six years Brouwer's senior. He had defended a dissertation on enumerative geometry one year before Brouwer's doctorate. After being accepted as a 'privaat docent' in Groningen, his career took him first to Delft (1907–1909), then to Groningen (1909–1916) and finally back to Delft (1916–1945). In 1906 he had been second on the list for a chair in mathematics in Amsterdam, when the vacancy left by Van Pesch was filled by Hk. de Vries. Schuh was a mathematician with a traditional expertise in geometry, analysis, number theory, statistics, etc. His publications dealt with almost all topics under the sun, but they did not earn him a particular niche in the Pantheon of mathematics. In Holland he was renown for his efforts to raise the standards of secondary mathematical education. Whole generations of mathematics teachers were raised on books of Schuh In addition he was a good expositor, who exercised his talents also in the more mundane papers and magazines. When still a third year student, the University of Amsterdam had crowned his prize essay in mathematics. By 1920, Schuh had published a considerable amount of papers on various subjects, for a large part in the area of applied mathematics, but also on the theory of algebraic curves. Korteweg did point out that 'Not all the papers on the list are of course of the same value. A smaller number has been written for mainly pedagogical reasons. Yet, also in these, the astuteness of the author, who has carried out a useful task for our future mathematicians with these papers, is to be praised'.[47]

In 1919, apparently, the general opinion was that Schuh would eventually make his mark (and, to be fair, compared to some of the incumbent members of the academy, he was quite an acceptable candidate) and so the lobbying for Schuh started. The candidacy of Schuh was supported by Kluyver, Kapteyn, Jan de Vries, Brouwer and Hk. de Vries. Some time after the start of this campaign, Brouwer must have felt that the Academy deserved a better candidate, and he proposed Denjoy. The recommendation of Denjoy sketches in flowing words his merits in 'almost all areas of function theory and set theory', but in particular his invention of the notion of 'totalization', which closed the chapter on the inversion of differentiability. This document was signed by Kluyver, Cardinaal, Hk. de Vries, Jan de Vries, Korteweg, Kapteyn, Lorentz and Brouwer, so it had a solid support. In the May meeting of the section for the physical sciences, neither Denjoy nor Schuh got enough votes to get elected. Six rounds of voting did not yield a positive outcome for either of them, Schuh being far ahead of Denjoy all the time.

In 1919 a leading foreign mathematician got elected: Brouwer had, with the support of Hk. de Vries and J. de Vries, put forward David Hilbert's candidacy. The recommendation took only half a page. For a man like Hilbert this was more than enough! Brouwer simply stated that

> David Hilbert, professor at the University of Göttingen, who has added many mathematical theories, of which some did not exist at all before, others had been in a chaotic situation during longer periods, as monuments of crystalline simplicity to the spiritual property of humanity. These theories are in particular concerned with invariants, algebraic number fields, integral equations, the principle of Dirichlet, the axiomatizing of geometry, arithmetic, and physics.
>
> It is not surprising that the Bolyai prize, which is awarded every five years to the scholar whose total oeuvre can be considered to have exerted the greatest influence on the general development of mathematics, and which was for the first time awarded in 1905 to Poincaré, has been presented for the second time in 1910 to Hilbert.

Of course, Hilbert was elected without any hesitation.In May 1919 he was appointed as a foreign member of the KNAW.

The Secretary of the Academy, Van Bemmelen, worried by the number of candidates for the ordinary membership, and the possible unpleasant discussions, had proposed a voluntary reduction of the list for the next year. The proposal was adopted, and in consequence Brouwer dropped the Denjoy proposal in favour of Schuh when Korteweg called a meeting of the mathematical members and advocated a reduction in line with Van Bemmelen's proposal.

---

[47] In his second recommendation of 1920.

When, however, some members of the Academy revoked the agreement, Brouwer did not feel any longer obliged to support Schuh, and he again pushed the candidacy of Denjoy. Another private reason for reconsidering the matter was the information that Denjoy's announced departure for Strasbourg had become uncertain and if realized at all, would take place much later. Brouwer's *volte à face* probably irked some of his fellow academicians, for on 27 March 1920 he distributed a printed pamphlet among the members of the section mathematics and physics, which contained the above arguments. It closed with the lofty statement that the withdrawal of the Denjoy candidacy in favour of Schuh was, in his eyes, a disgraceful injustice, which his conscience allowed him to support only if, and as long as, a considerable effect could be expected for the cause of preventing the greater injustice of unreasonable elections of academy members. As this justification had disappeared, he felt no longer free to suppress Denjoy's name on the list, as he was a man 'whose genius and international fame could not be compared to Schuh's many and all-round merits.'

The candidacy of Schuh was this time supported by Korteweg, J. de Vries, Kapteyn, Cardinaal, Hk. de Vries, Brouwer and Kluyver, the one of Denjoy by the same members plus Lorentz. Brouwer again wrote the letter of recommendation, in which he praised Denjoy as 'one of the most gifted younger representatives of the mathematical school of Paris', who had 'created, as an application of Cantor's transfinite induction process, the operation of *totalization*, of which one can say, that it, with its capacity for inverting, without exception, any differentiation, has ended a period of development of the theory of real functions.' Korteweg was the spokesman for the Schuh-appointment; he vigorously supported him, but to no avail.

Brouwer's eloquent arguments, and (one would like to think) Denjoy's superiority won the day. The correspondence contains a cordial letter from Denjoy in which he expressed his gratitude for the appointment and for Brouwer's efforts, which he conjectured behind the operation.[48]

This short episode may serve to illustrate the influence of Brouwer among his colleagues—it is no minor feat to succeed in getting a candidate elected against a gentleman's agreement of the majority! As to the comparison of Schuh and Denjoy, one is now inclined to side with Brouwer, but at the time the choice was not all that clear to the members concerned. Whatever the justification was, the outcome was painful for Schuh, who never was admitted to the Dutch Olympus of scholars.

8.6 Negotiations with Hermann Weyl

After having turned down the offers of Göttingen and Berlin in exchange for far-reaching promises of the City of Amsterdam, Brouwer started to contemplate the use he should make of the present opportunity to build a strong mathematics department. His explicit goal was to found a mathematical institute in Amsterdam that could compete with Göttingen, and there is no doubt that the concessions

---

[48]Denjoy to Brouwer 29 April 1920.

made by the City Council were concrete signs of his successful bargaining. Later in life Brouwer used to refer to 'the City Council's promise of a second Göttingen'. In this form no evidence has been found in the archives, but there is little doubt that the promises made in 1920 embody that what Brouwer used to call 'his Göttingen'. It is more than likely that in the discussions with the Mayor, a man of consequence in his twin role of President of the Board of Curators and of Mayor of Amsterdam, terms of this sort were used. His first attempt to raise the level of the Amsterdam mathematics department, was to offer a chair to Hermann Weyl.

Brouwer and Weyl met no later than the summer of 1912, during Brouwer's visit to Göttingen. There is a postcard from Brouwer to his friend Piet Mulder (5 June 1912), signed by Lize, Weyl, Weitzenböck and Bernstein, which documents the contact to the two. This card also solves a minor riddle in Brouwer's files. The archive contains a little notebook with notes on the problem of the schoolgirls. This problem was formulated in 1850 by Kirkman. It runs as follows:*Fifteen young ladies in a school walk out three abreast for seven days in succession: it is required to arrange them daily, so that no two shall walk twice abreast.* The problem was, of course, generalized to arbitray numbers of 'girls and days'. The required lists of arrangements are called Kirkman systems.

Apparently Brouwer was interested in the problem in his student years. At least, there is a post card from P.H. Schoute of 1 December 1903 in which the literature on the problem is listed. Mulder had been working on the problem for his Ph.D. thesis,[Mulder 1917], and this must have reawakened Brouwer's interest. Brouwer wrote to Mulder that 'For a couple of days I am taking part in the local routine, which is quite pleasant for an outsider. I have explained the15 schoolgirls to the company here. See you soon.' This is the only instance where Brouwer actually tried his hand on a combinatorial topic. Since there are no recorded results, we may assume that he did not pursue the matter. Most mathematicians have their taste and preferences, Brouwer was no exception, he was too much of a geometer to get seriously involved in other areas.

Hermann Weyl. Niedersächsische Staats- und Universitätsbibliothek Göttingen.

Weyl was thoroughly familiar with Brouwer's topological work, there is an explicit reference to Brouwer's work in the 'Idea of the Riemann Surfaces' (1913).[49] In the introduction he generously acknowledged his debt to Brouwer's innovations in topology: 'Three events decisively influenced the composition of my book: the

---

[49] *Die Idee der Riemannschen Flächen.*

fundamental papers on topology of L.E.J. Brouwer, dating back to 1909, ...'.[50]

Also, at the time of the uniformization conflict, Felix Klein had asked him to write a critical evaluation of the Brouwer–Koebe controversy,[51] so he was thoroughly familiar with Brouwer's mathematics. In July of that same year Weyl reported in the Göttingen Mathematics Society on Brouwer's work on the invariance of dimension and of domain. Weyl was four years younger than Brouwer, he had studied in Munich and Göttingen, and he was highly regarded by his teachers and fellow students for his balanced personality and his truly great insight in mathematics and physics; he was equipped, moreover, with an exquisite sense of beauty. Both Hilbert and Klein were greatly impressed by the young man. One might say that he was, among the bright students drawn to Göttingen, Hilbert's favourite son. In 1913 Weyl was appointed in Zürich, where he was safe from the horrors and the inconveniences of the war. Already before the war he had won instant fame at the age of twenty eight with the above mentioned brilliant exposition of the theory of Riemann surfaces and its applications.

In 1918 he had further enhanced his status as a leading researcher and expositor, with two books: *The Continuum* and *Space, Time and Matter*.[52] The latter was an exposition of the theory of relativity, to which Weyl had significantly contributed, and the first one was the formulation and execution of a program for a version of constructive mathematics, to be precise, of predicative mathematics. Weyl had, like so many of his contemporaries, been worried by the paradoxes that threatened to paralyse mathematics. However, whereas most mathematicians considered the paradoxes, roughly, as a newspaper-reader considers an earthquake in a far away exotic country—an interesting phenomenon but too far away to worry about, Weyl felt that indeed everyday analysis (the theory of real functions) was jeopardized. We will return to Weyl's foundational work later in this chapter (cf. p. 317)

After the war Brouwer and Weyl met again in Switzerland; during the summer vacation Brouwer stayed in the Engadin, where Weyl visited him.[53]

With a little bit of luck the occasion could have been grander. Hilbert had visited Switzerland just before Brouwer, as it appears from a postcard of the latter to Hilbert, saying how sorry he was to have missed him. It would have been of decisive importance for the foundations of mathematics if the three of them could have sat down and discussed the new developments! Hilbert would have had the story of Brouwer's new intuitionism first hand, and it could have changed the course of the coming foundational conflict. These are history's little quirks.

For a cosmopolitan person like Brouwer it was a relief to be able to travel again, even though the circumstances were not wholly without risk. In Wiessing's autobiography there is a small passage referring to Brouwer's experience on the way

---

[50] The two other events were Koebe's solution of the uniformization problem and Hilbert's treatment of the Dirichlet Problem.

[51] Weyl to Klein 6 May 1912.

[52] *Das Kontinuum; Raum, Zeit und Materie*.

[53] Cf. Brouwer to Fraenkel 28 January 1927.

back from Switzerland. Discussing the revolutionary events in Germany, Wiessing wrote 'Brouwer, the mathematician, returned yesterday from Switzerland. Everywhere along the Rhine he saw the revolutionaries with red bands around the arms and with guns in their hands, men who laid down the law.'[54]

The personal contact between Weyl and Brouwer must have opened Weyl's eyes for the deeper issues of constructive mathematics, for he immediately mastered the Brouwerian insights and started to present them in his own way. We will discuss Weyl's involvement in the intuitionistic reshaping of mathematics below, see section 8.8. Brouwer was very pleased and impressed by Weyl's grasp of the issues and his support.

At that time Weyl was a candidate for a chair in Berlin (cf. p. 300), and not much later Brouwer must have invited Weyl to Amsterdam. From the correspondence it becomes abundantly clear that Brouwer felt that he had found the right man to join him in Amsterdam. With a man like Weyl he was prepared to take on the mathematical establishment in an attempt to set mathematics straight! And thus he set himself to organize a chair for Weyl in Amsterdam.

The international contacts, including the crucially important visits to colleagues and universities, being resumed, Brouwer spent his next summer holiday after the war abroad once again; in August 1920 he breathed again the pure aire of the Harz mountains. On this occasion he must have bought a house in Bad Harzburg to invest Cor Jongejan's money.[55] This may be inferred from the fact that in April he still stayed at the hotel Lindenhof in Bad Harzburg, whereas in August Bertus, Lize and Cor were living in their own villa, called *Friedwalt*, in Harzburg.[56]

In a letter of 7 September 1920, written in Bad Harzburg, Brouwer discussed the position in Amsterdam with Weyl. The letter is significant, as it gives us a glimpse of Brouwer's views on the centre of mathematical activity that he had in mind.

> With respect to mathematics in Amsterdam, I am in no way pursuing the plan to organize there an intensive lecture and seminar activity, but only to bring together a circle of people, whose mathematical papers are mainly stimulating and controlling additional phenomena of their general spiritual development, that is of people who feel themselves more or less the thinking organ of the community and who frankly put the immediately concrete academic teaching practice on the second place for this vocation. (I view, indeed, the mathematical drive for knowledge—which is basically distinct from the joy of the solving of mathematical problems— the hallmark of the state of mind which guarantees in the most diverse moral and practical domains a wide and free view, which considerably surpasses the common view). I may add, that we mathematicians in Amsterdam have won in the last years a very high measure of academic freedom,

---

[54][Wiessing 1960], p. 33.
[55] Inherited from het father, see 252
[56] The villa actually is still there in the Krodothal.

and make use of it in the above sense.

Furthermore we are respected in our faculty (of the sciences), and our discipline is there regarded without scepticism. With the other faculties (maybe partially with the exception of the medical faculty) we have, to be sure, roughly the reputation of Bolsheviks.

As to assistants, I have one,[57] who administrates the reading room and who has worked out some of my courses. My colleagues have none and wish none. You can certainly have one as soon as you wish. Apart from the salary (which, by the way, will probably be raised again in the near future) there are no lecture fees. You can of course, if you prefer so, live in a suburb, like me, if necessary in Zandvoort at the North Sea, 30 minutes by train from Amsterdam. Rent and taxes will together make an amount of over 1000 guilders and less than 2000 guilders; in the suburbs both are considerably less than those in the city. My colleagues De Vries (Vossiusstraat 39), or even better, Mannoury (Koninginneweg 192), who has four children at the school going age, will be able to inform you accurately about schools. The schools in town are excellent. I have heard less praise about the ones in the suburbs, but there they certainly are also tolerable. There is even a German school in Amsterdam; about its quality nothing is, however, known to me.

As to your official language, you automatically have permission to teach for two years in another than the Dutch language; this permission is then, when necessary, extended year by year. Ehrenfest already lectured in Dutch in his second year; for Denjoy, however, it is already the fourth year that he lectures in French in Utrecht. At the end of this week I will be again in Holland and travel from there to Nauheim. So please write [...] to me in Laren. The courses start on the first of October.

Give my respect to your wife and accept with the greetings of my wife, a warm handshake from
>your
>Egbertus Brouwer

Weyl contemplated Brouwer's offer with more than routine interest. He had reached a point in his mathematical development which asked for radical decisions. He reported to the Board of the ETH in Zürich that

> At the meeting of the 'Naturforscher' at Nauheim, Professor Brouwer conveyed to me an offer from the University of Amsterdam. At the rejection of the offer from Berlin, Brouwer got permission for a sizeable extension of the Mathematics Department at the Amsterdam University; the position that was offered to me, is a new chair in function theory, to be founded as a consequence of this plan. It is above all my close relation to Brouwer—we work jointly for a revolutionizing and re-founding of analysis—which makes me consider the offer in earnest. I would in any

---

[57] J.H. Schogt, later succeeded by Belinfante.

case like to make myself familiar with the matters at the spot, before I make a decision, and thus I will travel this very vacation to Amsterdam.[58]

In spite of the attraction of a close cooperation with Brouwer, Weyl preferred to stay in Zürich, provided certain conditions were met. In negotiation with the authorities of the ETH Weyl indeed obtained all he wished.[59] In a letter of 20 January 1921 to the President he confirmed the rejection of the offer of Amsterdam. An important reason for declining the Amsterdam offer was Weyl's poor health. On account of his asthma he had to withdraw from time to time to the mountains with their clear air. As a matter of fact, the ETH granted him a regular vacation in the first months of the year.

Weyl announced his research plans, with a request for financial assistance. He hoped to be able to publish the first results soon: 'It concerns, on the one hand, impulses to a new foundation of the analysis of the infinite, the present foundation of which is in my opinion untenable; on the other hand, in close connection with that and in connection with the general theory of relativity, a clarification of the relation of the two basic notions with which modern physics operates, that of *field of action and of matter, the theories of which can at present by no means be joined together consistently*.'

To conclude our account of the postwar period of normalization, let us return to the vacant chairs in Amsterdam, Berlin and Göttingen. Brouwer had turned down both offers, strengthening his position at home. Weyl, similarly had preferred to remain in Zürich. He rejected the offers from Berlin, Göttingen and Amsterdam. In December 1920 he informed Brouwer of his decision. Brouwer was sincerely disappointed; like Hilbert he was much taken with the virtually universal young mathematician, who not only knew his way around in mathematics and physics, but also in philosophy, and who displayed an unbounded enthusiasm combined with a solid integrity.

> When your telegram arrived, my disappointment was very great, which surprised me and from which it really became clear to me how I would have loved to have you here.[60]

The letter makes it evident beyond doubt that the professional relation had been transformed into a mutual friendship, which may have gone through a long period of hibernation in the thirties and forties, but which was still strong enough to be rekindled at the end of Weyl's life.

Brouwer was for the moment at a loss. He wrote Weyl that he hoped to succeed in attracting 'another young personality with a really special reputation'. Bieberbach, he thought, would not qualify, 'as he is not sufficiently known outside the strict professional circle of mathematicians, and in addition he does not yet have

---

[58] Weyl to President of ETH 27 September 1920.
[59] cf. [Frey, G. 1992].
[60] Brouwer to Weyl 1 January 1921. ETH. HS.91:493.

a completely unchallenged name as an innovator.' For the chair offered to Weyl, Brouwer thought perhaps of Birkhoff, 'whom also the astronomers and the specialists in mechanics know, and who is especially considered as a star of the first magnitude, ...'

He ended his letter with a cordial

> Now my dear boy, here is to 1921 for you and your family. May the mountains bring you full health and vitality. I long now for a sojourn there: I do not feel well at all the last weeks, and every day, if I do not lay down for a few hours, I have a considerable temperature at night. In itself this does not worry me, because I have been through many such periods; should my situation remain the same much longer, then I will have to request a vacation, and I will have to go to Switzerland for some time. We would then see each other really soon.[61]

And thus ended the first attempt of Brouwer to create a mathematics department of the first magnitude.

8.7 Intuitionism and the *Begründungs* papers

It is about time to retrace our steps to the last year of the war; after a relatively quiet period of reflection on the foundations of mathematics, Brouwer returned to the debate, that had withered somewhat after the surge of interest following the paradoxes and Zermelo's axiom of choice. The comparative rest of the period 1914–1918 had given him an opportunity to think over the problem of infinity and the nature of mathematical objects, and thus the foundations of mathematics once again became the prime target of Brouwer's efforts. He had discovered that it made sense to consider infinite sequences of well-defined mathematical objects, such as natural numbers, which were *not* given by a law, and he had reached the viewpoint that these sequences, called *choice sequences*, were the answer to a host of questions in mathematics. The first observable result of these reflections was a substantial exposition of the new constructive mathematics: *'Founding mathematics independently of the logical theorem of the excluded middle. First part: general set theory'*.[62] This paper was published by the Academy in Amsterdam in 1918. It was followed by a second and third part, respectively in 1919 and 1923. This first paper (in German), in which the basics of Brouwer's intuitionistic mathematics was presented, was a purely *technical* paper; it contained no philosophical remarks. Neither, and this is surprising, did the paper contain the term 'intuitionism'! It is worth noting that in the 1920 paper *Intuitionistic Set Theory*, Brouwer for the first time adopts the name 'intuitionism' for his new program. His earlier use of the term 'intuitionistic' covered, as we have seen, a more amorphous collection of principles and methods. It certainly includes the French (semi-) intuitionists. On some occasions, notably his

---

[61] In March Brouwer asked and obtained a three month sick leave.

[62] *Begründung der Mengenlehre unabhängig vom logischen Satz vom ausgeschlossenen Dritten. Erster Teil, Allgemeine Mengenlehre.*[Brouwer 1918B]. We will refer to the papers of this series as the *Begründungs*-papers.

inaugural address *Intuitionism and Formalism*, Brouwer called his mathematics 'neo-intuitionistic', to distinguish it from its predecessors. The name 'neo-intuitionism' stuck for some time, for example the influential book *Einleitung in die Mengenlehre* of A. Fraenkel used the term.

There are a few confusing aspects about the title of the papers. In the first place, an uninformed reader might think to find a Cantorian, or maybe axiomatic approach à la Zermelo to set theory. In the second place he might think that the author advocates some sort of mathematics based on a restricted logic. In both cases he would be wrong; the sets (*Mengen*) that Brouwer introduced were made up of choice sequences subjected to specific conditions, and logic does not figure at all in the papers (nor is the principle of the excluded third mentioned!). In Brouwer's terminology there lingered the old aim of presenting an alternative to Cantor's set theoretic universe. Already in the dissertation Brouwer had listed the possible kinds of sets, and now he was carrying out a similar project for his new intuitionism.

One gets the impression that Brouwer on the one hand wanted to lay down the principles for a conceptually sound mathematics, but on the other hand did not want to estrange the mathematical community. Perhaps for that reason he cautiously stuck to the positive aspects of intuitionism; in particular he did not repeat the dissertation's attacks on all and sundry. Nor did he present his counterexamples, based on unsolved problems. Compared to his missionary dissertation, the *Begründung*-papers seemed a harmless intellectual exercise of another eccentric mathematician. If the leading exponents of non-constructive abstract mathematics got to see the papers at all, it probably would not have cost them a minute of their sleep.

Yet, the very first paper contained a revolutionary principle, that by itself refuted the practice of traditional mathematics (*classical mathematics*, as Brouwer started to call it). It seems highly unlikely, however, that anybody discovered the sting in Brouwer's new approach, since the papers were published as little separate monographs, and did not even appear in the Proceedings of the Academy, so potential readers were few. In short, the 'marketing' of Brouwer's new ideas was as low key and uncontroversial as only a learned treatise can be.

The first paper starts right away with the definition of the main concept, given in one sentence of 13 lines. In a later publication he repeated this definition and added dryly: *Dem Umstand dasz die Mengendefinition langatmig ist, ist leider nicht abzuhelfen.* (The circumstance that this definition of set is long-winded can unfortunately not be helped).[63]

A *set* (*Menge*) is what nowadays is called a *spread*; the terminology is rather unfortunate, since Cantor had already fixed the meaning of the term set as 'collection to a whole M of definite, well-differentiated objects m of our intuition or our thought', which has become the standard usage in mathematics, and which was already firmly entrenched in Brouwer's days.

---

[63] [Brouwer 1925A]. In the first presentation the definition is given in *one* sentence running over *eleven* lines!

The confusion was further increased by introducing the term *Spezies* (species) for what we would call, following Cantor, set, in Brouwer's words, a property that a mathematical object may have.

In discussing Brouwer's intuitionistic mathematics we will, as a rule, stick to the modern terminology, that is we will use 'spread' and 'set' for *Menge* and *Spezies*.

The definition of spread mystified almost all of Brouwer's contemporaries. Karl Menger in his 'My memories of L.E.J. Brouwer'[64] reported that set theoreticians like Hausdorff and Fraenkel failed to understand Brouwer's phraseology. Since the spread became the centre piece of Brouwer's mathematics, we will reproduce here the original definition:

> A *spread* is a *law*, according to which, if over and over an arbitrary number of the sequence $\mathbb{N}$ is chosen, every of these choices either generates a certain sign or nothing, or blocks the process and brings about the definitive destruction of the result, where for every n after each unblocked sequence of $n-1$ choices, at least one number can be indicated, which, if it is chosen as the nth number, does not result in blocking the process. Every sequence of signs that is generated in this way by the spread (which, thus can in general not be represented as finished) is called an element of the spread. We will denote the common mode of coming into existence of the elements of a spread M also as *the spread M*.[65]

The spreads and their elements make up the basic material of Brouwer's mathematics. It is convenient to visualize spreads by means of tree-pictures, Fig. 8.1.

The subsequent choices of natural numbers are indicated in the tree by the labels at the nodes. After each unchecked sequence of choices, at least one choice does not check the process, so below each node, at least one other node is eligible according to the spread(law). This means that, going down via the unchecked nodes, there are no finite (maximal) paths. In general, choice sequences are obtained as follows: unchecked choices generate signs; this means that after a finite number of choices $a_0, a_1, \ldots, a_n$, a particular object (sign) is generated; that is to say, we have an assignment of objects $s_{a_0,\ldots,a_n}$ to sequences $a_0, a_1, \ldots, a_n$. So the unchecked nodes have an extra label $s_{a_0\ldots,a_n}$. And the infinite sequences of these labels are the elements of the spread.

Imagine putting a cross at each checked node, then the *elements* of the spread can be visualized as the labelled infinite paths avoiding crossed nodes.

---

[64][Menger 1979].

[65] Eine *Menge* ist ein *Gesetz*, auf Grund dessen, wenn immer wieder ein willkürlicher Ziffernkomplex der Folge ζ gewählt wird, jede dieser Wahlen entweder ein bestimmtes Zeichen, oder nichts erzeugt, oder aber die Hemmung des Prozesses und die definitive Vernichtung seines Resultates herbeiführt, wobei für jedes n nach jeder ungehemmten Folge von $n-1$ Wahlen wenigstens ein Ziffern-komplex angegeben werden kann, der wenn er als n-ter Ziffernkomplex gewählt wird, *nicht* die Hemmung des Prozesses herbeiführt. Jede in dieser Weise von der Menge erzeugte Zeichenfolge (welche also im allgemeinen nicht fertig darstellbar ist) heisst ein *Element der Menge*. Die gemeinsame Entstehungsart der Elemente einer Menge M werden wir ebenfalls kurz als *die Menge* M bezeichnen.

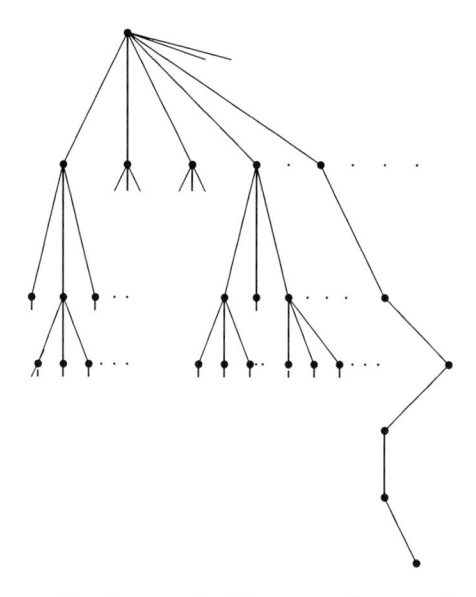

FIG. 8.1. underlying tree of a spread

The simplest example of a spread that comes to mind is obtained by allowing at each step every natural number and by assigning the last number a_n to the sequence $a_0, a_1, \ldots, a_n$. The result is the spread of all choice sequences of natural numbers. Of course, it is only a matter of convention to call the last mentioned sequences 'choice sequences of natural numbers'. One could also have considered the infinite sequences in the underlying tree as 'choice sequences of natural numbers'. The main idea is that one allows choice sequences to consist of elements of a previously given (or constructed) collection.

In the above definition the choice sequences are somewhat 'played down'; the spread appears as the primary notion, and the choice sequences behave as auxiliary objects. Since the choice sequences are the visible witnesses of a subjectivist approach to mathematics, Brouwer's particular presentation might well be the result of a deliberate policy to let sleeping dogs lie, that is not to invite the outcry and ridicule of a mathematical world that was undeniably cultivating a formalistic-positivistic (and definitely objectivistic) outlook.

The highlight of the first paper, although modestly hidden in the mass of new notions and techniques, is the proof of the non-denumerability of the universal spread, that is the spread in which no nodes are checked, and in which the assignment of signs to finite sequences of choices is the trivial one:

$$(a_0, a_1, \ldots, a_n) \longmapsto a_n.$$

This is the same spread which we called the spread of natural number choice sequences, and which may be denoted by $\mathbb{N}^\mathbb{N}$. In geometric terminology, the spread determined by this assignment is called the *Baire space*, a well-known topological structure.

Brouwer had used in his 1915–1916 course on set theory, Cantor's diagonal argument, but he had abandoned it a year later in favour of his own new argument, which now was unobtrusively slipped in amongst the other material.[66] In just ten lines a revolution was launched, which was to have marked consequences. Hardly anybody recognized the significance of the principle involved; it took a man like Hermann Weyl to recognize a revolution when he saw one.

Brouwer's argument runs as follows: suppose we have a function F that assigns a natural number to each choice sequence of natural numbers. Then the number $F(\alpha)$ must be completely known after a suitable initial segment of α has been generated; and this initial segment must already determine $F(\alpha)$. But then all sequences β with the same initial segment will get the same value as α. Hence F cannot be one–one.

The initial segment-argument, used above, is Brouwer's continuity principle in a nutshell. Together with some other principles, it would give an enormous power.

Observe that the crucial condition is the totality of F, that is to say, F must give an output for *all* inputs. A function like

$$F(\alpha) = \begin{cases} 0 & \text{if } \alpha \text{ is eventually constant} \\ 1 & \text{else} \end{cases}$$

does not qualify because we cannot tell for an arbitrary α whether it will become eventually constant. So we are not allowed to say that F is defined for all α. The intuitionist is not allowed to appeal to the higher authority of the principle of the excluded third, as the classical mathematician is.

The '*Begründung*'-papers contained most of the tools required for a systematic practice of constructive mathematics, to mention just a few:
the intuitionistic versions of
- real numbers, the continuum;
- elementary topology (of the plane);
- located (*katalogisierte*) sets;
- measure.

Most, if not all, refinements are dictated by the modesty of logic under an intuitionist regime. In the *Unreliability*-paper of 1908, Brouwer had drawn the consequences of his intuitionistic views; in particular he had indicated certain simple statements of the form 'there exists an object x such that …', one may think of 'there are infinitely many pairs of equal consecutive decimals of π', and 'there is a sequence 012…9 in the decimal expansion of π', which have neither been proved nor rejected.

For such statements A there is no reason to claim the principle of the excluded third: A or not A (in symbols $A \vee \neg A$). Likewise there is no reason to accept that

---

[66][Brouwer 1918B], p. 13, see also p. 240.

from the impossibility of the absence of a sequence $012\ldots9$ in π, the presence of such a sequence may be concluded, that is, the impossibility of the absence does not give slightest idea where this sequence might be among the decimals of π. Or, in general, $\neg\neg A \to A$, cannot be affirmed. The reader should realize that Brouwer did not say, in 1908, that $A \vee \neg A$ *cannot* be proved; he cautiously said that the principle is *as yet unreliable*.

In 1918 he had all the tools to go one step further and to show that $A \vee \neg A$ indeed cannot be proved. He drew this conclusion only much later; apparently even Brouwer could miss something that was staring him in the face.

The failing of the *double negation law* ($\neg\neg A \to A$), made it necessary to distinguish between a property and its double negation, for example, 'it is impossible that a is not an element of X' is not the same as 'a is an element of X'. In symbols $\neg\neg a \in X$ is distinct from $a \in X$. So, for example, where we had the notion 'the sets X and Y are *identical*', that is '$\forall a(a \in X \leftrightarrow a \in Y)$' ($X$ and Y have the same elements), Brouwer introduced a new notion 'X and Y are *congruent*' if $\neg\exists a(a \in X \wedge a \notin Y) \wedge \neg\exists a(a \in Y \wedge a \notin X)$.[67]

The restrictions on the logic required a whole list of similar refinements, of which some never gained any practical significance, but others have become standard material in intuitionistic (but also in general constructive) mathematics.

Brouwer's practice of naming notions and objects is highly idiosyncratic. Many suggestive names have never been adopted. On the other hand, he doggedly used long descriptions of colourless names for key concepts. For example, he preferred 'absurdity of A' to 'not A' (let alone $\neg A$, $\sim A$, or $\overline{A}$). He never introduced names for the positive versions of 'non-empty' or 'distinct'. Notions like 'bar-induction' and 'continuity principle' fared no better.

8.8 And Brouwer—that is the revolution

At the turn of the century there was a flurry of foundational activity in mathematics. The paradoxes had just appeared, and a good part of the mathematical community was wondering if there was not something basically wrong with the subject. Zermelo's introduction of the axiom of choice added to the confusion: what were the legitimate objects of the mathematical universe? After the heyday of Frege's *Grundgesetze*, Russell's *Principia*, Hilbert's *Heidelberg Lecture*, Poincaré's predicativity and Brouwer's early intuitionism, the general attention turned towards other areas, and the 'foundations' were left to the individual scholar who was willing to spend his time on yesterday's fashion.

As we have seen there were two well-respected mathematicians, who were not satisfied with the general state of mathematics: Brouwer and Weyl. While Brouwer was rewriting mathematics in Blaricum, Weyl had embarked on a similar project, resulting in his monograph *Das Kontinuum*. Weyl was one of those exceptional characters who combined an impeccable technical know-how with deep philosophical insight. In Göttingen he had become Hilbert's star pupil; but he had also joined

---

[67] A sniff of logic shows that this can be rewritten as $\forall a(a \notin X \leftrightarrow a \notin Y)$.

the circle around the philosopher Edmund Husserl, who had given a new impetus to the study of phenomenology. Like Brouwer, Weyl was in an excellent position to draw attention to the foundations of mathematics. His status as a mathematician was so unassailable that he could not be brushed aside as 'eccentric' or 'uninformed'.

The roads of Weyl and of Brouwer to an alternative mathematics were rather different. Whereas Brouwer was in a certain sense a radical, who wanted to guide mathematics back to its true sources (and thus to safeguard it from the excesses of Formalism and Cantorism), Weyl's first concern was to avoid the paradoxes that had appeared around the turn of the century. Comparing Weyl's and Brouwer's approaches to the foundations of mathematics, Weyl is the cautious revisionist, who wanted to find the safe kernel of traditional mathematics. Brouwer, like Samson, wanted to bring the whole building of mathematics down before erecting it again, whereas Weyl wanted a careful restoration.

Most mathematicians had their favourite villain, for some it was the comprehension principle, which caused the downfall of Frege by the hands of Bertrand Russell. For some it was the phenomenon of impredicative definitions, that undermined mathematical practice. The early adherents of Cantor's set theory belonged to the first group; they mostly accepted Zermelo's solution, consisting of a well chosen axiomatisation. Poincaré and Russell were the exponents of the second group; they focused their attention on the phenomenon of 'impredicative definition'. Such a definition defines an object in terms of the totality to which it belongs. The familiar example of an impredicative concept is the supremum of a bounded set of real numbers. One has to know the set of upperbounds before one can determine the smallest among them; so the supremum, that is the least upperbound, presupposes the set of all upperbounds, of which it is itself an element. And how can one define something that has to be known before it can be defined? Impredicativity was at the bottom of the paradoxes of Richard and Berry, therefore both Poincaré and Russell each had proposed a ban on impredicative definitions (the *vicious circle principle*).

Weyl took the issue of predicativity seriously by constructing a continuum consisting of real numbers which were definable in a particular predicative way: 'arithmetically definable numbers'. In *The Continuum* he gave a systematic development of the theory of real numbers and their functions along the lines of arithmetic definability. He saved, so to speak, the continuum by restricting it. In this sense his method was a big step forward compared to the ideas of the French semi-intuitionists, who did not go into the details of what they meant by 'definable in finitely many words'. *The Continuum* was Weyl's coherent answer to the anarchy that was threatening the realm of mathematics. His conception was not a mere technicality for escaping unpleasant phenomena; it was based on the philosophical insights of phenomenology. Phenomenology was his explicit point of departure.[68]

Hermann Weyl's monograph was a gem, it was far beyond anything produced in the foundations of analysis, it thoroughly treated the mathematical-logical aspects,

---

[68] Cf. [Weyl 1954].

and actually dealt with issues of analysis. Weyl's monograph could truly be called a mathematician's presentation, in contrast to, say, the works of Frege and Whitehead and Russell. 'The Continuum' was, on his own view, a technical realization of a phenomenological analysis of space (the real line). Weyl's monograph was considered to be rather forbidding, for example, Klein, thanking Weyl for a copy of 'The Continuum', wrote 'As you can imagine, it is difficult for me to make myself familiar with these abstract things, about which I would nonetheless like to have an opinion'.[69] The book was, as a matter of fact, much praised and cited, but it took the mathematical community some 60 years before it could and did read it properly. Only when Feferman, Kreisel and others started to investigate the foundational status of predicative mathematics, Weyl's creation was recognized for its elegance, depth and daring. Weyl's monograph distinguished itself from Brouwer's contemporary papers by its philosophical excursions. Modern readers will view Weyl's monograph as a pleasant and thorough exposition of the foundations of analysis, not in the last place because of Weyl's sincere effort to explain the philosophy behind his enterprise. In his own words: 'Although this publication aims above all at mathematical goals, I have not avoided *philosophical* questions, and I have not tried to dispose of them by that crude and superficial amalgamation of Sensualism and Formalism, which [combated with gratifying clarity by Frege in his *Grundgesetze der Arithmetik*] is still highly appreciated by mathematicians'.[70]

Weyl was not at all averse to radical positions and statements, even though he did not go all the way to a constructive mathematics. In particular he did not revise the logic in the *The Continuum*. He had independently reached a good many radical insights. Some of these have become famous in the wager between Weyl and his Zürich colleague Polya:

> within 20 years Polya and the majority of representative mathematicians will admit that the statements
>
> 1) *Every bounded set of reals has a precise supremum*
>
> 2) *Every infinite set of numbers contains a denumerable subset*
>
> contain totally vague concepts, such as 'number', 'set' and 'denumerable', and that therefore their truth or falsity has the same status as that of the main propositions of Hegel's natural philosophy.
>
> However, under a natural interpretation 1) and 2) will be seen to be false.[71]

Even Brouwer could not have outdone Weyl in provocation; his views were by no means identical with those of Weyl, but he heartily agreed with Weyl on the untenability of the classical principles.

When, after the long interruption of the war Brouwer and Weyl finally met again in the Engadin in the summer vacation of 1919, they discussed the founda-

---

[69] Klein to Weyl 19 April 1918.
[70] [Weyl 1918], p. IV
[71] 9 February 1918, cf. [Polya 1972].

tional problems of the day. Brouwer gave a private crash course on his new intuitionism, including the choice sequences. Weyl, as ever, was quick to grasp what was going on. He turned the new notions over in his mind, and soon he had mastered the topic. In the same year he gave a series of lectures with the title 'On the foundations of analysis' in Zürich.[72] The lectures contained, in fact, all the ideas of the subsequent paper; it is quite clear that Weyl had fully recognized the meaning of the new constructive views; from Gonseth's notes it appears that Polya heartily disagreed with Weyl's views, as may illustrated by the following small discussion:

> Polya: You say that mathematical theorems should not only be true, but also be meaningful. What is meaningful?
> Weyl: That is a matter of honesty.
> Polya: It is error to mix philosophical statements in science. Weyl's continuum conception is emotion.
> Weyl: What Polya calls emotion and rhetoric, I call insight and truth; what he calls science, I call symbol pushing (*Buchstabenreiterei*). Polya's defence of set theory ([that] one may one day provide meaning to these formulations) is mysticism.
> To separate mathematics, as being formal, from spiritual life, kills it, turns it into a shell. To say that only the chess game is science, and that insight is not, *that* is a restriction.[73]

Soon afterwards Weyl finished his beautiful provocative paper *Über die neue Grundlagenkrise der Mathematik*, [Weyl 1921]. In May 1920 it was finished and a copy was sent to Brouwer:

> Zürich, 6 May 1920
>
> Dear Brouwer,
>
> Finally I have sent off the long promised [object] to you. It should not be viewed as a scientific publication, but as a propaganda pamphlet, thence the size. I hope that you will find it suitable for this purpose, and moreover suited to rouse the sleepers; that is why I want to publish it. I would be grateful for your criticism and comments. Did I enclose everything that you let me have only as a loan? If not, please reclaim it; the lecture on *Formalism and Intuitionism*[74] was already in my possession in the old days; at that time I did not pay attention to it or understand it ...

One can imagine how pleased Brouwer was with Weyl's conversion; here was one of the foremost members of the new generation of mathematicians, who was not only a master in almost all traditional topics in mathematics, but who had an exceptional insight in philosophical and foundational matters. The fact that Weyl was Hilbert's best student, and was viewed as the crown prince of mathematics

---
[72] 2, 9, 16 December in the seminar of Fueter.
[73] Polya had advanced the view that research in set theory should not be restricted.
[74] [Brouwer 1913C], note the 'Freudian' permutation.

in Göttingen, may have contributed to the importance of this first convert, but one should keep in mind that the *Grundlagenstreit* had not yet started and that the relations between Brouwer and Hilbert were still free from personal and scientific antagonism. As the enclosed manuscript showed, Weyl wholeheartedly gave up his own foundational program, trading his arithmetically definable sequences for Brouwer's choice sequences, and rejecting the principle of the excluded third (PEM). Only a rough draft, full of crossed-out parts, has survived of Brouwer's reaction. Nonetheless, one can fairly well guess Brouwer's views from this draft, and from a few remarks, inserted in pencil in the margin of Weyl's manuscript. The first lines of Brouwer's draft run

> Your wholehearted assistance has given me an infinite pleasure. The lecture of your manuscript was a continuous delight, and your explanation, it seems to me, will also be clear and convincing for the public ... That we judge differently on some side issues, will only stimulate the reader. However, you are completely right in the formulation of these differences of opinion; in the restriction of the objects of mathematics you are in fact more radical than I am; however, one cannot argue about this, these matters can only be decided by individual concentration.

Brouwer had carefully studied Weyl's' manuscript and made some notes in the margin, but on the whole he was inclined to grant Weyl his own insights in the matter. He was critical here and there, for example, at the paragraph where Weyl seemed to pussyfoot in recognizing the scope of the new notions. Weyl had, for example, stated that choice objects were alright as individuals, but that one could not (or should not) quantify over them, 'It should be stressed once more that certain individual functions of that kind occur from case to case in mathematics, that general theorems are, however, never asserted about them. The general formulation of this notion is therefore only required if one is giving a justification of the meaning and methods of mathematics; for mathematics itself, the subject matter of its theorems, it is never considered at all'. Brouwer not only disagreed with this view, but he also upbraided Weyl for being too cautious:

> It seems to me that the whole point of your paper is endangered by the end of the second paragraph of page 34.[75] After you have roused the sleeper, he will say to himself: 'So the author admits that the real mathematical theorems are not affected by his considerations? Then he should no longer disturb me!' and turns away and sleeps on. Thereby you do our cause injustice, for with the existence theorem of the accumulation point of an infinite point set, many classical existence theorems of a minimal function, and also the existential theorems of the geodetic line without the second differentiability condition, loses its justification!'

Weyl distinguished in his paper two distinct views of the continuum, the *atomistic* one and the *continuous* one. In the first version the continuum is made up of

[75][Weyl 1921] p. 66.

individual real numbers which can be sharply distinguished. Weyl allowed in his program of *The Continuum* only specific (arithmetic) relations on natural numbers, and thus also restricting the class of Dedekind cuts. For this particular class of reals, Weyl accepted existence problems involving reals as meaningful:

> Only if we conceive the notion [of the continuum] in this way, which fixes and demarcates its extent, matters of existence concerning real numbers become meaningful. By this restriction of the notion, a bunch of individual points is picked from the flowing mash. The continuum is smashed to isolated elements, and the blending of all its parts is replaced by certain conceptual relations, based on the 'greater–smaller', between these elements. Therefore I speak of an atomistic conception of the continuum.'

The arithmetically definable continuum of 'The Continuum' has this atomistic character. Almost immediately Weyl added, however, that 'It has never been my opinion that the continuum given by intuition is a number system of Weyl; rather that analysis merely needs such a system for its constructions and does not have to worry about the 'continuum' poured in between. Indeed, already in *The Continuum*, Weyl pointed out that his atomistic version of analysis represented 'a *theory of the continuum* which has (over and above its logical correctness) to prove itself correct, much as a physical theory.'

After the discussion of the 'predicative', atomistic continuum, Weyl went on to give an exposition of Brouwer's continuum with choice sequences and choice real numbers. Weyl's style is beautiful and poetic, it is almost the opposite of Brouwer's style of presentation. The descriptions of the various new notions and the metaphorical presentation did more to popularize intuitionism than Brouwer's œuvre so far. Weyl introduced a number of slogans that caught the imagination of the otherwise not foundationally interested readers. Indeed, Weyl's 'New crisis' reads as a manifesto to the mathematical community. It uses an evocative language with a good many explicit references to the political and economic turmoil of the postwar period. The opening sentences castigate the complacency of the mathematical community, which had not paid any serious attention to the potential dangers of the various paradoxes of Cantor, Russell, Richard, . . .

> The antinomies of set theory are usually considered as border conflicts that concern only the remotest provinces of the mathematical empire and that can in no way imperil the inner solidity and security of the empire itself or of its genuine central areas. Almost all the explanations of these disturbances which were given by qualified sources (with the intention to deny them or to smooth them out) however, lack the character of a clear, self-evident conviction, born of a totally transparent evidence, but belong to that way of half to three-quarters attempts at self-deception that one so frequently comes across in political and philosophical thought. Indeed, every earnest and honest reflection must lead to the insight that the troubles in the borderland of mathematics must be judged as symptoms, in which what lies hidden at the centre of the superficially glittering and

smooth activity comes to light—namely the inner instability of the foundations, upon which the structure of the empire rests.

The readers of 1920 were all too familiar with the phenomenon; the German State with its skirmishes in the Baltic and its political instability was a painful reminder of the self-deception of politics. Weyl used the political metaphor with great dexterity, he compared the classical use of existential statements with the use of paper money:

> The point of view sketched above[76] only expresses the meaning which the general and existential propositions in fact have for us. In its light mathematics appears as a tremendous 'paper economy'. Real value, comparable to food products in the national economy, has only the direct, downright singular; general and all the existential statements participate only indirectly. And yet we mathematicians seldom think of cashing in this 'paper money'! The existence theorem is not the valuable thing, but the construction carried out in the proof. Mathematics is, as Brouwer sometimes says, more activity than theory (*mehr ein Tun als eine Lehre*).[77]

And the final clarion blast of Weyl, one that fired the imagination and fed the wrath of many a practising mathematician, rang through the next decade: 'For this order cannot in itself be maintained, as I have now convinced myself, **and Brouwer—that is the revolution!**'

Weyl in his paper addressed mainly two points: the notion of the continuum, that is the nature of the real numbers, and the meaning of logic, in particular the meaning of the quantifiers and the failure of the principle of the excluded third.[78] An analysis of Weyl's contributions is given elsewhere.[79] Here we will just look at the more superficial effects of the 'New Crisis' paper.

Real numbers are given by Weyl as infinite sequences of shrinking intervals, and thus the need arises to specify the notion of sequence further. Weyl allowed two sorts of sequences: those given by a law, standing for the individual points of the continuum, and the free choice sequences, which determine a continuum of 'becoming', or 'emerging', sequences (*werdende Folgen*),[80] that is sequences of intervals that are freely chosen and hence cannot in a predetermined way point at an element of the continuum. In the most literal sense of the word, these sequences remain 'in statu nascendi' forever. In Weyl's words:

---

[76] that is the construction of a specific object versus its non-effective existence.

[77] Cf. p. 105.

[78] We will indiscriminatingly use the terms 'principle of the excluded third', 'principle of the excluded middle'. Brouwer preferred 'principium tertii exclusi'. The acronym used here is *PEM*.

[79] [van Dalen 1995].

[80] The translation of the German *werdende Folge* is somewhat problematic. Various adjectives, such as 'becoming', 'developing', 'emerging', 'in statu nascendi' have been used in the literature.

> It is a first basic insight of Brouwer that the sequence which is emerging (*werdend*) through free choice acts, is a possible object of mathematical concept formation. Where the law φ, which determines a sequence up to infinity, represents the single real number, the 'choice sequence, restricted by no law in its freedom of development, represents the continuum.
>
> 'Brouwer's remark is simple but deep: here a 'continuum' arises in which, indeed, the single real numbers fit, but which itself does by no means dissolve into a set of completed existing real numbers, rather it is a medium of free becoming.

Weyl drew the conclusion that the true continuum consisted of the real numbers generated by choices and in addition those given by a law. Hence 'for all real numbers' meant for him 'for all choice real numbers', and existence could only mean 'there is a real number given by a law', for 'existence' must have the meaning 'I have a construction method here and now'. The consequence was that the negation of an existential statement over reals was not at all the same thing as the universal version of its negation. Hence there was not the faintest justification of the principle of the excluded third! But Weyl wanted more, he wanted to understand why the Principle of the Excluded Middle failed even for such pedestrian objects as natural numbers. For statements like $A(n)$, $\exists n A(n)$ is established if one finds a number n, during the consecutive testing of $A(0), A(1), A(2), \ldots$. The negation would then come to 'after running through all natural numbers no instance $A(n)$ has been established', but Weyl correctly rejects this as nonsense, $\forall n \neg A(n)$ can only be true because it is part of the essence of the notion natural number. He concluded that these statements do not act as each others negation. So this pleads against PEM. But, he continued,

> Although this speaks in favour of Brouwer, I am always thrown back to my old standpoint by the thought: if I run through the sequence of natural numbers and break off when I find a number with the property A, then this eventual breaking off does or does not occur; it is or is not the case, without wavering and without a third possibility [...] Finally I found for myself the magic word. An existential statement, for example, 'there is an even number', is not a judgement in the proper sense at all, which states a state of affairs; existential states of affairs are an empty invention of logicians.

Weyl termed such pseudo-statements 'judgement abstracts', to be compared to 'a piece of paper which announces the presence of a treasure, without giving away its place.' Similarly he provided a metaphor for the universal quantifier.

The 'New Crisis' paper is extraordinarily rich in ideas and claims; Weyl saw clearly what the Brouwerian revolution would bring. He drew some conclusions, that were to be established a couple of years later by Brouwer. For example, he asserted the 'indecomposability' of the continuum (which includes the connectedness):

'A true continuum is simply something connected in itself and cannot be split in separated pieces; that contradicts its nature.'

This is the first enunciation of the unsplittability of the continuum. However, Weyl does not prove the principle, but rather asserts its plausibility on conceptual grounds. Similarly, the prize-winner—the continuity of real functions—is announced without a proof:

'Above all, however, there can be no other functions at all on the continuum than continuous functions'

It is useful and necessary to correct the impression that Weyl thus proved the continuity of all real functions years before Brouwer. One should rather view Weyl's statement as an immediate corollary of his definition of real function. On his account an approximation of the value of a function was, by definition, determined by the approximation of the argument. And thus continuity was, so to speak, part of the definition of 'function'.

Another way to put it is: Weyl was convinced of for the unsplittability of the continuum and the continuity of the real functions on the basis of a *phenomenological analysis*. Subsequently this was confirmed by Brouwer's *proof*.

We have devoted so much space to the 'New Crisis' paper of Weyl as it has become one of the landmarks of the foundational debate in the interbellum, known as the '*Grundlagenstreit*'. While Brouwer toiled on to provide an impeccable foundation to mathematics, producing scholarly but hardly exciting papers (it was as if Brouwer tried to avoid even the semblance of that circumlocution that is so characteristic of cranks and self-proclaimed missionaries), Weyl had sounded a trumpet blast that woke many to the serious problems posed by mathematics.

8.9 Intuitionism, the Nauheim Conference

Brouwer, in the meantime, had put his foundational theories before the international mathematical forum at the *Naturforscherversammlung* at Nauheim, 1920. This meeting of the German scientists, which was more or less an act of defiance towards the Entente-dominated conference in Strasbourg, gathered the flower of German physics, mathematics and medicine. Brouwer gave his talk *Does every real number have a decimal expansion?*,[81] in the mathematics section in the afternoon of 22 September 1920. Hermann Weyl was also present. He lectured in the joint mathematics–physics section on 'Electricity and Gravitation'[82] in the morning of the next day. The conference brought together a large number of, mostly German, mathematicians for the first time after the war, to mention some: Bernstein, Bessel-Hagen, Bieberbach, Brouwer, Engel, Fricke, Hausdorff, Hensel, Kerékjártó, Koebe, Landau, Mie, E. Noether, F. Noether, Polya, Radon, Schoenflies, Schur, Szasz, Weitzenböck, Weyl, Wiener,

---
[81] *Besitzt jede reelle Zahl eine Dezimalbruchentwicklung?* [Brouwer 1921A].
[82] *Elektrizität und Gravitation*.

The list of speakers shows that the meeting was rather a 'family matter' for the Germans, no speakers were listed from the allied countries, and only three speakers from neutral countries: Brouwer (the Netherlands), Weyl and Polya (Switzerland). The title of Brouwer's talk, combined with Brouwer's reputation, was enough to draw attention. How far the actual talk diverged from the written text of the subsequently published paper[83] is hard to say. On the one hand speakers usually try to enliven their exposition with didactic remarks, possibly elicited by questions from the audience. On the other hand, Brouwer was notorious for ignoring his audience.

Group at the Nauheim Conference, 1920. In front: Hamburger; first row l.t.r.: Kerékjártó, Brouwer, Szász, Landau. Second row: Schur, Polya, Bessel–Hagen.

So the talk could have contained elucidating comments, but it could equally well have been a scholarly recitation of the contents of the manuscript. Even now the paper does not strike the reader as 'accessible', but in 1920 it must have found few readers. In itself the paper is a small piece of art. It succinctly treats the fine structure of the continuum. Starting from a constructive version of Dedekind cuts, Brouwer considered classes of reals which were in particular ways determined by their position with respect to the rationals. For example, the reals with decimal expansions or continued fractions appear in a natural way.

The audience must have been puzzled, a contemporary reaction can be found in a letter from the mathematician Robert Fricke to his uncle Felix Klein, in which he reported on the highlights of the conference:[84]

---

[83] [Brouwer 1921A].
[84] Fricke to Klein 28 September 1920.

Of course, Brouwer has to be taken very seriously, but he essentially is incomprehensible. Landau proposed, in view of Brouwer's contribution, to introduce a section for pathological mathematics and to incorporate it in the medical section'.

Whatever the audience might have thought of Brouwer's talk (which was clear, but not simple) one thing was clear: Brouwer had presented his program, and his criticism, in an international setting; he could no longer be ignored.

It is fairly evident that most observers considered what Brouwer had to offer a drastic austerity program. Indeed the Nauheim talk did not discuss the new notion of choice sequence, but remained rather neutral, and those who had not seen Brouwer's papers in the Amsterdam Proceedings, or the introductory paper *Intuitionistic Set Theory* (1919) in the home journal of the German mathematicians, would be rather confirmed in their opinion that constructivism only promised restrictions. The Set Theory–paper, however, clearly spelt out the new notions of 'choice sequence' and 'spread'. But since Brouwer did not give applications, it is no wonder that the readers could not extract any benefits from the paper. It did, on the other hand, contain a criticism of Hilbert's dogma of the solvability of mathematical problems (that is every mathematical problem can be solved or rejected). Brouwer had already in his 'unreliability' paper, [Brouwer 1908C], identified the Principle of the Excluded Third and Hilbert's Dogma, adding that 'there was no hint of a proof' for either of them. Here he repeated the identification, remarking that PEM

> can be assigned none but a scholastic and heuristic value, such that theorems, in the proofs of which its applications cannot be avoided, lack every mathematical content.[85]

Footnote 4 to this paper goes further into the general belief in the principle of the excluded third:

> In my opinion both the axiom of solvability and the principle of the excluded third are false, and the belief in these dogmas has been historically caused by the fact that people abstracted classical logic from the mathematics of subsets of a fixed finite set, and then ascribed to this logic an existence a priori, independent of mathematics, and that they finally wrongfully applied this logic to the mathematics of infinite set, on the ground of this alleged *a priori*.

This rejection of the principle of the excluded third was the inevitable conclusion of the objections to 'logical arguments', put forward in Brouwer's letter to Korteweg of 23 January 1907, cf. p. 99. The above passage (which is repeated over and over again in later papers and talks) pin-points, so to speak, the disastrous consequences of the unjustified step from finite to infinite.

Although the paper showed a glimpse of the underlying philosophical motivations, Brouwer was reticent to provide full explanations. He admitted to have used

---

[85][Brouwer 1920J].

'old methods' in his 'philosophy-free' papers, but claimed at the same time to have endeavoured 'to deduce only such results, of which I could hope, that they could, after carrying out a systematic construction (*Aufbau*) of intuitionistic set theory, find possibly in a modified form a place in the new doctrine, and maintain a significance.'

It is, indeed, not hard to find places in the topological papers, where proof by contradiction is used, or the second number class, but practice has born out Brouwer's claim to a surprising extent. Of course, one has to allow for suitable constructive modifications, see, for example, the intuitionistic version of the fixed point theorem, which Brouwer published more than forty years after the fixed point theorem itself.[86] The above claim was not an empty boast. Brouwer had enough visionary power to distinguish the constructive content of mathematical developments!

Before we leave the topic of intuitionism, let us see what the first assault at the fortress of mathematics had yielded. In the first two *Begründungs*-papers, dealing respectively with *General Set Theory* and *Point Sets* a rich collection of tools was laid out. The first paper contained the basic concepts of *spread(Menge)* and *set (Spezies)*, the various *inclusion* and *coincidence* relations for sets, the notion of *equivalence* (*gleichmächtigkeit*), and its weaker variants, notion of *finite* and *infinite*, various notions of *denumerability*, the constructive theory of *order* and *well-ordering* (including *ordinals*).[87]

A number of these notions have reappeared in different guises at various places in later years, for example, the refinements of the notion of denumerable were to play an important role in recursion theory, where similar distinctions must be made.

Some of the most common 'countability' notions are summed up (in modern language) as:

- S is *denumerable* if there is an injection $f : S \to \mathbb{N}$,
- S is *numerable* if it is denumerable, and if the range of f is decidable (that is '$\forall n (n \in ran f \vee n \notin ran f)$,
- S is *enumerable* if there is a surjection $f : \mathbb{N} \to S$.

For Brouwer's contemporaries, who were not in the least familiar with constructive mathematics, it must have been almost impossible to see that the distinctions made any sense. Even today it takes some time before students in an introductory course grasp the subtleties of constructive mathematics, so one can imagine that Brouwer's early papers, which were sparingly provided with didactic examples, found few readers!

The second paper introduced the basic concepts, needed for elementary analysis and topology. It contained likewise a number of novelties:

- a constructive notion of 'point of the plane',

---

[86] [Brouwer 1952A].
[87] N.B. Brouwer's ordinals were constructive, and hence countable.

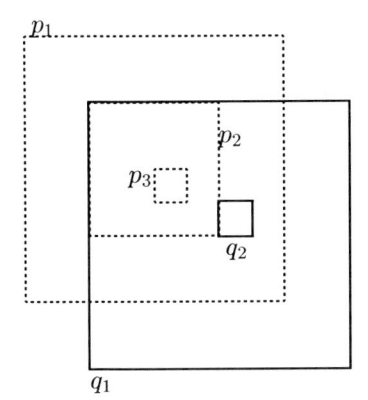

FIG. 8.2. Apartness for points

- a positive inequality relation, the apartness (*örtlich verschieden*),
- limit point, boundary point, closure,
- located sets (*katalogisiert*), measurable, content.

Some of the above mentioned notions are fairly straightforward, that is one has to make the proper constructive choices in the definitions. Some, on the other hand, are peculiar to constructive mathematics, and their significance disappears if one uses classical logic. One such notion is *apart*; a number of definitions of this notion have been given in the past. We will sketch Brouwer's original definition here:

Consider the plane divided in squares κ_1 with integer corner points, and subdivide each square in four congruent squares $\kappa_2, \ldots$ A λ_i-square is a square consisting of four κ_{i+1} squares. A point is defined as a choice sequence of λ-squares (that is λ_i-squares for certain i's) such that each square lies properly inside its predecessor.

Now, let P and Q be defined by sequences of squares $(p_i)_i, (q_i)_i$, then P is *apart* from Q if some p_i and q_j do not meet, see Fig. 8.2.

We note in passing that Brouwer did not introduce a symbol for apartness, neither here, nor later.[88] Strictly speaking he could do without, since in the presence of a metric, apartness is the same as having a (strictly) positive distance.

The other notion that plays an essential role in analysis and topology is that of a '*located set*'. In general subsets (say of the plane) can be rather elusive, for example, $\{x|A \vee \neg A\}$, where A is some unsolved problem, for example the Riemann hypothesis,[89] has rather strange features; one cannot say that it contains a point, but it

---

[88] Heyting introduced in his dissertation (1925) $A\omega B$ (*A deviates from B*) for apartness. In [Heyting 1936] he replaced ω by $=||=$, and since [Heyting 1941] the symbol $\#$ is generally accepted. Warning: the Bishop school uses '$\neq$' instead.

[89] Some people worry about the moment this problem A will be solved. There is no need to: there is an inexhaustible stock of unsolved problems; one can always take another one.

cannot be empty, for that would imply $\neg(A \vee \neg A)$, which is a contradiction.

The set $\{x|(x = 0 \wedge A) \vee (x = 1 \wedge \neg A)\}$ offers similar problems: one cannot say that either 0 or 1 belongs to it, but it cannot be empty. It is a subset of a finite set (namely of $\{0, 1\}$) but it cannot be said to be finite (for then it has exactly one element, and we should be able to tell which one).

Evidently, such sets make life miserable in mathematics. We want reasonably well-behaved sets, without requiring them to be decidable, because decidability is a scarce commodity in mathematics! Brouwer introduced *located sets* for the purpose.

Although there are now more convenient definitions, we will reproduce Brouwer's original definition: a point set X is given by a set of sequences of λ-squares. It is no restriction to assume that all n-th elements of all sequences have the same length, say $2^{1-\mu_n}$. X is called located (*katalogisiert*) if for any sequence $\sigma_1 < \sigma_2 < \sigma_3 < \cdots$ of natural numbers, and for each n, there is a finite set of S_n of squares λ_{μ_n} such that (i) squares that do not belong to S_n, are not squares of points in X, and (ii) for each $m \geq 0$ and each square λ_{μ_n} of S_n, there is a square λ_{μ_n} of X, within a distance less than 2^{μ_n} of λ_{μ_n}.

This definition, popularly speaking, says that a located set can be arbitrarily well approximated by finite sets of squares. A somewhat more manageable definition is: X is located if for any two squares q_1 and q_2, such q_2 is in the interior of q_1 without touching the boundary, we have $q_2 \cap X = \emptyset$ or $q_1 \cap X$ is inhabited.[90] Equivalently: X is located if for each point P, the distance $d(P, X)$ exists.

8.10 The failure of the Institute for Philosophy

Any reader who consults Brouwer's list of publications, will be struck by the fact that this same man who stunned the mathematical world with his tremendous outpour of substantial papers before the war, now took his time and carried out his revolution in instalments. One possible, and plausible, explanation is that the world had started taking its toll. Whereas during the years 1908–1913, Brouwer spent almost all his time at research, now he not only had the duties that came with a professorship, but he had accumulated over the years a large number of extra activities. He was an editor of the *Mathematische Annalen*, a job he took extremely seriously. He studied the papers that were submitted to him thoroughly, corresponded with the authors, suggested and discussed improvements, etc. The time spent on editorial business was considerable, and Brouwer refused to cut a corner here and there; he had on principle all papers that were submitted to him copied (in typing or by hand), so that no author should put the wool over his eyes. In the pre-Xerox days that was an awfully time-consuming procedure! Furthermore, he was seriously involved in the organization and development of the philosophical activity, known as Signific Circle.

Generally speaking, Brouwer's lifestyle did not at all conform to the accepted idea of the single-minded researcher. He enjoyed his leisure, had an active social

---

[90] A is *inhabited* if it contains an element, the positive analogue of non-empty.

life, travelled a great deal in Europe, and corresponded with all and sundry. Cultural events also took up their share of the spare hours. He regularly frequented the concerts in the Concertgebouw, or even in The Hague, where his friend Peter van Anrooy was the conductor of the *Residentie Orchestra*. Indeed, the outburst of energy during his first topological period seems rather uncharacteristic!

All this time, while Brouwer carried out his professional duties, designed a new mathematics, refereed papers, and resumed his topology activities, he conscientiously played his role in the small group of (mostly amateur) philosophers, gathered around Van Eeden and Mannoury.

During the war, a good deal of activity had been going on a national level. Now that the international contacts could be re-established, the thought about a Forte Kreis-like institution surfaced again.

In 1918 the future of Significs, as a scientific- and an academic-discipline started to look somewhat rosier. In any case, the group started to attract some attention. Mannoury, who had been promoted to an ordinary chair (in analytic and descriptive geometry, mechanics and philosophy of mathematics) at the University of Amsterdam started a course in 'Mathematical Significs'. Frederik van Eeden, who had acquired a good deal of respect (and attracted a lot of abuse on the way!) was invited to give a talk on *Intuitive Significs* in the Auditorium of the University of Amsterdam. The small group consisting of Van Eeden, Brouwer, Mannoury, Henri Borel, Jacob Israel de Haan got together regularly to discuss business of the institute or to listen to an exposition of a fellow member. The company varied slightly, as new members were introduced from time to time. Usually they left after some time, and so there was basically a hard core of the above mentioned. From 1919 onwards Father Jac. van Ginneken also attended the meetings.

Van Eeden's diary contains a number of short notes on the meetings of the group, accompanied by incidental comments on the individual members. He had slowly come to see himself as the centre of an exclusive intellectual universe and considered others as convenient and necessary supporters. There is no doubt that he was quite genuinely fond of his friends. Brouwer, especially, was close to his heart, although he was not blind for the shortcomings of the brilliant, impulsive mathematician. An exception was Henri Borel, who reciprocated the dark suspicions of Van Eeden. The mutual dislike increased over the years, so that finally Borel refused to see Van Eeden any longer. In numerous letters to Brouwer, he complained about the perfidy of Van Eeden. The letters that have survived, suggest that Borel and Brouwer formed a jolly couple; Borel addressed Brouwer as 'Joost' (for what reason is not clear) and signed as *Borreltje*, which in Dutch stands for a glass of *Jenever* (gin). Since only Borel's letters have been preserved, we actually do not know what Brouwer thought about all this jocularity. There is a strong suggestion, however, that the pair of them were not averse to an occasional binge and a bit of philandering.

On 3 January 1919, Joop de Haan left the group; he emigrated to Palestine. De Haan had lived a surprisingly active life so far. He was a talented poet and author, albeit of a somewhat shocking sort. Most of his acquaintances would agree that he was querulous and neurotic; Brouwer called him hysterical and unbalanced.[91] In spite of his pronounced sexual inclinations he married Johanna van Maarseveen, a medical doctor. After a period of rejection of the faith of his forefathers, he eventually became an orthodox Jew and a fervent Zionist. As a logical conclusion, he exchanged his country for Palestine, leaving his wife behind. After his emigration he remained faithful to the signific cause. As a correspondent for the newspaper *Het Handelsblad*, he wrote about a wide variety of topics, including, for example 'Hebraic Significs'. Once in Jerusalem, he discovered that the Zionist claim was not as unproblematic as he had thought. When he started to write critically about the Zionist program, he became the target of abuse and threats. His hardly veiled homosexual practice did not much to endear him to his Jewish countrymen. Eventually he was executed on 30 June 1924 by a member of the Hagana.[92]

The International Institute for Philosophy never lived up to its name, as far as the predicate 'international' was concerned. A number of foreign scholars and artists was nominated for the membership of the International Academy for Practical Philosophy and Sociology,[93] but in the end only *one* foreign member joined the Academy (Eugen Ehrlich).

International traveling was in 1919 still a hazardous affair, trains were not for the faint-hearted, it took a compulsive traveller like Brouwer, to visit the former Forte Kreis members in Berlin, and to re-establish the contacts that had laid dormant during the war. True, the Gutkinds had managed to come and visit Van Eeden in his Walden, but they had not met the whole significs group. In particular they had missed Brouwer. And now, prodded by Borel, who could not stop praising the charms of Lucia Gutkind, Brouwer traveled as a kind of emissary of the significists to Berlin. Borel would have come along, were it not for the miserable conditions of the German railways. He had heard about the horrors of wagons where one had to stand for 14 hours, where the toilets were blocked by masses of luggage, so that one could not pee for 12 to 14 hours.[94]

The more adventurously inclined Brouwer survived the uncomfortable trip by train to Berlin; at his place of destination, still undecided about the Berlin chair (see p. 301), he asked Schoenflies to advise him on such mundane matters as housing and financial arrangements.[95] It appears that he was rather pessimistic about the general situation in Germany. In particular for a Dutchman, the economic perspective of life in Germany seemed rather bleak. Schoenflies' reply (if any) is unknown.

The planned visit to the Gutkinds proved more complicated than expected. The couple was away when Brouwer arrived. They wrote a worried letter to Borel,

---

[91] Van Eeden, diaries 22 January 1919.

[92] The assassin was interviewed after 67 years in a Dutch television program on 29 October 1991.

[93] The Institute was the Foundation supporting the Academy. For the structure of the organization, see p. 266 and [Schmitz 1990].

[94] Borel to Brouwer 18 December 1919.

[95] Brouwer to Schoenflies 29 December 1919.

fearing that they had missed Brouwer altogether. In the end everything worked out, Brouwer was immediately taken in by Erich and Lucia. The Gutkinds wrote enthusiastically to Van Eeden, 'Brouwer is with us and he will bring you our greetings. We get along wonderfully and we are enormously pleased with him'. Mrs. Gutkind added 'It was a great pleasure to meet Brouwer, ... '[96] A comparision of the mystical writings of Brouwer and Gutkind easily convinces the reader that the two were of the same disposition. It is not surprising that they got along wonderfully; although the future drove them geographically apart (the Gutkinds emigrated, like so many, to America), the friendship closed in Berlin lasted for a lifetime.

Erich came from a Jewish family. He was deeply religious and had good contacts with the leading Jews in Berlin, among others Gustav Landauer, Martin Buber and Walter Benjamin. Both Gutkinds were deeply interested in spiritual and cultural matters. They offered the intellectual environment in which Brouwer felt comfortable, and which engendered a lasting spiritual relationship.

We have already seen that, philosophically speaking, Gutkind was not really close to Van Eeden. Brouwer clearly offered the reception and response that Gukind had been waiting for. From Borel's letters, we can infer that Brouwer repeatedly visited the Gutkinds or met them somewhere in Germany, sometimes in the company of Borel. Reading Borel's letters one cannot suppress the thought that Brouwer must have made a strange impression on these travels. In one of Borel's letters we read, for example,[97]'I just want to tell you that I have returned, and that I did indeed find a room in the hotel Windsor in Cologne. I got the impression that they held you for a burglar or a Bolshevik, for I came in with an enormous aplomb [...] and immediately got the 'only room still available', for only 22 German marks (*Reichsmarken*). [...] I had no more bad luck after you disappeared so mercilessly. You look too much like one of the

Erich and Lucy Gutkind (Courtesy of the University Library Amsterdam).

---

[96]Gutkind to Van Eeden 15 January 1920.

murderers of Erzberger, or like a friend of Lenin or so'.[98]

The contacts between Brouwer and Borel were probably mostly based on the admiration of the latter for the former. Borel's letters are filled with plans for meetings and apologies for broken appointments. Among the assorted gossip, there are all kinds of complaints about people and things that did not conform to Borel's ideas. He was particularly critical of Van Eeden, whom he considered an enormous fake.

All the time, meetings of the signific group went on fairly regularly; The arrival of father Van Ginneken had introduced an element of Roman Catholic orthodoxy. This in itself did not cause problems, but it took all of Mannoury's intermediating capacities to keep the ensuing conceptual conflicts from flaring up.

In 1920 Barend Faddegon, a specialist in Sanscrit and Indian philosophy, joined the Institute, but by 1921 he had already left. A linguist, A. Verschuur, also had a short-lived interest in significs. He attended the meetings from 1921 to 1922.

The Institute for Philosophy did not flourish, as its founders had hoped; the international cooperation was not forthcoming, and in Holland the activities of the significists were watched with detached curiosity. The public at large mainly identified significs with Van Eeden, who was its main propagandist. But since Van Eeden's ideas were not representative, and were in due course coloured by his newly acquired sympathy for the Roman Catholic Church, his publications[99] did not do much for significs as a scientific enterprise.

In view of the failure to set significs on its way as a recognized discipline, the conspicuous lack of harmony, and, last but not least, the financial problems, associated with the publication of the *Mededelingen* (Communications), the remaining members, Brouwer, Van Eeden, Van Ginneken[100] and Mannoury, decided to dissolve the Institute (23 March 1922) and the supporting association,[101] and to found a new institution, the *Signific Circle*. This circle was founded in an atmosphere of happy informality at Walden on 21 May 1922,—'At the lawn beneath the chestnut trees [...] we founded our Signific Circle'.[102]

And so the stage was prepared for the twenties, Brouwer had got himself involved in a large number of projects (or affairs, one could say): the intuitionistic revolution, significs, editorial business of the *Mathematische Annalen* and the struggle against the boycott of German scientists. One major event of the twenties was not foreseen by Brouwer: his renewed involvement in topology.

---

[96] Gutkind to Van Eeden, 15 January 1920.

[97] Borel to Brouwer 20 September 1921.

[98] Brouwer was in a boyish way rather proud of the epithet 'Bolschevist', it would go too far to attach any political significance to this fact. Let us add that there were some unconfirmed rumours that Brouwer met Lenin in Zürich.

[99] Cf. *Het roode lampje*, (the red lamp), [van Eeden 1921].

[100] Still a visitor and not a full member.

[101] Although legally it was only 'suspended'.

[102] Diary Van Eeden, 22 May 1922.

Altogether, the auspices for the decade were not favourable, although Brouwer might not have viewed it that way. But for a man with a full-time academic position, the punishment for spreading himself too thinly could and would prove disastrous.

9

POLITICS AND MATHEMATICS

9.1 The *Conseil* and the boycott of Germany

Resuming the pre-war routine might have been simple for scientists in the neutral countries; for the citizens of the belligerent nations it turned out to be far from easy. The war had conjured up all those nationalistic feelings that before the outbreak of the conflict were quietly on their way out in the scientific community. In particular in mathematics, the otherwise abundant national differences, or animosities, had been virtually absent. Communication was free and open; international conferences, for example in Paris (1900), Heidelberg (1904), Rome (1908), were the meeting places of a rich variety of mathematicians from all nations. The personal relations between the leading mathematicians had more or less guaranteed the absence of malicious or secretive manœuvres. The war had not, perhaps, changed the professional respect for colleagues across the dividing line of the Central Powers and the Entente, but it had fostered a collective hysteria that in some cases overruled the personal common sense of scientists. Although mathematics, in this respect, did not suffer as much as other disciplines, even in this privileged domain of quiet contemplation unrestrained emotions surfaced, as we will see in this chapter.

In spite of the relative sober-headedness of the scientific community, there had been nationalistic tendencies in certain scientific quarters. As a rule, the leading scientists who frequently travelled abroad and who cultivated an extensive circle of correspondents were fairly immune from bombastic nationalistic feelings, but the larger scientific community was not always invulnerable where the virus of nationalism was concerned. We have seen that high-minded thinkers like the members of the *Forte Kreis* reverted to a 'right or wrong, my country'-attitude as soon as the war broke out, how much less could one then expect from the rank-and-file intellectual? In Europe the *Wilhelminische* empire suffered in particular from the tragicomic illusion of grandeur; it cultivated a cultural missionary feeling to a dangerous degree. It is not hard to find traces of irritation with this German sense of superiority. For example, one of France's top mathematicians, Emile Picard, published an essay with the title *The History of the Sciences and the Pretentions of German Science*[1]

Where the exhibitions of nationalistic pride had so far been just an annoying or ridiculous phenomenon, which one could shrug off as immature behaviour, the

---

[1] *L'Histoire des Sciences et les Prétentions de la Science Allemande*, Revue des Deux-Mondes, 1915. Cf. [Picard 1922].

outbreak of the war suddenly transformed the latent or not so latent national feelings into a dangerous mixture of aggression and self-pity. And so we see that, while the world at large showed every inclination to pin the blame for the war on Germany, the German scientists and intellectuals closed ranks and protested as one man against the real and imagined prejudices in the foreign press reports and in foreign political reactions. The violation of Belgium's neutrality, the sack of Leuven, and the barbaric treatment of Belgian and French civilians had indeed provided the international press with an abundance of material to write about.[2]

The German scientific community, following the general opinion in the country, reacted to the adverse publicity with a flurry of declarations. The German university teachers published an *Erklärung* on 7 September 1914. It was followed on 4 October 1914 by the fateful *Aufruf an die Kulturwelt* (Call to the Cultural World).[3] This declaration, known as the 'Declaration of the 93', was the source of a good deal of negative publicity. Brouwer had criticized it after its publication in the Academy meeting, cf. p. 248. The list of signatories of this document is a dazzling display of the flower of German science. Many of the leading scientists of the beginning of the century had lent their names for the defence of the national honour. Outside Germany, the declaration met with incredulity and indignation. The scientists who had signed were viewed with suspicion and contempt. We may note that among the large number of prominent physicists, chemists and representatives of the humanities who had signed the declaration, there was only one mathematician: Felix Klein. As so often happens, some were cajoled into signing and some had not even seen the statement they were signing.[4]

The scholars in the allied camp were not slow to publish *their* views; a group of English scholars replied to the 'Call' in an open letter to the Times of 21 October. Members of the Oxford Faculty of Modern History published a scholarly monograph 'Why we are at war'.

Whereas the war had a sobering effect on many of the directly concerned, in particular the young men in the trenches, many German scholars stuck to their nationalistic position. All during the war a strong conservative and annexionistic section of the academic community squarely supported the war efforts and ideology, opposed only by a small band of scientists.[5]

It is a well-known fact that wars and conflicts tend to radicalize the various positions, and that had the unfortunate effect of hardening the determinedness of the victors to 'settle the bill'.

Indeed, at the end of the war there was a feverish activity at the side of the allied scientists. Not surprisingly, the most vehement denunciation of the Central Pow-

---

[2] See [Tuchman 1962].

[3] For these and other details the reader is referred to [Schroeder-Gudehus 1966] and [Kellermann 1915].

[4] Klein, for example, had not seen the text and was under the impression that he supported a 'protest against foreign lies'. C.f. [Schroeder-Gudehus 1966], p. 76.

[5] There is a considerable literature on the role of German scholars during the war, see for example [Schwabe 1969], [Krockow 1990].

ers and their scientists came from the French and the Belgians, who had suffered horrific civilian and material losses, quite apart from their military losses. On various occasions the French and Belgian scientists vented their bitterness. Some of these statements have been preserved in the records of academic institutions and in newspapers. The leading and well-respected French mathematicians Emil Picard and Paul Painlevé[6] figured prominently on the post-war scientific-political scene. They spoke for a vast majority of Belgian and French scientists, who had experienced the war in person or in their families.

Picard, as the permanent Secretary of the Académie des Sciences, sternly rejected all contacts with German colleagues: 'The Academy judges that personal relations between scientists of the two groups of belligerents are impossible until the reparations and expiations, which have been made necessary by the crimes that have placed the Empires of the Central Powers under the ban of humanity, allow them to take again part in the concert of the civilized nations' (31 September 1918).

At the meeting of the Academy of 21 October 1918 he added: 'In order to restore the trust, without which fruitful collaboration is impossible, the central empires must renounce the political methods, the application of which has led to the atrocities that have filled the civilized nations with indignation.' Painlevé, doubtlessly, expressed the feelings of his colleagues and his countrymen, when on 2 December 1918, before that same audience, he expressed his views that the past war presented the picture of a desperate life-and-death duel between two concepts of civilization, without even the possibility of dialogue. 'It was about learning if science will be a means for liberating and ennobling mankind, or an instrument for its enslaving'.

He denounced the German scientific community in a fierce indictment: '..., at the other side of the Rhine, science was a gigantic enterprise, where a whole nation stubbornly continued to manufacture with a patient servility the most formidable killing machine that ever had existed.'[7]

One should of course not construe Painlevé's statements as a call for pacifism or non-violence. On the contrary, he was one of the advocates of the enlisting Science in the cause of warfare. At the session of the Academy of 7 January 1918, he elaborated on to the contributions of his learned colleagues and their students to the eventual defeat of the enemy. 'Your mission', he said, 'is the research of scientific truth, on which neither time, death, nor the human passions have any hold.

---

[6] Painlevé (1863-1933) had a double career as a mathematician and a politician. In 1910 he was elected as a deputy for the 5th 'Arrondissement', and in 1916 he was re-elected. He was minister of education (*Ministre de l'Instruction et de Invention intéressant la Défense National*) from 1915–1916, minister of war (1917) and in 1917 he became prime minister. He was president of the Chamber of Deputies in 1924 and in 1925 he again was prime minister. In 1925 he was, for an interim period, minister of finance, followed by a longer period as minister of war (1925 to 1929) and finally he served as minister of aviation (1930-1933). He was, in the meantime, incredibly active in innumerable projects, for example, at the request of the Chinese Government, he reorganized the Chinese railway system (1920). Where Painlevé was, so to speak, politically predestined to fight German influence, Picard had suffered a personal loss during the war: his son fell at the front.

[7] C.R. 167, (1918) p. 800.

Under the worst upheaval your reason will not deviate from its inflexible rules. But in the beleaguered city of Syracuse, Archimedes applied the rigorous correctness of geometry to the construction of gigantic catapults; who is therefore the scholar whose spirit will remain deaf to the call of his country (*patrie*) in danger'.

No doubt Painlevé would have viewed the military scientific projects on the other side of the Rhine in a totally different light. He would probably have argued that there is no symmetry in the situation of the aggressor and the victim.

For the French and Belgians the remedy was quite obvious: never again should German scientists be misused by national and military authorities. Since, in the instance of the Great War, if you did not live in Germany or Austria, there was a good reason for blaming the Central Powers, it escaped the politicians and the general public that the involvement of science in military projects was a general issue, not restricted to the Central Powers. The warning voices of those thinkers or politicians who managed to rise above a narrow nationalist viewpoint, Bertrand Russell and Romain Rolland, for example, were easily drowned.

The deep bitterness of the victorious but ravaged nations can be observed in many official documents and in the general press. The scholars of the city of Lille, for example, announced that they would not take part in any activity involving contact with Germans, stating that ' . . . one is forced to recognize that in a very general way, with only very rare exceptions, the German heart is closed to noble, generous or human feelings.'

As soon as the arms had been laid down, the Entente scientists started to formulate rules for a new Science organization, in which there was to be no place for the Germans. From 9 to 11 October 1918 representatives from Great Britain, France, the USA, Belgium, Japan, Serbia, Brazil, Italy and Portugal met at the *Conference of Inter-allied Scientific Academies* in London. The 'London resolutions' aimed at an organization of international science, while at the same time isolating Germany and Austria. The next meetings of the Conference were in Paris in November 1918, and in Brussels, 18–28 July 1919.[8] The general atmosphere of these meetings was not encouraging for open international cooperation; the French and Belgians were firmly resolved to keep the Germans out, and they competently and effectively manipulated the proceedings to that effect. Even though the official wordings were 'matter of fact', a good deal of sabre-rattling went on in and out of the conference halls. Generally speaking, the leading personalities representing the scientific community of the Entente nations saw themselves as the counterparts of the political representatives at the Peace Conference at Paris. They did not hesitate to let political motives prevail over the scientific ones.

The minutes of some Entente academies contained language that would embarrass most of the present readers. The victors were quite frank about their goals, and they did not mince their words. Only academics from Entente countries partic-

---

[8]One can find some records of the post-war scientific organization in *Ens. Math.* 20(1918) and 21(1920).

ipated in the discussions: ' Already now new societies—recognized as useful for the progress and applications of science—will be founded by the states that are at war with the Middle European Powers, possibly with the admission of the neutrals.'[9] Proposals for excluding the Germans were circulating, e.g. 'the Middle Europeans should be forced by the peace treaty to withdraw from all international scientific bodies'.[10]

A large-scale boycott of German scientists was envisaged. The Belgian astronomer Lecointe had already, in 1917, formulated a number of points in his memorandum for the Belgian government. In order to keep wavering colleagues in line, he proposed a black list of those who failed to observe the boycott rules, accompanied by punishments such as exclusion from public office and striking from honour lists.[11]

The influential Picard had also called for an exclusion of the 'nation that had placed itself beyond humanity', with the words 'There is too much blood, and there are too many crimes, separating us.'[12]

The meeting in Brussels saw the founding of the *Conseil International des Recherches*, with member countries Belgium, Brazil, U.S.A., United Kingdom, Australia, Canada, New Zealand, South Africa, Italy, Japan, Poland, Portugal, Rumania, and Serbia. Countries other than the Central Powers, were allowed to join, provided they subscribed to the rules of the London Resolution, and in particular joined the boycott of Germany.

The practical consequences of the allied activities in the field of science were varied, and differed from discipline to discipline. Some Associations were rather strict. For instance at the International Tuberculosis Conference in Lausanne (Switzerland) German was not admitted as a conference language, so that the Basel newspaper drily observed that it was made impossible for the delegates from the German speaking part of Switzerland to speak their own mother tongue in their own country.[13]

The actual boycott of the Central Powers concentrated on a number of issues: exclusion of Germany from international conferences, a ban on the German language in scientific discourse, a reallocation of central bureaux, institutes and councils to countries of the allied part of the world and the termination of the German monopoly of bibliographical and review journals.

As was to be expected, the boycott was answered by German countermeasures; a number of organizations took care of the interests of German scientists and scientific institutions. Among those, the *Reichszentrale für naturwissenschaftliche Berichterstattung* was a very effective one. It was led by Karl Kerkhof, a civil servant from the Bureau of Measures and Weights. Kerkhof organized the flow of scientific in-

---

[9] Ac. Royale de Belgique, Bulletin de la classe des sciences, 1919, p. 63.
[10] Picard, cf. C.R. 21 October 1918, vol. 107, p. 570.
[11] [Schroeder-Gudehus 1966], p. 107.
[12] Ibid.
[13] Basler Nachrichten 17 August 1924.

formation into Germany, and the converse flow of information from Germany to other countries. Thus the Germans actively organized their own counter-boycott in response to the Entente boycott.

This short excursion into post-war politics of science may help the reader to understand some of the events Brouwer got mixed up in. At the end of the war Brouwer was equally disposed towards the Allies and the Germans: his letters to Hilbert, Klein and Denjoy show that he wished both sides well. And the visit that he made with his friend Frederik van Eeden to the American Consul (cf. p. 285), on which occasion he advocated a council of scientists as the platform for peace talks, clearly demonstrated his trust in the rationality and goodwill of his fellow scientists. Although Brouwer's relations with his French colleagues had been somewhat less intense than those with his German colleagues, and in spite of his brush with Lebesgue, he fostered no animosity towards the French. It seems plausible that he kept himself informed of the developments at the Paris and Brussels meeting, which would not have been difficult, as the information would anyhow reach the Dutch Academy; and there is no doubt that he violently disagreed with the intended boycott of Germany. He soon found an occasion to demonstrate his feelings.

The *Conseil International des Recherches* had set itself the goal of creating new international organizations for the various branches of science, free from German influence and German members. The mathematicians had not been among the first to comply with the wishes of the leaders of the *Conseil International des Recherches*, who tended to reign over the scientific world in the style of the old absolute monarchs. In the absence of an international body, the French national committee of mathematics, presided over by the inflexible Picard, had taken the initiative by organizing, in September 1920, an international conference of mathematics in Strasbourg. Brouwer was approached by the secretary, G. Koenigs, of the committee with the request to provide a list of Dutch mathematicians. The choice of Strasbourg could not very well have been a coincidence; the city had repeatedly changed hands between the French and the Germans . In 1871 the Germans had annexed the town, and in 1918 Strasbourg and the Alsace changed hands again. The organization of an international conference in a contested (or newly acquired) town has always been an excellent means of demonstrating the new status quo, and of rubbing the adversaries' noses in the dirt.

9.2 The Nauheim conference and Intuitionism

The Germans, barred as they were from this international meeting, organized their own conference at exactly the same time in Nauheim. The *Naturforscherversammlung* at Nauheim opened on 20 September 1920, two days before the conference in Strasbourg. The organization of a congress of the size of the *Naturforscherversammlung* was a major operation so shortly after the war. Nauheim, a modest spa, north of Frankfurt a.M., had volunteered to house the participants (free of charge!)[14]

---

[14]The reader is referred to [Forman 1986] for more detailed information on the Nauheim Conference.

and to provide meals at reasonable prices. The town and its inhabitants took great pains to make the 2500 participants welcome. Indeed, the visitors could easily have imagined themselves in a pre-war country, with pre-war conditions! The Nauheim Congress is memorable mainly for the confrontation between the proponents and opponents of the young theory of relativity, culminating in a debate at the end of the Relativity Theory session, between two Nobel Prize winners, Einstein and Lenard. The topic of the theory of relativity was a hot item in Germany; it had met with fierce opposition from conservative quarters, and the debate had attracted interest beyond the world of scholars. Einstein had already been the focal point of a tumultuous meeting in Berlin, and the Nauheim meeting became a battle ground for the various undercurrents in physics, with a considerable political-cultural bias.

Compared to this spectacular event in the centre of the congress, the mathematical section was solid but not surprising. As we have seen, Brouwer presented his talk with the provocative title '*Does every real number have a decimal expansion?*';[15] it treated the fine structure of the reals in an intuitionistic setting. Just like his earlier papers this was a scholarly exposition with hardly a controversial note, cf. p. 325.

At the time of the Nauheim Conference, Brouwer had already published several papers on his new intuitionism. Most of these were scholarly and dry: it seems almost as if Brouwer was avoiding controversy at all cost. There is one mild exception: the paper *Intuitionistic Set Theory* of 1920, published in the *Jahresbericht der Deutschen Mathematiker Vereinigung*. In this paper Brouwer gave an exposition of the principles and ideas of intuitionism and also discussed his objections to the current views on the foundations of mathematics. Hilbert is mentioned in connection with his 'Axiom of the solvability of each problem', known as Hilbert's dogma. Brouwer pointed out that Hilbert's dogma and the principle of the excluded third were equivalent—as he had claimed already since 1908, and that they were obviously false.

In view of later developments it is, of course, of some importance whether Brouwer did or did not provoke the formalists through personal attacks or biased statements. The modern reader will agree that even the 'Intuitionistic Set Theory'—the only pre-Grundlagestreit paper that mentioned Hilbert—is a model of objectivity and courteous academic presentation.

The reference to Hilbert was neither provocative, nor aggressive, just factual. His comments on the principle of the excluded third, on the other hand, might have offended its staunch believers.

One might well ask if the Nauheim Lecture was the opening of the foundational dispute between Hilbert and Brouwer, now known as the *Grundlagenstreit*.

In all fairness one cannot but conclude that the answer is 'no'. The Nauheim lecture (or at least its published version) is a scholarly exposition that would not excite the most sensitive natures. There is no mention of either Hilbert or the excluded third. Of course, one does not know what Brouwer actually said in his

---

[15]The positive answer is usually taken for granted; even Brouwer saw fairly late that there was a problem. It was as late as 1919, that he explicitly asked himself if π had a decimal expansion.

lecture, but even if he did elaborate the criticism of traditional mathematics, verbal comments are soon erased from the collective memory. It is not clear what the impact of Brouwer's lecture was. As we have seen in Fricke's letter (p. 326) the more humorously inclined members of the audience did not see it as a serious challenge.

One would guess that Weyl, who at that time had adopted Brouwer's program, listened with approval, but apart from Fricke's reaction, little is known about the impact of Brouwer's message.

As we have seen in the preceding chapter, the situation radically changed when Hermann Weyl threw in his lot with Brouwer; the study of Brouwer's foundational papers, and doubtless the personal conversations, had convinced Weyl that his own program did not touch the heart of the problems in the foundations of mathematics. Weyl was not the person to practise his mathematical–philosophical convictions quietly in a corner. Often, when a new insight was mastered, he presented the outcome in grand and sweeping formulations to the scientific world.

In the provoking paper *The new Crisis in the Foundations of Mathematics*,[16] Weyl forcefully expounded the intuitionistic principles and criticism. The paper was widely read and some of its slogans became the watchwords of the twenties and thirties, cited by friend and foe. After a try-out at the mathematics colloquium at the ETH in Zürich on 2, 9 and 16 December 1919, he lectured on the topic in Hamburg on 18 July 1920.

Weyl published his 'New Crisis' paper in the *Mathematische Zeitschrift*, thus, no doubt with good reasons, bypassing the *Mathematische Annalen*, where Hilbert might possibly have raised objections, which would have, if not stopped, at least delayed the publication. One should note, however, that the paper, although written in a provocative style, did not mention Hilbert or his formalism. Weyl altogether avoided personal attacks. Nonetheless there could be no doubt as to his allegiance; he mentioned with approval, for example, Kronecker's views. The 'New Crisis' paper captured the imagination of the contemporary mathematician because of its imaginative use of language and its provocative claims. It was to dominate the discussion for some time to come.

Weyl's exclamation, '*Brouwer—that is the revolution!*', became a rallying cry for some, and a threat to others. Weyl's battle cries were to haunt the mathematical community in Germany for years to come, and doubtless they angered Hilbert, who saw his favourite student desert him for a dangerous eccentric. Weyl did by no means intend to give up mathematics for the study of its foundations. His interests were too wide ranging to keep him fettered to one particular topic for a long period. Already at the Nauheim Conference, he lectured on mathematical physics, and his greatest work in mathematics was still to come, but not in the foundations of mathematics.

The intuitionistic ideas were certainly not ignored, but on the whole poorly understood. As Weyl had put it, it was considered more as 'interesting news from the border provinces', than as fascinating and worrying developments one had to learn

[16] [Weyl 1921].

about. In Göttingen things were taken seriously, in so far that Bernays and Courant gave talks at the Göttingen Mathematical Society on 1 and 8 February 1921, 'On the new arithmetical theories of Weyl and Brouwer'.[17] One may note here the adjective 'arithmetic'. From Brouwer's and Weyl's papers it should have been clear that the 'new' theories were not arithmetical; somehow it took a long time before even prominent mathematicians understood the true nature of Brouwer's program!

The general misconception, which survived into the late thirties, was that Brouwer (and hence Weyl) was a constructivist of the kind that only recognized the finite combinatorial part of mathematics. Even before Brouwer's new program, this would have been a doubtful proposition, but after 1918 nobody could have defended that view. Judging from the remarks in papers from the period, one can but guess how the general mathematical public got its information on Brouwer's ideas. The reception of Brouwer's ideas must have been slow; one explanation is that, before 1918, hardly any material was in print in the internationally accepted journals. Up to that time the dissertation and its 'appendix', the unreliability paper, gave the most extensive exposition of the foundational views, albeit in Dutch. Furthermore there were the Rome paper on the possible cardinalities and the translation of the inaugural address. In addition there was the review of Schoenflies' new version of the *Bericht*. One had to be a discerning reader to draw the right conclusion from the material! It was certainly tempting to view Brouwer simply as another Kronecker, that is, a constructivist with a bias towards the discrete. Although Brouwer was quite explicit in his dissertation about the unique irreducible nature of the continuum, it was not easy to explain the significance for ordinary mathematics. It is therefore not surprising that the more abstract, infinitistic features of his mathematics escaped a lot of readers. Even Bernstein, who was a personal friend of Brouwer, classified Brouwer, together with Poincaré, Richard, Borel and Lindelöf, as a finitist.[18]

One wonders, was Brouwer just poorly understood, or did he keep his foundational convictions to himself before 1918? The latter seems improbable in view of Brouwer's personality. We know at least that Hilbert was the one-man audience of an exposition on the foundations of mathematics (including the language levels) in 1909 at their joint walks in the dunes of Scheveningen, cf. p. 130.

9.3 The Denjoy conflict

Soon after the Nauheim Conference, Brouwer was plunged into a political debate that a more cautious man would have sought to avoid. After the conference, Blumenthal visited Brouwer in Laren and used the occasion not only to discuss general matters, but also to ask the expert opinion of the Utrecht Professor Arnaud Denjoy on a paper submitted for the *Mathematische Annalen*, of which Blumenthal was the managing editor. The first symptoms of a new conflict, which was to drag

---

[17] One would guess that Bernays' views on Brouwer and Weyl were along the lines of his paper *On Hilbert's thoughts on the founding of mathematics*, JDMV 1921.

[18] [Bernstein 1919], *Cantor's set theory and finitism*, JDMV 1919.

on, and which was mostly fought by only one participant, announced itself in a letter from Denjoy to Blumenthal—*chez M. Brouwer, à Laren*.[19]

In this letter Denjoy explained to Blumenthal that in spite of his fond recollection of their contacts during the Rome Conference in 1908, the time for a renewal of personal relations had not yet come. Instead of a brief refusal, the letter contained a list of arguments for avoiding contacts with the former enemy:

> The visible causes of a renewal of the conflict between our two countries have been far from eliminated. One must have seen the state of utter powerlessness of certain regions in the north and the north-west of France, and have measured the effort of work and of credit necessary for their restoration, to explain that the French people, being affected so strongly in their means of production, will not consent to bear that burden alone and to spare their enemies of yesterday, who are less sorely tried.
> Should Germany be preparing itself to elude its obligations, should France see itself forced to make it comply with force, then the initiative already taken by a French scholar to relax his reserves with respect to a German colleague before the permitted hour [...] would at this moment seem full of inconsistency and frivolity.

As soon as the French government thought that it had received from the German government guarantees of goodwill, and had judged them sufficient to elevate Germany again to the rank of normal nations, Denjoy would have no basic objections to renewing his relationships with those German scholars who had not displayed provoking behaviour by clamorous publicity. Until that moment, however, Denjoy chose to stick to the instructions that the policy of his country dictated to him.

The letter of Denjoy illustrates the bitterness that prevailed after the war. At the time of the writing of the letter, the peace treaty with Germany had been concluded in Versailles, but that had not automatically restored the pre-war situation. The size of the reparation payments had been set in June, but the actual implementation of the payments was to give rise to endless trouble, including a number of occupations of German territory by the French and Belgians, intended to guarantee the payments. It is an accepted historical fact that some of the allied nations displayed an understandable but short-sighted vindictiveness towards the former adversary. As a partial explanation, it should be pointed out that the Belgians and the French had borne the brunt of the almost apocalyptic destruction and killing of the Great War.

Denjoy had adopted a formal position; he stated that his behaviour was dictated by the official French position. Today that seems rather harsh, but one cannot ignore the atmosphere of the Great War and its emotional impact. One wonders if Denjoy himself was a wholehearted supporter of the strict measures of the French government, or whether he was hiding behind a convenient excuse? Whatever his

---

[19] Denjoy to Blumenthal 4 October 1920.

views were, he soon had no choice but to stick to his guns.

Brouwer was no sooner informed of this cold rejection of Blumenthal's request than he took the side of the insulted party. Characteristically, he did not write an avuncular letter which might have had a soothing influence, but instead dispatched a coolly argued indictment of the misguided behaviour of a professor in the service of the Kingdom of the Netherlands. In view of the friendly relationship between Brouwer and Denjoy, one would perhaps have expected an attempt to arrive at an understanding that would allow for some sort of compromise. But such a thing was not in line with Brouwer's strict code of behaviour. Misconduct of this sort could not be tolerated!

To avoid possible misunderstanding, it should be pointed out that Brouwer did not engage in this somewhat quixotic conflict because he had a bone to pick with Denjoy or because there was some hidden animosity. The correspondence prior to the Blumenthal incident had been cordial; nothing had foreshadowed this abrupt stern reaction. As a matter of fact Brouwer explicitly mentioned in a later letter to the Minister of Education that he had never detected any chauvinistic tendencies in Denjoy, and that, on the basis of personal experience, a civil reception of Blumenthal had been taken for granted.[20]

The relationship between Brouwer and Denjoy had been one of mutual respect and understanding. Denjoy appreciated Brouwer's topological fame, and Brouwer had enthusiastically welcomed the young analyst in Dutch mathematics. The correspondence preceding the conflict showed all the signs of a beginning friendship. The nationality of Denjoy played no role at all. In general Brouwer treated his German colleagues and Denjoy on strictly the same footing.

Arnaud Denjoy.
Courtesy M. Loi.

On the occasion of the Versailles peace treaty Brouwer had not only sent letters to Klein and Hilbert, but also to Denjoy, who had acknowledged this the same day: 'Infinitely many thanks for your sympathetic thought to congratulate me with the great events which at the moment take place in the history of my country. [...]'

Denjoy's policy was not without its nuances. On one occasion he expressed his regret at not being able to meet Carathéodory, when Brouwer told him of Carathéodory's stay in Holland.[21] He excused himself on the grounds of an overloaded schedule. He remembered Carathéodory well from the Rome Conference, he said, which he had attended with Montel and other colleagues. It is, of course,

---

[20] Cf. page 15 of an open letter of Brouwer to the Minister of Education, 27 September 1922.
[21] Denjoy to Brouwer 14 October 1919.

possible that Denjoy did not wish to meet Carathéodory, but did not want to give offence either. Anyway, Brouwer could not have foreseen Denjoy's reaction to Blumenthal's request. In the above mentioned letter to the Minister of Education, Brouwer gave a complete exposition of the whole affair, explaining that Blumenthal had been staying with him in October 1920 for the purpose of discussing business matters of the *Mathematische Annalen*. Blumenthal had a manuscript of H. Hake, which touched on the area of research of Denjoy; he felt that it would be good to consult Denjoy, but apparently he was hesitant, on account 'of the position adopted by several French mathematicians after the fall of Germany with respect to German colleagues'. Brouwer advised Blumenthal to contact Denjoy, as Denjoy in his role of Dutch civil servant and ordinary member of the Dutch Royal Academy, was to be considered as a Dutch rather than French mathematician, and hence would not harbour any grudge against Blumenthal. Furthermore, the position of Denjoy in the Kingdom of the Netherlands (as a professor, he had been appointed directly by Her Majesty the Queen; professors were so-called *Kroon docenten*, Crown-teachers) would in any case guarantee a courteous treatment of Blumenthal. In particular one might expect so in view of the rule of the Academy which said that 'the Academy was intended as a means of union between the scientists of the Netherlands and those of other countries'. Blumenthal, thus encouraged, had approached Denjoy, with the above result.

The possibility that Denjoy had a personal grudge against Blumenthal seems highly unlikely; Blumenthal was generally well liked, and no personal feud between the two men at that moment is known. Blumenthal had served in the army, but it is hard to say whether Denjoy was aware of that fact.

Denjoy's reaction to Blumenthal's request must probably be seen in the light of the enhanced national consciousness at a time of international conflict. This, in combination with the strict guidelines of the Conseil and the leading French mathematicians, may have forced him to do things that one would regret in a more sober mood.

When, so unfortunately, Blumenthal was rudely rebuffed, Brouwer's never failing sense of justice dictated the logical next move in this sad tragi-comedy. He wrote a letter to Denjoy stating that:

> You will no doubt have realized the consequences of this incident for our mutual relationship, since the duties of hospitality oblige me to consider behaviour towards my guest as touching me personally.[22]

He added that his opinion of the political task of scholars ('in particular of ourselves, academicians of the neutral countries') was diametrically opposed to that of Denjoy.

This short note opened the sluices of Denjoy's eloquence, almost as if he had been waiting to lecture the Dutch on their neutrality. Three days later he spelled out his views on the rights and duties of hospitality, but above all on the behaviour of the Dutch scientists. After rejecting Brouwer's view on the rules of behaviour towards someone's guests, he went on to say that if Brouwer had written a letter to

---

[22] Brouwer to Denjoy 17 October 1920.

him from Nauheim (the place of 'the congress organized by the Germans and coinciding with the one in Strasbourg, which was of no less scientific interest than the first mentioned') he would have replied in his usual cordial way, but that the German hosts would have been mistaken to think that these feelings could be assumed to extend to them.

The letter goes on to state, in a fairly moderate tone, that once the suspicions against the Germans had laid themselves to rest, Brouwer's efforts for a reconciliation would be welcome, but that he would not find things all that easy. Moreover, he warned Brouwer not to try to impose his sympathies for persons of one of the sides on those of the other side. 'The French', he said, 'have no taste for orders, not to give and even less to take.' Denjoy repeated that once the suspicion against the Germans was officially lifted, Brouwer's attempts to re-unite the former enemies was perfectly in order. He was quite willing, for example, to put his reprints at the disposal of German scientists,[23] but he thought it was premature to try to bring authors belonging to the two nations together.

Denjoy had in particular been upset by Brouwer's phrase 'we, academicians of the neutral countries'; he discerned a tendency to subject him, as a member of the Dutch Academy, to obligations incompatible with his French citizenship, and this he resolutely declined to accept. Even stronger, if there had been obligations of the sort Brouwer suggested (or worse), then the choice between the membership of the Academy and his simple citizenship of France would not have been difficult.

'The general opinion', he wrote, 'lends to scientists, more perhaps than to the majority of individuals, businessmen, for example, a kind of national character which must make those scholars very circumspect, when it is a matter of lending their persons to conciliatory actions which could be criticized by sensible patriots.'

Denjoy's letter is typical for the curious situation in which many scholars found themselves after the war. They were obliged to follow the official directions in so far as international scientific contacts were concerned. What seems surprising, in the case of Denjoy, is that there is no trace of regret at the situation; he seems heartily to support the French position on contacts with colleagues from the Central Powers. One would expect at least some signs of reticence, where personal and scientific matters are concerned. Let us, however, not judge too hastily here: Denjoy may have had private reasons for his attitude concerning the Germans, or he may have simply been swept along by the popular sentiment after the capitulation.

The above letter is in perfect accord with the official position of the French, as laid down and pronounced at the meetings of the *Conseil International des Recherches*: no contact with Germans and Austrians, until satisfactory promises have been given and reparations made. It lacks the harsh rhetoric of, for example, Picard, but it also lacks personal feelings towards those German colleagues that Denjoy must have known personally.

---

[23] As he already had done at Brouwer's request.

Brouwer's reply, a week later, was brief.[24] He did not see the purpose of a continuation of the discussion, for example, on hospitality. He could not resist pointing out that the hospitality of the conference in Strasbourg was offered him, because of the coincidence of his place of birth. As to the role of academics of neutral countries, he quoted the minutes of the particular meeting of the Academy (31 October 1914), at which Brouwer had commented on the role of scientists in the context of the war, in response to the 'Declaration of the 93'. He merely wished to draw Denjoy's attention to the rules for ordinary members of the Academy (in contrast to those for corresponding or foreign members) which stipulated that the Academy was to be a *means for uniting scholars of the Netherlands and those of other countries.* According to Brouwer, Blumenthal's scientific questions to Denjoy were justified on the grounds of this rule. He was surprised, he wrote, by Denjoy's interpretation of the rules regarding ordinary members of the Dutch Academy, but he wished to abstain from comments, since 'in the end it only concerned Denjoy and the Dutch government'. The letters so far showed some emotions, but no hostility. The flowery declarations of respect were certainly not insincere. Denjoy had ended his letter of 20 October 1920 with the words: 'There should not be any doubt that the sentiments expressed in this letter do not diminish in the least the great regard with which the mathematicians consider you and with which I associate myself without reserve.' Brouwer answered in kind: 'It is superfluous indeed, my dear colleague, to say that on the one hand I infinitely regret the circumstances that put a distance between me and a man of your value, and on the other hand that the circumstances diminish in no way the feelings of respect I have for you.'

Denjoy swiftly hit back, as if stung by a viper.[25] He was not, he said, going to consult the rules of the Academy; he would be surprised if the Academy forced its members to do the things Brouwer had mentioned. The Dutch government would not chase a Frenchman from its Academy just because he had declined to receive a German, if only because an incident of this sort would give too much pleasure to the enemies of the good relations between France and the Netherlands. The letter also contained a curious passage which hinted at a poorly veiled disapproval of the position of a small nation that did not take part in the war, that did not suffer, yet ventured to express its opinion in international affairs touching on the consequences of the war.

> The French opinion—the only one which, ultimately, concerns me (my letter to Blumenthal has elucidated this point)—the French opinion is already rather unfavourably towards Holland. It feels very clearly that in the eyes of certain people here, the disappearance of France and its civilization would have been a minor accident. It would not have been a pity if the world had become German. The aggression of 1914, the German crimes, for four years on land and at sea; these are cute sins. It serves no purpose whatsoever to bolster with my compatriots this (indeed inexact)

---

[24] Brouwer to Denjoy 27 October 1920.
[25] Denjoy to Brouwer 29 October 1920.

belief that this is what the Dutch think.

The last part of the letter refers to something that apparently was incomprehensible to Denjoy, that is Brouwer's reference to the 'accident of his birth place'. Denjoy had hit on a problematic issue: almost all speakers mean by the phrase 'coincidence of my birthplace' something like, 'if I had been born in China, I would have been a loyal Chinese, but since I was born in Greece, I am a loyal Greek'. From Brouwer's lips it meant something else; as we have seen in the case of his review of Van Eeden's anti-nationalism brochure, Brouwer felt himself a true internationalist. Nationalism, for him, was one of the basic red herrings of social training. Possibly, Brouwer and Denjoy had discussed this topic before, or else Denjoy was an extremely keen reader and observer. Denjoy sketched in most eloquent words the ties that bind people to their nation, and that made him a Frenchman for better or worse. The eulology has a sincerity and beauty that official addresses usually lack, and so I cannot resist the temptation to reproduce a translation here.

> Your letter informs me that with respect to Holland you only recognize vague ties created by the coincidence of a birth place. Don't you exaggerate your indifference? If the Belgians or the English had invaded the Netherlands, wasting and destroying the richest region, from Rotterdam to Amsterdam, killing three hundred thousand young men, maybe you would experience enough repulsion towards the aggressor to feel yourself Dutch.
>
> Your obligations towards the fortuitous country of your birth forbid you to be a participant at the Strasbourg Congress and allow you to be in Nauheim. I would have understood, if you had equally turned down both the German and the French invitation. Your duties towards Holland imply singular kinds of strictness and of compromise.
>
> In addition to the kinds of countries that you observe, one also counts the country of affinity, a category whose existence you will find difficult to deny.[26]
>
> But above all, you forget the country of nationality. Belonging to a nation brings burdens, but also benefits. It is for each man both a matter of honour and of wisdom to join a nation in some destiny.
>
> I dare to congratulate myself for uniting in one single country my countries of nationality, affinity and birth.
>
> It is not by chance that I come from Gascogne. My forebears have lived without interruption in that corner of France for many generations; it would have been accidental if I had been born somewhere else. I don't hesitate to feel myself a member, and a very humble member, of one and the same family with the human beings who have made my language, which is incomparably superior to all others for its rigour, its precision, its immaterial strength, totally vigorous as it is. That language is perfectly suitable for providing expressions with a certain spiritual sense, which I

---

[26] Brouwer had referred to the country of birth and of activity. (letter of 27 October 1920).

believe to observe in myself. And it lends itself poorly for translating the confused mental dispositions which offend my nature, and in which many foreign spirits find their pleasure.

I recognize myself in the aversion of the French intellect to crude authorities, charlatanism and the appeal to childish curiosity.

Among the dominant traits that belong most particularly to the French people, there is this one, that gives me a strong inclination (*goûte*) to resist opposing characters. I don't know people where one can observe more careful self-criticism and more reluctant self-admiration. There is no other nation on which noble arguments have more effect, and base arguments have less grip. One cannot make this people go to war in the hope that it will get them rich. All these affinities determine my impression that I am not a Frenchman by accident.

I am touched by your respect; I never asked that much. Less respect for me and less antipathy for my country would have satisfied me better.

The above letter is an example of the patriotism that was fostered by the war, and that was—contrary to what Denjoy might have thought— not a French prerogative. Some of the claims made by Denjoy were traditional exaggerations that one should not take seriously; the superiority, for example, of the national language has been claimed by a rich variety of nations. There is, however, an undertone that seems to accuse Brouwer and the Dutch of an anti-French disposition. There seems to be little evidence for such a claim; until 1920 Brouwer published in the *Comptes Rendus*,[27] and his personal relationships with French mathematicians was affected by neither the war, nor the brief Lebesgue conflict. The circumstance that his contacts after 1920 were largely German can be explained by the fact that the developments in the foundations largely by-passed France. Privately, he was a frequent visitor to France. As for the Dutch, there was indeed a certain amount of pro-German feeling among the population, but this was certainly not a dominant feature.

Brouwer considered the above letter totally unsatisfactory, he could not share Denjoy's views, and he could not find a coherent answer to his remark concerning the rules of the Academy. He returned the letter[28] with a few lines:

Dear Sir,
The lines below, which I am truly unable to grasp, I do not wish to keep.
I urgently beg you to direct no more letters to me.
In the meantime be assured of my sincere esteem.
L.E.J. Brouwer

In the typed copy of the letter which is in the Brouwer archive, Brouwer had added in the margin a remark to the effect that Denjoy had misinterpreted his reference to the relationship between the Academy and its members. Indeed Denjoy's answer

---

[27] Only in 1950 Brouwer resumed his habit of submitting papers to the *Comptes Rendus*.
[28] But not before making some copies.

is hard to understand; Brouwer had intended to say that the relationship between a member and the Academy only concerned the member and the government of the Netherlands. He wrote in the margin

> This quotation is incorrect and must *perhaps* explain the—for me—incomprehensible sequel. If you really think that the government of the Netherlands has nothing to do with the way in which rules, made by her, are interpreted by the civil servant concerned, then I will not quarrel with you. What I wanted to say, was no more than that apart from the civil servant concerned and the government, *certainly no third party* should be involved, and you apparently agree to that.

So Brouwer wanted to say that the relationship between Denjoy and the Dutch Academy (viz. the government) did not concern the French government.

This was indeed the end of the exchange of letters between Brouwer and Denjoy; it was not, however, the end of the affair. Brouwer felt that a regular member of the Dutch Academy, irrespective of his nationality, had to observe the rules. He feverishly tried to get the Academy on his side. All through the year 1921 he compiled a Denjoy file, and hoped to activate the Academy or even better, the Minister. He did not stand quite alone in this respect. Apparently there was a growing irritation with the demeanour of Denjoy, and Brouwer, already displeased in 1920 with the fact that Denjoy could still not lecture in Dutch after four years, decided to force the language issue in the Academy.

He must have been rather pessimistic about the outcome of this particular action when he wrote in a letter of 23 September 1921 to various colleagues asking them to sign an enclosed letter protesting at the behaviour of Denjoy. He told his fellow members that Denjoy considered himself rather an *agent de pénétration pacifique* in the service of the French government, than their colleague. If the addressees found the letter unfriendly with respect to Denjoy, he added, they should realize that Denjoy in his function in Holland refused to obey the rules of the Academy, but on the contrary accepted instructions issued by the government in Paris.

Brouwer had no illusions concerning the reaction of the Academy 'which is already halfway in the hands of French imperialism'. He just wanted to encourage 'those who found the responsibility for this new mocking of our native country too heavy, to remove it by means of this protest.'

Hk. de Vries, who was one of the recipients of the letter, was in fact upset by its sharp tone. Like Brouwer, he disliked the way Denjoy operated in Holland; he spoke of 'that Frenchman, who despises everything here, except our money (*onze rijksdaalders*)'. He pointed out to Brouwer that the legal basis for an action against Denjoy's choice of language was extremely narrow. De Vries' advice had little effect, Brouwer just carried on, regardless of the consequences and the remote possibility of success.

In a letter to the chairman of the physics section of the Academy, Brouwer denied the personal nature of the exchange. In his opinion Denjoy's invectives were more directed at the institution behind Brouwer (that is the Academy and the international community of scientists). Anyway he said, Denjoy blew his horn *pour la Galerie Parisienne*.[29]

The *Algemeen Handelsblad* of 29 October reported the meeting of the physics and mathematics section of the Academy and repeated verbatim a letter of Brouwer to the section's board. The letter explained the reason for Brouwer's absence from this particular meeting with the words:

> Since for part of the agenda of the ordinary meeting of the section, according to the convocation, another language than Dutch has been chosen as a medium, and, in the opinion of the undersigned, in public meetings of mandatories of the Dutch Government no departure from the official language of the Kingdom of the Netherlands should be allowed without previously obtained permission, it is impossible for the undersigned to attend the present meeting.

Brouwer had already, before the meeting, informed F.A.F.C. Went, the chairman, of his intention. Should Denjoy again address the meeting in French, then he, Brouwer, would demand that the above message of his be read in the meeting. And to make sure that his gesture would not go unobserved, he informed Went that the press had received a copy in advance. In itself this was no breach of confidence, as the press had free access to the documents of the public meeting anyway. He wrote to Went that

> my respect for the Dutch Government and for the national character of the Academy forbids me to take part in sanctioning by my presence an improper act, such as the use of a foreign language by an official appointed four years ago, is in my eyes. I will give a copy to the press.

A month later the *Algemeen Handelsblad* needed no prompting by Brouwer to report the meeting.

Royal Academy of Sciences

> In the present meeting of the section of the Academy for mathematics and physics, professor Brouwer, after the reading of the minutes, asked for the reading of a note which he had sent to the Section.
>
> The chairman, professor Went, said that the particular note had been discussed in the extra-ordinary meeting.
>
> Professor Brouwer pointed out that the letter was directed to the ordinary meeting, and that he maintained his request to have it read. Only one member[30] seconded professor Brouwer, upon which he demanded permission to make a statement.
>
> The secretary, professor Bolk, pointed out that professor Brouwer's desire for publicity had already been fulfilled, because he had published the

---

[29] Brouwer to Went 13 November 1921.
[30] Hk. de Vries.

note in the *Algemeen Handelsblad*.[31]

Thereupon professor Brouwer requested that it should be recorded in the minutes and in the report of the ordinary meeting, that he had been refused the right to read a message to the ordinary meeting.

Thus the matter was closed.

Faced with the flat refusal of the Academy to hear his case, Brouwer decided to go directly to the press; he sent a letter to the editor of the *Algemeen Handelsblad*, dated 27 November 1921, explaining to the public the issue of the 'official language'. He made it clear that he wished to be absolved from sharing the responsibility for allowing the unauthorized use of foreign languages in meetings of the Academy. The refusal of the meeting to take notice of this abuse, 'which constituted with respect to the undersigned a violation of the elementary rights of a minority', could only leave the readers of the reports of the Academy with the incorrect impression that the national character of the Academy could be disregarded without protest.

On 19 December Brouwer again sent a letter to the editor of the *Algemeen Handelsblad*; this time to correct an inaccuracy that had slipped in. He had erroneously claimed in his earlier letter to the editor, that the readers would be informed of the lapse of the Academy in the minutes of the meeting of 29 October. According to recent new regulations the records of meetings, etc. would however be published in the yearbook. The above mentioned incorrect impression, caused by 'the strangulation of the protest of the undersigned', would thus be noted 'by the international readership' at a later moment.

The effect of Brouwer's actions must have been limited; even at the best of times the Dutch were not very sensitive to national honour; anything that smelt of pompousness was easily and happily ignored.

In the meantime the fears of the board of Utrecht University, that Denjoy would consider Utrecht an intermediate station in his career turned out to be well-founded. In 1922 he left Holland to become *Maître de Conférences* in Paris, and so the person who had caused the conflict no longer figured in Dutch mathematics.

Denjoy's mathematical influence in Holland had been stimulating indeed. He gave courses and seminars which were attended by students from all over the country. He even went so far as to inquire with his colleagues, whether they had any clever students who might be interested in working for a Ph.D. thesis with him! Brouwer sent his own student Belinfante to Utrecht to attend the courses of Denjoy. Van der Corput and T.J. Boks were assistants of Denjoy; Boks, H. Looman and J. Ridder were his Ph.D. students in Utrecht. There is no evidence that Denjoy ever reconsidered the matter after the last letter was exchanged. Not so, however, Brouwer. When it became known that Denjoy was about to leave and thereby automatically would be made a corresponding member of the Academy, Brouwer made a final attempt to bring him to justice. He wrote a long letter to the Minister of

---

[31] strictly speaking this was not the case. Brouwer's note was printed as a letter to the editor, but reproduced within the report of the meeting of 29 October 1921.

Education, Arts and Sciences, accompanied by a complete dossier.[32]

This letter summed up the events, starting from the request for advice by Blumenthal. Somewhat bitterly, Brouwer pointed out that Denjoy's role in the matter resembled more 'that of an agent of the French Government in a territory under French sovereignty, than that of a Dutch civil servant in an independent Kingdom of The Netherlands'. On the basis of Denjoy's interpretation of the rules of the Academy and their applicability, and the real or seeming disregard for the obligations of Academy members with respect to the Dutch Government, Brouwer concluded that the automatic transfer of the ordinary membership to a corresponding membership should not be granted, at least until Denjoy should have given the Dutch Government the necessary explanation and satisfaction. He expressed his hope that the Minister would act accordingly. The Minister did what a politician automatically does: he asked the Academy for advice; the Academy appointed a committee, which duly reported, through H.A. Lorentz, its chairman, the findings to the Academy on 27 January 1923. The report stated that Denjoy had, in his dealings with Blumenthal, been guided by his feelings as a Frenchman, rather than by the circumstance that he was a member of the Academy.

Uebersetzung.
Als Manuskript gedruckt.

Unberücksichtigt gebliebenes
Schreiben von Dr. L. E. J. Brouwer
an
Seine Exzellenz den Minister für Unterricht,
Kunst und Wissenschaft
vom 27. September 1922.
Symptomatisches zu einer Gefährdung der niederländischen Staatshoheit.

Open letter to the Minister of Education

> It would certainly have pleased us if it had been otherwise and thus could have led to some reconciliation. All the same, we are of the opinion that the recovery of good relationships between scholars of mutually estranged countries, which we hope to be the case in the future, would not be furthered if we acted with regard to Mr. Denjoy, according to the ideas of Mr. Brouwer, and demanded an explanation.

The Academy advised the Minister accordingly. This left Brouwer far from satisfied. In another letter to the Minister he once more explained the issue, which he considered grossly misrepresented in the committee report.[33] With admirable perspicuity (which would have found a more rewarding goal in mathematical research) he pointed out that the treatment of Blumenthal was completely irrelevant; what mattered was Denjoy's attitude with respect to the Academy, a member of the

---

[32] Brouwer to Minister of Education 27 September 1922.
[33] Brouwer to Minister 12 February 1923.

Academy, the Dutch Government and the Dutch people. He, understandably, did not share the views of the committee, and suggested that the Minister ask for the desired explanation, and terminate Denjoy's membership in case of refusal. The dismissal of members of the Academy, by the way, was far from simple (it even required the Queen's signature). The Minister, in view of the modest importance of the problem, simply followed the advice of the Academy.

The Academy must have been rather displeased that Brouwer, in spite of the report of the committee, had thought it necessary to approach the Minister directly. At the meeting of 27 January, Brouwer defended his action in the (closed) meeting of the Academy.

This was the end of this rather quixotic episode in Brouwer's career; his emotional reaction to the insult of Blumenthal, coupled with his indignation at the treatment of the defeated Central Powers by, of all people, his fellow scientists, had carried him further than he might have foreseen. More than the earlier conflicts, this affair gave him the reputation of quarrelsomeness. He even went so far as to have his petition to the Minister printed at his own cost, including a German translation, with the ominous subtitle 'Symptomatic observations on the jeopardizing of the Sovereignty of The Netherlands'.[34]

The whole Denjoy affair and the way it was handled by the Academy left Brouwer emotionally exhausted. It incapacitated him for months; in a letter to Mrs. Ehrenfest he confessed that ' ... since the Dutch Academy of Science has been pulling the loot wagon of the Parisian Shylock gang as the n-th yoke of oxen (and allows the members who do not swallow this humiliation to be abused by the Shylock lackeys) I have fallen into such a state of disillusion and apathy that most of the incoming letters are not answered by me'.[35]

Denjoy's career did not suffer through this intermezzo; perhaps, on the contrary, it was furthered. Among the many honours that befell him during his career, there was the *Legion d'honneur*, awarded in 1920.[36] The motivation contained the following passage: 'Scientist of very great value. Was sent as a professor to the University of Utrecht; has rendered through the success of his teaching the greatest services to the French propaganda. Was torpedoed twice while rejoining his post ... [37] In a way the award confirmed the worst suspicions of Brouwer: the French government acted as if Denjoy was posted in the Netherlands, much as a soldier or diplomat is sent to a foreign outpost. The officials simply chose to ignore the fact that Denjoy had gone through an ordinary application procedure, and was appointed as an ordinary civil servant by the Dutch authorities. Perhaps the wording was chosen so as to teach the Dutch a lesson; it seems more likely that the French authorities

---

[34] *Symptomatisches zur Gefährdung der Niederländische Staatshoheit.*
[35] Brouwer to Mrs. Ehrenfest, 26 April 1922.
[36] *Chévalier*–1920, *Officier*–1935.
[37] Extract of the citation: *Savant de très grande valeur. Envoyé comme professeur à l'université d'Utrecht, a, par le succès de son enseignement, rendé les plus grands services à la propagande Francaise. A été torpillé deux fois en rejoignant son poste* ... (2 October 1920).

could not see much difference between posting a professor in the Netherlands and a diplomat somewhere in Central Africa.

Brouwer was not intrinsically, or by disposition, a German supporter. There is little doubt that had the war taken another course and ended with victory of the Central powers, the peace conditions would have been equally harsh and short-sighted. In that case Brouwer would undoubtedly have campaigned for the other side. As it was, the present constellation of the international powers and executives tended to drive Brouwer more and more into the arms of the underdog—and of the German nationalists.

While Brouwer was conducting his campaign against Denjoy, his national mathematics association, the *Wiskundig Genootschap*, was wooed from its neutral position by the *Conseil International des Recherches*. The *Wiskundig Genootschap* was invited in April 1921 to join the *Conseil*; on the annual meeting on the thirtieth a letter from the *Conseil*, written by G. Koenigs, to that effect was read to the members. And when the meeting showed itself not averse to joining the *Conseil*, the board was authorized to settle the matter. And so that same afternoon, the board took steps to join the *Conseil*; from the minutes it appears that the motives for this hasty act were not wholly idealistic. There was a certain fear that the *Conseil*, that is to say, the French, would revise the international allocation of the review journals after the eclipse of the Germans, and the Dutch published a mathematical review journal, the *Revues Semestrielles*, which was eyed by some parties abroad with a keen interest. The society argued that nobody would rob a member of the *Conseil* of its review journal. And so the secretary duly wrote a letter of application on 11 May.

Brouwer's opposition was conspicuously absent, why? The most likely explanation is that he was not aware of the goings-on. He had been on a holiday trip to Italy in April, where he visited Assisi and Florence with his wife. Right after the Italian tour he must have gone on to Germany, for it appears from a letter from Brouwer to Schoenflies that he was in the Harz in the beginning of May. Brouwer's presence in the Harz is not as surprising as it seems; already at an early age his health problems had been the cause of parental concern, and he had been an incidental visitor to health spas, among them the institution of Dr. Just in Jungborn, not far from Harzburg. Indeed some of the pre-war letters to Hilbert were dispatched from Just's Institution. After the war he resumed his old habit, and the letter to Schoenflies in effect was written in Just's institution, where Brouwer had sought refuge for health reasons. One would be hard pressed to find a more idyllic and healthy area than the Harz, which had an additional advantage—it was close to Göttingen, and many prominent mathematicians spent their vacations or weekends in the Harz, for example Cantor, like Klein, was a habitué of Hahnenklee.

For the Mathematical Society, Brouwer's absence meant a quick decision and no painful debate; no one, and certainly not the board, would have looked forward to doing battle with Brouwer, who was a fearsome opponent and who could, like the legendary warriors of the past, cut down a numerous opposition single-handedly.

It is thus one of the minor ironies of history that the Dutch Mathematical Society, which counted among its members the most determined neutral opponent of the *Conseil International des Recherches*, inconspicuously joined it. One is reminded of decisions taken 'when father is away'!

The innocent cause of the Denjoy affair, Otto Blumenthal, had quietly disappeared from the scene. Brouwer, who was sincerely fond of Blumenthal, did some lobbying for him when Bieberbach left Frankfurt for Berlin. In glowing words he recommended Blumenthal to Schoenflies, as a man who in matters of 'universal mathematical knowledge, working capacity, helpfulness, and on top of that honesty and decency hardly finds his equal among our colleagues'.[38] He praised Blumenthal lavishly for carrying on the management of the *Mathematische Annalen*, which owed its prominence largely, according to Brouwer, to Blumenthal's unselfish efforts.

> That Klein and Hilbert have, nonetheless, never helped him to a university chair, I can only explain by Machiavelli's maxim: 'The first duty of kings is ingratitude', in addition to which the excessive modesty of Blumenthal (who never blew his own horn) has played a role.

Schoenflies' reply was sympathetic, but he regretted that he could not do much; Bieberbach had built up an excellent reputation as a Ph.D. adviser, with a wealth of scientific inspiration, and the faculty wished to find a comparable man. Max Dehn eventually became Bieberbach's successor in Frankfurt, and Blumenthal remained in Aachen. Brouwer was, of course, disappointed; he wrote that Blumenthal had exercised a considerable influence on himself: 'many of my papers I would never have written without him'.[39]

9.4 Weitzenböck's appointment in Amsterdam

Gradually, the spirited and brilliant mathematician–philosopher Brouwer had become entangled in the web of academic traditions. Had he been primarily a researcher in the pre-war years, by the time the war ended, Brouwer had, whether he liked it or not, become a part of the establishment, with all the obligations that eventually reduce a free man to a slave of the circumstances. During the years 1915 and 1916 he had served the Dutch Mathematical Society as a chairman, to be succeeded by his close friend and ally Mannoury. The Amsterdam Academy and the International Academy for Philosophy also laid a claim to his time and energy. And, of course, the faculty required his attention; one cannot escape the impression that Brouwer took these institutions somewhat more seriously than was good for him and for them.

Like so many theoretical disciplines, mathematics had no place of its own, just a faculty room in the central building of the Oudemanhuispoort. For Brouwer, who was familiar with the facilities in other places, this was far from satisfactory, but official signs of discontent have not been preserved in the minutes of the faculty.

---

[38] Brouwer to Schoenflies 17 January 1921. Schoenflies was Rector in Frankfurt at the time.
[39] Brouwer to Schoenflies 14 May 1921.

Brouwer was surprisingly faithful in attending the faculty meetings, but before the twenties he was not conspicuous. His activities were rather low-key: the minutes record his ideas and comments sparingly. For example, after Korteweg's retirement Brouwer and Hk. de Vries advocated to change the vacancy of the extra-ordinary chair into a lecturer's position (24 October 1918) and in February 1919 he is on record as requesting a typewriter for the use of the library assistant. The offers from Göttingen and Berlin were of course major events in the otherwise quiet faculty life. The minutes faithfully record the 'feelings of joy that Mr. Brouwer has resisted the lure of foreign parts' (17 March 1920).

In spite of his tendency to lose his audience in long monologues, Brouwer's teaching was appreciated by his students. The course material was always impeccably presented.

The records of Brouwer's courses show that his classes were well attended.[40] Some of his early followers already appear before 1920: the roll of the course 'Elementary Mechanics', 1915/16 contains the name of Maurits Belinfante, and in 1918–1919 the name of Arend Heyting appears on the list of the course 'non-euclidean and axiomatic geometry'. Both were to become the first intuitionistic followers of Brouwer. They had been preceded by Brouwer's first Ph.D. student, B.P. Haalmeijer, who wrote a dissertation on a topological topic.[41]

Brouwer used the prestige of foreign offers of such considerable weight to bargain with the Curators of the University for substantial concessions. The permission to hire Hermann Weyl must have been one of them, but when Weyl declined the offer, Brouwer did not press his luck and settled for a lecturer. The minutes of the meeting of 23 February 1921 record that Brouwer had found a suitable candidate, Dr. Roland Weitzenböck, ordinary professor in number theory and the theory of invariants at Graz (Austria). The choice of Weitzenböck is difficult to explain. Brouwer knew many excellent mathematicians abroad, and there seemed no special reason to prefer Weitzenböck. At least it shows that Brouwer wished to widen the spectrum of expertise in Amsterdam: the new candidate brought new blood and new topics.

Weitzenböck was born 26 May 1885 in Kremsmünster (Austria). He studied at the Military Academy in Mödling and obtained his doctor's degree in Vienna. From 1911 to 1912, he studied in Göttingen and he wrote his *Habilitationsschrift* in Vienna. Subsequently, he taught mathematics in Graz, and got an offer from Prague just before the war. After the war he was appointed in Prague as an extra-ordinary professor. After being promoted to ordinary professor, he got an offer from Graz, where he was teaching at the Institute of Technology. He was still in Graz when the Amsterdam Faculty showed an interest in him. Although he was in active military service during the war, he had kept up his mathematical production; he was a regular speaker at mathematics meetings. He must have impressed Brouwer by his

---

[40] Only a few of the attendance lists have been put by Brouwer in his private files. There is no central registration of student attendance.

[41] *Bijdragen tot de theorie der elementairoppervlakken* (Contributions to the theory of elementary surfaces) 28 November 1917.

specific expertise. Brouwer and Weitzenböck knew each other already in 1912, as confirmed by a postcard from Brouwer in Göttingen to the Dutch mathematician P. Mulder.

Evil tongues had it that Brouwer had an ulterior motive for Weitzenböck's appointment: to annoy and keep out J.A. Schouten (who had a chair in Delft at the Institute of Technology). Weitzenböck's appointment followed on 11 October 1921 and since then he remained at Amsterdam.

The appointment of Weitzenböck was no easy routine matter; for economy's sake the Curators had already suggested that future lecturers could have a shared position with some gymnasium or high school. This made it harder to attract foreign experts, and even Dutch mathematicians might think twice before accepting such a position. Brouwer, however, had found his candidate: he informed the faculty in the meeting of February 1921, that Dr. R. Weitzenböck, at present ordinary professor in number theory, theory of forms and the theory of invariants at Graz would be a good choice. The faculty, in turn, put out its feelers to the Curators of the university, but the Board of Curators was somewhat sceptical about the proposal. It wanted to know why no Dutch mathematicians were considered for the position. A month later the faculty informed the Curators that their question of 19 March, whether competent Dutch mathematicians could be found to fill the position of lecturer, had to be answered in the negative. It went on to propose Weitzenböck formally as a candidate for the position. The curators were, however, not to be convinced that easily; they reminded the faculty in their letter of 18 April, that at the time of Weyl's candidacy, the name of Wolff was mentioned. They pointed out that on that occasion one professor and two lecturers were claimed, and that in connection with the lecturers' positions two Dutchmen were mentioned. The faculty replied that, although the mentioned mathematicians showed promise, they did not qualify for the position. Wolff, they added, was already a professor in Groningen.

The faculty admitted that it could not dispel the fear of the Curators that Weitzenböck, who was praised sky-high, could at any moment decide to accept a position elsewhere. But, they went on, in view of the excellence of Weitzenböck, and his extensive record of publications, the faculty considered the possibility to entrust in due time Weitzenböck with the teaching of 'Analytic Functions', so that he would eventually have a task comparable to the one proposed for Weyl. Thus the faculty was willing to offer an ordinary chair to Weitzenböck, while giving up the claim to the two lecturer positions.

This was in a sense exactly the opposite of what the curators had in mind. They did not mind so much appointing a man like Weitzenböck, but they were worried that such a man would foster expectations of a future promotion from lecturer to full professor, something they expressly wanted to avoid. Hk. de Vries tried to put the curators at ease in his letter of 7 April. He made it clear that at the moment Weitzenböck was invited for the lecturer's position only. His teaching would be exclusively in the areas of his specialization, to wit: 'Vector Mathematics' and 'Theory of Invariants'. The faculty might possibly ask him if he would be interested to make himself familiar with the topics of the second lecturer's position, and thus qualify himself for an ordinary chair. In that case the faculty would forgo the second lecturer's position. Brouwer had in the meantime talked to Weitzenböck, and found him more than willing to come to Amsterdam. The only real problem was that the cost of moving to Amsterdam was forbidding. Since the curators had firmly refused to pay the removal expenses, Brouwer had suggested that Weitzenböck should, for a short period, keep both jobs in Graz and Amsterdam, thus saving enough to move to Amsterdam. De Vries closed his letter with the expectation that hiring Weitzenböck would be a step in the direction of their common goal: 'to make Amsterdam the mathematical centre of the Netherlands'

After some more negotiating the curators accepted the proposal of the faculty. On 12 May they submitted Weitzenböck's name to the City Council for an appointment, and on 6 June the Rector was able to inform the Academic Senate that Weitzenböck would be appointed as a lecturer. The final appointment followed on 11 October of the same year. This appointment was, for the time being, the end of the expansion desired by Brouwer.

9.5 Kohnstamm and the Philosophy of Science curriculum

The Amsterdam Faculty had boasted an excellent staff of professors; J.D. Van der Waals, P. Zeeman, Hugo de Vries and D.J. Korteweg were the early stars, and Brouwer was a worthy second-generation member. Among the members of the faculty there was a former student and associate of J.D. van der Waals,[42] Ph. Kohnstamm, co-author with Van der Waals of a textbook on Thermodynamics. Kohnstamm was appointed in 1908 as the successor of Van der Waals, but gradually his interest in physics was overshadowed by other interests. We have met his name before, he was the editor of the *Tijdschrift voor Wijsbegeerte*,[43] who had handled Brouwer's 'Unreliability'-paper in 1908. Kohnstamm's comments at that time would have done little to increase Brouwer's respect for his colleague's philosophical insights, (cf. p. 108).

Kohnstamm, a man of considerable charisma, eventually gave up his chair in physics, and accepted a special chair for pedagogy in Amsterdam.[44] In his function as a professor in the faculty of the exact sciences, he actively promoted the study of philosophy in the faculty of mathematics and physics. At that time philosophy, as a major subject, belonged to the literary faculty, and Kohnstamm tried to get philosophy, in particular the part that was oriented towards the sciences, accepted as one of the majors in the science faculty. He felt that the philosophy of the sciences had come of age, and that it was time to incorporate it into the academic curriculum; the proper place for the philosophies of the various disciplines seemed to him the

---

[42] The Nobel prize winner, not to be confused with his son of the same name.
[43] Journal for Philosophy.
[44] There were three kinds of professorships: the ordinary one, which was a full time job, the extra ordinary one, which was usually a part-time appointment for some specialism, and finally the special one, which usually was a small part-time affair with scant remuneration, paid by some society for the furthering of the interest in ... These societies ranged from religious groups to para-psychologists.

corresponding faculty (called 'the faculty of mathematics and physics', but also containing chemistry, biology, astronomy, ...). Brouwer was fervently opposed to the whole enterprise, he was (correctly, one would guess) worried about the dilution of the content of the traditional curriculum.

The procedure for such a change in the academic curriculum was long and wearisome, but apparently Kohnstamm knew how to speed up the official machinery, and so we see that in a surprisingly short time the proposal reached the Academic Statutes.

The matter was brought up in the fall of 1919, when the official body for discussing matters of curriculum went over the major subjects of the faculties. Kohnstamm handled the whole affair as a born politician; at the meeting of the faculties in Utrecht, he was one of the spokesmen of the Amsterdam faculty. The faculty had bound him to a negative advice, to be conveyed to the other delegates. In the general confusion, when the meeting discussed this minor detail of 'philosophy in the science faculty', Kohnstamm managed to promote the introduction of the subject! The mix-up resulted in a general impression that philosophy had been accepted. When it finally dawned that there had not been a majority for philosophy, the positive recommendation had already been sent to the Minister.

Brouwer violently opposed Kohnstamm's move: he agitated where he could, in the faculty and in public, and indeed in January 1921 the united faculties had decided to advise against the proposal, but by a polished manoeuvre, stretching logic to its breaking point. Kohnstamm convinced the delegates that a vote against the proposal automatically implied the rejection of an older rule, which formulated the teaching qualification of students in philosophy. In this tragi-comedy of error and calculation, the proposal eventually reached the Minister. In spite of Brouwer's attempts to change the mind of Minister De Visser—he enlisted the help of his fellow significist Van Ginneken—the study of philosophy was added to the science curriculum. The Minister, a member of one of the Protestant parties, was only vaguely aware of the purport of the proposal. He introduced the plans in parliament[45] with the words:

> For I have learned that lately a scientific (*natuurwetenschappelijk*) discipline has come up, especially in physics, which, if I may say so, wants to see, more than before, that emphasis is layed on the psyche of plants and animals. And which, to stick to today's terminology, tries rather to do justice to a more idealistic direction than to the biased materialistic direction. As a couple of the representatives of that movement here in our country I would like to mention: Prof. Kohnstamm at Amsterdam and Prof. Buytendijk at the Free University.[46]

It is clear that the Minister was thinking of something rather different from what we nowadays call the philosophy of the sciences: he probably thought of something related to psychology (as the names of Kohnstamm and Buytendijk suggest). He was

---

[45] Meeting of Parliament, 25 May 1921.
[46] The Calvinist universtity at Amsterdam.

a member of the Protestant party, and he probably thought that anything opposed to the evil of materialism was automatically a good thing.

Brouwer was not inclined to let the Minister get away with his curious motivation. As it happened, he had just joined the editorial board of a new literary magazine, *De Nieuwe Kroniek*.[47] This provided an excellent opportunity to write a biting comment on the proposal to adopt 'philosophy' as a major in the Science faculty. It is not known how Brouwer got into the literary–artistic circle that ran the magazine, but he clearly had no objections to try his hand at this unexpected opportunity.

In a few lines Brouwer took the plan and its motivation to pieces; he forcefully argued that there was no such thing as an independent curriculum (*leervak*) philosophy in mathematics and physics.

> As all measures that are impervious to reasonable realization, the one of the minister will only have effect as a pretext for abuses; in this case the more so because of the subsidiary meaning of the word 'philosophy' in the mouth of eloquent superficiality, which is the expression of its reluctance to submit to any control of intelligence or factual knowledge.

Even less appreciation was shown for the minister's amateur opinion, or for his choice of representatives:

> And it is incomprehensible, that the minister, who has no expertise in the field, and who should find it more important to inspire confidence than astonishment, felt that it sufficed to communicate his personal motives, and did not feel obliged to name the authorities, on the excellent authority of whom he has ignored the unanimous advice which he had received from the competent and involved faculties. The mentioning of the names of the gentlemen Kohnstamm and Buytendijk has no significance in this context, since Mr. Buytendijk is a medical man, who never showed any mathematical or physical competence, whereas Mr. Kohnstamm, to be sure, has a place among the members of the faculties of mathematics and physics as an extraordinary professor, but who does not belong there, to those whose opinion one can make prevail over that of a large majority of colleagues, without affronting the latter and disgracing oneself.[48]

Objectively and factually, Brouwer was completely right. There was no urgent need to start a philosophical education in the science faculties, and there were (apart from Brouwer and Mannoury) no acceptable candidates for such a project. One may well surmise that Kohnstamm was promoting a private pet project, and that Brouwer saw through the manoeuvre. He was furious when Kohnstamm succeeded. At one instance he snapped at Kohnstamm, when the matter was discussed in the faculty: 'In this matter you are an ignoramus, and you have to abstain from an opinion!' Again, Brouwer's cool arguments were a front for highly emotional outbursts.

---

[47] The New Chronicle.
[48] [Brouwer 1922B].

Van Eeden recorded in his diary (15 June 1920) that 'Bertus was again fierce in sympathy and antipathy. Furious at Kohnstamm, ...'

Kohnstamm defended himself in an educational magazin with arguments that avoided the heart of the issue, and that breathed an atmosphere of reasonable reflection, suggesting that Brouwer was an isolated malcontent. This battle was lost by Brouwer, mostly because his opponent, Kohnstamm, had his campaign smoothly organized, and it seems likely that as a philosopher of the traditional kind, he was better able to mobilize the sympathy of the average philosophy amateur, than the forbidding Brouwer, who did not spare his readers.

9.6 The New Chronicle

Brouwer's temporary platform, *The New Chronicle*, was a magazine that brought together artists and scientists. The editors were Brouwer, Frans Coenen (an author and journalist), F. Fisher, J.F. Staal (an architect) and Matthijs Vermeulen (a composer).

Brouwer's contributions to this journal were modest. He probably attended the meetings of the editorial board. He was, as a matter of fact, surprisingly conscientious in matters of that sort. He did not belong to the class of the perpetual absent. But in writing, there is little to call attention to; apart from two articles dealing with the philosophy issue and Kohnstamm, he contributed a short review of a Dutch translation of the Russian book of statutes of labour.

In this review Brouwer discussed the following principles of the Soviet law:

1. Only the state has the right of inheritance.
2. Man and woman have the same legal position.
3. All healthy individuals from 26 to 50 years of age, are, by authority of the state, entitled to labour, in accordance to their ability, and to a recompense which guarantees their welfare.
4. All other individuals (including children) will be cared for by the state, even without their labour.

He accepted the first two without hesitation. The third and fourth principles he met, however, with considerably reservation. Brouwer's commentary is rather penetrating, especially if one takes into account the fact that in the first years after the Russian revolution intellectuals were inclined to give the new rulers the benefit of the doubt in almost every way. Keeping in mind the ultimate collapse of the Soviet experiment, which was viewed as the new paradise by the young idealists in the Western world, it is rewarding to reread Brouwer:

> The first two principles mean doing away with decayed institutions, obscuring selection and family life, and can hence be unconditionally welcomed with approval. The enforcement of the third and fourth principles, however, will then only meet their goals (putting an end to the needlessly cruel pauperization of the socially weak) if with respect to the connected abolition of private property of real estate, means of production and transport, the utmost reservation is practised. For, one of the deepest rooted

desires of civilized man aims at freedom, freedom of employment: unhindered unfolding of talents; private freedom: undisturbed possession of house and household goods, unhampered choice of company. And only the hope of freedom is capable of evoking in the individual that intense energy for work, through which alone a nation can gather riches, sufficient to bring affluence and civilization to broad sections of the population. But hope for freedom, in a society without slavery, means hope for money, and hope for money is only possible in the case of a market for goods and labour which is at least partially free. By the total abolition of this market, i.e. the total militarizing of society, the social efficiency of labour would be reduced to a minimum and universal poverty would be created. With respect to the equilibrium between state enterprise and private enterprise thus to be pursued, the following demands have to be made: In the first place, the higher revenues of private enterprise compared to that of state enterprise, should benefit the state not only indirectly, but also directly, and indeed by means of strongly progressive taxes. In the second place, luxury enterprises should be forbidden as long as a high general affluence has not been reached. In the third place, neither the wages in the state enterprises, nor the taxes on private enterprise, should be raised to such an extent that the possibility of existence of the latter is endangered. In particular, one should leave private trade the freedom of action, since only that can bring the goods in time and undamaged into the hands of the consumers. One should even keep up *malafide* trade as an indispensable domain of escape for the activity of crooks and hustlers, whose perpetual birth in a normal percentage is after all guaranteed by the biological–statistical laws. Woe to the state that would block the possibility of trading to these gentlemen, and would force them to enter, all of them, politics!

Brouwer's review is interesting, as it shows that he was not blind to the idea of a reasonable state interference with society and trade, but he did not choose to forego his private philosophical views for a fashionable political idealism.

Extreme and unrealistic ideas were seriously considered by Brouwer. There is a passage in Van Eeden's diary in which Brouwer's views on social reform were recorded. According to Brouwer, two things were necessary for that purpose:

1. Abolishment of the anarchy in marriage and procreation, under the supervision of society
2. Conferring a temporary value to money. Thus the value of a bank note is fixed for a limited period, after which it becomes worthless.

It certainly seems that ideas came first with Brouwer, the realizability was often left out of the considerations.

In view of the recent decline and fall of the Soviet Empire, Brouwer's comments are all the more modern, but at the time of writing, his more progressive friends would not have found his views overly appealing. One might say that Brouwer's views were the paradigm of common sense, but one should also remember that in progressive circles common sense is often a scarce commodity.

The New Chronicle was a short lived magazine; it soon expired, thereby ending Brouwer's short fling with the literary world.

10

THE BREAKTHROUGH

10.1 The Signific Circle

Although the political affairs of the post-war period laid a disproportionate claim to his time, Brouwer still managed to think about the more academic matters. There was his concern for the constructive content of mathematics and, on a smaller scale, his interest in the psycho-linguistic topics of significs.[1] Looking at the list of Brouwer's publications, one sees that an outburst of productivity in 1919 and 1920, was followed by a slack period, with no publications at all in 1922.

Brouwer blamed politics for the temporary lull in his scientific activity. In a postcard to Schoenflies of 1 September 1923, he wrote 'my own work has rested completely for three years, because my strength is almost fully taken up by the fight against our annexation by France, which is diligently promoted by the Lorentz clique.'

This claim may seem exaggerated, but we must realize that Brouwer was not a slick operator who could combine science and political manoeuvring without winking an eye. He was always emotionally involved in his fights against injustice. The conflicts sapped his energy to a dangerous degree; his work suffered perceptibly at times of crisis. So it is plausible that the Denjoy episode and the 'Conseil' problems seriously affected his functioning as a mathematician. After Brouwer's letter to the minister (cf. p. 346) the Denjoy case was more or less closed, and there were no urgent extra-mathematical matters preventing Brouwer's return to his research.

This research consisted of his intuitionistic mathematics, but in addition there was his involvement in the signific enterprise. The latter was in the hands of a small group of Dutch scholars, the most prominent among them was the celebrated author–utopist–psychiatrist–linguist Frederik van Eeden, but the driving force behind the signific enterprise was Gerrit Mannoury, self-styled mathematician, philosopher and Marxist.

Walden, the former colony (commune) of Van Eeden, was one of the favourite meeting places, the company also met at other places, such as the pharmacy of Brouwer's wife, Mannoury's place or Brouwer's house in Blaricum.

Schmitz gives in his book *De Hollandse Significa* a detailed report of the activities of the International Institute for Philosophy, its meetings and discussions, its principles and conflicts. For the present it may suffice to say that the discussion

---

[1]The section on significs makes extensive use of Walter Schmitz' book, *De Hollandse Significa*, [Schmitz 1990]. The reader will find a wealth of information in its pages.

never produced the required consensus, at least not to a degree necessary for a program. Even Mannoury and Brouwer, the two members of the project who were close friends, differed on the topic of 'language'. Mannoury was keenly aware that Brouwer's concept of language reform was too specialized to attract supporters on a desired scale.

Things did not become simpler when a newcomer made his entrée. The old group, more or less the founding fathers of the International Institute for Philosophy, had been reinforced in November 1919 by the linguist Jacques van Ginneken, a powerful and charismatic man. Van Ginneken (1872–1945) was one of the shining examples of Roman Catholic emancipation, a man with an almost inexhaustible store of energy and a wealth of ideas and plans; he became a Jesuit in 1895 and, after writing a much admired dissertation on the foundations of psycho-linguistics, devoted himself to the conversion of the Netherlands to Catholicism. He gained prominence in Catholic Netherland where he was, among other things, the key figure in organizing the mass movement *De Graal* (The Grail) for girls and women.

His authority in linguistics was recognized in 1923 when the newly founded Catholic University at Nijmegen made him a professor.

Right from his introduction into the company of significists, Father van Ginneken occupied an important, if somewhat puzzling, place in the group. He introduced an element of dogmatism into the discussion which had been absent so far. Whereas the older members, of course, had their convictions and did not lose an opportunity to defend them, their discussions were in general ruled to a high degree by the weight of arguments. Van Ginneken's Catholicism did not agree well with this tradition, but since his presence as an authority was appreciated, Mannoury had to exert himself in devising compromises. The very first act of the circle, the formulation of its basic principles, already called for such a compromise.

Back in 1919, Mannoury had put forward a system of language levels, which was adopted on 18 February of that year at the board meeting of the Institute for Philosophy.

The reader should not get any grand ideas at terms like 'Board of the Institute' and of the various committees mentioned in the prospectus of the Institute for Philosophy, since in practice the four friends, Borel, Brouwer, van Eeden and Mannoury, were the major actors who kept the enterprise going. The whole enterprise had something of an old-fashioned comedy, where the cast of a small travelling company had to play at least three parts each. So when we read that the board accepted Mannoury's language levels, we should not think of a proposal put on the agenda and subsequently put to a vote by the members. Everything was on a small cosy scale. The four of them would spend months debating the issues involved, changing a little here, polishing a bit there, and finally reaching a satisfactory formulation.

The group had accepted the general premise that 'language is never able to represent or render any art of reality adequately, but that the so-called meaning of the words only depends on the effect which the speaker has in mind, or which the hearer undergoes'. The readers of Brouwer's *Life, Art and Mysticism* will not be surprised by this viewpoint. He must, however, allow for the fact that these ideas were, so to speak, in the air, in the words of the report: 'they were nowadays generally accepted'. The immediate task, formulated in the founding statement of 21 January 1919, was the compilation of one or more dictionaries for the 'ground words'. This should eventually lead to a clarification and purification of the exchange of thoughts.

The language levels were in this context an appropriate classification, intended to import some system into the amorphous domain of language. The opening statement was followed by a further specification, reproduced below.[2]

The content of the document is rather Brouwerian in tone and intention, but this is not particularly surprising in view of the relationship between Brouwer and Mannoury.

The distinction between the language levels with respect to social understanding.

The *ground language* need not presuppose any dynamic relationship of will to other individuals, and if it does, that relationship can be one of friendship as well as one of enmity.

In the *mood language*, on the contrary, the mutual right of life of speaker and listener is recognized, though not further regulated, and hence loneliness is objected to, and enmity stylised.

In *daily language*, a measure of unanimity of the speaker and listener is obtained, which excludes almost all militancy, and a strong limitation of the possibility of misunderstanding, by admitting to the understanding only those elements of life that are expressed in generally recognized human needs of a peripheral nature.

In *scientific language*, this selection of elements of life has gone so far that only those are admitted which are inherent in the assumption of an objective outer world (common to all individuals).

Finally, in the *symbol language*, only elements of life are considered which are covered by intellectual categories (common to all individuals) so that an almost complete exclusion of misunderstanding is reached.

The document was signed by the trustees of the International Institute for Philosophy: Mannoury, Brouwer, Borel and Van Eeden.

We have presented the formulation of the language levels in full, because they were adopted as a basis for further work. It is also a typical instance of the Mannoury–Brouwer approach to language. Further discussions of the language levels is to be found in the document *Signific Language Research*, reproduced in [Schmitz 1990] p. 415.[3]

---

[2] 18 February 1919. See [Schmitz 1990], p. 421.

[3] A detailed treatment of the history of significs would take us too far afield, so we will stick to the aspects that are somehow relevant for Brouwer's life and work. The book of Schmitz (on the basis of his Habilitationsschrift) is recommended for full information.

The discussions dragged on, and although some progress was made, the institute never became operative. In particular the lack of international response proved to be fatal. In passing it should be mentioned that the absence of financial support was another reason for putting the institute out of its misery. The brochures and mailings exhausted whatever slender means there had been. The institute in fact did not manage to rise above the level of a debating club of bright academics. The end of the project did not come as a surprise to its members.

When it was clear that the *International Institute for Philosophy* was not viable, it was dissolved by the remaining members on 23 February 1922. Not willing to give up their hopes for a signific revision of society and philosophy, the hard core, Brouwer, Van Eeden, Van Ginneken and Mannoury, almost immediately founded a new group: the *Signific Circle* (*Signifische Kring*). Mannoury was appointed chairman, and a little later, on 25 June Brouwer became its secretary.

The new group, called the *Signific Circle* at van Eeden's instigation, decided to present itself to the world in a common declaration of its founders.[4] The formulation of this declaration was no improvement over the founding declaration of the short lived Institute. As we all too often experience: too many cooks spoiled the meal. Comparing the new program to its predecessor, one notes that under influence of Van Ginneken, the language levels lost their prominent role, and 'language act' had become the central term. Furthermore, Van Ginneken introduced statistical and experimental considerations into the program.

The new Circle was no longer after ultimate perfection and unanimity; therefore it was considered fitting to add the personal views of its members to the joint declaration. As to be expected, the formulation of the declaration involved a good deal of discussion, and gradually it became clear that Van Ginneken could not and would not endorse in all respects the common views that had been put forward in the past. Not surprisingly, his personal statement, submitted in the fall of 1922, contained a good measure of provocation. Considering himself the missionary soul of the emerging Catholic progressives, he did not try to hide his convictions and prejudices. Brouwer's expectation and hope that the significists would rise above their respective groups of reference as a 'neutral humanitarian community' did not meet with Van Ginneken's approval. On the contrary, he wrote to his fellow members that:

> The undersigned does not mind the inadequacy of language very much, as long as he directs himself to his equals, his close friends and relations, in short the members of his confessional group. And in this way the difference between me and my co-signatories has to be explained, I think, because I feel that affinity of soul in my private circle far stronger than they do.

This somewhat parochial and in a sense uncooperative viewpoint irked Brouwer. On 24 November 1922 Brouwer confessed in a letter to Mannoury that he had

---

[4]Cf. [Schmitz 1990], p. 423.

given up all hope:

> In the contribution of Van Ginneken I read for the first time a formally pronounced, ruthless negation of the only thing that attracts me in significs: the hope on the creation of linguistic social reform tools, independent of all existing (and in my opinion mostly obsolete) groupings, and indeed by people who would rise in a neutral-humanitarian community above their respective groups. It is true that this view has been pushed more and more to the background, but I have patiently allowed this; in the first place counting on my learning capacity in the matter, and in the second place in the expectation that the community I hoped for, would in spite of all difficulties, eventually come about and function.
>
> This expectation I have to give up definitely, after the experience that one of my fellow members draws inspiration from the rejection of my (unchanged) principle. And the consequence of that can be no other than my resignation from our circle. I am even of the opinion that it would smell of dishonesty or lack of character, if under these circumstances I took part in the publication of our joint manifesto, knowing that it will be followed by Van Ginneken's words.
>
> In spite of the above, I have the feeling that there is something that binds the four of us to each other, than to others, but *je ne sais quoi* seems not to admitted to the domain of conscious reality.

Mannoury's superior statemanship and his genuine personal warmth must have prevented the imminent collapse of the circle. Brouwer did not quit and he delivered his personal statement, be it with a considerable delay, on 25 March 1924. He stressed his fundamental views:

> For the undersigned, significs does not so much consist of a practice of linguistic criticism as:
>
> 1. the tracing of the affect-elements, into which the cause and action of words can be analysed. By this analysis, the affects that touch on the human understanding can be brought closer to control by the conscience.
> 2. the creation of a new vocabulary, which also for the spiritual tendencies of life of man yields access to their reflective exchange of thoughts and, as a consequence, to their social organization.
>
> For the realization of the part of the program mentioned under 1, cooperation cannot be dispensed with: for, numerous affect complexes can only be analysed through the catalysing action of philosophical discussion between those not of the same mind.
>
> Also with respect to the creative labour mentioned under 2, I have for a long time believed in the great importance of cooperation. However, I have become more and more convinced that this higher task of significs can be accomplished only by the utmost concentration of mind of the individual.

Thus Brouwer remained true to his first conception. He was confident that via introspection and exchange of thought with, in particular, people of different sociocultural convictions and backgrounds, the affects which underlie words, could be traced and subsequently named, so that people could be more aware of them and that they could thus guide the conscience. His view, in fact, differed from those of the others (in particular Van Eeden and Mannoury) in that he assigned the social component of language a far more modest role.

The first practical enterprise which had to illustrate significs in action, was the creation of an *Encyclopaedia of Significs*. The Circle composed a list of prospective authors: Faddegon, de Haan, Eugen Ehrlich, Revesz, Bouman, Van der Plaats, Grünbaum, Bertrand Russell, Uhlenbeck, Husserl, Spengler, Peano, Couturat, Henri Borel, Brouwer, Van Eeden, Van Ginneken and Mannoury.

The Encyclopaedia project, formulated by Mannoury, meant a clean cut with the older ideas of social reform *à la* Van Eeden and Brouwer. As with so many plans of the significists, this one also came to nought. The envisaged authors, when approached, were not interested. After some more half-hearted initiatives, the Signific Circle, too, foundered. The last project, the publication of a survey of the various viewpoints, *The Signific Dialogues*, was undertaken in 1924. Its publication followed in 1937.[5]

Mannoury sadly accepted that the Signific Circle was doomed to sterility. In particular, Van Ginneken's objections to Mannoury and his communism played a role in the general ineffectiveness of the group. In 1924 Van Ginneken left the Signific Circle. He may have distrusted Mannoury's influence, as might be inferred from a letter of Van Ginneken to Van Eeden of 26 December 1924: 'Although the person of Mannoury remains very sympathetic to me, I feel that I have to break with him as a party member, since his principles force him to malign my principles'.[6] In view of the virulent anti-religious ideas of the more crude Marxists, van Ginneken may have had some grounds to keep his distance, but in the case of Mannoury he was sadly mistaken. Mannoury was, in spite of his occasional political rhetoric, an extremely tolerant person.

When the Circle showed no prospect of healthy activity, Mannoury proposed in 1926 to dissolve the Signific Circle. This marked the end of an ambitious enterprise, one that failed on account of ordinary human weaknesses, and an unworldly approach to matters of organization. The group had been too heterogeneous to succeed in the face of real differences of opinion. There have been attempts to compare the Signific Circle with the Vienna Circle, and indeed there are a number of similarities. But on the whole the Signific Circle lacked the coherence and academic homogeneity necessary for influencing the development of science and philosophy. In spite of their literary reputation, Van Eeden and Borel did not match the intellectual potential of Brouwer and Mannoury. Moreover, the goal that the significists

---

[5] [Brouwer 1937].
[6] Cf. [Fontijn 1996], p. 550.

had set themselves, a socio-linguistic reform, was so ambitious and vulnerable to squabble, that it was almost doomed to fail. Significs, as such, was promoted further by Mannoury, who gave numerous talks and published a number of books and papers, but even his efforts could not secure it a place among the popular philosophical topics, nor in the academic curriculum.[7]

Van Ginneken played more than one role in our history. He was also instrumental in the conversion of Van Eeden to Catholicism. His attempts to convert Borel and Brouwer too did not meet the same success.[8] In his younger years Van Eeden had been rather critical of religion in general, but in his sixties, he changed his views and became, guided partly by Van Ginneken and his second wife, Truida, a faithful son of the Church.

Although there was no longer any social or political suppression of the Catholics, the church still had solid militant and expansionist ideas. It considered the conversion of Van Eeden as one of its spectacular 'catches' and decided to make the most of it. Van Eeden was baptised and received into the church on 18 February 1922, in the presence of Van Ginneken and Brouwer. At the occasion of the baptism, Van Eeden had to make a solemn retraction of most of his earlier publications, which were in part anti-religious. The books and plays that were not compatible with the views of the Catholic Church were displayed on a separate table. Van Eeden raised each book separately abjuring it. It made Brouwer feel sick to witness the procedure.[9]

Van Eeden's perception of Brouwer's reaction can be found in his diary (18 February 1922): 'Brouwer, too, was moved'. Reports of events of a personal nature have the inevitable tendency to contradict each other. Louise's recollection of the proceedings and Brouwer's reactions may very well be compatible with Van Eeden's observation.

Van Ginneken and Brouwer were present at the occasion, much to the indignation of Henri Borel, who wrote bitterly to Brouwer 'So you stood there as a friend and brother of that liar and faithless one, that fraud! [...]. Would Frederic Paulus now also condone the murder of Giordano Bruno and so many noble minds?'[10]

Van Eeden became 'more Catholic than the Pope', a phenomenon not uncommon with converts. He thoroughly reconsidered his views on religion and revoked part of his earlier views. A year after his baptism he was the star of a meeting, organized by the Catholic Church in the Concertgebouw of Amsterdam (22 March 1923) at which occasion he ardently testified to his new-found religion.

Van Eeden's conversion had not been the result of a sudden impulse; like a beleaguered fortress he had been made ready for the final attack. Truida, his second wife, had already embraced the Catholic religion two years earlier: she was received

---

[7] As far as I know, only Mannoury, Van Dantzig, and Vuijsje taught courses in significs.
[8] Borel eventually became a Catholic at the end of his life.
[9] Interview Louise Peijpers.
[10] Borel to Brouwer 18 February 1922.

into the fold on 3 April 1920 and van Eeden's two children had been baptised on 23 August 1920. The influence of Truida on Van Eeden's decision had been considerable.

Brouwer's stepdaughter Louise Peijpers was baptised together with the children of Van Eeden. She had left home somewhere in the middle of the nineteen-tens[11] and entered a home in the Hague, run by Father van Ginneken. Under the guidance of Van Ginneken and his helpers, she became a devout Catholic, and expressed the wish to become a nun. Indeed, she entered a convent, but after a trial period her wish was denied. It appears that the strict rules and the obligatory menial duties did not agree with Louise's view of life. Eventually she became a Catholic, but the process took her more than eight years. According to Louise, Brouwer and Van Ginneken got to know each other through her.[12]

Later in life Louise became exceedingly religious, bordering on mania. She joined the church of the anti-Pope, and in her old age lived a secluded, pious life that was rather similar to that of mediaeval religious ladies, be it that she combined the love for the church with a deep and bitter hatred of her stepfather.

Louise as a novice.

Already before his conversion, Van Eeden had occasionally gone into retreat in the St Paulus Abbey in Oisterwijk.[13] At his invitation, Brouwer accompanied him a couple of times. In fact, the Catholic practice of retreat may have appealed to Brouwer, because of the similarity with his own practice of mystical introspection. But there certainly was also a good deal of plain curiosity involved. Brouwer never missed an opportunity to attend mystical religious or pseudo-religious manifestations. For example, later in life he went with friends to hear Krishnamurti.[14]

The role of Van Eeden in Brouwer's life became smaller as Brouwer was more and more absorbed by his mathematical and academic duties. They continued to meet occasionally. The few cards and letters that were exchanged suggest a cordial relationship, but the close contact, the walks and visits, had gone.

In 1923 Van Eeden apparently entertained the idea that he might be considered a candidate for the Nobel Prize; he asked Brouwer to write a recommendation.[15] There is no trace of such a recommendation, so we cannot ascertain whether Brouwer complied with the request. In view of the standard procedures of the No-

---

[11] 'Ran away', as she put it.

[12] Interview Louise Peijpers.

[13] A village in the province of North-Brabant.

[14] Interview C. Emmer, one of Brouwer's family doctors.

[15] Van Eeden to Brouwer 28 June 1923. Mannoury and Van Ginneken were also approached with the same request.

bel committee, Van Eeden's idea seems strange; it rather illustrates his self-esteem.

On Van Eeden's seventieth birthday, in 1930, a Liber Amicorum was presented to him, with contributions from friends. The list contains a fair number of prominent artists and scientists. There is no contribution from Brouwer, who is listed as one of the organizers.

Life at home was for Brouwer as hectic as ever during the early years of the twenties. He travelled, invited colleagues to stay, and carried out his extensive and complicated research and correspondence from the idyllic village Blaricum. Cor Jongejan had become a permanent member of the entourage, she acted as a general factotum, typing letters and manuscripts, copying large manuscripts which Brouwer handled for the *Mathematische Annalen*, went shopping, did jobs around the house, assisted Lize in the pharmacy, and at the same time was a companion to Brouwer. In these busy years Lize fell seriously ill. She turned out to have pernicious anaemia. In order to get the best possible prospect of recovery, she retired for some time to a health clinic in Antwerp.

10.2 Intuitionism—principles for choice sequences

The scientific activity of Brouwer concentrated more and more on the further development of his new intuitionism. Of course, he still handled the topology papers for the *Mathematische Annalen* and therefore automatically kept himself informed of the progress in that area, but his personal research was completely devoted to foundational matters. He had set out no more, no less, to rebuild mathematics along constructive, that is intuitionistic lines. In 1921 Hermann Weyl had taken his side and enthusiastically proclaimed the Brouwerian revolution, but, apart from a few isolated contributions, he hardly took part in the realization of Brouwer's grand design. In practice, Brouwer had to reform mathematics single-handedly.

His basic papers of 1918 and 1919 had provided the basis for concrete mathematical theories, and in 1923 a substantial part of the intuitionistic theory of real valued functions was developed in the paper *The Founding of the theory of functions independent of the logical proposition of the excluded third. First part, continuity, measurability and derivability*.[16].

In the two earlier papers of the same series, the basic concepts, such as spread, real number, and function, had been introduced, and in addition Brouwer had added to the arsenal of constructive mathematics the fundamental tool of *finitary spread* (*finite Menge*, later called *fan*) which played the role of the constructive substitute for compact sets. Basically, a fan is a set of choice sequences over a finitely branching tree. By reducing questions about sets to questions about finitely branching spreads (fans), Brouwer intended to exploit more fully the intuitionistic viewpoint. A major advance in the theory of real functions was presented right at the

---

[16] *Begründung der Funktionenlehre unabhängig vom logischen Satz vom ausgeschlossenen Dritten. Erster Teil, Stetigkeit, Messbarkeit, Derivierbarkeit*, [Brouwer 1923A]

beginning of the paper on real functions. The first three theorems showed beyond a shadow of a doubt that intuitionistic analysis was not just the poor sister of classical analysis.

1. *A continuous function whose domain coincides with a finite point-spread is uniformly continuous.*

2. *Every function whose domain coincides with a closed located point set is uniformly continuous.*

3. *Every full function is uniformly continuous.*[17]

Although at first sight it is marvellous that all real functions are continuous, even an intuitionist would like to have things like step functions. Mathematical practice just begs for functions of that sort. Brouwer realized this, and he earnestly tried to find intuitionistic versions of discontinuous functions, which, of course, could not be everywhere defined. He did not find a wholly satisfactory solution, as one can read in his paper on the Domain of Functions.[18] Brouwer's letter of 1920 to Hermann Weyl confirms his pre-occupation with the problem of discontinuous functions.[19]

Brouwer soon realized that his continuity theorem required a more substantial proof than the one given in the 1923 paper. In March 1924 he submitted a new paper to the Academy with the title, *Proof that every full function is uniformly continuous*.[20] This paper, and its subsequent improvements, contained the major breakthrough of the new intuitionism. Until this time intuitionistic mathematics had been a high-spirited adventure, but with little to show—except new notions. An outsider could thus easily have convinced himself that Brouwer's program was purely negative: a banning of certain principles (notably the principle of the excluded third). It is true that Brouwer's constructivism of the first period, before the full-scale adoption of choice sequences, mainly held the ideas and tools for straightforward constructive mathematics (much like the work of Bishop in the sixties). Brouwer's second Ph.D. student, Belinfante, wrote a dissertation in this vein on a constructive theory of series [Belinfante 1923]. It was followed by a long series of papers on constructive function theory.

In 1924 Brouwer had reached the watershed; he could show the mathematical world results that proudly set intuitionism aside from classical mathematics. In other words, Brouwer started to produce theorems that classical mathematicians could not possibly prove. From now on classical and intuitionistic mathematics were recognized to be mutually incompatible. Brouwer used to express this as 'classical mathematics is contradictory'. The impact on classical mathematicians was negligible; even if they had been able to appreciate the finer details, they would

---

[17] In this paper Brouwer restricts his attention to functions on [0, 1]; 'full' means 'everywhere defined on [0, 1]'. In general, 'full stands for 'everywhere defined on the natural domain'.
[18] [Brouwer 1927B].
[19] Cf. [Dalen, D. van 1995].
[20] *Beweis dass jede volle Funktion gleichmässig stetig ist*, [Brouwer 1924D2].

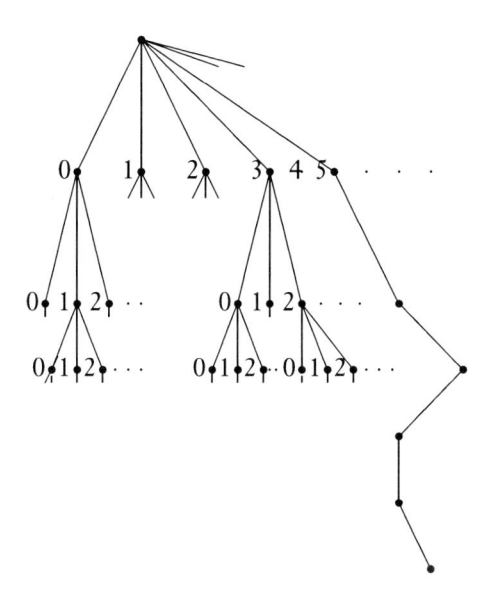

FIG. 10.1. spread

have shrugged their shoulders. Why should Goliath pay heed to David?

In the following pages I will attempt to give a home-and-garden view of the ideas involved in the proof of the continuity theorems.

Since we are dealing with choice sequences of natural numbers, we have to introduce some notations for choice sequences and their (finite!) initial segments. The latter are nothing but finite sequences of natural numbers. We will use Kleene's notation, which is more flexible and convenient than Brouwer's.

We consider the universal tree of all finite sequences of natural numbers. Sequences are denoted by bold face numerals, $\mathbf{n} = \langle n_0, \ldots, n_{k-1} \rangle$, and numerical functions by greek symbols, $\alpha, \beta, \ldots$ An initial segment $\langle \alpha(0), \ldots, \alpha(k-1) \rangle$ of α is denoted by $\overline{\alpha}(k)$, in particular $\overline{\alpha}(0) = \langle \, \rangle$ is the empty sequence. The functions α can be viewed as infinite paths in the tree (that is, infinite sequences of natural numbers): in other words, Brouwer's choice sequences, see Fig. 10.1.

Now suppose that we are given a set S of nodes of the tree (that is to say, a set of finite sequences) such that each infinite path contains at least one element of S. One can say, picturesquely, that S cuts off all the infinite paths. The question is now: what do we know about such a set? The classically trained mathematician is thoroughly familiar with this question: in the case of 0-1-sequences, the infinite paths are the points of the *Cantor Space*. The points of the set S can be viewed as an open covering of the Cantor space, and by compactness we know that a finite subset of S already suffices to cut off all infinite paths. In analysis (infinitesimal calculus) the phenomenon goes by the name of the *Heine–Borel theorem*.

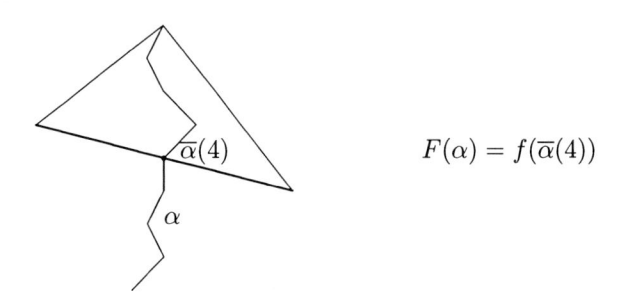

FIG. 10.2. Value of $F(\alpha)$ determined at the bar by f

For Brouwer it was a matter of great urgency to find 'nice' subsets of S that already suffice to cut off all infinite paths. The structure of the functions from the universal spread (that is the spread of all infinite paths through the universal tree) to natural numbers depended on the answer. The link between 'function from choice sequences to natural numbers' and 'path-cutting sets' is provided by Brouwer's continuity principle (cf. p. 243) This principle states that for a function $F : \mathbb{N}^{\mathbb{N}} \to \mathbb{N}$ (where $\mathbb{N}^{\mathbb{N}}$ stands for the set of all choice sequences of natural numbers) there is a set of nodes B such that for each α there is an initial segment $\overline{\alpha}(k)$ in B; the values of F are determined by these initial segments in B. A set like B is called a *bar*, and the paths are *barred* by B. In logical notation:

$$\forall \alpha \exists k (\overline{\alpha}(k) \in B)$$

This bar is used to determine the outputs of F, that is, $F(\alpha)$ is determined by the node $\overline{\alpha}(k)$ of B, see Fig. 10.2. To be precise, there is a function f, acting on nodes, such that $F(\alpha) = f(\overline{\alpha}(k))$.

The behaviour of the function F is thus determined by B and f. It was Brouwer's *tour de force* to determine the structure of the bars B.[21]

There are no notes, no letters, no pieces of scratch paper that shed any light on the process of this brilliant achievement. In the case of some of his other discoveries there are traces of his preparatory research. (The Analysis Situs paper was the result of painstaking exploration of point set topology, made necessary by Schoenflies' *Bericht* with its mistakes.) In the case of the recognition of the legitimacy of choice sequences, we know of his wavering, his return to point set theory. But in this case there are no indications of any research, of any experiments. It is not unlikely that the insight came in a flash. Louise Peijpers recalled that at one particular occasion when Brouwer was working in the garden, he suddenly saw the solution to one of

---

[21]This is not an obvious generalization of the compactness of Cantor space. For in that case we find that a finite subset of B will do. But the generalization from finite to infinite is far from unique.

his mathematical problems at the moment that the sounds of a street organ that played on the road, reached him. It is not known which problem it was, but it is said to be one of his major achievements. The story illustrates that Brouwer was no stranger to these sudden flashes of illumination.

One thing is certain: Brouwer had the tools necessary for the job. In the first of the *Begründungs papers* he had devoted many pages to well-ordered sets. And well-ordering was the key to the problem of characterizing bars.

In classical mathematics a set is well-ordered if it is ordered, and every nonempty subset has a first element. In intuitionistic mathematics this notion is no good, for even the set $\{0, 1, 2, 3\}$ is not well-ordered according to this definition.

One can see this by considering the set $A = \{n = 0 | R \vee \neg R\}$, where R is an open problem, for example the Riemann Hypothesis. Clearly $A \subseteq \{0, 1, 2, 3\}$; and if $A = \emptyset$, then there is no number $n \leq 0$, such that the condition $R \vee \neg R$ is true, so it must be false. Now a disjunction is false if both parts are false. This means that $\neg R$ and $\neg\neg R$ are true. But this is impossible, and therefore A cannot be empty. So, does A have a first element? If so, it must be 0. But then $0 \in A$ and hence $R \vee \neg R$, which is an open problem! The upshot is that we cannot indicate a first element, and this means that $\{0, 1, 2, 3\}$ is not well-ordered. Clearly this is not what we had in mind!

Brouwer recognized the problem and decided to use another, more constructive, definition. He went back to Cantor's first ideas about well-ordering and defined well-ordered sets as those sets that could be obtained by means of the following process: start with a one-element set and make new well-ordered sets by putting together finite or denumerable sequences of (smaller) well-ordered sets.

This process is perfectly constructive, and it yields many well-ordered sets. Here are some examples:

1. $\{a_0\}$
2. $\{a_0, a_1, a_2\}$ obtained from $\{a_0\}, \{a_1\}, \{a_2\}$ by putting them in a linear order
3. $\{a_0, a_1, a_2, a_3, \ldots, a_n, \ldots\}$ obtained from the denumerably many sets $\{a_i\}$
4. $\{a_0, a_1, a_2, \ldots, b_0, b_1, b_2, \ldots\}$ obtained from the two well-ordered sets $\{a_0, a_1, a_2, \ldots\}$ and $\{b_0, b_1, b_2, \ldots\}$
5. $\{a_0, a_1, a_2, \ldots, b_0\}$ obtained from $\{a_0, a_1, a_2, \ldots\}$ and $\{b_0\}$

As the reader will see, the actual nature of the elements a_i, b_j is irrelevant. We could just as well have used dots:

(1) • $(= 1)$
(2) • • • $(= 3)$
(3) • • • • · · · · · $(= \omega)$
(4) • • • • · · · · · • • • · · · · · $(= \omega + \omega = \omega \cdot 2)$
(5) • • • · · · · · • $(= \omega + 1)$

Traditionally, these well-ordered sets are called *ordinals* or *ordinal numbers*, and the

FIG. 10.3. Ordinal trees

operation of putting together ordinals is thought of as a sum. So $3 = 1 + 1 + 1, \omega = 1 + 1 + 1 \cdots + 1 + \cdots (\Sigma a_i,$ with $a_i = 1)$.

Note that if we use the dot notation, both ω and $1 + \omega$ look alike:

$$\omega = \bullet \bullet \bullet \cdots, \qquad 1 + \omega = \bullet \quad \bullet \bullet \bullet \cdots$$

Hence we identify them.[22] There are specific rules in set theory about how to identify and order ordinals. Here it will suffice to consider them in our naive fashion.

Brouwer noted that the ordinals matched certain trees, in accordance with their construction process. The set 1 above corresponds to a tree with one node, and a composite ordinal corresponds to all the corresponding trees lined up under a single top node:

Let us call the corresponding trees, as shown in Fig. 10.3, *ordinal trees*.

Under the construction process different trees may belong to the same ordinal, for example $1 + 2$ and 3 both correspond to the same ordinal, but that is not a serious problem. The intuitive idea of the ordering of the ordinal is to run through the bottom nodes from left to right. From a mathematical point of view the ordinals and their ordinal trees are the same things in different disguises. Let us, for convenience, call such a bar, consisting of the bottom nodes of an ordinal tree, an *ordinal bar*.[23]

Now Brouwer's astonishing insight was that *a bar always contains a complete set of bottom nodes of an ordinal tree*.[24]

The above insight goes by the name *Bar Theorem*. It reflects rather bleakly on Brouwer's didactic gifts that this result remained nameless until after 1945. He gave it its name in his lecture at the Canadian conference talk in 1953.

He did, however, give a name to a special case of the theorem. Suppose we deal with a fan, that is with finitely branching tree, then the ordinal trees will all be finite, and hence, if in a fan all paths are cut off by a bar, this bar has to be *finite*! In particular one can determine a longest node in the bar, which means that there is an upper bound to the length of the nodes on top of the bar, see Fig. 10.4.

Let us give this theorem a more formal formulation. We use $B(n)$ for 'n is on the bar B'. Then 'every path α is cut off by B' is rendered as

$$\forall \alpha \exists x B(\overline{\alpha}(x))$$

---

[22] Note that $\omega + 1 \neq 1 + \omega$; addition of ordinals is therefore not commutative.

[23] In his original proofs Brouwer considered thin bars, i.e. bars in which no two nodes are comparable (one above the other).

[24] There are actually some conditions to be put on the bar, such as monotonicity or decidability, cf. [Troelstra–van Dalen 1988].

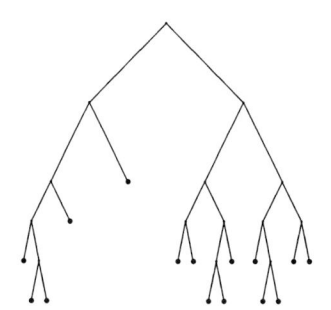

FIG. 10.4. A bar in a fan

The conclusion above says that there is a maximal length z, such that each path is cut off at the latest at length z. This reads $\exists z \forall \alpha \exists x \leq z B(\overline{\alpha}(x))$.

Now the famous *fan theorem* is

$$\forall \alpha \exists x B(\overline{\alpha}(x)) \to \exists z \forall \alpha \exists x \leq z B(\overline{\alpha}(x)$$

The fan theorem is an immediate corollary of the bar theorem. Brouwer was rightfully proud of the fan theorem, which was the main tool in his subsequent papers. In recognition of its importance, Brouwer gave it a name, albeit a rather uninspired one: *the fundamental property of fans* (Haupteigenschaft der finiten Mengen). The name *fan theorem* appears for the first time in his post-war publications.

It is an interesting historical curiosity that the fan theorem preceded its better known contrapositive: König's infinity lemma.[25] This lemma says that a fan with infinitely many nodes contains an infinite path. The infinity lemma is not constructively valid. An interesting observation is that this refutes the popular idea that constructive proofs and theorems are always posterior upgradings of classical theorems.

The proof of the bar theorem (as we will call it) was a complicated and mysterious affair. It is to be doubted that at the time anybody but Brouwer understood it, partly because few mathematicians were willing to devote much time to a project that met blank incomprehension in professional circles, but also because the only competent fellow revolutionary, Hermann Weyl, had left constructive mathematics.

Brouwer's proofs rest on an analysis of the data that go into the theorem, in this case a proof Π that each choice sequence (path) hits the bar. Brouwer wanted to apply some form of induction to obtain his result, so he introduced auxiliary subproofs $\Pi_\mathbf{n}$ of Π showing that 'all choice sequences through the node $\mathbf{n}$ hit the bar'.

The gist of the argument is that we can conclude from the proof $\Pi_\mathbf{n}$ that the little bar below $\mathbf{n}$ is an ordinal bar; having (by induction hypothesis) a suitable

[25][König 1926]

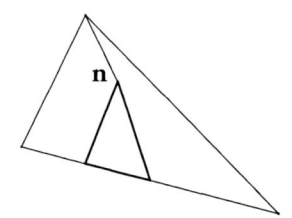

FIG. 10.5. Bar under **n**

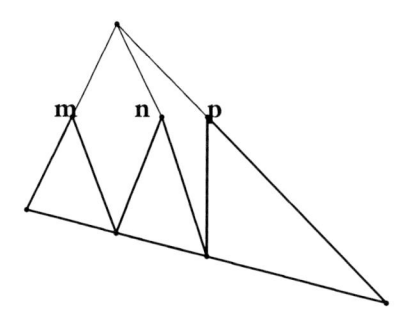

FIG. 10.6. joining bars

collection of **m**'s available for which there is a proof π_m. We may 'pull up' this knowledge until finally $\Pi_{\langle\rangle}$ (that is Π itself) is reached. Then the whole bar is an ordinal bar.

Now two things are clear:
(1) if **n** is in the bar then the little bar below **n**, that is a bar consisting of one node, is an ordinal bar (corresponding to the ordinal 1),
(2) if all nodes immediately below **n** have ordinal bars below them, then **n** itself has an ordinal bar below it.

So, starting at the bottom, one travels all the way up in the tree until the top node is reached. And then we have established that the bar is an ordinal bar.

The above sketch does not do justice to Brouwer's argument; rather, it gives the reader an impression of the underlying idea. The 'pulling up' trick in the proof is a special way of carrying out transfinite induction; it now goes by the name of *Bar Induction*.[26]

Since Brouwer was involved in actual constructive mathematics—not axiomatic reasoning on the basis of some principles (in this case the *Continuity Principle* and *Bar Induction*) he had to justify his argument conceptually. So he presented his argument as a constructive procedure based on the given data. Thus he had to analyse the given proof that each choice sequence hits the bar. Doing so he isolated the

---

[26] Cf. [Kleene–Vesley 1965], [Kreisel–Troelstra 1970], [Troelstra–van Dalen 1988].

basic proof steps (1) and (2) above and argued that each such proof would be replaced by a canonical proof consisting of basic steps only. This canonical proof then enabled him to use bar induction.

In the early literature there are no comments on this proof. It is safe to assume that few, if any, of the experts in the foundations of mathematics had grasped Brouwer's argument. In the thirties some of the members of the Göttingen school, in particular Gentzen and Bernays, started to consider the fan theorem, but rather the result than the proof.[27]

Brouwer returned to the bar theorem in the same year, with a couple of further comments on the proof. In 1926 he submitted an extended version of his results to the *Mathematische Annalen* as a contribution to the Riemann volume (at the occasion of the hundredth anniversary of Riemann's birthday). This paper *On the domains of functions*[28] contained, in addition to the updated proof of the basic theorem, a number of foundational remarks. One of the more surprising statements concerns the nature of intuitionistic mathematical proof. In footnote 8, Brouwer casually remarked that 'intuitionistic proofs were constructed by means of the two generating operations from nil-elements [that is, immediately given insights] and elementary inferences, given immediately by intuition'. Thus, according to Brouwer, proofs are well-ordered sets of steps. In particular, he admitted infinite proofs. This looks stranger than it is, for 'infinite' means 'intuitionistically infinite'. And in that sense infinite remains potential.

In all of Brouwer's papers the bar theorem and the fan theorem were immediately followed by the *continuity theorem*, which states that 'all real functions are locally uniformly continuous'. Here, too, Brouwer could not fall back on older classical proofs, for the simple reason that this theorem was classically false. Brouwer proved the theorem by reducing functions on the reals to functions on choice sequences of nested intervals. The proof, once you have seen it, is not difficult, but it requires a certain degree of maturity to put it together for the first time.

It is fair to say that Brouwer's experience in topology and analysis gave him a headstart in constructive mathematics; especially in his particular infinitary constructive mathematics that dealt with infinite objects, such as choice sequences. Brouwer's career as an intuitionist was indeed intimately connected with his career as a topologist. A person without this penchant for the infinite and for topology would probably have embraced a more combinatorial or finitist constructivism, like Kronecker's.

Brouwer's own view of his breakthrough can be gathered from the titles of the first papers; he considered the local uniform continuity of the real functions the highlight of the new intuitionism. In this instance he showed a keen appreciation of the advertisement value of his work: the statement 'all real functions are continuous' was far more likely to shock the reader than some scholarly title, involving choice

---

[27] Gentzen, being a specialist in countable ordinals, must have recognized the significance of Brouwer's ordinals and ordinal trees. His first consistency proof of arithmetic was based on Brouwer's ideas, cf. [Gentzen 1969]

[28] *Über Definitionsbereiche von Funktionen*, [Brouwer 1927B].

sequences or transfinite induction. The reader expecting provoking details must, however, have been seriously disappointed!

The 'domains of functions' paper contained, as the title implicitly promised, also a study of intuitionistic versions of discontinuous functions which are total in the classical sense. The results of this part are rather inconclusive. Brouwer notes that neither the measure theoretic, nor the logical approaches are satisfactory.

The paper contains one real curiosity: in the introductory part the notion '*negative continuity*' is introduced, presumably as a didactic tool to introduce the reader to the new results. A function is called negatively continuous if it has no discontinuities. Brouwer's argument, unfortunately, showed less than he promised. Instead of showing that a total function on the continuum is negatively continuous, he only established a weak result: we cannot prove that there are discontinuous functions.

He must have discovered this lapse fairly soon, for in his Berlin lectures, which he gave in the first months of 1927, he had changed the presentation. There he showed that the existence of discontinuous functions implied a certain omniscience principle.[29]

Apart from the didactic reason for incorporating the 'negative continuity theorem', there may have been a more worldly one: to put on record that Brouwer had conjectured the continuity theorem before anybody else.[30] The 1927 paper contains a passage to that effect.

> On the basis of this theorem, mentioned often since 1918 in lectures and in conversations, which is an immediate consequence of the intuitionistic viewpoint, the conjecture of the validity of the following theorem, claiming much more, has become plausible. I succeeded with its proof, however, much later.[31]

It is no exaggeration to put the ideas that led to the continuity theorem in the same class as the basic ideas of Brouwer's new topology (for example the mapping degree). These two complexes of ideas are definitely the high water marks in Brouwer's career.[32]

One would expect that a similar period of feverish activity with its outpouring of papers would follow. Alas, the times had changed: Brouwer was no longer the unfettered researcher of the early years. Moreover, he was slowed down by developments in topology which called for his attention (cf. ch. 11), and by the *Grundlagenstreit*, which started to assume unpleasant proportions (cf. ch. 14).

Whatever the cause may have been, it is a fact that the breakthrough was not followed by a flurry of new results. The published fall-out in this period of the new

---

[29] Omniscience principles are certain non-constructive statements; the prime example is the principle of the excluded third. Brouwer introduced these principles in the twenties, cf. [Brouwer 1992]. The term 'omniscience principle' was introduced by Bishop.

[30] Cf. p. 240

[31] [Brouwer 1927B], p. 62.

[32] Freudenthal referred to the pair of innovations, in an obituary of Brouwer, as 'the man who went through the wall twice with his head'.

insights is restricted to a few papers: on intuitionistic dimension theory and on an intuitionistic form of the Heine–Borel theorem.[33]

The first paper gave a metric characterization of intuitionistic compact spaces and introduced intuitionistically the notion of dimension. It reproved his 1913 result, $\mathbb{R}^n$ is n-dimensional, in an intuitionistic setting. The second paper contained the correct intuitionistic formulation of the Heine–Borel theorem and its proof. The Heine-Borel paper is a clear instance of progress in the intuitionistic program; as late as 1923 Brouwer had branded the general Heine–Borel theorem (in its classical formulation) as false, but within three years he presented an intuitionistically correct version.[34]

This seems a very meagre harvest for such a powerful new method. The situation is, however, not as bad as it looks. The Berlin Lectures presented a few more applications:

- the indecomposability (*Unzerlegbarkeit*) of the continuum;

- the natural ordering of the continuum is the only ordering of the continuum;

- the existence of the supremum of a continuous function on $[0, 1]$.

The indecomposability of $\mathbb{R}$ is a most surprising property of the reals; it says that one cannot split $\mathbb{R}$ into two disjoint sets.[35] It is the mathematical expression of the intuitive conviction that $\mathbb{R}$ is extremely closely knit. For example, the pair of sets $(-\infty, 0]$ and $(0, +\infty)$ is not a decomposition, since we are missing out on those points for which it is unknown, whether they are on the left or on the right of 0; so $(-\infty, 0] \cup (0, +\infty)$ is a subset of $\mathbb{R}$ but not the whole of $\mathbb{R}$. Thus Brouwer's indecomposability theorem can be viewed as the mathematical expression of this surprising coherence of the continuum.

The continuum may be compared to those lines produced by syrup on a pancake; one cannot split a syrup line in two parts without losing some particles. Something will always stick to the knife.

In classical mathematics the splitting of the continuum is the result of a logical trick: the principle of the excluded third (PEM). This principle tells us, for example that each real number is either rational or irrational, although we have not the faintest idea how to test whether a real number is rational. On the strength of the PEM the classical mathematician claims that $\mathbb{R}$ is split into the rationals and irrationals. To pursue the metaphor: the classical continuum can be viewed as the frozen intuitionistic continuum; all points are nicely fixed in place, and there are no more uncertainties like 'where exactly is this point?'

---

[33][Brouwer 1926B1, 1926C2].

[34]The classical Heine–Borel theorem says that if a bounded closed subset of $\mathbb{R}$ has a covering by a collection of arbitrary open sets, then it is already covered by a finite subcollection. Brouwer added an extra condition: the given set should also be located (cf. p. 330).

[35]More precisely: if $\mathbb{R} = A \cup B$ and $A \cap B = \emptyset$, then $A = \mathbb{R}$ or $A = \emptyset$.

The indecomposability of the continuum was proved by Brouwer directly from the fan theorem in his Berlin Lectures; it was mentioned in his *Intuitionistic Reflections on Formalism* and again in *The Structure of the Continuum*.³⁶ In the first mentioned paper Brouwer used the indecomposability property to refute the generalized version of the principle of the excluded third: $\neg \forall x \in \mathbb{R}(x \in \mathbb{Q} \vee x \notin \mathbb{Q})$. The indecomposability is, of course, an immediate corollary of the continuity theorem.³⁷

Let $\mathbb{R} = A \cup B$, with $A \cap B = \emptyset$. Define $f(x) = \begin{cases} 0 \text{ if } & x \in A \\ 1 \text{ if } & x \in B \end{cases}$

Then f clearly is a total function, and hence it must be continuous. But there are only two continuous functions with values 0 or 1, namely the constant 0 and the constant 1 function. If f is the first one, then, by definition, $\mathbb{R} = A$; in the second case $\mathbb{R} = B$.

Brouwer never published a proof of the theorem or referred to it in print after the above mentioned papers. This may not have been intentional; he planned to publish the Berlin Lectures, but somehow the project aborted. The manuscript was found among Brouwer's papers, and it was finally published in 1992. His Cambridge Lectures also contain a proof of the indecomposability theorem.

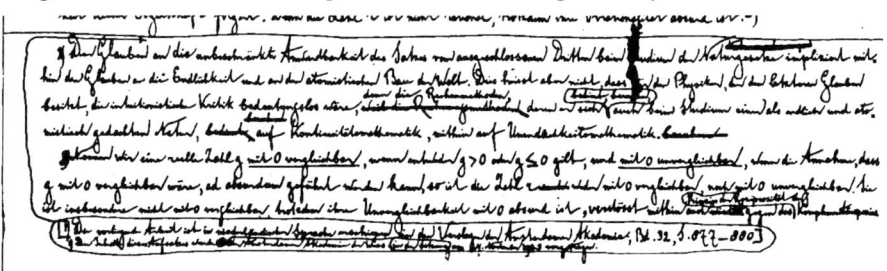

A draft of the indecomposability of the continuum.

The extra power and insight that the new methods had put at Brouwer's disposal should ideally have been the starting point for a real school of intuitionistic mathematics. However, as much as Brouwer would have liked to give intuitionism its rightful place in the world of mathematics, he lacked the leadership qualities and ambition of, say, Hilbert. There was enough of the mystic left in him, to resist the temptation of a career as theorem prover or super manager. Instead of devoting all his efforts to the promotion of the true conceptual mathematics, he divided his time and energy between travelling, talking, quarrelling, lecturing, keeping an eye on the pharmacy, and a hundred more things.

---

³⁶ *Intuitionistische Betrachtungen über den Formalismus*, [Brouwer 1928A1], *Die Struktur des Kontinuums*, [Brouwer 1930A].

³⁷ Cf. [Heyting 1956].

10.3 Intuitionism in the *Mathematische Annalen*

He was not idle, nor did he neglect his intuitionistic research, but the driving impetus was missing. Nonetheless, he was deadly serious about promoting intuitionism. He carefully upgraded his *Begründungs* papers and submitted a series of three basic papers to the *Mathematische Annalen* under the title *On the foundations of intuitionistic mathematics*.[38] In the absence of any records it is hard to say who handled the papers, and who accepted them. It is not impossible that Brouwer himself had a hand in the procedure, but on the other hand he did not need tricks of that sort; his name was sufficient guarantee for quality. Even Hilbert could not have objected to these papers of his adversary; they were totally devoid of provocation. As usual, the papers were a paragon of respectable mathematics. The negative aspect was wholly absent. No names were mentioned; they refer to neither the principle of the excluded middle, nor classical mathematics. Brouwer presented his material with a superior disregard for the competing, and indeed prevailing, view. A prisoner, serving his years in isolation and being presented by a charitable institution with a copy of Brouwer's three expository papers, would not even have been aware that there was an alternative mathematics.

The new series lent a large measure of respectability to intuitionism; the general opinion of the *Mathematische Annalen* was so high that mediocre authors or impostors were not supposed to get access to its pages. The series also showed that, at a time when Hilbert was still issuing new programmatic statements, Brouwer had already worked out a substantial piece of mathematics.

The papers were an improvement upon the original papers published by the Amsterdam Academy.[39] In the intervening years, some loose ends had been discovered and corrected; in addition, Brouwer's views on the subject had ripened.

Brouwer did carry on his program, but in a rather eclectic way. The post-'continuity' papers deal with a number of topics that rather varied in importance. There was, for example, a paper that further refined the various equality and inequality relations between sets.[40] The results are instructive in so far as they illustrate the complications one may expect when dealing with sets; they also show that the intricacies of intuitionistic logic were no obstacle for Brouwer. To be fair, it must be said that the problems were somewhat esoteric, out of the mainstream. At the same time the contents must have provided a powerful deterrent for the curious among the mathematical community. Putting these concoctions of negations and double negations in the shopwindow was not the right policy to attract customers.

Together with his Ph.D. student B. de Loor, he published an intuitionistic proof of the fundamental theorem of algebra,[41] and a few months later he added the

---

[38] *Zur Begründung der intuitionistische Mathematik*, [Brouwer 1925A,1926A,1927A], submitted 20 June 1924, 14 March 1925, 28 November 1925. Note that there is a certain ambiguity in the title. It could be translated as 'On the foundations ...' or as 'On the founding of ...'. In view of the fact that Brouwer opted for 'foundations' (*grondslagen*) in the title of his dissertation, I have stuck to 'foundations'.

[39] [Brouwer 1918B,1919A,1923A].

[40] [Brouwer 1923C1, 1924F1, 1925E, 1927D].

[41] [Brouwer 1924C2].

finishing touch to the fundamental theorem by proving the general case.

The fundamental theorem of algebra is an old family heirloom of mathematics. It says, roughly, that the complex numbers are rich enough to solve their own equations. It is well-known that the simpler number systems have their limitations when it comes to solving equations. Linear equations can be solved in the rational numbers, but not always in the integers; for instance $2x = 1$ has no integer solution. In the reals some equations, but not all, of higher degree have roots. Equations of odd degree always have a solution; but an $x^2 + 1 = 0$ has no real solution. In order to solve this equation, the complex numbers, with the new number $i(= \sqrt{-1})$ were introduced. Here the story stops: every algebraic equation $x^n + a_{n-1}x^{n-1} + \cdots a_1 x + a_0 = 0$, with complex coefficients, has a complex solution. Gauss was the first person to prove the fundamental theorem of algebra, after a number of prominent predecessors had failed.

The fundamental theorem of algebra is one of the most famous *existence theorems*. It states that there is a number α such that $\alpha^n + a_{n-1}\alpha^{n-1} + \cdots + a_0 = 0$.

The usual proofs one finds in the textbooks proceed by contradiction; hence from an intuitionistic point of view they only show that it is impossible that there is no such α. Therefore it was a matter of some urgency to give an effective procedure for constructing a root of the equation. This was accomplished in 1924 by Brouwer and De Loor and in the same year, independently, by Hermann Weyl.[42]

Brouwer saw, of course, that the theorem, as formulated above, was not fully general; it was based on the assumption that the leading coefficient was 1. Classically this is not a problem. If the leading coefficient a_n is distinct from 0, one just divides all coefficients by a_n. Intuitionistically this does not work, since it may be unknown whether a_n is *apart* from 0. The improved version of the fundamental theorem asserts that $a_n x^n + \cdots + a_1 x + a_0 = 0$ has a solution if at least one of the coefficients $a_n, \cdots, a_1$ is apart from 0. A matching characterization in terms of a factorization in linear factors was also provided.[43]

There are just two more papers of Brouwer to be mentioned. One of them is a refined analysis of the nature of spreads,[44] and the other contains a marvellous theorem on the intuitionistic theory of ordering.

Order is one of the basic tools of mathematics; without it, life would be very difficult, if not impossible. Now, the basic structures in mathematics—the integers, the reals—have a natural ordering, which in classical mathematics satisfies the *trichotomy law* : $x < y \vee x = y \vee y < x$.

---

[42][Weyl 1924].

[43]*Intuitionistische Ergänzung des Fundamentalsatzes der Algebra*, [Brouwer 1924E2], see also [van Dalen 1985], [Troelstra–van Dalen 1988] p. 434. The story does not end here. A more refined analysis in sheave models showed that without the hidden assumption of the countable axiom of choice the fundamental theorem fails (Fourman and Hyland), cf. [Troelstra–van Dalen 1988], p. 794.

[44][Brouwer 1928B2].

In intuitionistic mathematics this law is valid for the natural numbers, the integers and the rationals, but, as Brouwer showed in 1923,[45] the law fails for the real numbers. He formulated this as 'the continuum is not ordered'. In his writings he insisted that orderings should be *total*, in the sense that for distinct elements, a and b, either $a < b$ or $b < a$ holds.[46] His well-known counterexamples show that this is impossible for the ordering of the reals. Nowadays in constructive mathematics, one drops the requirement of totality, and adopts some weaker laws.[47] Somehow, Brouwer had not yet made the final step to free himself from the time-honoured traditions of classical mathematics. Either unconsciously or expressly, he wished to stay as close as possible to the classical notions (at least in a number of instances). So, when he had observed that the relation $<$ fell far short of its classical counterpart, he decided to introduce a finer relation.

In order to give an idea of the intricacies, we will briefly look at the ordering of the reals.

Assuming that we have already ordered the rationals (which is unproblematic, as $a = b$ and $a < b$ are decidable for rational a and b) and assuming that the real numbers a and b are given by Cauchy sequences,[48] we let $a < b$ stand for $b - a > 0$, that is

$$\exists k \exists n \forall p (b_{n+p} - a_{n+p} > 2^{-k})$$

In other words, if 'eventually' the b_i's are larger than the a_i's plus a fixed positive rational number.

For this order one easily shows the following:[49]

$$a = b \rightarrow \neg a < b$$
$$a < b \rightarrow a \neq b$$
$$a < b \wedge b < c \rightarrow a < c$$
$$a < b \rightarrow c < b \vee a < c$$

Writing $a \leq b$ for $\neg b < a$ (not to be confused with the classical $a < b \vee a = b$) we see that the following holds:

$$a = b \rightarrow a \leq b$$
$$a \leq b \wedge b \leq a \rightarrow a = b$$

But we still do not get $a \leq b \vee b \leq a$.

---

[45] [Brouwer 1923B2].

[46] In formal notation: $a \neq b \rightarrow a < b \vee b < a$ (note that $a \neq b$ stands for $\neg a = b$). This is, intuitionistically, not the same as the trichotomy property.

[47] In particular the trichotomy law is replaced by the following law $a < b \rightarrow c < b \vee a < c$. Heyting deserves the credit for cleaning up the admittedly confused situation in the case of ordering, as left by Brouwer.

[48] Sequences of rationals which satisfy the Cauchy criterion: $\forall k \exists n \forall p (|a_n - a_{n+p}| < 2^{-k})$. Observe that in constructive mathematics one works with effectively given ϵ (and δ). This explains why we have $< 2^{-k}$ instead of $< \epsilon$.

[49] Cf. [Troelstra–van Dalen 1988], p. 256.

Let us call $<$ the *natural ordering* on $\mathbb{R}$. Brouwer's examples showed that $<$ is far from trichotomous. He reacted to this phenomenon by introducing a finer ordering:
$$a \prec b := \neg a > b \wedge a \neq b$$
or in the above notation: $a \leq b \wedge a \neq b$.[50]

He called the new relation $\prec$ a *virtual ordering* and his treatment of ordering in the *Mathematische Annalen*[51] is based on this notion.

Since the trichotomy law did not hold for virtual orderings, cf. [Brouwer 1926A], p. 455, the question arose if there is an even finer ordering (that is an ordering defined for more pairs) which is trichotomous?

Brouwer settled the problem in a beautiful note *Virtual and inextensible order*,[52] which was published not in the *Mathematische Annalen*, but in the *Journal für Reine und Angewandte Mathematik*, also known as *Crelle's Journal*, or even as '*Crelle*'.[53] In this paper he introduced the notion of *inextensible order*, which is an ordering relation on a set with the property that; if for a, b the relation $a < b$ is consistent with the given ordering and equality on the set, then $a < b$ is already the case in the given ordering ('what is not forbidden must happen'); similarly for $a = b$. Nowadays one would rather call this a *maximal ordering*.

The main result of the paper is a gem: an ordering is virtual if and only if it is maximal. This fact, which is rather isolated in the body of Brouwer's œuvre, is one of the encouraging events in the development of intuitionistic mathematics. Newcomers to the field usually start to feel that the stern conditions of intuitionism prevent almost any regularity or structure but for the most trivial situation. Indeed, this was the general feeling in the mathematical world: no doubt, intuitionism is a commendable, and perhaps the right, approach to mathematics, but its rules are so strict that one can only expect scattered *ad hoc* results. Finsler, in his inaugural address in Cologne in 1922, spoke for the majority of his colleagues:[54]

> Such [undecidability] assumptions may by themselves lead to very interesting investigations; an *exact* science cannot very well be based on them; apart from the great complications this would bring, many of the best secured results would have to be abandoned.

Brouwer's former ally, Hermann Weyl, also had 'second thoughts'. His initial enthusiasm of the 'New Crisis' paper:

---

[50] In Brouwer's notation, the natural ordering was denoted by $\lozenge$, and the double negation ordering by $<$. It is not difficult to see that $a \prec b \leftrightarrow \neg\neg a < b$.

[51] [Brouwer 1926A], p. 453.

[52] *Virtuelle Ordnung und unerweiterbare Ordnung*.

[53] [Brouwer 1927C]. In view of Brouwer's long association with the *Mathematische Annalen*, the choice of journal is not quite obvious. It is possible that the relationship with Hilbert caused Brouwer to look for another place to publish, but that seems an unlikely motive. After all, the paper is of the inoffensive, scholarly sort. It is more plausible that Brouwer was invited to submit a paper to *Crelle* on the occasion of its centenary jubilee.

[54] [Finsler 1925].

> The new conception, as one sees, brings very far-reaching restrictions in comparison with the generality that virtually disappears into vagueness, to which an analysis, developed so far, has made us accustomed the last few decades. We must again learn modesty. We wanted to storm the heavens, and we have only piled cloud on cloud that cannot support anybody who tries in earnest to stand on them. That what is saved could at first sight appear so insignificant, that the possibility of analysis is questionable—this pessimism is, however, unfounded, [...]

gradually gave way to more pessimistic views:

> Mathematics attains with Brouwer the highest intuitive clarity; his doctrine is idealism in mathematics analysed to the extreme. But with pain the mathematician sees the larger part of his towering theories fall apart. [...]
> Brouwer succeeds in developing the beginnings of analysis in a natural way, retaining much closer contact with intuition than was achieved before. But, progressing to higher and more general theories, one cannot deny that finally a hardly bearable awkwardness results from the fact that the simple principles of classical logic are not applicable. And with pain the mathematician sees that the larger part of his tower, which he thought was joined with strong blocks, dissolves into smoke.[Weyl 1928].

One can imagine that where Weyl's frustration was of the charitable variety, the majority of practising mathematicians would have little patience with a doctrine that promised hard labour and fewer results. The theorem about inextensible orders demonstrates, however, convincingly that there is more structure and regularity in intuitionistic mathematics than meets the eye. Clearly, intuitionistic mathematics is *not*, as many thought it to be, the loosely connected remnant of results of the powerful traditional mathematics. There is a good deal of structure, albeit more sophisticated than what people were used to.

The feature that makes this paper on virtual order stand out in Brouwer's work is that it is a general model-theoretic result. Most of the time, Brouwer's papers dealt with concrete, real life situations; in this case Brouwer considered arbitrary ordered sets. The 'virtual = inextensible' result is a striking piece of model theory *avant la lettre*.[55]

Heyting reported in the Collected Works (p. 596) that Brouwer was not quite certain of the correctness of the proof of the theorem. There is a note of 1933 in which he reconsiders the proof. One of the confusing features of Brouwer's paper was his convoluted use of language; one has to find a suitable way of reading the paper. Martino has, however, found a clean interpretation that completely vindicates the theorem.[56]

Brouwer's mathematical output in the twenties contains a few more interesting

---

[55] Cf. [Martino 1988].
[56] Ibid., cf. [Brouwer 1992], p. 10, 11.

items, such as an intuitionistic proof of the Jordan curve theorem[57] and a number of results in the Berlin Lectures. The proof of the Jordan curve theorem was in itself not such a miracle, but it was important to Brouwer for a couple of reasons. It confirmed Brouwer's deep-rooted feelings that a considerable part of topology was constructive after all, once one found the proper formulations and proofs. Moreover, it showed that intuitionism did not reduce mathematics to abject poverty, as so many thought.

The Berlin Lectures, in addition to the positive achievements, contained a list of counterexamples to well-known classical theorems, such as the existence of a maximum of a (uniformly continuous) function on [0,1], the Bolzano–Weierstrass theorem, the intermediate value theorem, the incidence axioms of projective geometry and the fixed point theorem. In each case Brouwer indicated, however, an intuitionistic version, mostly in an ϵ-formulation. For example, for a topological mapping f of the unit disk in itself there is for each $\epsilon > 0$ a point x at a distance less than ϵ from $f(x)$.

10.4 Beyond Brouwerian counterexamples

There is one particular innovation in Brouwer's writings of the twenties that deserves special attention: *the fleeing properties* and *the pendulum number* (*Pendelzahl*). Recall that in Brouwerian (counter-) examples certain sequences are constructed on the basis of some algorithm and a criterion. Here is such an example:

Let $R(n)$ stand for 'the decimal digits of π at places $n - 9$ through n are all 9'. For each n we can test $R(n)$ by simply calculating the digits of π up to the n-th place;[58] we can thus effectively check if the last 10 decimals are nines.

Now we use $R(n)$ to define a real number as follows:

$$a_n = \begin{cases} 2^{-n} & \text{if } \forall k \leq n \neg R(k) \\ 2^{-k} & \text{if } k \leq u \text{ and } R(k) \text{ and } \forall p < k \neg R(p) \end{cases}$$

In other words, we write down simultaneously the decimal expansion of π and powers of 2^{-1}, see Fig. 10.7; as long as we have not encountered a sequence of 10 nines in the expansion, we take one half of the previous choice, but as soon as we have found a string of 10 nines, we keep repeating the last number.

It is not hard to see that a_n converges to a real number, say a. This a can be used to refute some classical laws; for example, $a = 0 \vee a \neq 0$ cannot be asserted, for that would mean that *at present* we have a proof that either there will never be a sequence of 10 nines, or that it is impossible that there is no such sequence. But we have no evidence supporting this, so we have no right to assert that $a = 0 \vee a \neq 0$.

---

[57][Brouwer 1925B].

[58]That this can actually be done, was shown in [Brouwer 1921A1]. Most mathematicians at the time would have considered this completely superfluous on abstract grounds. The intuitionist, however, needs a proof. Brouwer had jotted in the margin of his private copy of 'Mathematics, Truth and Reality' the question 'does that exist?', at the place where the decimal expansion of π was exploited. Note, by the way, that the precise number of digits in the sequence figuring in the Brouwerian counterexamples is immaterial. One can always take a very large string.

$\pi =$	3,	1	4	...	9	9	9	9	9	9	9
$a_n:$	2^{-1}	2^{-2}	2^{-3}	...	2^{-k}	2^{-k-1}	.	.	.	.	.
	9	9	9	7	3						
	.	.	2^{-k-10}	2^{-k-10}	2^{-k-10}	...					

FIG. 10.7. A Brouwerian counterexample

Neither can we assert that $a < 0 \lor a = 0 \lor a > 0$. Indeed, it is clear that a is not negative, so let us look at $a = 0 \lor a > 0$. This holds if there is no 'string of nines', or if we can indicate a place where 10 nines will occur—which we cannot.

The special feature of this example is that it is completely algorithmic, so the number a is constructed by a law.

After Brouwer had come to the conclusion that choice objects are not only legitimate but even indispensable, he had reached the point of view that the distinction between classical and intuitionistic real numbers was that classical real numbers were completely fixed in advance, that is given by a law. This is a difficult point, for it raises questions about lawlike sequences. What does one mean by a law, and would the classical mathematician be prepared to give such a law? The term 'pre-determinate' seems more apt in this case. The classical reals are fixed by the magic of the principle of the excluded third, not by honest calculation.

Intuitionistic reals could, on the other hand, also be given by a choice process. In the Berlin Lectures and the Vienna Lectures, he called the continuum of all choice reals the *full continuum*, and the continuum of all lawlike reals the *reduced continuum*. Both forms of the continuum were treated side by side.

The introduction of the reduced continuum was not a passing fad. Brouwer kept the distinction up to the end of his career (cf. [Brouwer 1952B], [Brouwer 1981A], p. 41). In [Brouwer 1954B] Brouwer called the lawlike part of the continuum the *classical continuum*, thus affirming the view that classical reals could be considered from an intuitionistic viewpoint. He did not add any comments, but we may be fairly certain that he merely wanted to express his conviction that choice phenomena did not belong to classical mathematics, not that the lawlike reals have classical properties. Roughly speaking, one might say that Brouwer wanted to indicate certain reals that rather resembled the classical reals.

The early Brouwerian counterexamples thus referred to the reduced continuum. In the late twenties Brouwer cast his counterexamples in a somewhat more abstract form. Instead of fixing the property $R(n)$ he only required that

(i) $\forall n(R(n) \lor \neg R(n))$ (that is $R(n)$ is decidable) and
(ii) neither $\exists n R(n)$ nor $\forall n \neg R(n)$ has been shown.

For such an R he carried out similar constructions as above; Brouwer considered the construction sufficiently important to give names to the key notions: the least number n (if any) for which $R(n)$ holds is the *critical number* of R. R itself was called a *fleeing property*, and the real number constructed by means of R carried the evocative name of *pendulum number*.[59] Wittgenstein, who attended Brouwer's Vienna Lectures, refused to admit the pendulum number as a proper mathematical object.[60]

Brouwer wanted, however, to move on to non-lawlike Brouwerian counterexamples. He did so in his Berlin Lectures [Brouwer 1992] p. 31. The method was later called the method of the *creating subject*,[61] but until 1948 it remained nameless. We will illustrate the new version of Brouwerian counterexamples:

Let A be a mathematical statement for which neither A nor $\neg A$ has been proved. Define
$$a_n = \begin{cases} 2^{-n} & \text{as long as neither } A \text{ nor } \neg A \text{ has been established} \\ 2^{-k} & \text{if } k \leq n \text{ and between the choice of } a_{k-1} \text{ and } a_k \\ & \text{either} A \text{ or } \neg A \text{ has been established.} \end{cases}$$

Clearly, the sequence (a_n) converges to a real number, say a. Again we cannot assert that $a = 0 \vee a \leq 0$.

The above example is taken from the Berlin Lectures (p. 31). It produces, like the example based on π, a weak counterexample. Only after the Second World War, Brouwer returned to the counterexamples based on the creating subject; by then he had found a way to strengthen the method so that actual contradictions could be produced, instead of mere 'it cannot be shown that'. The remarkable fact remains that Brouwer waited for 20 years before he published these 'extended counterexamples'.[62]

In his postwar publications he clarified the ideas behind the procedure. The example above, reformulated in the terminology of 1948, runs as follows:

$$a_n = \begin{cases} 2^{-n} & \text{as long as the creating subject has experienced neither} \\ & \text{the truth nor the falsity of } A \\ 2^{-k} & \text{if } k \leq n \text{ and the creating subject has experienced the} \\ & \text{truth of } A \text{ or of } \neg A \text{ between the choice of } a_{k-1} \text{ and } a_k. \end{cases}$$

What had to be read between the lines in 1927, is spelled out in 1948 in provoking detail. The full extent of Brouwer's subjective approach to mathematics is hidden in this procedure. The creating subject produces choices depending on its mental experiences with respect to a particular statement. It was as far as Brouwer could go in his intuitionism.

There are few reports on Brouwer's private reactions to all these new developments of the twenties. It is beyond doubt that he was proud of his fan theorem and the continuity theorem, but perhaps the discovery of the creating subject and its possibilities did not leave him untouched either. There is a letter from the signifi-

---

[59]*Pendelzahl* of R, cf. [Brouwer 1929A].

[60]Cf. [Wittgenstein 1984].

[61]In the literature, it is known as the *creative subject*. However, Brouwer's naming seems much more apt, keeping in mind the intended meaning.

[62]We will return to this method at the proper place to show how Brouwer cleverly exploited the logical possibilities of the creating subject.

cist Henri Borel to Brouwer, right after the Berlin Lectures, in which he mentioned that "Gutkind wrote me that 'you had made an enormous discovery', which 'had to do with the foundations of logic' ".[63] Taking into account the timing of the statement, there are two candidates for this discovery: the 'virtual–inextensible' theorem, or the creating subject. (To be fair, it could have been the continuity theorem or the indecomposability of the continuum just as well, but that belong rather to mathematics than to logic.) Under the circumstances, the creating subject would be a good candidate for this discovery, assuming that Brouwer had referred to a recent event. Whichever may have been the alleged discovery, we may rest assured that Brouwer considered the creating subject as an important step in the further development of intuitionism. In general, Brouwer was not given to bouts of self-congratulation, so we may safely conclude in this case that Brouwer had expressed his own true view on the recent developments in intuitionism. [64]

The investigations into more general choice sequences, depending on the subjective behaviour of the individual mathematician, probably did not yield satisfactory results, for Brouwer did not publish any of the material that he lectured on in Berlin until after World War II.[65] It is, of course, possible that Brouwer anticipated an outcry over this openly subjective mathematics and that, fearing a reversal of the generally tolerant opinion of the mathematical world, he suppressed the material for strategic reasons. This does not seem plausible; like the late Samson, he was willing to face the Philistines at any time and any place. It seems more likely that he was not satisfied with the results obtained so far. Indeed, the strong results of the post-war years were well within his reach, and we cannot exclude that he was in the possession of the decisive results, and that he lost interest for reasons that had little to do with mathematics but, as we will see, everything with the power structure in the mathematical community.

10.5   Fraenkel's role in intuitionism

Brouwer's intuitionistic mathematics found by no means the response that was necessary for a wholesale adoption. As he directed his missionary activities mainly towards the German mathematicians, it is in Germany that one has to look for the first signs of appreciation. The first convert was Hermann Weyl, but a second one was not easily forthcoming. No matter how one thought about the foundations of mathematics, giving up the existing practice for an arduous life of hard labour in the house of intuitionism was going too far for most mathematicians.[66]

---

[63] Borel to Brouwer 26 March 1927.

[64] One cannot exclude the possibility that Brouwer referred to the bar theorem and the fan theorem. This certainly changed the face of logic, be it that Kleene and Kreisel had to act as midwives some fourty years later.

[65] With the exception of his Vienna lectures, Brouwer 1929, but there the uninitiated reader could easily miss its significance.

[66] The reception of Brouwer's intuitionism in the crucial years before 1930 will be described in the dissertation of Dennis Hesseling [Hesselin 1999, Hesseling 2002].

Intuitionism found a sympathetic reception with the German mathematician Adolph Fraenkel,[67] a specialist in the young discipline of set theory. Fraenkel was a man with vivid foundational interests. He had written a famous monograph *Introduction to Set Theory*[68] (1919) while serving in the German army during World War I. His name lives on in Zermelo–Fraenkel set theory, an extension of Zermelo's set theory. Fraenkel (and independently Skolem) had added a powerful existence axiom that allowed the introduction of larger sets and infinities. Furthermore, he was the first person to attack successfully the independence problem for the axiom of choice; the Fraenkel–Mostowski models are the lasting witnesses of his pioneering work.

Fraenkel's wife came from Holland, and the couple cultivated strong ties with that country. While staying with his in-laws in 1921, Fraenkel visited Brouwer in Laren, and a cordial relationship was established. He used Brouwer's library and attended his lectures in Amsterdam during the break between the (German) winter and summer term of 1923. The discussions with Brouwer stimulated his interest in intuitionism:

> It was, among other things, very interesting for me to observe the fresh life of intuitionism, already pronounced dead from many quarters; inside me it still ferments with these questions.[69]

At that time Fraenkel was preparing a new and extended edition of his *Introduction to Set Theory*. He had discussed the book with Brouwer, who had read the proofs and commented on the material (probably the chapter on intuitionism, but most likely the entire text, for Brouwer's interest in set theory had not waned). Brouwer and his wife visited the Fraenkels in Marburg during the Annual Meeting of the German Mathematical Society in 1923. When the second edition of Fraenkel's book appeared, Mrs. Fraenkel personally presented a copy to Brouwer, and Brouwer gratefully acknowledged its receipt, expressing his expectation that 'The book will exert [...] in very wide circles, both of mathematically and epistemological interested readers, an intensive and beneficial effect'.[70]

Fraenkel was a brilliant expositor, and so, when he gave a series of lectures on set theory in Kiel in 1925, the publisher Teubner was only too happy to publish the lectures as a book with the title *Ten Lectures on the Foundations of Set Theory*.[71] In 1926 the proof sheets had already been sent to the author, who once more asked Brouwer to go over the proofs and to suggest changes where necessary. And here, suddenly, a conflict arose. When Fraenkel, after three weeks, told Brouwer that there was no more time to make changes, the latter reacted as if stung by a bee:

---

[67] Who later in life, for obvious reasons, changed his name into Abraham.
[68] *Einleitung in die Mengenlehre*.
[69] Fraenkel to Brouwer 18 April 1923.
[70] Brouwer to Fraenkel, 15 December 1923.
[71] *Zehn Vorlesungen über die Grundlegung der Mengenlehre*.

What kind of a wizard do you think I am, that I could in the middle of term, with my time almost completely occupied otherwise, study a book of more than 100 pages so thoroughly that I could bear the responsibility for suggestions of changes!

He went on to say that on account of this disastrous haste, the only thing left for him to do was to write a critical review '... where I would indeed (in particular with respect to intuitionism) have to correct a great deal, but maybe it is really good that I have an occasion to dispose of the false information on intuitionism that is offered to the public.'

Brouwer's wrath was mainly caused by the sections about intuitionism. Regardless of a possible revision, he demanded some immediate changes, among others the insertion of a list of intuitionistic publications, '... among which there are, after all, so far, the only publications existing in the field of intuitionism, apart from Heyting's dissertation, that "do not just talk, but also create" '.[72] In the absence of Fraenkel's original text, it is hard to say what exactly incited Brouwer's anger, but given the correspondence and the final text, one can make an educated guess. Fraenkel, in line with the prevailing tradition (encouraged by Hilbert) had probably described intuitionism as a more or less continuous foundational program, initiated by Kronecker, carried on by Poincaré, and taken over by Brouwer. Commentators found it often difficult to distinguish Kronecker's program, which was in the most literal sense an arithmetization of mathematics (in particular of analysis and algebra) from Brouwer's intuitionism.

A.A. Fraenkel. Courtesy Mrs. M. Fraenkel.

It was even harder for the general mathematician to keep Brouwer's intuitionism and the French semi-intuitionists apart. The latter group was far from homogeneous; it held strongly diverging views on foundational matters. It was not uncommon to present Brouwer's intuitionism as a hybrid offshoot of the programs of Kronecker and the French semi-intuitionists. The popular view of Brouwer's program was however mistaken: the influence of Kronecker on Brouwer is negligible. Neither in Brouwer's dissertation nor in his inaugural address is Kronecker mentioned. In the Berlin Lectures there is a reference to Kronecker in the historic introduction, but in the Vienna Lectures his name does not appear. Taking into account Brouwer's basic ideas, one has to admit that the programs of Kronecker and Brouwer had preciously little in common. The resemblance of the two programs was largely accepted on Hilbert's authority.

---

[72] Brouwer to Fraenkel 21 December 1926.

As for the French intuitionists, one would be hard pressed to extract a coherent program from their publications. Their unreserved acceptance of classical logic, made them and Brouwer strange bedfellows. It is quite possible that Fraenkel had, under the influence of the prevailing opinion, fallen a victim to similar crude simplifications.

It is not easy to interpret Brouwer's comments and complaints; some of the objectionable expressions had probably been deleted by Fraenkel in reaction to the arguments of Brouwer. There is, for example, a passage in Brouwer's letter which suggests that Fraenkel had lamented the lack of consistency in Brouwer's nomenclature in the treatment of the Cantor–Bendixson theorem (called 'the fundamental theorem of Cantor'): 'every closed set is the disjoint sum of a countable set and a perfect set'. In the second *Begründungs*-paper,[Brouwer 1919A], Brouwer had introduced a number of constructive notions in the topology of the plane, partly for the purpose of proving an intuitionistic version of the Cantor–Bendixson theorem. One of those was an analogue of the derived set of a set, yielding the notion of *Abbrechbare Punktmenge* (deconstructible point set). The resulting version of the Cantor–Bendixson theorem was accordingly formulated in terms of *Abbrechbare Punktmengen*. In *Intuitionistic Set Theory*, however, Brouwer declared that in the *Begründungs*–paper 'the consequences of intuition were not completely clear in my mind'; for that reason he felt obliged to revise his earlier statements.

One can safely say that it was not a trivial matter to keep track of some of Brouwer's notions and theorems. For example, in the review of Schoenflies' *Bericht*, Brouwer had pronounced the Cantor–Bendixson theorem trivially true.[73] But in his *Intuitionistic Set Theory*, he retracted his words and declared that it was a task for intuitionistic mathematics to classify the sets for which the theorem holds.

Fraenkel, not surprisingly, was struck, it seems, by the evolution of the Cantor–Bendixson theorem in intuitionism. His manuscript probably contained some statement to the effect that this was the consequence of a 'gradual refining' of the definition. In Brouwer's eyes this was such a flagrant misconception of the development of intuitionism that he immediately set out to lecture Fraenkel:[74]

> Dear Fraenkel,
>
> That Cantor's fundamental theorem is obvious for totally *Abbrechbare* sets and false for general point sets has nothing to do with a 'gradual refinement' of the fundamental notions, but only with the fact that the intuitive initial construction of mathematics (which, where it occurs with my predecessors, nowhere exceeds the denumerable) was conceived by me at first (1907) as totally *Abbrechbare*, finitary spread; next as a totally *Abbrechbare* (not necessarily finitary) spread, and finally as a spread *tout court*; but in the introductory phase it was always denoted as just 'spread'. One cannot keep introducing new notions; therefore I have always called

---

[73] [Brouwer 1914].
[74] Brouwer to Fraenkel 12 January 1927

my intuitionistic basic construction of mathematics, when it required an extension, 'spread',[75] just as before.

Indeed, a couple of months ago, another such extension became necessary, as you can read in my note *Intuitionistic introduction of the notion of dimension*. Also, after this extension, some so far 'obvious' theorems will turn out to be 'false', without lending warnings of yours, as in the footnote under consideration, the least justification. Should you stick to this insulting and hollow insinuation, even after my urgent request and urgent advice to strike them out, then the competent reader (as I claim to be myself) can only read this as a declaration of war on me. I ask myself in vain what grounds I may have given you. Forgive me that I write sharply and clearly, but I will have to do this also in public; and then it should not be said that I have not called your attention to the consequences of the above-mentioned statement of yours, and warned you.

With friendly greetings, your Brouwer.

It is difficult to explain this sudden outburst of Brouwer. Of course, he was already irritated by the undue haste of the publisher, but his extremely defensive reaction must have had deeper grounds.

Before we go further into the remaining part of the correspondence let us point out that Brouwer did not harbour any grudge against Fraenkel. On the contrary, he had been impressed, and somewhat flattered, by the young man's interest in his foundational views. Moreover he could not have failed to notice that Fraenkel possessed the valuable gift that he himself lacked, or maybe did not have the patience to cultivate: the gift of exposition. A man like Fraenkel could bring intuitionism to the attention of the mathematical community in a way which was much more suited to rouse interest and sympathy than Brouwer's uncompromising, stern papers. At the time of the correspondence, the development of intuitionism was in a critical phase; solid and competent mathematicians were more and more inclined 'to give intuitionism a chance'. But this required from Brouwer more than just being right, it required a measure of convincing that was alien to Brouwer's way of thinking. The process of patient indoctrination did not appeal at all to him. He was quite willing to lecture on intuitionistic mathematics and to show other mathematicians where they went wrong, but he had no taste for the more didactic–propagandistic side. Much later Heyting attempted to regain the lost ground, but even he could not elevate intuitionistic mathematics above the level of an interesting specialism in the corner of dungeons of the mathematics castle. In the twenties there was, however, still hope that intuitionistic mathematics might become an alternative to the existing practice, and Fraenkel was more than anybody else at that time equipped, and mentally suited, to influence the mathematical opinion. Hence Brouwer's high hopes of the influence of Fraenkel's monographs.

[75] i.e. *Menge*.

As we have seen, the relationship between Brouwer and Fraenkel had been warm and unproblematic. Fraenkel admired Brouwer, and Brouwer valued the opinion of the bright younger man. Under the circumstances Fraenkel did not know what to make of this stern letter. He must have ascribed it to some accidental irritation, and written to Brouwer in that spirit. In his letter he probably commented, again, on Brouwer's 'dynamical' presentation of the Cantor–Bendixson theorem, and told Brouwer that such change of notions 'easily led to mistakes'. This was more than Brouwer, who was in the middle of his course on intuitionism in Berlin, could bear; he answered with a long letter full of bitter complaints.[76] In spite of his utter disappointment by Fraenkel's reaction, he had managed to convince the publisher Teubner that more time should be allowed for the proof reading and the corrections. As Brouwer's list of complaints is rather significant for his state of mind at the time of writing, let us have a look at some of them.

He started by urgently requesting Fraenkel not to assist in the 'expropriation, which the German-speaking review journals practised on him, by making me share with Poincaré, Kronecker and Weyl, what is my exclusive personal and spiritual property'. It is quite true that his commentators always referred to his work in connection with the aforementioned three persons; for instance, Hilbert with his immense authority, suggested that Brouwer was something like a mathematical reincarnation of Kronecker. In order to straighten out these misunderstandings, Brouwer sent Fraenkel a German copy of a paper which he had written for the *Revue de Métaphysique et de Morale* (but which never appeared) in which he presented his view on the status of intuitionism.[77]

He went on to dictate to Fraenkel a few corrections, which were partly adopted. The reader who knows Brouwer's style will be amused by these sudden interruptions of Fraenkel's narrative by the solemn Brouwerian formulations.[78] In addition he asked Fraenkel to insert a number of bibliographic references (all to himself).

Finally he proposed to replace 'opinion of the radical intuitionist' by 'opinion of Brouwer' (p. 156 top); for, he explained, 'this opinion is, even when it has been repeated since then by others, nonetheless my spiritual property'.

The letter ended on a sad, almost pathetic note:

> According to a statement of Schopenhauer, there will be practised against each innovator, by the automatically appearing opposition, at first the strategy of (factual) ignoring *totschweigen*), and after the failure of this strategy, that of priority theft. Should this also apply in my case, I am convinced that you do not belong to my enemies, that on the contrary you harbour the wish— and after learning the above—will help to make the above-mentioned strategy against me as little successful as possible.

---

[76] Brouwer to Fraenkel 28 January 1927.

[77] No manuscript of the paper has been found.

[78] See [Fraenkel 1927], p. 35, starting from 'Die zweite'; p. 48, line -8 to -3. p. 50, line -9 to -3 (changing 'Urintuition' to 'Intuition'). Cf. [van Dalen 1997B].

Finally I beg you to believe that the purely factual content of this letter is accompanied only by benevolent and friendly feelings towards you.[79]

It is most likely indeed that Brouwer sincerely wished to avoid a fight. His relationship with the Fraenkels was friendly enough, and although Fraenkel may have had difficulties grasping Brouwer's unusual theories, he definitely was a scholar of recognized integrity.

There is one more letter from Brouwer to Fraenkel, which shows that their relationship had survived this emotional storm. Fraenkel produced many more books after the 'Ten Lectures on Set Theory' and each of them contained an exposition of intuitionism. It is a fact that Fraenkel's books introduced many readers all over the world to the idea that there was more in this world than set theory. At the end of his life, he was even preparing, together with Bar-Hillel, a new edition of his last version of *Foundations of Set Theory*. Above all, Fraenkel should be praised because he conscientiously strove to incorporate the basic ideas of mathematics in his books, including Brouwer's ideas on intuitionism.

10.6 Heyting's first contributions

The reconstruction of 'everyday' mathematics, that is geometry, algebra, probability theory, topology and so on, was a project that figured on Brouwer's agenda, but he could not really bring himself to the toil of routine labour. What he urgently needed was a group of young disciples who could and would revise mathematics in a systematic way. In this respect he was only marginally successful. Although his classes on foundations of mathematics were well attended, few students were willing to devote themselves to the task that Brouwer envisaged. There were two notable exceptions: M.J. Belinfante and Arend Heyting. The first one completed his doctoral examination (M.Sc.) on 16 February 1921, and the second on 22 May 1922. Belinfante took up intuitionistic mathematics and published a series of papers on the theory of infinite series and on the theory of complex functions. He was Brouwer's assistant from 1921 to 1925. In 1925 he became a *privaat docent* at the University of Amsterdam, where he lectured on intuitionistic topics. Those who attended his lectures described them as exceedingly dull and clerically precise. During the Second World War, Belinfante, a Jew, was deported. Euwe, in his Obituary, lamented his tragic fate: 'Belinfante had more opportunities to bring himself to safety than many others, for as a Portuguese Jew he was in a preferential position for some time, so that he had ample opportunity thoroughly to prepare his disappearance. But Belinfante was not the kind of person to do so. Being a hundred percent correct and reliable person, he had difficulties—even in this case—to distrust given promises. Advice from friends was of no influence: he stuck to his decision and only remarked that others had so much less than he had. He even wanted to help! A big child, but at the same time a great man!' He ended his life in

---

[79] Brouwer to Fraenkel 18 January 1927.

Theresienstadt. The physicist A. Pais was in the same category. He was fortunate enough to escape the fate of some of his fellow Portuguese Jews, who had been given a so-called reprieve.[80]

Heyting, the second student who embraced constructive mathematics, became the best-known intuitionist after Brouwer. When Heyting expressed his wish to engage in research in intuitionism, Brouwer suggested the investigation of the axiomatization of intuitionistic projective geometry as a suitable topic for a dissertation. Although this may seem a strange topic for an intuitionist (for it deals more with the linguistic description of a part of mathematics than with the subject matter proper) it was a plausible one. Hilbert's monumental *Foundations of Geometry* had set the tone in geometry, and Brouwer had been lecturing on that topic right from the beginning of his career; a tradition he pursued until the forties. It does not seem to have worried Brouwer that this Ph.D. student made axiomatic investigations. He was sufficiently open-minded to recognize the practical and theoretical value of the axiomatic method as such, as long as one refrained from using unwarranted logical principles or tools and from drawing unwarranted conclusions. After all, he held officially the chair in (among other things) axiomatics! Of course, he denied the axiomatic method any 'creative' role.

Arend Heyting. Courtesy Lien Heyting.

In line with the tradition of the day, Heyting left the university after passing his final examinations. He became a high school teacher in Enschede, a town close to the German border. Notwithstanding his duties as a full-time teacher (luxuries such as research grants or part-time jobs were unheard of) he managed to complete his dissertation in three years. Heyting conducted his research in relative isolation; for some time he had no contact with his Ph.D. adviser. According to Heyting, Brouwer usually left his few Ph.D. students largely to their own resources. When

---

[80]The story of the Portuguese Jews is complicated. Arguments were offered to show that they were not Jewish in the normal sense. Part of the group was given permission to stay in Holland for the time being. Eventually they were transported to Theresienstadt, where some managed to survive the war, but a large number was sent to Auschwitz, with the expected fatal consequences, cf. [Presser 1965].

he did not hear from his student Heyting, he simply assumed that he had given up. And so one day Heyting found out to his surprise that Brouwer had assigned the same research topic to someone else! When he wrote to Brouwer about this, Brouwer replied 'I have not heard from you, so I thought that nothing had come of it'.[81] Heyting, however, persevered, and finished his dissertation three years after his final examination; on 27 May 1925 he defended his dissertation in the central auditorium of the University at the traditional *Oudemanhuispoort* before the Faculty of Mathematics and Physics.

Heyting's dissertation *Intuitionistic Axiomatics of Projective Geometry* was a beautiful piece of work indeed. He was the first to explore the inaccessible intricacies of intuitionistic axiom systems. In particular, he had to solve the far from trivial problems of formalizing the specific positive notions that made intuitionism stand out from classical practice. For geometry one needs a relation like Brouwer's 'apartness' (*örtlich verschieden*) in order to give a reasonable version of the familiar axioms. For example, 'through any two distinct points there passes exactly one line'; this axiom cannot be copied literally, because one cannot determine the requested line if the points are not clearly distinguishable. To be more precise, if we cannot put a positive distance between two points, then we have no means to find the direction of the line passing through the two points.[82] Similarly, there has to be a positive equivalent of the notion 'a point is not on a line'. In analytic geometry one can reduce apartness for points to apartness for real numbers, but in axiomatic geometry one is not allowed to use number systems; one has to find intrinsic properties of 'apart'!

Heyting formulated three simple axioms for the notion of apartness that have ever since been an indispensable part of intuitionistic practice:

(i) $a \# b \rightarrow b \# a$
(ii) $a = b \leftrightarrow \neg a \# b$
(iii) $a \# b \rightarrow c \# a \vee c \# b$

The first axiom is rather obvious, but the second and the third captured the essence of 'being apart from'. The third axiom says that 'apart' is a global property: if a and b are far apart, then any c is far apart from at least one of them. Phrasing it negatively: two things cannot be far apart if there is something close to both of them.[83] The second axiom relates '=' and '#'. It gives a very specific meaning to 'apart', not just as the negation of 'close to', but (in terms of 'distance') as 'having at least a distance 2^{-n} for a suitable n'. Brouwer had indeed shown earlier that in the case of the points in the plane, 'not apart' coincides with 'is equal to'. Once he had found the correct formulations, Heyting developed a true mastery in the manipulation

---

[81] Oral communication A. Heyting.

[82] Think of the co-ordinate version of projective geometry: it is not sufficient for the solving of two linear equations that the determinant is distinct from 0. It has to be apart from 0!

[83] The formulation is un-intuitionistic, but it may suggest the background of the property: the contrapositive of (iii) is the ordinary transitivity of equality.

of the axiomatic machinery. His dissertation is full of clever axiomatic constructive arguments.

In the same period a Ph.D. student from abroad, B. de Loor, had been attracted by Brouwer's fame. He had come to Amsterdam from South Africa to study with Brouwer, and in 1925 he defended his dissertation, *The fundamental theorem of algebra*. We have already seen that Brouwer and De Loor, published an intuitionistically correct version of the fundamental theorem of algebra in 1924. On the basis of this experience De Loor was expected to compose a dissertation. In 1925 the relationship between Brouwer and De Loor suddenly turned sour. The latter was offered the position of lecturer in Pretoria (South Africa). There was, however, a condition attached to the offer: a doctorate was required. So, to oblige him, a few corners were rounded: in exchange for the acknowledgement of the fact that most of the work for the thesis was done by Brouwer and Belinfante, he would be awarded the doctorate 'on grounds of sufficient assimilation by him of a dissertation appearing under his name, but completely composed by others'. The faculty would then graciously grant the doctor's title 'in the interest of cultural relations between the Netherlands and South Africa'. When the dissertation was presented, to the annoyance of Brouwer, no such acknowledgement was to be found. When at the end of the Ph.D. ceremony it was Brouwer's duty to deliver a short laudatio on the work of the young doctor, he publicly castigated De Loor for his breach of academic mores, an understandable but slightly curious action since he had entered into the unusual deal in the first place.[84]

From a mathematical point of view Brouwer's program was showing some real progress: hard theorems were proved. The general conviction that intuitionism was a synonym for poverty was firmly refuted (although the mathematical community at large never bothered to base its opinion on these facts). Objectively speaking, the prospects for intuitionism were fairly rosy.

---

[84]Minutes of the faculty meeting of 11 March 1925.

11
BIBLIOGRAPHY OF BROUWER'S WRITINGS

Here we give list of Brouwer's writings that have, in one form or another, appeared in print. The list may very well be incomplete—there may be letters to the editor, reviews, brochures that have escaped our attention, but it covers probably most of the output of Brouwer. The minor publications have not been listed in the bibliographies of the Collected works. In some cases the present coding of the papers differs from the coding of Freudenthal and Heyting. We have stuck, as far as possible, to the convention that a paper should be classified according to the year printed on the journal volume. In many cases this does not correspond to the actual date of publication. If the the present code differs from the code of the Collected Works, the *Freudenthal–Heyting Verzeichnis* is added after the code. An annotated bibliography of Brouwer's publications, [Dalen, D. van 1997], can be found on the web at www.phil.uu.nl/preprints.html, preprint no 175.

Two of the publications listed here are probably not by Brouwer. The contributions [1905C] and [1905D] to the *Studenten Weekblad Delft* have the appearance of clever imitations of Brouwer's style. Whereas [1904D] and [1904D] are signed 'L. van der Zee', [1905C] and [1905D] are signed 'J. van der Zee' (see p. 8). It is possible that Aldert, Brouwer's youngest brother, who studied in Delft, is the author of the 'imitations'.

Brouwer's Publications

1904A1 Over de splitsing van de continue beweging om een vast punt O van R_4 in twee continue bewegingen om O van R_3's. *KNAW Verslagen*, 12: 819–839.

1904A2 On a decomposition of a continuous motion about a fixed point O of S_4 into two continuous motions about O of S_3's. *KNAW Proc*, 6: 716–735.

1904B1 Over symmetrische transformatie van R_4 in verband met R_r en R_l. *KNAW Verslagen*, 12: 926–928.

1904B2 On symmetric transformation of S_4 in connection with S_r and S_l. *KNAW Proc*, 6: 785–787.

1904C1 Algebraïsche afleiding van de splitsbaarheid der continue beweging om een vast punt van R_4 in die van twee R_3's. *KNAW Verslagen*, 12: 941–947.

1904C2 Algebraic deduction of the decomposability of the continuous motion about a fixed point of S_4 into those of two S_3's. *KNAW Proc*, 6: 832–838.

1904D Een Blijde Wereld? (A joyous world?) *Stud. Weekblad Delft*, 6.10.1904: 8.

1904E No title. *Propria Cures*, 15.10.1904.

1904F Bolland in Delft. *Stud. Weekblad Delft*, 17.11.1904: 58.

1904G Over Moraal. (On morality) *Propria Cures*, 16, 19.11.1904: 110.

1904H Troostgronden. (Grounds of Solace) *Propria Cures*, 16, 31.12.1904: 157.

1905A [FHV: 1905] *Leven, Kunst en Mystiek*, (Life, Art and Mysticism). 99 pp. Waltman, Delft.

1905B Over Moraal (Excerpt). *Stud. Weekblad Delft*, 9.1.1905: 137–138.

1905C Wat is Wijsheid? (What is Wisdom?) *Stud. Weekblad Delft*, 19.10.1905: 20.

1905D Over Wiskunde. (On Mathematics) *Stud. Weekblad Delft*, 30.11.1905: 75.

1906A1 Meerdimensionale vectordistributies. *KNAW Verslagen*, 15: 14–26.

1906A2 Polydimensional vectordistributions. *KNAW Proc.*, 9: 66–78.

1906B1 Het krachtveld der niet-Euclidische, negatief gekromde ruimten. *KNAW Verslagen*, 9: 116–133.

1906B2 The force field of the non-Euclidean spaces with negative curvature. *KNAW Proc.*, 9: 116–133.

1906C1 Het krachtveld der niet-Euclidische positief gekromde ruimten. *KNAW Verslagen*, 15: 293–310.

1906C2 The force field of the non-Euclidean spaces with positive curvature. *KNAW Proc.*, 9: 250–266. (Corr. in 1909H2.)

1907A1 *Over de grondslagen der wiskunde.* (On the Foundations of Mathematics) Dissertation, Amsterdam. 183 pp. + 21 theses.

1907A2 *Over de grondslagen der wiskunde.* Maas en Van Suchtelen, 1907. Amsterdam. 183 pp. (Commercial version of 1907A1.)

1908A Die mögliche Mächtigkeiten. In *Atti IV Congr. Intern. Mat. Roma III*: 569–571.

1908B Over de grondslagen der wiskunde. *Nieuw Arch. Wiskunde*, 8: 326 – 328.

1908C De onbetrouwbaarheid der logische principes. (The unreliability of the logical principles) *Tijdschrift voor Wijsbegeerte*, 2: 152–158.

1908D1 Over differentiequotiënten en differentiaalquotiënten. *KNAW Verslagen*, 17: 38–45.

1908D2 About difference quotients and differential quotients. *KNAW Proc.*, 11: 59–66.

1909A *Het wezen der meetkunde. Rede, uitgesproken 12 October 1909.* (Inaugural lecture, privaat docent). 23 pp.

1909B Die Theorie der endlichen continuierlichen Gruppen unabhängig von den Axiomen von Lie. In *Atti IV Congr. Intern. Mat. Roma*, II: 296–303.

1909C Die Theorie der endlichen kontinuierlichen Gruppen, unabhängig von den Axiomen von Lie, I. *Math Ann*, 67: 246–267.

1909D Over de niet-Euclidische meetkunde. *Nieuw Arch. Wiskunde*, 9: 72–74.

1909E Karakteriseering der Euclidische en niet-Euclidische bewegingsgroepen in R_n. In *Handelingen van het Nederlandsch Natuur- en Geneeskundig Congres*, vol. 12: 189–199.

1909F1 Over één–ééneenduidige, continue transformaties van oppervlakken in zichzelf. *KNAW Verslagen*, 17: 741–752.

1909F2 Continuous one–one transformations of surfaces in themselves. *KNAW Proc.*, 11: 788–798.

1909G1 Over continue vectordistributies op oppervlakken. *KNAW Verslagen*, 17: 896 – 904.

1909G2 On continuous vector distributions on surfaces. *KNAW Proc.*, 11: 850–858. (Corr. in 1910D2.)

1909H1 Over één–ééneenduidige, continue transformaties van oppervlakken in zichzelf, II. *KNAW Verslagen*, 18: 106–117.

1909H2 Continuous one–one transformations of surfaces in themselves, II. *KNAW Proc.*, 12: 286–297. (Corr. in 1911H2.)

1910A1 Over continue vectordistributies op oppervlakken, II. *KNAW Verslagen*, 18: 702 – 721.

1910A2 On continuous vectordistributions on surfaces, II. *KNAW Proc.*, 12: 716–734.

1910B1 Over de structuur der perfecte puntverzamelingen. *KNAW Proc.*, 18: 833 – 842.

1910B2 On the structure of perfect sets of points. *KNAW Proc.*, 12: 785–794.

1910C Zur Analysis Situs. *Math. Ann.*, 68: 422–434.

1910D1 Over continue vectordistributies op oppervlakken, III. *KNAW Verslagen*, 19: 36–51.

1910D2 On continuous vectordistributions on surfaces, III. *KNAW Proc.*, 12: 171–186.

1910E Beweis des Jordanschen Kurvensatzes. *Math. Ann.*, 69: 169–175.

1910F Über eineindeutige, stetige Transformationen von Flächen in sich. *Math. Ann.*, 69: 176–180. (Corr. in 1910L, 1919F.)

1910G Berichtigung. *Math. Ann.*, 69: 180. (re 1909C.)

1910H Die Theorie der endlichen kontinuierlichen Gruppen, unabhängig von den Axiomen von Lie, II. *Math. Ann.*, 69: 181–203.

1910J Sur les continus irréductibles de M. Zoretti. *Annales Ec. Norm. Sup.*, 27: 565–566.

1910L Berichtigung. *Math. Ann.*, 69: 592. (re 1910F.)

1910M G. Mannoury, Methodologisches und Philosophisches zur Elementarmathematik. [review] *De nieuwe Amsterdammer*, 28.8.1910.

1911A G. Mannoury, Methodologisches und Philosophisches zur Elementarmathematik. Haarlem 1909. [review] *Nieuw Arch. Wiskunde*, 9: 199–201.

1911B1 Over de structuur der perfecte puntverzamelingen, II. *KNAW Verslagen*, 19: 1416–1426.

1911B2 On the structure of perfect sets of points, II. *KNAW Proc.*, 14: 137–147.

1911C Beweis der Invarianz der Dimensionenzahl. *Math. Ann.*, 70: 161–165.

1911D Über Abbildung von Mannigfaltigkeiten. *Math. Ann.*, 71: 97–115. (Corr. in 1911M, 1921E.)

1911E Beweis der Invarianz des n-dimensionalen Gebiets. *Math. Ann.*, 71: 305–313.

1911F Beweis des Jordanschen Satzes für den n-dimensionalen Raum. *Math. Ann.*, 71: 314–319.

1911G Über Jordansche Mannigfaltigkeiten. *Math. Ann.*, 71: 320–327. (Corr. in 1911N.)

1911H1 Over één–éénduidige, continue transformaties van oppervlakken in zichzelf, III. *KNAW Verslagen*, 19: 737–747.

1911H2 Continuous one–one transformations of surfaces in themselves, III. *KNAW Proc.*, 13: 767–777. (Corr. in 1911J2.)

1911J1 Over één–éénduidige, continue transformaties van oppervlakken in zichzelf, IV. *KNAW Verslagen*, 20: 24–34.

1911J2 Continuous one–one transformations of surfaces in themselves, IV. *KNAW Proc.*, 14: 300–310.

1911K Sur une théorie de la mesure. A propos d'un article de M. G. Combebiac. *Enseign. Math*, 13: 377–380.

1911L Sur le théorème de M. Jordan dans l'espace à n dimensions. *Comptes Rendus*, 153: 542–543.

1911M Berichtigung. *Math. Ann.*, 71: 598. (re 1911D.)

1911N Bemerkung zu dem Aufsatz von L.E.J. Brouwer: "Über Jordansche Mannigfaltigkeiten". *Math. Ann.*, 71: 598. (re 1911G.)

1912A *Intuïtionisme en Formalisme. Rede bij de aanvaarding van het ambt van buitengewoon hoogleraar in de wiskunde aan de Universiteit van Amsterdam op Maandag 14 October uitgesproken door Dr. L.E.J. Brouwer.* Clausen, Amsterdam. 32 pp.

1912B Beweis des ebenen Translationssatzes. *Math. Ann.*, 72: 37–54.

1912C Zur Invarianz des n-dimensionalen Gebiets. *Math. Ann.*, 72: 55–56.

1912D Über die topologischen Schwierigkeiten des Kontinuitätsbeweises der Existenztheoreme eindeutig umkehrbarer polymorpher Funktionen auf Riemannschen Flächen (Auszug aus einem Brief an R. Fricke). *Nachr. Gött.*: 603–606.

1912E1 Over omloopcoëfficiënten. *KNAW Verslagen*, 20: 1049–1057.

1912E2 On looping coefficients. *KNAW Proc.*, 15: 113–122.

1912F Sur l'invariance de la courbe fermée. *Comptes Rendus*, 154: 862–863.

1912G Über die Singularitätenfreiheit der Modulmannigfaltigkeit. *Nachr. Gött.*: 803–806.

1912H Über den Kontinuitätsbeweis für das Fundamentaltheorem der automorphen Funktionen im Grenzkreisfall. *Jahresber. Dtsch. Math. Ver.*, 21: 154–157.

1912K1 Over één–éénduidige, continue transformaties van oppervlakken in zichzelf, V. *KNAW Verslagen*, 21: 300–309.

1912K2 Continuous one-one transformations of surfaces in themselves, V. *KNAW Proc.*, 15: 352–360.

1912L Beweis der Invarianz der geschlossenen Kurve. *Math. Ann.*, 72: 422–425.

1912M Sur la notion de 'classe' de transformations d'une multiplicité. In *Proc. V Int. Congr. Math. Cambridge 1912, II*: 9–10.

1913A Über den natürlichen Dimensionsbegriff. *J. Reine Angew. Math.*, 142: 146–152. (Corr. in 1924M.)

1913B1 Eenige opmerkingen over het samenhangstype η. *KNAW Verslagen*, 21: 1412–1419

1913B2 Some remarks on the coherence type η. *KNAW Proc.*, 15: 1256–1263.

1913C Intuitionism and Formalism. *Bull. Am. Math. Soc.*, 20: 81–96.

1914 A. Schoenflies und H. Hahn. Die Entwickelung der Mengenlehre und ihrer Anwendungen, Leipzig und Berlin 1913. [review] *Jahresber. Dtsch. Math. Ver.*, 23: 78–83.

1915A1 Opmerking over inwendige grensverzamelingen. *KNAW Verslagen*, 23: 1325–1326.

1915A2 Remark on inner limiting sets. *KNAW Proc.*, 18: 48–49.

1915B Over de loodrechte trajectoriën der baankrommen eener vlakke eenledige projectieve groep. (On the orthogonal trajectories of the orbits of a one parameter plane projective group) *Nieuw Arch. Wiskunde*, 11: 265–290.

1915C Meetkunde en Mystiek. (Geometry and Mysticism, review) *De nieuwe Amsterdammer*, 4.12.1915.

1916A L.E.J. Brouwer, eenige opmerkingen over het samenhangstype η. *Jahrbuch Fortschr. Math.*, 44: 556 (Auto-review).

1916B E. Borel, Les ensembles de mesure nulle. [review] *Jahrbuch Fortschr. Math.*, 44: 556–557.

1916C J.I. de Haan, Rechtskundige significa en haar toepassing op de begrippen: 'aansprakelijk', 'verantwoordelijk', 'toerekeningsvatbaar'. (Legal significs and its applications to the notions 'liable', 'responsible', 'accountable', review) *Groot Nederland*, 14: 333–336.

1916D Mr. Israël de Haan's Rechtkundige Significa. (Mr. Israël de Haan's Legal Significs, review) *De nieuwe Amsterdammer*, 9.9.1916.

1916E Een vraag aan de Belgische Regeering. (A question to the Belgian Government) *De nieuwe Amsterdammer*, 99, 16.9.1916.

1916F Luchtvaart en photogrammetrie. (Aviation and Photogrammetry) *Avia*, 6: 29-30, 122-124, 223-225.

1917A1 [FHV: 1917A] Addenda en corrigenda over de grondslagen der wiskunde. (Addenda and Corrigenda to the foundations of mathematics) *KNAW Verslagen*, 25: 1418–1423.

1917A2 [FHV: 1917A] Addenda en corrigenda over de grondslagen der wiskunde. *Nieuw Arch. Wiskunde*, 12: 439–445.

1917B1 Over lineaire inwendige grensverzamelingen. *KNAW Verslagen*, 25: 1424–1426.

1917B2 On linear inner limiting sets. *KNAW Proc.*, 20: 1192–1194.

1917C Luchtvaart en photogrammetrie. *Het Vliegveld*, 142–144, 165–167.

1917D Anti-nationale Literatuur. (Anti-national Literature) *De nieuwe Amsterdammer*, 3.2.1917, 110.

1918A Voorbereidend Manifest. (Preparatory manifesto) *Mededelingen van het Internationale Instituut voor Wijsbegeerte*, 1: 3–12.

1918B Begründung der Mengenlehre unabhängig vom logischen Satz vom ausgeschlossenen Dritten. Erster Teil, Allgemeine Mengenlehre. *KNAW Verhandelingen*, 5: 1–43.

1918C Über die Erweiterung des Definitionsbereiches einer stetigen Funktion. *Math. Ann.*, 79: 209–211. (See 1919E.)

1918D Lebesguesches Mass und Analysis Situs. *Math. Ann.*, 79: 212–222.

1918E Beweging van een materieel punt op den bodem eener draaiende vaas onder den invloed der zwaartekracht. (Motion of a particle on the bottom of a rotating vessel under the influence of the gravitational force) *Nieuw Arch. Wiskunde*, 12: 407–419.

1918F Voorbereiding van Hoogleeraarsbenoemingen. (Preparation of the appointment of professors) *De nieuwe Amsterdammer*, 27.7.1918.

1918G Intuïtieve significa. (Intuitive Significs) *Propria Cures*, 9.9.1918.

1918H Aan de Ned. Burgerij. (To the Dutch Citizenship) *De nieuwe Amsterdammer*, 30.11.1918 (Joint letter to the editor).

1919A Begründung der Mengenlehre unabhängig vom logischen Satz vom ausgeschlossenen Dritten. Zweiter Teil, Theorie der Punktmengen. *KNAW Verhandelingen*, 7: 1–33.

1919B *Wiskunde, Waarheid, Werkelijkheid.* (Mathematics, Truth, Reality) Noordhoff, Groningen. 65 pp.

1919C Signifisch Taalonderzoek. Onderscheid der taaltrappen ten aanzien van de sociale verstandhouding. (Signific Investigation of Language. Distinction of the language levels with respect to the social understanding) *Mededelingen van het Internationale Instituut voor Wijsbegeerte*, 2: 5–32.

1919E Nachträgliche Bemerkung über die Erweiterung des Definitionsbereiches einer stetigen Funktion. *Math. Ann.*, 78: 403. (re 1918 C.)

1919F Berichtigung. *Math. Ann.*, 79: 403. (re 1910F)

1919G Énumération des surfaces de Riemann regulières de genre un. *Comptes Rendus*, 168: 677–678. Errata. *C.R.* 168: 832.

1919H Énumération des groupes finis de transformations topologiques du tore. *Comptes Rendus*, 168: 845–848. Errata *C.R.* 168: 1168; 169: 552.

1919J Sur les points invariants des transformations topologiques des surfaces. *Comptes Rendus*, 168: 1042–1044. Errata *C.R.* 169: 552.

1919K Sur la classification des ensembles fermés situés sur une surface. *Comptes Rendus*, 169: 953–954.

1919L1 Over één–éénduidige, continue transformaties van oppervlakken in zichzelf, VI. *KNAW Verslagen*, 27: 609–612.

1919L2 Über eineindeutige stetige Transformationen von Flächen in sich, VI. *KNAW Proc.*, 21: 707–710.

191M1 Opmerking over de vlakke translatiestelling. *KNAW Verslagen*, 27: 840–841.

1919M2 Remark on the plane translation theorem. *KNAW Proc.*, 21: 935–936.

1919N1 Over topologische involuties. *KNAW Verslagen*, 27: 1202–1203.

1919N2 Über topologische Involutionen. *KNAW Proc.*, 21: 1143–1145.

1919P1 Opsomming der periodieke transformaties van de torus. *KNAW Verslagen*, 27: 1363–1367.

1919P2 Aufzählung der periodischen Transformationen des Torus. *KNAW Proc.*, 21: 1352–1356.

1919Q1 Opmerking over meervoudige integralen. *KNAW Verslagen*, 28: 116–120.

1919Q2 Remark on multiple integrals. *KNAW Proc.*, 22: 150–154.

1919R Luchtvaart en photogrammetrie, I. *Nieuw Arch. Wiskunde*, 7: 311–331.

1919S Über die periodischen Transformationen der Kugel. *Math. Ann.*, 80: 39–41.

1919T Hoogleeraarsbenoemingen. *De nieuwe Amsterdammer*, 1.2.1919.

1919U Corrections. *Bull. Am. Math. Soc.*:135. (re 1913C.)

1919V Antwort von L.E.J. Brouwer *Mededelingen van het Internationale Instituut voor Wijsbegeerte*, 2: 33–34.)

1920A1 Over de structuur van perfecte puntverzamelingen, III. *KNAW Verslagen*, 28: 373–375.

1920A2 Über die Struktur der perfekten Puntmengen, III. *KNAW Proc.*, 22: 471–474, 974. (Erratum same vol. p. 974.)

1920B1 Over één-ééndudige, continue transformaties van oppervlakken in zichzelf, VII. *KNAW Verslagen*, 28: 1186–1190.

1920B2 Über eineindeutige, stetige Transformationen von Flächen in sich, VII. *KNAW Proc.*, 22: 811–814. (Corr. in KNAW Proc.29 (1926), p. 1133.)

1920C Énumération des classes de transformations du plan projectif. *Comptes Rendus*, 170: 834–835. Errata. *C.R.* 170: 1295.

1920D Énumération de classes de représentations d'une surface sur une autre surface. *Comptes Rendus*, 171: 89–91. Errata. *C.R.* 171: 830.

1920E Über die Minimalzahl der Fixpunkte bei den Klassen von eindeutigen stetigen Transformationen der Ringflächen. *Math. Ann.*, 82: 94–96.

1920F Luchtvaart en photogrammetrie, II. *Nieuw Tijdschr. Wiskunde*, 8: 300–307.

1920G1 Over één-ééndudige, continue transformaties van oppervlakken in zichzelf, VIII. *KNAW Verslagen*, 29: 640–642.

1920G2 Über eineindeutige, stetige Transformationen von Flächen in sich, VIII. *KNAW Proc.*, 23: 232–234.

1920H DE NIEUWE AMSTERDAMMER. *De nieuwe Amsterdammer*, December 1920.

1920J [FHV: 1919D1] Intuitionistische Mengenlehre. *Jahresber. Dtsch. Math. Ver.*, 28: 203–208.

1920K Letter to the members of the Mathematics and Physics section of the KNAW. Privately printed pamphlet.

1921A1 [FHV: 1921A] Besitzt jede reelle Zahl eine Dezimalbruch-Entwickelung? *Math. Ann.*, 83: 201–210.

1921A2 [FHV: 1921A] Besitzt jede reelle Zahl eine Dezimalbruch-Entwickelung? *KNAW Verslagen*, 29: 803–812.

1921A3 [FHV: 1921A] Besitzt jede reelle Zahl eine Dezimalbruch-Entwickelung? *KNAW Proc.*, 23: 955–964.

1921C Opmerking over de bepaling van alle complexe functies, voor welke $f(z) = \varphi(|z|)$ is. (Remark on the determination of all complex functions, for which $f(z) = \varphi(|z|)$ holds) *Chr Huygens*, 1: 352–354.

1921D Aufzählung der Abbildungsklassen endlichfach zusammenhängender Flächen. *Math. Ann.*, 82: 280–286.

1921E Berichtigung. *Math. Ann.*, 82: 286. (re 1911D.)

1921F Berichtigung. *Math. Ann.*, 82: 286. (re 1911 E.)

1921G Sovjet wetten. (Soviet laws, review) *Nieuwe Kroniek*, 1: 2,3.

1921H Wis- en Natuurkunde en Wijsbegeerte. (Mathematics, Physics and Philosophy) *Nieuwe Kroniek*, 18.6.1921, 1: 1–2.

1921J [FHV: 1919D2] Intuïtionistische Verzamelingsleer. *Kon Ned Ak Wet Verslagen*, 29:797–802.

1922A [FHV: 1919D1] Intuitionistische Mengenlehre. *KNAW Proc.*, 23: 949–954.

1922B Wis- en Natuurkunde en Wijsbegeerte. *Nieuwe Kroniek*, 2, 11.2.1922: 3–5.

1922C Wetenschap, humaniteit, Frankrijk en Nederland. (?) (Science, humanity, France and the Netherlands)

1922D1 Schrijven van Dr. L.E.J. Brouwer aan Zijne Excellentie den Minister van Onderwijs, Kunsten en Wetenschappen. (Letter of Dr. L.E.J. Brouwer to his Excellency the Minister of Education, Arts and Sciences). *Privately printed brochure.*

1923A Begründung der Funktionenlehre unabhängig vom logischen Satz vom ausgeschlossenen Dritten. Erster Teil, Stetigkeit, Messbarkeit, Derivierbarkeit. *KNAW Verhandelingen* 1e sectie deel XIII, 2: 1–24.

1923B1 Over de rol van het principium tertii exclusi in de wiskunde, in het bijzonder in de functietheorie. (On the role of the principium tertii exclusi in mathematics, in particular in function theory) *Wis- en Natuurkundig Tijdschrift*, 2: 1–7.

1923B3 Die Rolle des Satzes vom ausgeschlossenen Dritten in der Mathematik. *Jahresber. Dtsch. Math. Ver.*, 32: 67.

1923C1 Intuïtionistische splitsing van mathematische grondbegrippen. *KNAW Verslagen*, 32: 877–880.

1923D1 Over het natuurlijke dimensiebegrip. *KNAW Verslagen*, 32: 881–886.

1923D2 Über den natürlichen Dimensionsbegriff. *KNAW Proc.*, 26: 795–800.

1924A1 Over de toelating van oneindige waarden voor het functiebegrip. *KNAW Verslagen*, 33: 41.

1924A2 Ueber die Zulassung unendlicher Werte für den Funktionsbegriff. *KNAW Proc.*, 27: 248.

1924B1 Perfecte puntverzamelingen met positief irrationale afstanden. *KNAW Verslagen*, 33: 81.

1924B2 Perfect sets of points with positive-irrational distances. *KNAW Proc.*, 27: 487.

1924C1 (With B. de Loor) Intuïtionistisch bewijs van de hoofdstelling der algebra. *KNAW Verslagen*, 33: 82–84.

1924C2 (with B. de Loor) Intuitionistischer Beweis des Fundamentalsatzes der Algebra. *KNAW Proc.*, 27: 186–188.

1924D1 Bewijs dat iedere volle functie gelijkmatig continu is. *KNAW Verslagen*, 33: 189–193.

1924D2 Beweis dass jede volle Funktion gleichmässig stetig ist. *KNAW Proc.*, 27: 189–193.

1924E1 Intuïtionistische aanvulling van de hoofdstelling der algebra. *KNAW Verslagen*, 33: 459–462.

1924E2 Intuitionistische Ergänzung des Fundamentalsatzes der Algebra. *KNAW Proc.*, 27: 631–634.

1924F1 Bewijs van de onafhankelijkheid van de onttrekkingsrelatie van de versmeltingsrelatie. *KNAW Verslagen*, 33: 479–480.

1924G1 Opmerkingen aangaande het bewijs der gelijkmatige continuïteit van volle functies. *KNAW Verslagen*, 33: 646–648.

1924G2 Bemerkungen zum Beweise der gleichmässigen Stetigkeit voller Funktionen. *KNAW Proc.*, 27: 644–646.

1924J1 Opmerkingen over het natuurlijk dimensiebegrip. *KNAW Verslagen*, 33: 476–478.

1924J2 Bemerkungen zum natürlichen Dimensionsbegriff. *KNAW Proc.*, 27: 635–638.

1924K1 Over de n-dimensionale simplex ster in R_n. *KNAW Verslagen*, 33: 1008–1010.

1924K2 On the n-dimensional simplex star in R_n. *KNAW Proc.*, 27: 778–780.

1924L Zum natürlichen Dimensionsbegriff. *Math. Zeitschr.*, 21: 312–314.

1924M Berichtigung. *J. reine angew. Math.*, 153: 253. (re 1913A.)

1924N [FHV: 1923 B2]. Über die Bedeutung des Satzes vom ausgeschlossenen Dritten in der Mathematik insbesondere in der Funktionentheorie. *J. reine angew. Math.*, 154: 1–8.

1924O Unberücksichtigt gebliebenes Schreiben von Dr. L.E.J. Brouwer an Seine Exzellenz den Minster für Unterricht, Kunst und Wissenschaft. *Privately printed brochure.*

1925A Zur Begründung der intuitionistischen Mathematik I. *Math. Ann.*, 93: 244–257. (Corr. in 1926A.)

1925B1 Intuitionistischer Beweis des Jordanschen Kurvensatzes. *KNAW Proc.*, 28: 503–508.

1925B2 Intuïtionistisch bewijs van de krommenstelling van Jordan. *KNAW Verslagen*, 34: 657.

1925C (with R. Weitzenböck, Hk. de Vries) Rapport over de verhandeling van P. Urysohn 'Mémoire sur les multiplicités cantoriennes. Deuxième partie: les lignes cantoriennes'. *KNAW Verslagen*, 34: 516–517.

1925D [FHV: 1924H] Zuschrift an dem Herausgeber. *Jahresber. Dtsch. Math. Ver.*, 33: 124.

1925E [FHV: 1923C2] Intuitionistische Zerlegung mathematischer Grundbegriffe. *Jahresber. Dtsch. Math. Ver.*, 33: 251–256.

1926A Zur Begründung der intuitionistischen Mathematik II. *Math. Ann.*, 95: 453–472.

1926B1 Intuïtionistische invoering van het dimensiebegrip. *KNAW Verslagen*, 35: 634–642.

1926B2 Intuitionistische Einführung des Dimensionsbegriffes. *KNAW Proc.*, 29: 855–873.

1926C1 De intuïtionistische vorm van het theorema van Heine-Borel. *KNAW Verslagen*, 35: 667–668.

1926C2 Die intuitionistische Form des Heine-Borelschen Theorems. *KNAW Proc.*, 29: 866–867.

1926D1 Over transformaties van projectieve ruimten. *KNAW Verslagen*, 35: 643–644.

1926D2 On transformations of projective spaces. *KNAW Proc.*, 29: 864–865.

1926E (with R. Weitzenböck, Hk. de Vries) Rapport omtrent de ter opneming in dewerken der Akademie aangeboden verhandeling van P. Alexandroff en P. Urysohn 'Mémoire sur les espaces topologiques compacts'. *KNAW Verslagen*, 35.

1927A Zur Begründung der intuitionistischen Mathematik III. *Math. Ann.*, 96: 451–488.

1927B Über Definitionsbereiche von Funktionen. *Math. Ann.*, 97: 60–75.

1927C Virtuelle Ordnung und unerweiterbare Ordnung. *J. reine angew. Math.*, 157: 255–257.

1927D [FHV: 1924F2] Zur intuitionistischen Zerlegung mathematischer Grundbegriffe. *Jahresber. Dtsch. Math. Ver.*, 36: 127–129.

1928A1 Intuitionistische Betrachtungen über den Formalismus. *KNAW Verslagen*, 36: 1189.

1928A2 Intuitionistische Betrachtungen über den Formalismus. *Die Preussische Akademie der Wissenschaften. Sitzungsberichte. Physikalische-Mathematische Klasse*, 48–52.

1928A3 [FHV: 1928A2] Intuitionistische Betrachtungen über den Formalismus. *KNAW Proc.*, 31: 374–379.

1928B1 Beweis dass jede Menge in einer individualisierten Menge enthalten ist. *KNAW Verslagen*, 36: 1189.

1928B2 Beweis dass jede Menge in einer individualisierten Menge enthalten ist. *KNAW Proc.*, 31: 380–381.

1928C1 Zur Geschichtsschreibung der Dimensionstheorie. *KNAW Verslagen*, 37: 626.

1928C2 Zur Geschichtsschreibung der Dimensionstheorie. *KNAW Proc.*, 31: 953–957. (Corr. in *KNAW Proc.* 32: 1022.)

1929A Mathematik, Wissenschaft und Sprache. *Monatshefte für Mathematik und Physik*, 36: 153–164.

1929B Herinnering aan C.S. Adama van Scheltema. (In memory of C.S. Adama van Scheltema) In *Ter herdenking van C.S. Adama van Scheltema*. Em. Querido, Amsterdam, p. 69

1929C Erratum. *KNAW Proc.*, 32: 1022. (re 1928C2)

1930A *Die Struktur des Kontinuums* [Sonderabdruck]. Vienna.

1930B A. Fraenkel, Zehn Vorlesungen über die Grundlegung der Mengenlehre. [review]. *Jahresber. Dtsch. Math. Ver.*, 39: 10–11.

1931 Ueber freie Umschliessungen im Raume. *KNAW Proc.*, 34: 100–101

1933A1 [FHV: 1933] Willen, Weten, Spreken. (Will, Knowledge, Speach) *Euclides*, 9: 177–193.

1933A2 [FHV: 1933] Willen, Weten, Spreken. In *"De uitdrukkingswijze der wetenschap, kennistheoretische voordrachten aan de Universiteit van Amsterdam (1932–33)"* Noordhoff, Groningen. 43–63.

1934 Erratum. *Euclides*, 11. (re 1933A1.)

1937 (with F.J. van Eeden, J. van Ginneken, G. Mannoury) Signifische dialogen. (Signific dialogues) *Synthese*, 2: 168–174, 261–268, 316–324.

1939A Zum Triangulationsproblem. *Ind. Math.*, 1: 248–253.

1939B Theodor Vahlen zum 70. Geburtstag. *Forschungen und Fortschritte*, 15: 238–239.

1939C (with F. van Eeden, J. van Ginneken, G. Mannoury), *Signifische Dialogen*. J. Bijleveld, Utrecht, p. 65.

1941A D.J. Korteweg. (Obituary). *Euclides*, 17: 266–267.

1941B D.J. Korteweg. (Obituary). *KNAW Verhandelingen*, 266–267.

1942A Zum freien Werden von Mengen und Funktionen. *Ind. Math.*, 4: 107–108.

1942B Die repräsentierende Menge der stetigen Funktionen des Einheitskontinuums. *Ind. Math.*, 4: 154.

1942C Beweis dass der Begriff der Menge höherer Ordnung nicht als Grundbegriff der intuitionistischen Mathematik in Betracht kommt. *Ind. Math.*, 4: 274–276.

1946A Synopsis of the Signific Movement in the Netherlands. *Synthese*, 5: 201–208.

1946B1 Toespraak van Prof. Dr. L.E.J. Brouwer en antwoord van Prof. Dr. G. Mannoury. *Jaarboek der Universiteit van Amsterdam 1946-1947*.

1946B2 [FHV: 1946B] Address delivered on September 16th, 1946, on the conferment upon Professor G. Mannoury of the honorary degree of Doctor of Science. *Jaarboek der Universiteit van Amsterdam 1946-1947*.

1947 Richtlijnen der intuïtionistische wiskunde. (Guidelines of intuitionistic mathematics) *Ind. Math.*, 9: 197.

1948A Essentieel negatieve eigenschappen. (Essentially negative properties) *Ind. Math.*, 10: 322–323.

1948B Opmerkingen over het beginsel van het uitgesloten derde en over negatieve asserties. (Remarks on the principle of the excluded third and on negative assertions) *Ind. Math.*, 10: 383–387.

1948C Discussion. (Following "Les conceptions mathématiques et le reel"). In *Symposium de l'Institut des Sciences théoriques. Bruxelles 1947. Actualités scientifiques et Industrielles*, Paris. Hermann: 31–60.

1949A De non-aequivalentie van de constructieve en negatieve orderelatie in het continuum. (The non-equivalence of the constructive and the negative order relation in the continuum) *Ind. Math.*, 11: 37–39.

1949B Contradictoriteit der elementaire meetkunde. (Contradictority of elementary geometry) *Ind. Math.*, 11: 89–90.

1949C [FHV: 1948C] Consciousness, Philosophy and Mathematics. *Proceedings of the 10th International Congress of Philosophy, Amsterdam 1948*, 3: 1235–1249.

1950A Remarques sur la notion d'ordre. *Comptes Rendus*, 230: 263–265.

1950B Sur la possibilité d'ordonner le continu. *Comptes Rendus*, 230: 349–350.

1950C Discours final. Les méthodes formelles en axiomatique. *Colloques Internationaux du C.N.R.S. Paris 1950*, 36: 75. (See also the discussions in this volume.)

1951 On order in the continuum, and the relation of truth to non-contradictority. *Ind. Math.*, 13: 357–358.

1952A An intuitionist correction of the fixed-point theorem on the sphere. *Proceedings of the Royal Society London*, 213: 1–2.

1952B Historical background, principles and methods of intuitionism. *South African Journal of Science*, 49: 139–146.

1952C Over accumulatiekernen van oneindige kernsoorten. (On accumulation cores of infinite core species) *Ind. Math.*, 14:439–441.

1952D Door klassieke theorema's gesignaleerde pinkernen die onvindbaar zijn. (Fixed cores which cannot be found, though they are claimed to exist by classical theorems) *Ind. Math.*, 14: 443–445.

1954A Point and Spaces. *Canadian Journal for Mathematics*, 6: 1–17.

1954B Addenda en corrigenda over de rol van het principium tertii exclusi in de wiskunde. (Addenda and corrigenda on the role of the principium tertii exclusi in mathematics) *Ind. Math.*, 16: 104–105.

1954C Nadere addenda en corrigenda over de rol van het principium tertii exclusi in de wiskunde. (Further addenda and corrigenda on the role of the principium tertii exclusi in mathematics) *Ind. Math.*, 16: 109–111.

1954D Ordnungswechsel in Bezug auf eine coupierbare geschlossene stetige Kurve. *Ind. Math.*, 16: 112–113.

1954E Intuïtionistische differentieerbaarheid. (Intuitionistic differentiability) *Ind. Math.*, 16: 201–203.

1954F An example of contradictority in classical theory of functions. *Ind. Math.*, 16: 204–205.

1955 The effect of intuitionism on classical algebra of logic. *Proceedings of the Royal Irish Academy*, 57: 113–116.

1956 [FHV: 1952B2] Voorgeskiedenis, Beginsels en Metodes van die Intuïtionisme. *Tydskrif vir Wetenskkap en Kuns*, 186–197.

1975 *Collected Works 1. Philosophy and Foundations of Mathematics.* (ed. A. Heyting). North-Holland Publ. Co., Amsterdam.

1976 *Collected Works 2. Geometry, Analysis Topology and Mechanics.* (ed. H. Freudenthal). North-Holland Publ. Co., Amsterdam.

1981A *Brouwer's Cambridge Lectures on Intuitionism.* (ed. D. van Dalen) Cambridge University Press, Cambridge. 109 pp.

1981B *Over de Grondslagen der Wiskunde.* (ed. D. van Dalen) Mathematisch Centrum, Amsterdam. 267 pp. (Dissertation, Correspondence Brouwer–Korteweg, reviews, 1908C.)

1983B *Brouwer. Lezioni Sull' Intuitionismo.* (ed. D. van Dalen, transl. S. Bernini) Boringhieri, Torino. 113 pp.

1984 *Carel Steven Adama van Scheltema & Luitzen Egbertus Jan Brouwer. Droeve snaar, vriend van mij (Sad string, friend of mine). Brieven.* (ed. D. van Dalen). De Arbeiderspers. Amsterdam. 173 pp.

1985 'Amsterdam' van C.S. Adama van Scheltema [review] (ed. R. Delvigne). *Juffrouw Ida*, 11: 41–44.

1992 *Intuitionismus.* (ed. D. van Dalen) Bibliographisches Institut, Wissenschaftsverlag, Mannheim.

1996 Life, Art and Mysticism (ed. W.P. van Stigt) *Notre Dame Journal of Formal Logic*, 37: 381-429.

CONTENTS OF VOLUME 2

1. The Fathers of dimension
2. Years of triumph and tragedy
3. The *Grundlagenstreit*
4. The war of the frogs and the mice
5. The thirties
6. War and occupation
7. After World War II. The decline of power
8. The battle for the Compositio mathematica
9. The restless emeritus

REFERENCES

[Alexander 1922] J.W. Alexander. A proof and extension of the Jordan–Brouwer separation theorem. *Trans Am Math Soc*, 23:333–349.
[Alexander 1924] ———— An example of a simply connected surface bounding a region which is not simply connected. *Proc. Nat. Ac. Sc.*, 10:8–10.
[Alexandroff, P. and Hopf, H. 1935] Alexandroff, P. and Hopf, H. *Topologie I*. Springer Verlag, Berlin.
[Alexandrov 1972] P.S. Alexandrov. Poincaré and Topology. *Russ Math Surveys*, 27:157–166.
[Antoine 1921] L. Antoine. Sur l'homéomorphie de deux figures et de leurs voisinages. *J Math*, 4:221–325.
[Arzela 1896] C. Arzela. Sull'essistenza degli'integrali nelle equazioni differenziali ordinarie. *Mem. Ac. Sci. Bologna*, 5:131–140.
[Baire 1907a] R. Baire. Sur la non–applicabilité de deux continus à n et $n + p$ dimensions. *Comptes Rendus*, 144:p.318–321.
[Baire 1907b] ———— Sur la non–applicabilité de deux continus à n et $n + p$ dimensions. *Bull Sc Math*, 31:94–99.
[Bashkirtseff 1888] Marie Bashkirtseff. *Journal de Marie Bashkirtseff*. Charpentier et Cie, Paris.
[Beardon 1984] A.F. Beardon. *A Primer on Riemann Surfaces*. Cambridge University Press, Cambridge.
[Behnke, H, and Sommer, F. 1955] Behnke, H, and Sommer, F. *Theorie der analytische Funktionen*. Springer Verlag, Berlin.
[Belinfante 1923] M.J. Belinfante. *Over oneindige reeksen*. PhD thesis, Amsterdam.
[Bellaar-Spruyt 1903] C. Bellaar-Spruyt. *Leerboek der Formeele Logica. Bewerkt naar de dictaten van wijlen Prof. Dr. C.B. Spruyt door M. Honigh*. Vincent Loosjes, Haarlem.
[Bernstein 1919] F. Bernstein. Die Mengenlehre George Cantors und der Finitismus. *Jahresber. Dtsch. Math.Ver.*, 28:63–78.
[Bockstaele 1949] P. Bockstaele. Het intuïtionisme bij de Franse wiskundigen. *Verhandelingen van de Koninklijke Vlaamse Academie van Wetenschappen*, XI (32).
[Boersen 1987] M.W.J.L. Boersen. *De Kolonie van de Internationale Broederschap te Blaricum*. Historische Kring Blaricum.
[Bolland 1897] G.J.P.J. Bolland. *Aanschouwing en Verstand. Gedachten over Continua & Discreta in Wiskunde en Bewegingsleer*. Adriani, Leiden.
[Bolland 1904] ———— *Zuivere Rede en Hare Werkelijkheid*. Adriani, Leiden.
[Bonger 1929] W.A. Bonger. Scheltema en het Socialisme (persoonlijke herinneringen). In *Ter herdenking van C.S. Adama van Scheltema*, pages 46–68, Amsterdam. Querido.
[Borel 1898] É. Borel. *Leçons sur la théorie des fonctions*. Gauthier–Villars, Paris. 2nd ed. 1914, 3rd ed. 1928.
[Borel 1908] ———— Sur les principes de la théorie des ensembles. In G. Castelnuovo, editor, *Atti IV Congr.Intern.Mat. Roma*, volume 1, pages 15–17, Accad Naz Lincei. Roma.
[Browder 1976] F.I.(ed.) Browder. *Hilbert's Problems. Proc. of Symposia in PURE MATHEMATICS, 2 vols*. AMS, Providence, Rhode Island.
[Cantor 1892] G. Cantor. Ueber eine elementare Frage der Mannigfaltigkeitslehre. *Jahresber. Dtsch. Math.Ver.*, 1:75–78.

[Cartan 1914] E Cartan. Les groupes réels simples finis et continus. *Ann. École Normale*, 31: 263–355.
[Caspary 1883] F. Caspary. Zur theorie der Thetafunctionen mit zwei argumenten. *J Reine Angew Math*, 94:74–86.
[Castelnuovo 1909] G. Castelnuovo, editor. *Atti IV Congr.Intern.Mat. Roma*, Accad Naz Lincei. Roma.
[Dalen, D. van 1981] Dalen, D. van. L.E.J.Brouwer en de eenzaamheid van het gelijk. *Vrij Nederland*, pages 3–23.
[Dalen, D. van (Ed.) 1984] ——— *Droeve snaar, vriend van mij. De correspondentie tussen Brouwer en Adama van Scheltema*. De Arbeiderspers, Amsterdam.
[Dalen, D. van 1985] ——— Eine Bemerkung zum Aufsatz "Der Fundamentalsatz der Algebra und der Intuitionismus" von H. Kneser. *Arch. Math. Logic*, 25:43–44.
[Dalen, D. van 1995] ——— Hermann Weyl's Intuitionistic Mathematics. *Bull. Symb. Logic*, 1:145–169.
[Dalen, D. van 1997] ——— A bibliography of L.E.J. Brouwer. *Logic Group Preprint Series*, 175.
[Dalen, D. van 1999] ——— *Luitzen Egbertus Jan Brouwer*, in'History of Topology (ed. I.M. James)' pages 947–964. Elsevier, Amsterdam.
[Dalen, D. van 2000] ——— Brouwer and Fraenkel on Intuitionism. *Bull.Ass.Symb.Logic*, 6:284–310.
[Dalen, D. van 2001] ——— *L.E.J. Brouwer en De Grondslagen van de wiskunde*. Epsilon, Utrecht.
[Dantzig 1957] D. van Dantzig. Gerrit Mannoury's significance for Mathematics and its Foundations. *Nieuw Arch Wiskunde*, 5:1–18.
[Delvigne 1985] Rob Delvigne. L.E.J. Brouwer over C.S. Adama van Scheltema. *Juffrouw Ida*, 11:41–44.
[Dieudonné 1989] J. Dieudonné. *A History of Algebraic and Differential Topology, 1900–1960*. Birkhäuser, Basel.
[Dijkstra 1986] Bram Dijkstra. *Idols of Perversity. Fantasies of feminine Evil in Fin–de–siècle Culture*. CUP, Cambridge.
[Eeden 1911] E. Gutkind Eeden, F. van. *Welt–Eroberung durch Helden–Liebe*. Schuster & Loeffler, Berlin.
[Eeden 1921] ——— *Het rode lampje*. Versluys, Amsterdam. 2 vols.
[Eeden 1971] ——— *Dagboek 1878–1923 (ed. H.W. van Tricht)*. Tjeenk Willink, Culemborg. 4 vols.
[Emmerik 1991] E.P. van Emmerik. *J.J. van Laar (1860–1938). A Mathematical Chemist*. PhD thesis, Universiteit van Amsterdam.
[Everdingen 1976] E. van Everdingen. *Zestig Jaar Internationale School van Wijsbergeerte*. Van Gorcum, Assen.
[Finsler 1925] P. Finsler. Gibt es Widersprüche in der Mathematik? *J Math*, 34:143–155.
[Fontijn 1990] J. Fontijn. *Tweespalt. Het leven van Frederik van Eeden tot 1901*. Querido, Amsterdam.
[Fontijn 1996] ——— *Trots verbrijzeld. Het leven van Frederik van Eeden vanaf 1901*. Querido, Amsterdam.
[Forman 1986] P Forman. Il Naturforscherversammlung a Nauheim del settembre 1920: una introduzione alla vita scientifica nella Republica de Weimar. In A. Rossi G. Battimelli, M. de Maria, editor, *La ristrutturazione delle scienze tra le due guerre mondiali*, volume 1, pages 59–78, Rome. La Goliardica.

[Fraenkel 1927] A. Fraenkel. *Zehn Vorlesungen über die Grundlegung der Mengenlehre.* Teubner, Leipzig. Reprinted by the Wissenschaftliche Buchgesellschaft Darmstadt, 1972.
[Franks 1992] J. Franks. A new proof of the Brouwer plane translation theorem. *Ergod. Th. and Dynamic Systems*, 12:217–226.
[Freudenthal 1954] H. Freudenthal. Leibniz und die Analysis Situs. In *Homenaje a Millias–Vallicrosa*, volume 1, pages 611–621, Barcelona. Consejo Superior de Investicaciones Cientificas.
[Freudenthal 1957] ———— Zur Geschichte der Grundlagen der Geometrie. Zugleich eine Besprechung der 8. Aufl. von Hilberts "Grundlagen der Geometrie". *Nieuw Arch Wiskunde*, 5:105–142.
[Freudenthal 1975] ———— The cradle of Modern Topology, according to Brouwer's Inedita. *Historia Mathematica*, 2:495–502.
[Freudenthal 1979] ———— Een manuscript van Brouwer. In *Tweehonderd Jaar Onvermoeide Arbeid. Tentoonstellingscatalogus.*, volume 2, pages 43–55, Amsterdam. Mathematisch Centrum.
[Freudenthal 1984] ———— A Bit of Gossip: Koebe. *Math. Intelligencer*, 6:77.
[Frey, G. 1992] U. Stammbach Frey, G. *Hermann Weyl und die Mathematik an der ETH, Zürich, 1913–1930.* Birkhäuser, Basel.
[Gentzen 1969] G. Gentzen. The consistency of elementary number theory. In M.E. Szabo, editor, *The Collected Papers of Gerhard Gentzen*, pages 132–213, Amsterdam. North-Holland.
[Gray 2000] J.J. Gray. *The Hilbert Challenge.* Oxford University Press, Oxford.
[Gutkind 1919] E. Gutkind. Einwand von Erich Gutkind. *Mededelingen Int Inst Wijsbegeerte*, 2: 33.
[Gutkind 1930] ———— Von Freundschaft. In *Liber Amicorum Dr. Frederik van Eeden aangeboden ter gelegenheid van zijn 70ste verjaardag 3 april 1930*, pages 68–69, Amsterdam. Maatschappij tot verspreiding van goede en goedkope literatuur.
[Gutkind 1910] Volker (E. Gutkind). *Siderische Geburt. Seraphische Wanderungen vom Tode der Welt zur Taufe der Tat.* Karl Schnabel, Berlin.
[Haan 1916] J. de Haan. *Rechtskundige Significa en hare toepassingen op de begrippen: "aansprakelijk, verantwoordelijk, toerekeningsvatbaar".* PhD thesis, Amsterdam.
[Hadamard 1910] J. Hadamard. Sur quelques applications de l'indice de Kronecker. In J. Tannéry, editor, *Introduction à la théorie des fonctions, 2e ed.*, volume 2, pages 437–477.
[Hahn 1921] H. Hahn. *Theorie der reellen Funktionen, I.* Springer, Berlin.
[Heijerman, Erik 1986] M.J. van der Hoeven Heijerman, Erik. *Filosofie in Nederland. De Internationale School voor Wijsbegeerte als ontmoetingsplaats 1916–1986.* Boom, Meppel.
[Hesseling 1999] D. Hesseling. *Gnomes in the fog. The reception of Brouwer's intuitionism in the 1920s.* PhD thesis, Utrecht.
[Hesseling 2002] ———— *Gnomes in the fog. The reception of Brouwer's Intuitionism in the 1920s.* Birkhäuser, Basel.
[Heyting 1936] A. Heyting. Intuïtionistische Wiskunde. *Mathematica B*, 5:105–112.
[Heyting 1941] ———— Untersuchtungen über intuitionistische Algebra. *Kon Ned Ak Wet Verhandelingen*, 18:36 p.
[Heyting 1956] ———— *Intuitionism, an Introduction.* North-Holland, Amsterdam.
[Heyting 1994] L. Heyting. *De wereld in een dorp.* Meulenhof, Amsterdam.
[Hilbert 1900] D. Hilbert. Mathematische Probleme. *Nachr Gött*:253–297.
[Hilbert 1902] ———— Ueber die Grundlagen der Geometrie . *Math Ann*, 56:381–422.
[Hilbert 1905] ———— Über die Grundlagen der Logik und der Arithmetik . In *Verhand-*

lungen des Dritten Internationalen Mathematiker-Kongresses in Heidelberg vom 8. bis 13 August 1904, pages 174–185, Leipzig. Teubner.
[Hilbert 1909] ——— Grundlagen der Geometrie. Teubner, Leipzig–Berlin.
[Hölder 1924] O. Hölder. Die Mathematische methode. Logisch erkenntnistheorethische Untersuchungen im Gebiete der Mathematik, Mechanik und Physik. Springer, Berlin.
[Hopf 1966] H. Hopf. Ein Abschnitt aus der Entwicklung der Topologie. Jahresber. Dtsch. Math.Ver., 68:182–192.
[Hurwitz 1892] A. Hurwitz. Über algebraische Gebilde mit eindeutigen Transformationen in sich. Math Ann, 41:403–442.
[Jahnke 1904] E. Jahnke. Bemerkung zu der am 27. Februar 1904 vorgelegten Notiz von Herrn Brouwer "Over een splitsing van de continue beweging om een punt O van R_4 in twee continue bewegingen om O van R_3's". Kon Ned Ak Wet Verslagen, 12:940–941.
[Johnson 1979] Dale M. Johnson. The Problem of the Invariance of Dimension in the Growth of Modern Topology, Part I. Archive for History of Exact Sciences, 20:97–188.
[Johnson 1981] ——— The Problem of the Invariance of Dimension in the Growth of Modern Topology, Part II. Archive for History of Exact Sciences, 25:85–267.
[Kellermann 1915] H. Kellermann (Ed.). Der Krieg der Geister: eine Auslese deutscher und ausländischer Stimmen. Alexander Dunker Verlag, Weimar.
[Kleene, S.C. 1965] R.E. Vesley Kleene, S.C. The Foundations of Intuitionistic Mathematics especially in relation to Recursive Functions. North–Holland, Amsterdam.
[Klein 1890] F. Klein. Zur Nicht–Euklidische Geometrie. Math Ann, 37:544–572.
[Klein 1923] ——— Zum Kontinuitätsbeweise des Fundamentaltheorems. In E. Bessel-Hagen R. Fricke, H. Vermeil, editor, Gesammelte Mathematische Abhandlungen III, volume 3, pages 731–741, Berlin. Springer Verlag.
[Klein 1927] ——— Vorlesungen über die Entwicklung der Mathematik im 19. Jahrhundert II. Springer, Berlin.
[Koebe 1914] P. Koebe. Zur theorie der der konforme Abbildung und Uniformisierung (Voranzeige). Ber. Math. phys. Kl. Sächs. Akad. Wiss. Leipzig, 66:67–75.
[Koebe 1936] ——— Wesen der Kontinuitätsmethode. Deutsche Mathematik, 1.
[König 1926] D. König. Sur les correspondances multivoque des ensembles. Fund. Math., 8:114–134.
[Korteweg 1905] D.J. Korteweg. Huygens' sympathische uurwerken en verwante verschijnselen in verband met de principale en samengestelde slingeringen die zich voordoen, als aan een mechanisme met één enkele vrijheidsgraad twee slingers bevestigd worden. Kon Ned Ak Wet Verslagen, 13:413–432.
[Kreisel, G. 1970] A.S. Troelstra Kreisel, G. Formal systems for some branches of intuitionistic analysis. Ann Math Log, 1:229–387.
[Krockow 1990] C. Graf von Krockow. Die Deutschen in ihrem Jahrhundert, 1890-1990. Rowohlt, Hamburg.
[Kühnau 1981] R. Kühnau. Paul Koebe und die Funktionentheorie. In Schumann, H. Beckert, H., editor, 100 Jahre Mathematisches Seminar der Karl–Marx–Universität Leipzig, pages 183–194, Leipzig, VEB Deutscher Verlag der Wissenschaften.
[Lebesgue 1911a] H. Lebesgue. Sur la non–applicabilité de deux domaines appartenant respectivement des espaces à n et $n + p$ dimensions (Extrait d'une lettre à M.O.Blumenthal). Math Ann, 70:166–168.
[Lebesgue 1911b] ——— Sur l'invariance du nombre de dimensions d'un espace et sur le théorème de M. Jordan relatif aux variétés fermées. Comptes Rendus, 152:841–843.
[Lebesgue 1921] ——— Sur les correspondances entre les points de deux espaces. Fund.

Math., 2:256–285.
[Mannoury 1898a] G. Mannoury. Lois cyclomatiques. *Nieuw Arch Wiskunde*, 4:126–152.
[Mannoury 1898b] ———— Sphères de seconde espèce. *Nieuw Arch Wiskunde*, 4:83–89.
[Mannoury 1900] ———— Surface–images. *Nieuw Arch Wiskunde*, 4:112–129.
[Mannoury 1907] ———— Review. Over de Grondslagen van de Wiskunde. *De Beweging*, 3:241–249.
[Mannoury 1937] ———— De Schoonheid der wiskunde als signifisch probleem. *Synthese*, 2:197–201.
[Mannoury 1947] ———— *Handboek der Analytische Significa. Geschiedenis der Begripskritiek.* Kroonder, Bussum.
[Mannoury 1948] ———— *Handboek der Analytische Significa.Hoofdbegrippen en Methoden der Significa. Ontogenese en Fylogenese van het verstandhoudingsapparaat.* Kroonder, Bussum.
[Martino 1988] E Martino. Brouwer's Equivalence between Virtual and Inextensible Order. *Hist. and Phil. of Logic*, 9:57–66.
[Mauthner 1906] F. Mauthner. *Beiträge zu einer Kritik der Sprache (3 vols.).* J.G. Cotta, Stuttgart.
[Menger 1979] K. Menger. *Selected Papers in Logic and Foundations, Didactics, Economics.* Reidel, Dordrecht.
[Mulder 1917] P. Mulder. *Kirkman–Systemen.* PhD thesis, Groningen.
[Nevanlinna 1953] R. Nevanlinna. *Uniformisierung.* Springer Verlag, Berlin.
[Nielsen 1920] J. Nielsen. Über fixpunktfreie topologische Abbildungen geschlossener Flächen. *Math Ann*, 81:94–96.
[Noether, E. 1937] J. Cavaillès Noether, E. *Briefwechsel Cantor–Dedekind.* Hermann, Paris.
[Otterspeer 1995] W. Otterspeer. *Bolland. Een Biografie.* Bert Bakker, Amsterdam.
[Peano 1890] G. Peano. Démonstration de l'intégrabilité des équations différentielles ordinaires. *Math Ann*, 3:182–228.
[Peano 1895] ———— *Formulaire de mathématiques.* Turin. Last volume (4) appeared in 1903.
[Phragmén 1885] E. Phragmén. Über die Begrenzung von Continua. *Acta Math*, 7:43–48.
[Picard 1922] E. Picard. *Discours et Mélanges.* Gauthier–Villar, Paris.
[Poincaré 1881] H. Poincaré. Mémoire sur les courbes définies par une équation différentielle. *J Math*, 7:375–442.
[Poincaré 1882] ———— Mémoire sur les courbes définies par une équation différentielle. [II]. *J Math*, 8:251–296.
[Poincaré 1885] ———— Mémoire sur les courbes définies par une équation différentielle. [III]. *J Math*, 1:167–244.
[Poincaré 1887] ———— Sur les résidus des intégrales doubles. *Acta Math*, 9:321–380.
[Poincaré 1895] ———— Analysis situs. *Journ. Ecole Polytechnique*, 1:1–123.
[Poincaré 1899] ———— *Les méthodes nouvelles de la mécanique céleste. III.* Gauthier–Villars, Paris.
[Poincaré 1910] ———— *Sechs Vorträge über ausgewählte Gegenstände aus der reinen Mathematik und mathematische Physik.* Teubner, Leipzig.
[Poincaré 1912] ———— Sur un théorème de géométrie. *Rendiconti Palermo*, 33:375–407.
[Polya 1972] G. Polya. Eine Erinnerung an Hermann Weyl. *Math. Zeitschr*, 126:296–298.
[Presser 1965] J. Presser. *Ondergang. De Vervolging en Verdelging van het Nederlandse Jodendom.* Vols. 1,2. Martinus Nijhoff, Den Haag.
[Richard 1905] J. Richard. Les principes des mathématiques et le probleme des ensembles. *Revue générale des sciences pures et appliquées*, 16:541. (transl. in [Heijenoort 1967]).
[Ritter 1898] Dr. P.H. (ed.) Ritter. *Eene Halve Eeuw. 1848–1898. Nederland onder de Regeering*

van Koning Willem den Derde en het regentschap van Koningin Emma door Nederlanders beschreven. I & II. J.L. Beijers, J. Funke, Amsterdam.
[Rogers, H. jr. 1967] Rogers, H. jr. *Theory of Recursive Functions and Effective Computability*. McGraw–Hill, New York.
[Schmitz 1985] H.W. Schmitz. Tönnies' Zeichentheorie zwischen Signifik und Wiener Kreis. In L. Clausen and V. Borries, editors, *Tönnies heute. Zur Aktualität von Ferdinand Tönnies*, pages 73–93, Kiel. Mülau Verlag.
[Schmitz 1990] ——————— Frederik van Eeden and the introduction of significs into the Netherlands: from Lady Welby to Mannoury. In H.W. Schmitz, editor, *Essays on significs. Papers presented on the occasion of the 150th birtday of Victoria Lady Welby (1837–19120)*, volume 23, pages 219–246, Amsterdam/Philadelphia Philadelphia. John Benjamins.
[Schmitz 1990] ——————— *De Hollandse Significa*. Van Gorcum, Assen.
[Schoenflies 1903] A. Schoenflies. Über den Beweis eines Haupttheorems aus der Theorie der Punktmengen. *Nachr Gött*, 21–31.
[Schoenflies 1904] ———————. Beiträge zur Theorie der Punktmengen. II. *Math Ann*, 59: 129–160.
[Schoenflies 1905] ———————. Über wohlgeordnete Mengen. *Math Ann*, 60:181–186.
[Schoenflies 1906] ———————. Beiträge zur Theorie der Punktmengen III. *Math Ann*, 62: 286–328.
[Schoenflies 1908] ———————. *Die Entwicklung der Lehre von den Punktmannigfaltigkeiten. II*. Teubner, Leipzig.
[Schouten 1914] J.A. Schouten. *Grundlagen der Vektor– und Affinorenanalysis*. Teubner, Leipzig.
[Schroeder-Gudehus 1966] B. Schroeder-Gudehus. *Deutsche Wissenschaft und internationale Zusammenarbeit, 1914–1928. Ein Beitrag zum Studium kultureller Beziehungen in politischen Krisenzeiten*. Université de Genève. Imprimerie Dumaret & Golay, Geneve.
[Schwabe 1969] K. Schwabe. *Wissenschaft und Kriegsmoral. Die deutschen Hochschullehrer und die politischen Grundfragen des ersten Weltkrieges*. Musterschmidt Verlag, Göttingen.
[Stigt 1979] W.P. van Stigt. The rejected parts of Brouwer's dissertation on the Foundations of Mathematics. *Historia Mathematica*, 6:385–404.
[Stigt 1990] ——————— *Brouwer's Intuitionism*. North Holland, Amsterdam.
[Stuurman, F. 1995] H. Krijgsman Stuurman, F. *Family Bussiness. On dictionary projects of H. Poutsma (1856–1937) and L.E.J. Brouwer (1881–1966)*. Stichting Neerlandistiek, VU Amsterdam, Amsterdam.
[Tietze 1914] H. Tietze. Über Funktionen die auf einer abgeschlossenen Menge stetig sind. *J Reine Angew Math*, 145:9–14.
[Tönnies 1899] F. Tönnies. Philosophical Terminology (I–II). *Mind*, 8:289–332, 467–491.
[Tönnies 1900] F. Tönnies. Philosophical Terminology (III). *Mind*, 9:46–61.
[Tönnies 1906] ——————— *Philosophische Terminologie in psychologisch-soziologischer Ansicht*. Thomas, Leipzig.
[Troelstra 1982] A.S. Troelstra. On the origin and development of Brouwer's concept of choice sequencechoice seq. In D. van Dalen Troelstra, A.S. (eds.), *The L.E.J. Brouwer Centenary Symposium*, pages 465–486, Amsterdam. North–Holland.
[Troelstra, A.S. and D. van Dalen 1988] Troelstra, A.S. and D. van Dalen. *Constructivism in Mathematics, I, II*. North-Holland Publ. Co., Amsterdam.
[Tuchman 1962] B.W. Tuchman. *The guns of August*. Bantam books, New York.
[Ular 1902] Alexandre Ular. *Le Livre de La Voie et la ligne–droite de LAO–TSË*. Éditions de la Revue Blanche, Paris.
[Weyl 1918] H. Weyl. *Das Kontinuum. Kritische Untersuchungen über die Grundlagen der Analysis*.

Veit, Leipzig.
[Weyl 1921] ——— Über die neue Grundlagenkrise der Mathematik. *Math. Zeitschr*, 10: 39–79.
[Weyl 1924] ——— Randbemerkungen zu Hauptproblemen der Mathematik. *Math. Zeitschr*, 20:131–150.
[Weyl 1928] ——— Diskussionsbemerkungen zu dem zweiten Hilbertschen Vortrag über die Grundlagen der Mathematik . *Abh Math Sem Univ Hamburg*, 6:86–88.
[Weyl 1954] ——— Erkenntnis und Besinnung (Ein Lebensrückblick). *Studia Philosophica, Jahrbuch der Schweizerischen Philosophischen Gesellschaft*, 1954.
[Wiessing 1960] H. Wiessing. *Bewegend Portret*. Moussault, Amsterdam.
[Willink 1998] B. Willink. *De Tweede Gouden Eeuw. Nederland en de Nobelprijzen voor natuurwetenschappen, 1970–1940*. Prometheus, Amsterdam.
[Wittgenstein 1984] L. Wittgenstein. *Philosophische Untersuchungen*. Surkamp, Oxford. first ed. Blackwell.
[Yoneyama 1917] K. Yoneyama. Theory of continuous set of points. *Tôhoku Math. Journal*, 12: 43–158.
[Young, W.H. 1906] G.C. Young Young, W.H. *The theory of Sets of Points*. Cambridge University Press, Cambridge.
[Zermelo 1904] E. Zermelo. Beweis dasz jede Mengen wohlgeordnet werden kann. *Math Ann*, 59:514–516.
[Zermelo 1908] ——— Neuer Beweis für die Möglichkeit einer Wohlordnung. *Math Ann*, 65:107–128.
[Zoretti 1911] L. Zoretti. Review of A. Schoenflies " Entwicklung der Lehre von den Punktmannigfaltigkeiten. II". *Bull Soc Math France*, 35:283–288.
[Zorin 1972] V.K. Zorin. On Poincaré's letter to Brouwer. *Russ Math Surveys*, 27:166–168.

INDEX

a priority of space and time, 214
Aachen, 162, 182
accessibility, 142, 210
Ackermann-Teubner prize, 181
Adama van Scheltema, C.S. (1877–1924), 14–16, 23–40, 51–53, 55, 56, 65, 76, 83, 108, 119, 130, 199–208, 217, 219, 220, 282
Adama van Scheltema, C.S. on Brouwer, 25
additional lecturer, 217
Adler, A. (1870–1937), 260
administration of the pharmacy, 202
air
   photography, 281
   reconnaissance, 276
   cartography, 281
Aldert (1886–1973), 4, 55, 65, 155, 201, 202, 224, 225
Alexander, W.J. (1888–1971), 166, 171, 176, 177, 291
Alexandrinius, 40, 201
algebraic
   functions, 234
   number fields, 305
Algemeen Handelsblad, 287, 353
algèbre de la logique, 81
American consul, 285, 341
Amsterdam as mathematical centre of the Netherlands, 361
Amsterdam Fraternity (*Amsterdamse Studenten Corps*), 13
Amsterdam Mathematical Society, 46
anaemia, 29
analysis situs, 124, 128, 141, 142, 148, 154, 155, 214, 300
   book on, 255
   paper, 128, 143, 162, 209, 210, 229, 231, 378
analytic geometry course, 214, 219
anarchy
   in word formation, 249
   of procreation, 250
annuity for Mrs. de Holl, 199
Anrooy, P. van (1879–1954), 331
Anschauungs-world, 70, 287
anti-nationalistic literature, 249
anti-Pope, 374
Antoine's set, 128
apart, 329
apartness, 329, 403

Apollinius, 40, 201
appeal to the world of culture, 248, 337, 349
approximating polygon, 142
Aristotle's logic, 106
Ariëns Kappers, C.A. (1877–1946), 289
armistice, 286
Assisi, 357
asthma, 311
astronomical phenomena, 97
astronomy, 97
Atheneum Illustre, 12
atomistic conception of the continuum, 322
automorphic functions, 178, 191, 192, 284
axiom of choice, 86, 113, 238, 312, 317, 396
axiom of the solvability of each problem, 342
axiomatic
   foundations, 100
   geometry, 84
   method, 105
   systems, 105
axiomaticians, 105
axiomatizing of geometry, arithmetic, and physics, 305

Baire classes, 169
Baire space, 316
Baire, R., 163, 165, 167, 169, 175, 176
Bakhuis Roozeboom, H.W. (1854–1907), 27
ban of the German language, 340
bar, 378, 380, 381
   induction, 382
   theorem, 381, 383
Barrau, J.A. (1873–1946), 119
Bashkirtseff, Marie (1860–1884), 36
Bazel, K.P.C. de (1869–1923), 260
becoming sequences, 323
Begründungs-paper, 281, 313
Belgian
   forced labour, 276
   walking tour, 201
Belinfante, M.J., 354, 359, 401
Bellaar-Spruyt, C. (1842–1901), 106
Benjamin, W., 333
Bericht, 134, 143, 210, 239, 292
Berlin, 301, 302, 311, 332
   lectures, 395, 397
   offer from, 302, 303, 306, 310
   University, 300
Bernays, P. (1888–1977), 130, 260, 344

Bernstein, F. (1878–1956), 86, 118, 184, 185, 192, 204, 213, 274, 292, 325, 344
Bessel-Hagen, E. (1898–1946), 325
Bewegungsgruppen*, 140
Bianchi, L. (1856–1928), 34
Bieberbach, L. (1886–1982), 184, 190, 192, 311, 325, 358
Birkhoff, G.D. (1884–1944), 150, 265, 292, 312
Bjerre, P.C. (1876–1964), 246
Blaricum, 28, 52, 59, 64, 129, 195, 202, 225, 245, 255, 282, 302, 317, 375
Blaschke, W. 1885–1962), 255
Bloch, E. (1885–1977), 260
Bloemers, H. (1880–1947), 267
Blumenthal, O. (1876–1944), 159–164, 166–168, 171, 182, 184, 185, 204, 291, 294, 296, 344, 346, 347, 358
Bockwinkel, H.B.A. (1881–1961), 274
Bohl, P., 291
Bolland, G.J.P. (1854–1922), 56, 57, 64, 65, 259
Bolshevik, 310, 333
Bolyai prize, 305
Bonger, W.A. (1876–1940), 30
Boomstra, 274
border conflicts, 322
Borel, E. (1871–1956), 162, 169, 203, 204, 221, 235–238, 241, 344
Borel, H. (1869–1933), 244, 255, 258, 260, 263, 267, 301, 331, 368, 372
Bosch, 280
boundary point, 329
boycott of Germany, 340
Breughel, P., 205
Brew, A Potent, 26
Brouwer
  appointed to KNAW, 222
  arrest for vagrancy, 34
  chairman of the Dutch Math. Soc., 358
  editor of the Mathematische Annalen, 228, 232
  entering UVA, 13
  health clinic, 62
  health freak, 29
  in Paris, 155
  president of the Dutch Mathematical Society, 221
  privaat docent, 153, 206–208, 213
  the failed lecturer appointment, 159
  theorems, 172, 177
Brouwer and
  Blaricum, 52
  Göttingen, 213, 226, 227, 284, 307
  health, 26, 62
  his mother-in-law, 199
  Louise, 77

  socialism, 34
  the Academy, 218, 219, 222, 305, 352
  the Amsterdam promise, 303, 306
  the Comptes Rendus, 176
  the Dutch Math. Soc., 221, 227
  the Mathematische Annalen, 212, 228, 232, 293, 295, 298
  the pharmacy, 55, 198
  Van Eeden, 247
Brouwer on
  appointments of professors, 289
  art, 205
  axiomatics, 105
  choice sequences, 239, 240, 314
  dimension, 177
  ends-to-means, 68, 91
  formalism, 109
  geometry, 48, 214
  himself, 194
  his dissertation, 194
  his non-constructive results, 328
  importance of words, 249
  intuition, 90, 102
  language, 71, 92
  logic, 99, 105
  mathematical thinking, 194
  mathematics, 46
  mysticism, 59, 65, 81, 287
  objectivity, 104
  physical measurements, 98
  politics, 207
  Russell, 96
  Scheltema's death, 26
  set theory, 113, 235
  sin, 81, 82
  teaching, 206, 309
  the continuum, 102, 114, 236, 385
  the importance of words, 266, 268, 270
  the PEM, 106, 240
Brouwer's
  Amsterdam appointment, 226
  auto-biography, 3
  bargaining with Amsterdam, 253, 302
  bargaining with the UVA, 359
  candidaats examination, 27
  chest complaint, 29
  conditions to the UVA, 302
  confirmation, 17
  courses, 214, 226, 234
  diligence at school, 11
  dissertation, 52, 89
  examinations, 11
  family life, 9, 62, 78, 251
  financial position, 224
  health problems, 29, 254

health rules, 63
height, 23
highschool, 5, 8, 11
hut, 52, 62, 64, 77, 202
inaugural lecture, 222, 236
inaugural lecture (privaat docent), 213
marriage, 55
military service, 22, 50
national feelings, 276
nose complaints, 62
profession of faith, 17
promotion, 119
recuperation, 27
research in topology, 215
research program for topology, 154
revolution, 323, 343
Rome talk, 142, 204
set [Menge] definition, 241
socialist inclinations, 34
stomach problem, 29
sweet tooth, 9
teaching, 214
travels, 203, 252
Brouwer, Egbertus Luitzens (1854–1947), 2
Brouwer, Hendrikus Albertus
  see Aldert (1886–1973), 4
Brouwer, Izaak Alexander
  see Lex (1883–1963), 4
  seeLex (1883–1963), 4
Brouwer–Korteweg correspondence, 86
Brouwer–Lebesgue controversy, 162
Brouwer–Poutsma, H. (1852–1927), 2
Brouwerian counterexamples, 107, 392
brown rice, 63
Bruggen, Carry van (1881–1932), 60, 249
Bruno, Giordano (1548–1600), 373
Brussels, 339
Buber's objections, 267
Buber, M. (18878–1965), 246, 247, 260, 267, 268, 333
Bussum, 59, 245
Buytendijk, F.J.J. (1887–1974), 362, 363
Böhme, Jakob (1575-1624), 68

campaign for the Liberals, 207
Canada, 9
candidaats examination, 27
Cantor's fundamental theorem, 149
Cantor's transfinite induction process, 306
Cantor, G. (1845–1918), 80, 86, 93, 105, 111–113, 118, 122, 125, 141, 151, 152, 209, 242, 313, 318, 357
Cantor–Bendixson theorem, 149, 398
Cantor–Bernstein theorem, 240
Cantor–Schoenflies topology, 145

Cantorism, 318
Carathéodory
  and Göttin-gen, 300
Carathéodory, C. (1873–1950), 204, 226, 232, 291, 294, 295, 300
Cardinaal, J. (1848–1922), 219, 222, 305, 306
Cartan, E. (1869–1951), 125, 204, 296, 299
Carus, P. (1852–1919), 270
Caspary, F, 50
Cassirer, E. (1874–1945), 260
Catechism class, 17
Cauchy sequence, 241
causal sequences, 91
Central Powers, 336
centralization, 82
Charon's pennies, 203
chess game as science, 320
choice, 238
choice sequence, 114, 241, 312, 314, 320, 322
Christian-anarchist, 61
Christina of Sweden, 208
Church, A. (1903–1995), 260
City Council of Amsterdam, 217, 219, 222, 225, 271, 290, 302–304, 361
clairvoyance, 51
classical
  continuum, 385
  mathematics, 313
CLIO, 14, 30
closed curve, 142, 144, 148
closed manifolds, 188
closure, 329
Coenen, F. (1866–1936), 60, 364
coincidence of birth place, 350
collected works of Christiaan Huygens, 43
Collegium Philosophicum (Bolland), 57
colony, 60, 245
Combebiac, G., 133
committee for prize essays, 221
comprehension, 240
comprehension principle, 318
Comptes Rendus, 166, 168–170, 175, 176, 351
Concertgebouw, 331, 373
Conference of inter-allied Scientific Academies, 339
conflict with Reiman, 259
conformal mapping, 187
congruent, 317
connected, 154
connectedness, 132, 145
Conseil International des Recherches, 340, 348, 357, 358
consistency, 109, 111
  of arithmetic, 93, 109, 111
  of the continuum hypothesis, 116

constructive mathematics, 308
content, 329
continuity, 124
  method, 187
  principle, 243, 316
  proof, 185
  theorem, 376
continuous mappings of the torus, 294
continuum, 102, 114, 324, 385, 389
  atomistic, 321
  continuous, 321
  hypothesis, 93
  problem, 115, 116
conversion of Van Eeden, 373
Corput, J.G. van der (1890–1975), 354
counter-boycott, 341
country
  of affinity, 350
  of nationality, 350
Courant, R. (1888–1972), 344
Couturat, L. (1868–1914), 81, 94, 372
Cranz–Hugershoff method, 281
creating subject, 394
creation of words, 267, 270
creative subject, 394
Crelle's journal, 168, 390
Critique of Pure Reason, 94
Crystal Structure, 140
cultural elite, 246
cum laude, 27, 36, 42, 51
curve, 210
cycle tour with Van Eeden, 264

daily diet, 63
Daniell-element, 98
Darboux, G. (1842–1917), 169, 203
De Amsterdammer, 287
De Donder, Th., 204
De Groene (Amsterdammer), 287
De Nieuwe Amsterdammer, 249, 264, 276, 287, 288
decimal expansion, 236
  free, 223, 237
  lawlike, 223
  of π, 107, 316, 392
  of reals, 325
declaration of spiritual values of human life, 266
Declaration of the 93, 248, 337, 349
decomposition of space, 301
Dedekind cuts, 322, 326
Dedekind, R. (1831–1916), 80, 151
definable in finitely many words, 318
definitions, 109
degeneration of forest, 69

Dehn & Heegaard, 134
Dehn, M. (1878–1952), 204, 291, 293, 358
Delft, 64, 65, 206, 217, 295, 304
Delft lectures, 65, 76
Den Helder, 27
Denjoy, A. (1884–1974), 274, 289, 292, 305, 306, 310, 341, 344, 346, 347, 350–352, 354, 357, 358, 367
denumerability, 328
denumerable, 112, 328
denumerably unfinished, 115, 116, 118
Descartes, R. (1596–1650), 123
Deutsche Mathematiker Vereinigung
  see DMV, 140
Deutschfreundlich, 301
diagonal argument, 242, 316
Dietzgen, J. (1818–1888), 30, 33, 34, 57
Dieudonné, J. (1906–1994), 172
differentiability condition, 94, 126
differential equation, 136
differential- and integral calculus for chemistry students, 227
dimension, 151, 152, 178
  definition of, 177
  natural notion of, 177
Dionysus, 40, 201
Dirichlet principle, 149
dispuut (fraternity), 14
dissertation
  of Belinfante, 376
  of De Haan, 256, 257
  of De Loor, 404
  of G. de Vries, 43
  of Heyting, 397, 403
  of Korteweg, 42
  of Nielsen, 293
  of Schouten, 295
  of Schuh, 304
dizziness, 7
DMV, 140, 182, 396
doctoraal examination, 36
doctoral examination of Lize, 199
doctorandus, 36
dogmatism, 20
domain, 142, 145
domination of nature, 90
Donatello (1386–1466), 31
double negation law, 317
Dresden, A, 223
Droste, J., 274
Du Bois-Reymond, P. (1831–1889), 237
Dutch
  1918–revolution, 286
  army, 277

Mathematical Society (*Wiskundig Genootschap*), 46, 140, 142, 159, 221, 227, 228, 271, 297, 357, 358
Royal Academy (see KNAW), 248
Dyck, W. (1856–1934), 292
Dürckheim, K. (1858–1917), 260

École Normale, 169
Economics, 75
education of the masses, 257
Eeden, F. van (1860–1934), 29, 60, 64, 66, 243–247, 249, 255, 257, 258, 260–264, 267, 269, 272, 282, 285, 287, 302, 304, 331, 334, 341, 367, 372
ego, 18, 74
Ehrenfest, P. (1880–1933), 253, 254, 292, 310
Ehrlich, E. (1854–1915), 270, 372
Einstein, A. (1879–1955), 295, 342
electrical machine of Van Marum, 208
electromagnetic field, 98
Emerson, R.W. (1803–1882), 36
emigration of De Haan, 332
emotion of beauty, 288
empirical space, 104
encyclopaedia
 of significs, 372
 project, 372
end–to–means, 69
Engadin, 308, 319
Engel, F. (1861–1921), 130–132, 161, 325
Enschede, 402
Entente, 336
enumerable, 328
epistemology, 70
equivalence, 328
Erlangen program, 122, 142, 153
Erzberger, M. (1875–1921), 334
ETH, 310
Euclid, 122
Euclidean space, 104
Euler's theorem, 123
Euler, L. (1707–1783), 123
exceptional genius, 26
existence, 105, 236, 237, 323, 324, 327, 388
 independent of experience, 104
 of decimal expansion of π, 392
 of discontinuous functions, 384
 of maximum value, 392
 of supremum, 319, 385
 theorem, 284, 301, 321, 323
exponentiation of sets, 113
externalizing, 82
extra credit for books, 302

faculty

 meeting, 226, 255, 269, 273, 302, 359, 360, 363
 politics, 270, 272
 room, 215, 358
faculty meeting, 221
Faddegon, B., 334, 372
faith in God, 17
fall through the Intellect, 68
fan, 375
fan theorem, 381, 383
fasting, 78
fatherland, 249
Fechner, G.T. (1801–1887), 96
Feferman, S., 319
fellow beings, 75
feminine logic, 73
Festschrift (Hilbert), 85
fine structure of the continuum, 326
finite combinatorial part of mathematics, 344
finitist, 344
Fisher, F., 364
fixed point, 133, 134, 136, 137, 149, 150, 156, 294, 301
 theorem, 139, 172, 328
fixed points, number of, 294
Flanders, 275
Flaubert, G. (1821–1880), 36
fleeing property, 392
Flemish
 higher education, 275
 University at Gent, 275
 university at Gent, 276
Florence, 203, 204, 357
Fokker, Anthony (1890–1939), 281
folding of surfaces, 43
formalism, 318
Formalism and Intuitionism, 320
Forman, P., 341
formula-free treatment, 214
Forte dei Marmi, 247
Forte Kreis, 247, 265, 285
fortune-teller, 202
foundations of
 a new poetry, 31
 a science, 70, 109
 geometry, 84, 86, 214, 235
 logic, 84, 395
 mathematics, 84, 86, 96, 105, 109, 121, 222, 285, 308, 312, 318, 395
 set theory, 84
four-dimensional geometry, 47
Fraenkel, A.A. (1891–1965), 313, 314, 396–401
Frankfurth a.M., 140
free
 choice sequences, 323

creation, 117, 195
freedom of contradiction, 117
Frege, G. (1848–1925), 318, 319
French
    imperialism, 352
    intellect, 351
    opinion of Holland, 349
freshman, 8
Freudenthal, H. (1905–1990), 48, 128, 130, 136, 154, 158, 168, 175, 189, 190
Fricke, R. (1861–1930), 185, 186, 191, 192, 230, 291, 325, 326
Friesland, 3
Fréchet, M. (1878–1973), 125, 274, 291
full continuum, 393
fundamental
    sequence, 241
    theorem of algebra from an intuitionistic viewpoint, 404

Galerie Parisienne, 353
Gascogne, 350
Gauss, C.F. (1777–1855), 123, 156
General C.J. Snijders (1852–1939), 277
general set theory, 234
Gent University, 275
Geometry and Mysticism, 287
German guarantees of goodwill, 345
German railways, 332
Germany's obligations, 345
Gesellschaft der Wissenschaften zu Göttingen, 284
Gimon, 36
gin drinking, 100
Ginneken, J. van (1877–1945), 331, 334, 362, 368, 370, 372, 373
gleichmächtig, 112
global behaviour, 136
gnomes, 200
goal–means, 69
Goethe, J.W. von (1749–1832), 31
Gooische Stoomtram, 61, 202
goose eggs, 62
Gorter, H. (1864–1927), 32, 60
grass eaters, 61
Graz, 359
Grenzkreis, 187
Groningen, 205–207, 225, 226, 304, 360
    system, 290
    University, 206, 226
    vacancy, 224
group, 126
groups of motions, 140
Grundlagenstreit, 321, 325, 342
Guinness Book of Records, 4
Gutkind's objections, 268

Gutkind, E. (1877–1956), 246, 247, 267, 268, 301, 333, 395
gymnasium, 4, 8, 194
Göttingen, 140, 162, 181, 190, 196, 213, 226–230, 252, 293, 294, 300–302, 304, 307, 308, 317, 344, 359
    library, 303
    Mathematical Society, 344
    offer from, 300, 306, 359
    University, 226
    vacancy, 226
Göttinger Gesellschaft der Wissenschaften, 183, 187
Göttinger Nachrichten, 185, 187–189

Haalmeijer, B.P., 359
Haan, J.I. de (1881–1924), 244, 255, 256, 258, 267, 272, 302, 331, 372
Haarlem, 7–9, 11, 13, 14, 16, 22, 27, 28, 208, 244, 255
Hadamard, J. (1865–1963), 137, 139, 154, 156, 176, 203, 204, 215
Hagana, 332
Hague Convention, 275
Hahn, H. (1879–1934), 204
Hahnenklee, 357
hairy ball theorem, 138
Hake, H., 347
Hamdorff, Jan (1860–1931), 60
Handelsblad, 332
Hankel, H. (1839–1873), 81
Hardy, G.H. (1877–1947), 204
Harst, van der, 280
Harwich, 274
Harz, 62, 213, 252, 357
Harzburg, 357
Hausdorff, F. (1868–1942), 171, 231, 291, 314, 325
HBS, 4, 7, 8, 11, 42, 251
heavenly bodies, 92, 94
Hecke, E. (1887–1947), 300
Heegaard, P. (1871–1948), 291
Hegel's natural philosophy, 319
Hegel, G.W.F. (1770–1831), 56, 96
Hegelian philosophy, 57
Heidegger, M., 260
Heidelberg
    address, 86, 108
Heidelberg
    Congress, 81
Hellinger, E. (1883–1950), 168
Helmholtz–Lie Raumproblem, 101
Hensel, K. (1861–1941), 325
herbal cures, 62
Herglotz, G. (1882–1953), 300
Heron, G.D., 265

Hesseling, D., 395
het Gooi ('t Gooi), 27, 59, 207, 245, 261
Heyting's dissertation, 403
Heyting, A. (1898–1980), 359, 389, 397, 401, 403
Heyting–van Anrooy, F.J. (1903–), 9
higher mechanics, 234
highschool examinations, 225
Hilbert
    member of the KNAW, 305
Hilbert's
    dogma, 105, 327, 342
    fifth problem, 101, 127
    problems, 126
    recommendation, 218
Hilbert, D. (1862–1943), 81, 93, 105, 108, 109, 126, 128–130, 135, 143, 148, 152, 154, 159, 160, 162, 165, 176, 181, 184, 187, 188, 191, 204, 209, 211, 212, 218, 219, 225, 230, 284, 300, 301, 305, 308, 311, 317, 320, 321, 341–343, 358
history of dimension theory, 129, 152, 178
Hoff, J.H. van 't (1852–1911), 13
Holdert, T. (1883–1979), 261
Holdert-van Syll, G.H. (1889–1983), 261
Holl, J. de (1834–1880), 53, 198
Holl, R.B.F.E.
    see Lize (1970–1959), 52
    see Lize (1970–1959), 53
Holl-Sasse, E.J.J. de, 53, 55, 198, 199
Holy Grail, 288
homeomorphism, 122, 154
homeopathic medicines, 62
homological methods, 177
homotopic changes, 139
homotopy, 171
honorary doctorate Mannoury, 45
Hoorn, 5–7, 11
Hopf, H. (1894–1971), 171
horned sphere of Alexander, 128
house
    in Blaricum, 199
    in the Harz, 252
house painter, 4
Hubrecht, A.A.W. (1853–1915), 219
Huet, W.G. (1869–1911), 29, 53
Huizen, 59
Human Rights, 75
Hurwitz, A. (1859–1919), 291
Husserl, E. (1859–1938), 318, 372
Huygens, C. (1629–1695), 1, 43
Hölder, O. (1859–1937), 296

Ibsen, H. (1828–1906), 36

idealism, 21
identical, 317
illusion of the woman, 72
immortality, 19
impredicative definition, 318
inaugural lecture
    of Brouwer, 153, 236, 253
    of De Haan, 258
    of Mannoury, 45, 272
inauguration of Queen Wilhelmina, 23
indecomposability of the continuum, 324, 385
indecomposable continua, 144
indicatrix, 156, 171
induction, 240
industry, 69
inextensible order, 390
infantry, 51
    barracks, 22
institute for language-reflection, 249
integral equations, 149, 305
intellectual elite, 39, 264
International
    Academy for Philosophy, 265, 267, 301, 332, 358
    Brotherhood, 61, 245
    Institute for Philosophy, 250, 266, 267, 269, 332, 334, 367–370
    Mathematics Congress at Rome, 116, 127, 201, 203
    School for Philosophy, 259
intuition of time, 91
intuitionism, 223, 281, 312
    new style, 281
Intuitionism and Formalism, 236
intuitionist, 237
intuitionistic
    dimension theory, 385, 399
    mathematics, 194, 282
    set theory, 312, 327, 342
intuitive
    continuum, 102, 114
    significs, 269, 331
invariance
    of dimension, 152, 154, 158–160, 165, 168, 169, 176, 192, 284, 301, 308
    of domain, 175, 182, 301, 308
    of the closed curve, 145, 176
invariant point, 133
invariants, 98, 305
invertibility of differentiability, 305
irrigation fields, 139
Israëls, Jozef (1824–1911), 59
Italian art, 204
Italy, trip to, 27, 34

Jaeger, F.M. (1877–1945), 277

Jahnke, E., 49, 200, 204
Jahrbuch über die Fortschritte der Mathematik, 130
Jahresbericht der Deutschen Mathematiker Vereinigung, 303, 342
Jerusalem, 332
job hunting, 208
Johnson, D.M., 152
Jongejan, Cor (1893–1968), 251, 252, 262, 298
Jordan curve, 124, 142, 145, 155
  theorem, 144, 146, 176, 182, 219
  theorem, generalization of, 284
Jordan theorem for n dimensions, 221
Jordan, C. (1838–1922), 124, 141, 162, 166, 168
Jordan–Schoenflies theorem, 142
Journal for Philosophy, 31, 108
Journal für Reine und Angewandte Mathematik, 168, 390
joy of the solving of mathematical problems, 309
Joyous World, 64
judgement abstracts, 324
Jungborn, 62
Just, Dr, 62, 357
Justice, 76
Jürgens, E., 152

Kamerlingh Onnes, H. (1853–1926), 13
Kant, E. (1724–1804), 81, 94, 97, 104
Kapteyn, J.C. (1851–1922), 219
Kapteyn, W., 201, 216, 219, 222, 305, 306
Kapteyn, W., death of, 274
Karlsruhe
  meeting, 181, 182, 191
  report, 192
karma, 22, 71–73
  loss of, 74
katalogisiert, 330
Kepler, J. (1571–1630), 50
Kerkhof, K., 340
Kerékjártó, B. (1898–1946), 292, 294, 325
kingship, 52
Klein
  and the Declaration of the 93, 248, 337
  bottle of, 294
  succession of, 226
Klein, F. (1849–1925), 50, 80, 86, 122, 162, 175, 179, 182, 184, 186, 187, 190–193, 219, 221, 226, 232, 284, 285, 293–296, 298–300, 308, 319, 326, 341, 357, 358
Klein–Fricke, 187
Kloos, W. (1859–1938), 31

Kluyver, J.C. (1860–1932), 205, 216, 219, 222, 253, 305, 306
KNAW, 48, 134, 136, 176, 218, 274, 276, 281, 297, 304, 305, 337, 347–349, 351–356, 358, 376, 387
Koebe affair, 291
Koebe, P. (1882–1945), 149, 175, 180–184, 187–193, 196, 204, 298, 325
Koenigs, G., 357
Kohnstamm, Ph. (1875–1958), 31, 108, 207, 361–363
Koninklijke Nederlandse Akademie van Wetenschappen
  see KNAW, 48
Korteweg's mathematical production, 43
Korteweg, D.J. (1848–1941), 35, 41, 42, 45, 47–51, 76, 80, 86–89, 93, 94, 96, 98, 99, 114, 127, 133, 136, 150, 176, 200, 201, 203, 205–209, 214–219, 221, 222, 224–229, 269, 271–273, 279, 285, 290, 295, 304–306, 327
Korteweg, retirement of, 253, 271
Korteweg–de Vries equation, 43
Koyré, A., 260
Kreisel, G. (1923–), 319
Kronecker's program, 397
Kronecker, L. (1823–1891), 156, 235, 243, 344, 400
Kroniek, 33
Kuenen, P.H., 219
König's lemma, 381
Königsberg, 140

Laar, J.J. van (1860–1938), 221
labelled infinite paths, 314
labour, 75
Lakwijk-Najoan, J.A.L. van, 198
Lambek, J., 10
Landau, E. (1877-1938), 190, 284, 301, 325, 327
Landauer, G. (1870–1919), 246, 267, 270, 333
Landgrebe, L., 260
landstorm (National Guard), 277
Langhout, W., 261, 262
Langhout-Vermey, Tine (1887-1964), 262
language, 71, 92
  daily, 369
  levels, 369
  of mathematics, 100
  of mood, 369
  scientific, 369
Laren, 59, 64
Larense School, 60
law, 314
law of attraction, 97

laws of astronomy, 92
leap from end to means, 69, 91, 107
leap-frog, 55
Lebesgue
    and invariance of dimension, 160, 166
    in Paris, 354
Lebesgue measure, 291
Lebesgue, H. (1875–1941), 159–170, 176, 195
Leck, van der B. (1876–1958), 60
Leclancher-element, 98
Lecointe, 340
lecturers' position, 302–304
legal significs, 256
Leibniz, G.W. (1646–1716), 123
Leiden, 12, 13, 56, 208, 271, 279
Leiden University, 12, 252, 253, 279
Leipzig, 181, 303
Lenard, P. (1862–1947), 342
length of a set, 149
Lenin (1870–1924), 334
letter to Fricke, 189
Leuven, 12, 337
levels in formalisation, 109
Levi-Civita, T. (1873–1941), 87, 204
Lex, 28, 55
    painter, 28
Liber Amicorum for Van Eeden, 375
liberal, 35
Lie groups, 94, 126, 128, 132, 133, 152, 209
Lie, S. (1842–1899), 80, 86, 125, 126, 130
Life, Art and Mysticism, 21, 257, 263
Life, Art and mysticism
    Van Eeden's review of, 264
Lily, 225, 232
limit point, 329
Lindelöf, E.L. (1870–1946), 344
linking manifold, 168
linking varieties, 166
Listing, J.B. (1808–1882), 124, 177
Lize (1870–1959), 52–55, 62, 79, 199, 200, 202, 213, 225, 232, 247, 279
Lize's final exams, 201
Lize's study of pharmacy, 56
Loa-Tse, 120
located, 329
logic, 71, 82, 84, 90, 92, 99–101, 105, 107–110, 116
logical reasoning, 99
Lokhorst, J. van (1878–1904), 14
London resolutions, 339
looping coefficients, 176
Loor, B. de, 387, 404
Lorentz transformations, 214

Lorentz, H.A. (1853–1928), 13, 197, 208, 214, 219, 228, 252, 253, 262, 272, 277–280, 295, 302, 305, 306, 355, 367
Lotze, R.H. (1817–1881), 96
Louis Napoleon (1778–1846), 48
Louise (1893–1980), 3, 14, 54, 77–79, 198, 199, 234, 250, 251, 373, 374, 378
love of my neighbour, 20
love for languages, 12
Lucca, 205
Luckenwalde, 190
Lupin, Arsène, 37, 282
lust for power, 71
Lüroth, J. (1844–1910), 152

Maarseveen, J. van, 332
malafide trade, 365
manifolds, 125
Mannigfaltigkeiten, 125
Mannoury
    privaat docent at UVA, 45
    successor of Korteweg, 273
Mannoury's duality theorem, 44
Mannoury, G. (1867–1956), 43, 45, 46, 108, 119, 120, 207, 243, 255, 260, 265, 266, 269, 271, 273, 274, 310, 331, 334, 358, 363, 367, 372
mapping degree, 155, 156, 171, 177
Marburg meeting (1923), 396
Martino, E., 391
Marxist, 64
mathematical
    drive for knowledge, 309
    experience, 96
    principles, 281
    reasoning, 99
    school of Paris, 306
    significs, 331
mathematics, 11
    and logic, 105
    in Holland, 216
    motivation behind, 82
    vacancy in Utrecht, 274
Mathematische Annalen, 127, 143, 209, 229, 232, 252, 255, 292, 295, 296, 298, 330, 343, 344, 383, 387, 390
Mathematische Probleme, 106
Mathematische Zeitschrift, 295, 343
Mauthner, F. (1849–1923), 246, 270
Mauve, A. (1838–1888), 15, 59
Mauve, R. (1878–1963), 14, 15, 62, 119, 245
maximal ordering, 390
maximum salary, 302
Mayor of Amsterdam, 254, 302, 303
measurable, 329

continuum, 114
measuring instruments, 92
Medemblik, 4, 5, 7
medical industry, 70
medium of free becoming, 324
Meerum-Terwogt, P.C.E., 58
Meister Eckehart (1260–1327), 68
Meltzer, Lieutenant, 280
Men of a Royal Mind, 246
Menge, 313
Menger, K. (1902–1985), 314
meta-mathematics, 109
middelbare acte, 42
Mie, G. (1868–1957), 325
Milano, 204
military
    aviation section at Soesterberg, 280
    dispensation, 277
    service, 23, 29, 277
Minister Lelie, 278
minister of education, 355, 362
Minkowski, H. (1864–1909), 149
Mises, R. von (1883–1993), 302
Mittag-Leffler, G. (1846–1927), 203, 265
Mondriaan, P. (1872–1944), 60
Montaigne, M. (1533–1592), 36
Morality, 59
Mother Goose, 5
Mulder, P. (1878–1963), 180, 307
MULO, 4, 7
Multatuli (1820–1887), 2
Municipal University, 12
museum of immovable truths, 46
musical education, 9
mystic experience, 20, 68
Möbius strip, 133
Méthode de Continuité, 181

Naber, H.A., 287
Naess, A. (1912–), 260
Naples, 205
natural
    notion of dimension, 168
    phenomena, 104
nature of geometry, 153, 214
Naturforscherversammlung, 341
Nauheim, 310
    Conference, 325, 341, 343, 348, 350
    Lecture, 342
negatively continuous, 384
neo-intuitionism, 313
Netto, E. (1846–1919), 152
New Chronicle, 363
New Crisis-paper, 320–325, 390
new foundational crisis in mathematics, 320

NEWTON, 13, 16, 17, 46, 47
Newton, I. (1642–1727), 50
Nielsen number, 294
Nielsen, J. (1890–1959), 293, 294
Nietzsche, F. (1844–1900), 40, 201
Nieuw Archief voor Wiskunde, 46
Nieuwe Rotterdammer Courant, 287
Nijmegen University, 368
no-separation theorem, 175
Noether, E. (1882–1935), 151, 325
Noether, F. (1884–1939), 325
non-Archimedean
    geometries, 101
    uniform groups, 101
non-Euclidean geometry, 80
Norlind, E, 267
number of fixed points, 294
number system of Weyl, 322
numerable, 328

objectivity, 104
objectivization of the world, 90
offer from
    Berlin, 300, 302, 303
    Göttingen, 300
    Leiden, 252, 255, 279
old-intuitionists, 238
omniscience principle, 384
open letter
    re Troelstra's coupe, 286
    to the Belgian Government, 276
open manifolds, 171
open-air bathes, 62
oppressed, the poor, 76
ordinal, 328, 379
ordinal bar, 380
Ornstein, L.S. (1880–1941), 221, 265, 267, 274, 297
Oud Leusden, 259
Oudemanhuispoort, 119, 215, 358
Overschie, 2, 3
Overtoom, 202

Paestum, 205
Painlevé, P. (1863–1933), 338, 339
Palestine, 331, 332
Pannekoek, A. (1873–1960), 30
paper money, 323
paradox
    of Berry, 318
    of Richard, 238, 318
paradoxes, 112, 318
parallel displacement, 87
parallelism of will, 257
paranimf, 119, 256

Paris, 29, 36, 155, 159, 162, 306, 336, 339, 352
Paris Peace Conference, 339
Pascal, B. (1623–1662), 36
Pasch, M. (1843–1930), 86
pathological mathematics, 327
paving principle, 160, 170, 182
Peano curve, 141, 142
Peano's existence theorem, 138
Peano, G. (1858–1932), 45, 105, 136, 265, 270, 372
Peijpers, A.L.E.
see Louise (1893–1980), 3, 54, 55, 63
Peijpers, H.F., 53
Peirce, C.S. (1839–1914), 245
Pels, Dina (1878–1929), 7, 54
PEM, 106, 108
Pendelzahl, 392
pendulum number, 392
Pennewip, 2
Pension Luitjes, 28
Pesch, A.J. van, 41, 51, 304
Pflaster Satz, 160
pharmacists exam Lize, 199
pharmacy, 53–55, 62, 197, 269
PHILIDOR, 14
Philistine, 76
philosophy, 56, 361
    in the science faculty, 362
    of the sciences, 362
photogrammetry, 277
Phragmen–Brouwer theorem, 146
physical time and space, 104
physically measurable quantity, 98
physicist, 92
physics, 92
piano playing, 78
Picard, E. (1856–1941), 165, 203, 338, 340, 341
Pieri, M. (1860–1913), 86
plagiarism, 185
plane
    curves, 142
    domains, 142
Poincaré's
    last theorem, 149, 232
Poincaré's closure trick, 191
Poincaré, H. (1854–1912), 81, 86, 87, 94, 112, 117, 118, 125, 136, 137, 139, 156, 172, 173, 175, 177, 179–181, 184, 185, 187, 188, 203, 204, 215, 218, 221, 236–238, 287, 292, 305, 318, 344, 397, 400
point sets, 234
political
    meetings, 34
    metaphor, 323

    task of scholars, 347
Polya, G. (1887–1985), 319, 325
Polya–Weyl wager, 319
polymorphic functions, 182
polynomial approximations, 171
Pompéiu, D., 148
possible cardinalities, 116, 127
potent brew, a, 264
potential
    field, 81
    theory, 84, 86
Poutsma, A. (1858–1941), 2
Poutsma, G., 9
Poutsma, Hendrik (1856–1937), 2
Prague, 359
pre-intuitionists, 235
predicative mathematics, 308
predicativity, 318
Pretoria, 404
primary notions, 257
primeval phenomenon, 91
principium
    contradictionis, 106
    tertii exclusi, 106, 107
    tertium non datur, 106
principle
    of the excluded third, 281
    of the excluded third, generalized, 386
    of Dirichlet, 305
    of the continuing self-isomorphism, 296
    of the excluded middle, 106
    of the excluded third, 106, 201, 312, 316, 321, 323, 324, 327, 342, 384
priority controversy (Jahnke), 49
privaat docent, 45, 208, 213
prize essay committee, 227, 228
Proceedings of the Academy, 134
process of wordless building, 106
profession of faith, 21
professor, appointments of, 288
projective
    geometry, 235, 402
    geometry course, 214, 219
projective plane, 294
promise of a second Göttingen, 307
promotion, 87, 88
Propria Cures, 8, 16, 30, 57, 269
provisor, 55, 198
pseudo-sets, 241
psychiatrist, 60
psychiatry, 244
psychologism, 99

quacks, 70
Quine, W. van Orman (1908–), 260

Radon, J. (1887–1965), 325
railway Medemblik–Hoorn, 5
Rang, F. (1864–1924), 247, 301
rash in the face, 29
Rathenau, W. (1867–1922), 247, 265, 267
reading room, 310
real number, 236
reduced continuum, 393
Rees, J. van (1854–1928), 61
reference library, 303
regular household, 225
Reichenbach, H. (1891–1953), 260
Reichszentrale für naturwissenschaftliche Berichterstattung, 340
Reiman, J.D. (1876–1957), 259
rejected parts of the dissertation, 99
relativity
  postulate, 214
  theory session in Nauheim, 342
religious matters, 18
religious-anarchist, 61
Remonstrant Church, 17
representations, 21
retreat, 374
return to studies, 51
Revesz, G. (1878–1955), 372
review of *Amsterdam*, 32
revolution
  Brouwer's, 323, 343
revolutionary events in Germany, 309
revolutionizing analysis, 310
Revues Semestrielles, 205, 227, 357
Richard's paradox, 239
Richard, J., 344
Ridder, J., 354
Riemann surface, 179, 182, 187, 291
Riemann volume, 383
Riemann's theory of algebraic functions, 148
Riemann, G.F.B. (1826–1866), 124, 133, 153
Riemann–Helmholtz Raumproblem, 124
Riemannian geometry, 299
rigid bodies, 92
Roland Holst-van der Schalk, H. (1869–1952), 30, 32, 60, 286
role of
  foundational research, 82
  logic, 100
  mathematics in physics, 97
  science, 90
  the woman, 73
Roll, Henriette (1863–1962), 262
Rolland, R., 247, 339
Rome, 203, 204
  conference, 152, 346
Rosenthal, A. (1887–1959), 184, 274

rotations in four-dimensional Euclidean space, 48
Royal Dutch Academy of Sciences, 48
Russell's nonsense, 96
Russell, B. (1872–1970), 80, 81, 86, 94, 96, 104, 105, 112, 222, 238, 245, 317, 318, 339, 372
Rutgers, J.G. (1880–1956), 274

Salomonson, W., 29
Sande Bakhuyzen, H.G. van de, 219
Scheler, M. (1874–1928), 247
Scheveningen, 128, 181, 213
Schmidt, E. (1876–1959), 146, 171
Schmitz, W.H., 265
Schoenflies, A. (1853–1928), 128, 140–143, 145, 152, 155, 176, 209–212, 214, 229–232, 239, 252, 265, 275, 276, 284, 291, 303, 325, 344, 357, 358
schoolgirls problem, 307
Schopenhauer, A. (1788–1860), 96
Schoute, P.H. (1846–1913), 201, 205, 216, 219, 222, 307
Schouten's direct analysis, 295
Schouten, J.A. (1883–1971), 87, 295, 296, 298–300, 360
Schroeder–Gudehus, B., 337, 340
Schröder, J.C., 276
Schuh, F. (1875–1966), 206, 216, 304–306
Schur, F. (1856–1932), 86, 325
Schwarz, H.A. (1843–1921), 181
Schweitzer, A., 260
science, 70
Science et Hypothèse, 94
Scientific committee of advice and research in the interest of public welfare and defence, 280
SDAP, 29
second number class, 112, 113, 140
semi-intuitionist, 238, 243, 312
seminar in pure mathematics, 226
seminar library, 255
sequence governed by induction, 240
set
  -theoretic topology, 125
  generating principles, 115
  theory, 84, 235
  well-ordered, 379
sexuality, 263
Siderische Geburt, 246
Siena, 205
signific circle, 330, 334, 370, 372
Signific Dialogues, 372
signifies, 243, 244, 246, 247, 255, 257, 258, 265, 269, 273, 288, 302, 304, 331, 332, 334, 367, 369, 371, 372

simplex, 171
simplicial approximation, 155, 171
sin, 81
Singer Museum, 60
Singer, William (1868–1943), 60
singular points, 139, 148
singularities, 191
siren song of Leiden, 255
Sissingh, R., 41
Sitter, W. de (1872–1934), 253
skating tour, 201
socialist, 34
   inclinations, 34
   reading society, 30, 34
socialistic
   degeneration, 75
   movements, 75
Soesterberg (air force), 280
solvability of mathematical problems, 105, 327
Sommerfeld, A (1868–1951), 162
Soviet law, 364
space, 18
space-filling curves, 152
species, 314
Spengler, O. (1880–1936), 372
Spezies (species), 314
Spil, Lily van der
   see Lily, 202
spiritistic seances, 262
sports, 8
spread, 313, 314, 328, 375
St. Jobs Foundation, 197
St. Jobsleen (foundation), 14, 88, 89
Staal, J.F. (1879–1940), 364
star of the mathematicians, 218
Stevin, S. (1548–1620), 1
sticker in Karlsruhe reprint, 191
Stieltjes, T.J. (1856–1894), 1, 290
Stokes' theorem in higher dimension, 87
Strasbourg, 306
   Congress, 325, 341, 349, 350
struggle for existence, 91
Struik, D.J. (1894–), 87
Studenten-weekblad, 65
Study, E. (1862–1922), 295
subjugation of nature, 92
Suchtelen, N. van (1878–1949), 245
surgery, 198
symbol pushing, 320
symbolic logic, 45
Szasz, O. (1884–1952), 325

Tagore, R. (1861–1941), 260, 270
Tak, P.L. (1848–1907), 32, 60
Tao, 268

Tarski, A. (1902–1983), 260
Taussky, Olga, 129
teaching load, 216
Technical University at Delft, 13
Tellegen, J.W.C. (1859–1921), 302
Tensor Calculus, 295
Teubner, 255, 303, 396
Teyler Museum, 208
Teyler Van der Hulst, 208
the Grail, 368
theatre plays of Scheltema, 282
theorem of Janiszewski and Mazurkiewics, 149
theoretical logic, 99, 100
theory
   of dimension, 178
   of functions, 226, 234
   of invariants, 360
   of order, 328
   of oscillations, 235
   of point sets, 218
   of relativity, 308
   of simple curves, 212
thermodynamics, 361
Thijsse, Jac. P. (1865–1945), 245
thing–in–time, 91
Thomae, J. (1840–1921), 152
Thoreau, H. (1817–1862), 61, 245
Thorn Prikker (1869–1932), 14
Tideman, B., 17
Tietze, H. (1880–1964), 291, 292
Tietze-extension theorem, 291
Tijdschrift voor Wijsbegeerte, 31, 108, 361
Tillich, P., 260
time, 18
Tolstoi (1828–1910), 61
topological
   group, non-Lie, 149
   mapping, 122, 142, 152, 154, 175
   mappings of spheres, 142
   notions, and their properties, 145
   transformation, 153
topology, 45, 124, 141, 214, 290
   of surfaces, 291, 292
   of the plane, 128, 133
   research program, 154
Torenlaan, 62
torpedo, 274
totalization, 305, 306
transfinite numbers, 112
transformation
   class, 177
   groups, 153
translation theorem, 133–136
tree, 377
tree climbing, 9

trichotomy, 388
triple tangents, 272
Troelstra, P.J., (1860–1930), 30, 286
trust in God, 19
truth, 18
two kingdoms, 38
two kings, 37
Tönnies, F. (1885–1936), 245, 265

unaccountability, 257
unauthorized change, 189
uniform differentiability, 127
uniformisation, 149
uniformization, 175, 178, 187, 191
    problem, 178, 180
    proof, 182
    theorems, 301
uniqueness conditions, 138
University of Amsterdam
    see UVA, 2
unreliability paper, 100, 108, 201, 316
unsolvable mathematical problems, 107
Unzerlegbarkeit, 385
Upward, A., 265
ur-intuition, 90, 102, 117, 118
Urysohn, P. (1898–1924), 127, 170
Utrecht, 265
    University, 271, 274, 354, 356
UVA, 12, 13, 41, 42, 45, 153, 216, 222, 226, 253, 264, 269, 271, 304, 331, 401
Uven, M.J. van (1878–1959), 215, 274

Vahlen, T. (1869–1945), 86
Valiron, G. (1884–1955), 274
Van der Waals surface, 43
Van Gogh Art gallery, 16
Vandermonde, A. (1735–1796), 124
varieties, 125
Veblen, O. (1880–1960), 124, 292
vector
    distribution, 136, 138
    field, 138, 139, 142, 156
    mathematics, 360
vegetarian, 28
vegetarianism, 28, 62
Veluwe, 27
Vermeulen, M. (1888–1967), 364
Veronese, G. (1854–1917), 50, 86
Versluys, W.A. (1870–1946), 204
Verzerrungssatz, 183, 189, 191
vicious circle principle, 239, 318
Vienna Circle, 372
virtual ordering, 390
Volker
    see Gutkind, E. (1877–1956), 246

Vries, Gustav de, 43
Vries, Hk. de (1867–1954), 78, 206, 216, 217, 219, 222, 224, 227, 271, 300, 304–306, 310, 352, 353, 359–361
Vries, Hugo de (1848–1935), 13, 219
Vries, J. de (1858–1938), 118, 216, 219, 222, 305, 306
Vuysje-Mannoury, C., 269

Waals jr, J.D. van der, 207, 219
Waals, J.D. van der (1837–1923), 13, 27, 41–43, 168, 197, 206, 228, 361
Wada, 144
Waerden, B.L. van der (1903–1996), 190
Walaardt Sacré, Major H., 280
Walden, 60, 245, 246, 260, 261, 269, 334
Weitzenböck lecturer at UVA, 361
Weitzenböck, R. (1885–1950), 296, 300, 307, 325, 359–361
Welby, Victoria Lady (1837–1912), 245, 246, 256
well-ordering, 328
    theorem, 112
Weltanschauung, 18
Weltgeist, 268
werdende Folgen, 323
Weyl
    and the UVA, 310
    on continuity of real functions, 325
    on existential statements, 324
Weyl, H. (1885–1955), 180, 213, 226, 291, 298–301, 307–312, 316–326, 360, 375, 390, 395, 400
WG see *Wiskundig Genootschap*, 46
wicker chair, 77
Wiener, H. ((1857–1939), 325
Wiessing, H. (1879–1961), 14–16, 30, 34, 47, 193, 286, 289, 309
Wijthoff, W.A., 274
Willink, B., 43, 99
winding number, 139
Winkler, C., 29, 219
Wiskundig Genootschap, 46, 140, 142, 159, 221, 227, 228, 271, 297, 357, 358
Wolff, J. (1882–1945), 271, 360
words of spiritual value, 267
words-of-power, 249
world, the sad, 66
world-state, 266
Woude, W. van der (1876–1974), 300
Würfelsatz, 189

Zandvoort, 310
Zee, Lau van der, 8, 58, 65
Zeeman, P. (1865–1943), 206, 216, 219, 254, 272, 277, 290

Zermelo, E. (1871–1953), 86, 112, 113, 125, 141, 149, 204, 312, 317, 318
Zernike, F. (1888–1966), 280
Zionist, 256, 332
Zoretti, L., 165

Zuiderzee, 4
Zur Analysis Situs, 143
Zyklosis, 176
Zürich, 311